PROCEEDINGS OF THE WORKSHOP ON

EFFECTIVE FIELD THEORIES OF THE STANDARD MODEL

PROCEEDINGS OF THE WORKSHOP ON

EFFECTIVE FIELD THEORIES OF THE STANDARD MODEL

DOBOGÓKŐ, HUNGARY AUGUST 22 – 26, 1991

EDITED BY

ULF-G. MEIßNER
Institut für Theoretische Physik
Universität Bern
CH-3012 Bern, Schweiz

World Scientific
Singapore • New Jersey • London • Hong Kong

Published by

World Scientific Publishing Co Pte Ltd
5 Toh Tuck Link, Singapore 596224
USA office: 27 Warren Street, Suite 401-402, Hackensack, NJ 07601
UK office: 57 Shelton Street, Covent Garden, London WC2H 9HE

Proceedings of the Workshop on
EFFECTIVE FIELD THEORIES OF THE STANDARD MODEL

ISBN-13 9789810210014
ISBN-10 9810210019

FOREWORD

Effective Lagrangian methods are a powerful tool to explore the symmetries and dynamics of the standard model and physics beyond it. This has become particularly visible through the many successes of chiral perturbation theory, which allows to make precise statements about the low energy behavior of QCD Greens functions. In this spirit, it appeared as a natural consequence to organize a meeting in this field and bring together its leading protagonists. This gathering took place at Dobogókö, Hungary, from August 22 to 26, 1991, in the form of a workshop under the title "Effective Field Theories of the Standard Model".

The workshop consisted of an introductory lecture delivered by Professor John Donoghue and a series of invited talks. The latter were grouped into the following categories: *Topics in chiral perturbation theory, non-leptonic weak interactions, baryons, finite temperatures and volumes, strongly coupled Higgs sector* and *various topics*. There were also a few shorter contributions embedded in the extended discussion sessions. The physics discussed covered phenomena at scales of a few MeV to some hundred GeV demonstrating the usefulness of the effective field theory methods. Here, I would like to take the opportunity to thank all speakers and participants for their valuable contributions to the workshop (part of which is found in the written material presented here).

It is my particular pleasure to thank Prof. Andras Patkós and his staff, Katalin Varga, Orsolya Újsághy and Antal Jakovác, for the excellent local organization which made this meeting truely memorable for all of us. The scientific organization was done in collaboration with Gerhard Ecker, Peter Hasenfratz and Andras Patkós. The workshop could not have taken place without the financial support from the Hungarian Academy of Sciences and the Swiss National Science Foundation. Their support is gratefully acknowledged.

Ulf-G. Meißner
Bern, January 1992

CONTENTS

Summary

INTRODUCTORY LECTURE

INTRODUCTION TO NONLINEAR EFFECTIVE FIELD THEORY

JOHN F. DONOGHUE
Department of Physics and Astronomy, University of Massachusetts
Amherst, MA 01003, USA

ABSTRACT

I summarize the general method of applying nonlinear effective field theory. After a illustrative example involving general relativity, the phenomenologically rich field of chiral perturbation theory is surveyed.

Even though our fundamental theories continue to be expressed in the framework of renormalizable quantum field theory, physicists know how to use certain nonlinear and nonrenormalizable field theories in a consistent and predicted fashion. In this talk I provide a brief introduction to and an overview of the use of such nonlinear effective field theories. The most complete example is that of chiral perturbation theory, but the theory of gravitation also has many similar aspects.

1. The Paradigm of Effective Field Theory

When we teach a course in quantum field theory, we typically give the following rules for building and applying a field theory:

1) Construct an action which is invariant under the desired symmetries. In the archetypical case of QED, the Lagrange density must be a Lorentz scalar, invariant under U(1) gauge invariance.
2) Keep only renormalizable interactions. This restricts the Lagrangian to terms of canonical dimension less than or equal to four. In the QED example, one drops Lorentz and gauge invariant terms such as
$$(F_{\mu\nu} F^{\mu\nu})^2 \ , \ \bar{\psi}\, \sigma^{\mu\nu}\, \psi\, F_{\mu\nu}$$
3) Quantize the theory and calculate scattering processes. When using perturbation theory, this consists of the calculation of tree and loop diagrams.
4) Determine the physical parameters from experiment and express the predictions in terms of the physical parameters. One must measure the charge and masses entering the theory, and the predictions of QED amount to relations between many experiments, all parameterized in terms of the physical (e, m).

Of these ingredients, #1, 3, 4 seem to be logically necessary, and one cannot imagine modifying these steps. However the issue of renormalizability is less obvious, and almost seems to be inserted more for the convenience of the physicist doing the calculation. It is true that for operators with $d > 4$, one finds divergences in all amplitudes when one calculates in perturbation theory and needs an infinite number of parameters to fully renormalize the theory. This would certainly be a nuisance, but is not inconceivable. Moreover, many (most?) people believe that the divergences are an artifact of our present incomplete theory/knowledge, and that we will ultimately arrive at a finite theory. If this is the case, why does Nature present us with a Lagrangian with all of the $d > 4$ operators set equal to zero?

It is interesting that another consistent and predictive paradigm exists. In this case steps #1, 3, 4 above are retained but #2 is replaced by:

2) Work at low energies and expand the theory in powers of the energy.

In these cases there is no need for the restriction to renormalizable interactions, as the energy expansion restricts the form of the Lagrangian instead. At a given order in the energy one needs to measure only a finite number of couplings to specify the theory. The predictions amount to relations among many experiments parameterized by these couplings. This pattern can be called effective field theory. The theory is "effective" because the procedure fails at high energy and the low energy theory must be replaced by a better high energy theory. While renormalizable theories can also suffer this fate (e.g., QED is replaced by $SU(2)_L \times U(1)_Y$ at high energies), the term is usually reserved for nonrenomalizable theories. By effective *field theory*, I also imply that one uses the full field theory formalism, excluding the simple tree level usages of effective Lagrangians.

2. Gravity as an Effective Field Theory

While I will devote most time to chiral perturbation theory, it is interesting to look at general relativity through the eyes of effective field theory. Here the symmetry is general covariance and the fields are the metric $g_{\mu\nu}(x)$ and matter fields. Objects which transform covariantly are components of the curvature $R^{\mu\nu\alpha\beta}$, $R^{\mu\nu}$, R, which are nonlinear functions of $g_{\mu\nu}$. The general action has an infinite number of terms

$$S = \int d^4x \sqrt{g} \{\Lambda_0 + F^2 R + \alpha R^2 + \beta R_{\mu\nu} R^{\mu\nu} + \gamma R^3 + \ldots\} \qquad (1)$$

where Λ_0, F^2, α, β are parameters or coupling constants (A third R^2 term $\sqrt{g}\,(R^{\mu\nu\alpha\beta} R_{\mu\nu\alpha\beta} + R^2 - 4R_{\alpha\beta} R^{\alpha\beta})$ is a total divergence and can be dropped from our discussion here.] Continuing to all powers of R leads to an infinite number of coupling constants. Since R itself contains fields to all orders, the theory is nonrenormalizable. However, fortunately for our existence, the theory makes a good low energy effective theory. R is constructed with two derivatives, schematically

$$R \sim \Box g + \nabla g\, \nabla g + O(\nabla^2 g^3) \ . \qquad (2)$$

In matrix elements, derivatives become factors of momentum, $\partial_\mu \sim p_\mu$, so that counting derivatives counts factors of the energy. Λ_0 has no derivatives and hence is $O(E^0)$, while F^2R is $O(E^2)$ and R^2 is $O(E^4)$. For low energy/small curvatures we have $R^2 << R$, so that we may truncate the Lagrangian to

$$L_g = \sqrt{g}\,[\Lambda_0 + F^2 R] \qquad (3)$$

which is general relativity with a cosmological constant.

We can work out the phenomenology to $O(E^2)$. The equation of motion is

$$R_{\mu\nu} - \frac{1}{2}\, g_{\mu\nu}\, R = \frac{1}{2F^2}\, (T_{\mu\nu} + \Lambda_0\, g_{\mu\nu}) \tag{4}$$

and one concludes that i) F is related to the Cavendish constant

$$F = \frac{1}{\sqrt{16\pi\, G_N}} = 0.17 \times 10^{19}\ \mathrm{GeV} \tag{5}$$

and ii) Λ_0 is unexpectedly small

$$\Lambda < 2.3 \times 10^{-48}\ \mathrm{GeV}^4 \tag{6}$$

such that we will set $\Lambda_0 = 0$.

What happens if we treat this seriously as a quantum field theory? Clearly loop diagrams, involving gravitons and matter fields, will be generated,[1] and one might expect problems due to the nonrenormalizability of the theory. In fact, there are no serious problems in perturbation theory. Consider quantum fluctuations about a tree level background field

$$g_{\mu\nu}(x) = \bar{g}_{\mu\nu}(x) + \frac{h_{\mu\nu}}{F}\ . \tag{7}$$

Expanding the Lagrangian, the terms linear in h vanish by the equation of motion and one finds[2]

$$L(\bar{g}, h) = \sqrt{\bar{g}}\, \Big\{ F^2 R + \frac{1}{4}\Big(D_\nu h^\beta_\alpha D^\nu h^\alpha_\beta - D_\mu h^\alpha_\alpha D^\mu h^\beta_\beta$$

$$+ 2\, D_\beta h^\beta_\alpha D^\Lambda h^\alpha_\lambda + D^\alpha h^\nu_\beta D_\nu h^\beta_\alpha \Big)$$

$$+ \frac{1}{8}\Big((h^\alpha_\alpha)^2 - 2h^\alpha_\beta (h^\beta_\alpha) \Big) \bar{R} + h^\nu_\beta h^\beta_\alpha \bar{R}^\alpha_\nu - \frac{1}{2}\, h^\alpha_\alpha h^\nu_\beta R^\beta_\nu + \ldots \Big\} \tag{8}$$

If one integrates out the quantum fluctuations to get the one loop effective action

$$e^{iZ(\bar{g})} = \int [dh]\, e^{iS(\bar{g},\, h)}\ . \tag{9}$$

The divergent terms can be picked out. For example, with the matter field being a minimally coupled scalar of mass m, one finds[2]

$$\Delta L_{\rm div}\, (\bar{g}) = \frac{\sqrt{\bar{g}}}{8\pi^2 (d-4)} \left\{ \frac{1}{2}\, m^4 - \frac{1}{6}\, m^2 \bar{R} + \frac{9}{720}\, \bar{R}^2 + \frac{43}{120}\, \bar{R}_{\mu\nu} \bar{R}^{\mu\nu} \right\}\ . \tag{10}$$

The divergent terms are themselves generally covariant, as they must be, and hence match terms already in the Lagrangian. The divergences can be absorbed into renormalized parameters Λ_0, F^2, α, β. Thus at one loop one has the action

$$S(g) = \int d^4x \sqrt{g}\Bigg\{\left[\Lambda_0 + \frac{m^2}{16\pi^2(d-4)}\right] + \left[\frac{1}{16\pi G_N} - \frac{m^2}{48\pi^2(d-4)}\right] R$$

$$+ \left[\alpha + \frac{9}{5760\pi^2(d-4)}\right] R^2 + \left[\beta + \frac{43}{960\pi^2(d-4)}\right] R_{\mu\nu} R^{\mu\nu}\Bigg\}$$

$$+ \text{finite}$$

$$= \int d^4x \sqrt{g}\left\{\Lambda_{ren} + \frac{1}{16\pi G_N^{ren}} R + \alpha^{ren} R^2 + \beta^{ren} R_{\mu\nu} R^{\mu\nu}\right\} + \text{finite} \tag{11}$$

One repeats the phenomenology for the renormalized parameters, obtaining the same results up to finite effects. One loop diagrams have required the presence of $O(E^4)$ Lagrangians, and higher loops bring in higher order coefficients.

The phenomenology can in principle be extended to order E^4 by considering the R^2 components of L. Although there is no positive data, this has been explored by Stelle.[3] For example the Newtonian potential is modified

$$V = -\frac{GM}{r}\left[1 - \frac{4}{3}e^{-m_2 r} + \frac{1}{3}e^{-m_0 r}\right] \tag{12}$$

with

$$m_0 = \frac{F}{[6\alpha - 2\beta]^{1/2}} \quad ; \quad m_2 = \frac{F}{\sqrt{\beta}} \; . \tag{13}$$

Laboratory measurements provide the best constraints on α, β restricting them to be less than $\sim 10^{28}$. If the world were such that we could measure these, one could apply the gravitational Lagrangian to $O(E^4)$.

The $R + R^2$ theory has soliton-like solutions[4] which generate inflation. These are not really consistent[5] with the use of gravity as a low energy effective theory as they require such a large curvature that the effects of the βR^2 Lagrangian is comparable to $F^2 R$.

The $R + R^2$ theory has problems as a fundamental theory. There are negative metric states, a lack of causality and an instability of flat space. However these effects are not problems as an effective theory. If one includes the lowest order constraints, the problems do not exist perturbatively.[6] That is, they are associated with the high energy aspects and do not show up in the low energy theory.

Many of the lessons of this example apply to other effective field theories, such as chiral symmetry. One organizes L by dimension and at low energy one needs only the low dimension Lagrangians. The theory can be renormalized order by order in the energy expansion, with one loop generating $O(E^4)$ terms. The phenomenology also proceeds order by order, and needs experimental input. "Soliton"-like solutions, which require comparable contributions from differing orders, lie outside of the energy expansion (the phrase "ineffective field theory "comes to mind for this). One lesson which does not generalize is the vanishing of the lowest order effect—the cosmological constant.

3. Chiral Symmetry

Effective field theory techniques are more practical in the application of chiral symmetry. I don't have the space here to develop the theory of chiral symmetry from the start,[7,8,9] so I would rather just describe how it works for mesonic processes. The fields in the theory are the Goldstone bosons [$\phi^i = \pi^i$ in SU(2) or π, K, η_8 in SU(3)], which are arranged in a matrix representation with a definite transformation under left and right chiral symmetries

$$U = \exp\frac{i\,\lambda^i\cdot\phi^i}{F}$$
$$U \to L\,U\,R^\dagger \tag{14}$$

with L in $SU(3)_L$, R in $SU(3)_R$ and F being related to the pion decay constant. Invariant Lagrangians can be formed from chiral invariants such as $\mathrm{Tr}(\partial_\mu U\,\partial_\nu U^\dagger)$. There are no invariant Lagrangians that can be formed without any derivatives since $\mathrm{Tr}(U^\dagger U)$ = constant.

The general pattern of effective field theories will generate a Lagrangian which is an expansion in the number of derivatives. There is in addition symmetry breaking due to the quark masses, which in turn generates masses for the Goldstone bosons. Each power of m_π^2 or m_K^2 counts as equivalent of two derivatives in the energy expansion. At order energy-squared, the Lagrangian has the form

$$L_2 = \frac{F^2}{4}\mathrm{Tr}(D_\mu U\,D^\mu U^\dagger) + \frac{F^2}{4}\mathrm{Tr}(M^2\,U^\dagger + U\,M^2) \tag{15}$$

where the meson mass matrix M^2 is related to the quark mass matrix by $M^2 = 2B_0\,m_q$ for some constant B. The phenomenology at $O(E^2)$ is governed by the pion decay constant F $= F_\pi = 92.3$ MeV, and the meson masses

$$m_\pi^2 = 2B_0\,\hat{m} \qquad \left(\hat{m} = \frac{m_u + m_d}{2}\right)$$
$$m_K^2 = B_0(\hat{m} + m_s)$$
$$m_\eta^2 = \frac{4}{3}\,m_K^2 - \frac{1}{3}\,m_\pi^2 \tag{16}$$

When this Lagrangian is used at tree-level one reproduces the results of the 1960's applications of current algebra.

In the past decade, the theory of chiral symmetry has been applied at order E^4, which is equivalent to say at the one-loop level. Here the ingredients[8] are the Lagrangian L_2, used both at true level and in one-loop diagrams, plus an $O(E^4)$ Lagrangian L_4 to be used at tree level. The general Lagrangian at $O(E^4)$ contains 3 operators in the limit of exact chiral symmetry (without sources), 5 involved with symmetry breaking, and a few more if external sources (such as the electromagnetic field) are added,

$$\mathcal{L}_4 = \sum_i L_i \, O_i \ . \tag{17}$$

Here L_i are low energy constants which parameterize the Lagrangian, and O_i are various combinations of the chiral matrix U, plus coupling to external sources. The classic Gasser-Leutwyler form[9] describes vector, axialvector, scalar and pseudoscalar sources. Some recent additions have been coupling to the metric tensor[10] $g_{\mu\nu}$ (in order to identify the energy momentum tensor) and the θ parameter[11] (in order to identify $F\tilde{F}$). In addition the axial anomaly[12] enters at $O(E^4)$, but does not contain any free parameters.

4. How well does it work?

The application of chiral symmetry are restricted to low energy processes. This encompasses only a finite number of processes, but, surprisingly, the analysis of many sectors of the theory is still not complete. I will describe two of the best studied segments, the radiative complex and the effects of the quartic derivatives, plus the still developing area of parameter free predictions. At this conference, D. Wyler[13] described our knowledge of quark masses, J. Bijnens[14] reviewed the anomaly sector, J. Kambor[15] described the extension of the chiral framework to weak nonleptonic processes, and several speakers[16] reported on the large field of applications to baryons. In the baryon sector, the low energy requirement is satisfied if the baryons are essentially static sources, interacting with low energy pions. A seemingly related approach, Skyrmions,[17] which are solitons of the chiral field, is in fact logically very different and is not part of the framework of effective field theory. The soliton curvature induces a large energy scale which does not obey the energy expansion and does not constitute a controlled approach .

The radiative complex of reactions includes a series of decays and amplitudes governed by two of the low energy constants

$$\begin{aligned} \mathcal{L}_{rad} &= i\, L_9 \, \mathrm{Tr}(L_{\mu\nu} \partial^\mu U \partial^\nu U^+ + R_{\mu\nu} \partial^\mu U^+ \partial^\nu U) \\ &\quad + L_{10} \, \mathrm{Tr}(L_{\mu\nu} U R^{\mu\nu} U^+) \end{aligned} \tag{18}$$

where $L_{\mu\nu}$ ($R_{\mu\nu}$) are left (right)-handed couplings of γ or $W^\pm$. Included are the pion form factor, radiative pion decays, pion polarizability and related kaonic processes. Of course, to proceed one must measure the low energy coefficients, so that it is only relations between decays which are predictions of chiral symmetry. The results[9,18] are shown in Table I. Aside from situations where there are experimental inconsistencies, the predictions work well.

Table 1. Chiral predictions and data in the radiative complex of transitions.

Reaction	Quantity	Theory	Experiment
$\gamma \to \pi^+\pi^-$	$<r_\pi^2>$ (fm^2)	0.44[a]	0.44 ± 0.02
$\gamma \to K^+K^-$	$<r_K^r>$ (fm^2)	0.44	0.34 ± 0.05
$\pi^+ \to e^+\nu_c\gamma$	h_V (m_π^{-1})	0.027	0.029 ± 0.017
	h_A/h_V	0.46[a]	0.46 ± 0.08
$K^+ \to e^+\nu_c\gamma$	$(h_V + h_A)$ (m_π^{-1})	0.038	0.043 ± 0.003
$\pi^+ \to e^+\nu_c e^+e^-$	r_A/h_V	2.6	2.3 ± 0.6
$\gamma\pi^+ \to \gamma\pi^+$	$(\alpha_E + \beta_M)$ (10^{-4}fm)	0	1.4 ± 3.1
	α_E(10^{-4}fm)	2.8	6.8 ± 1.4[b]
$\gamma\gamma \to \pi\pi$	α_E(10^{-4}fm)	2.8	2.2 ± 1.6[b]
$K \to \pi e^+\nu_c$	$\xi = f_-(0)/f_+(0)$	-0.13	-0.20 ± 0.08[b]
	λ_+(fm^2)	0.067	0.065 ± 0.005[b]
	λ_0(fm^2)	0.040	0.050 ± 0.012[b]

[a]Used as input.

[b]There are experimental inconsistencies in these measurements. We use average values for the λ_+, λ_0 rows.

A second "classic" area of application is in $\pi\pi$ scattering and $K \to \pi\pi e\nu$ decay. While these processes seem unrelated, they are in fact governed by the same parameters in the chiral Lagrangian

$$\begin{aligned} \mathcal{L}_{\pi\pi} = L_1[\mathrm{Tr}\,(D_\mu U\, D^\mu U^+)]^2 \\ + L_2\, \mathrm{Tr}(D_\mu U\, D_\nu U^+)\, \mathrm{Tr}(D_\mu U\, D^\nu U^+) \\ + L_3\, \mathrm{Tr}(D_\mu U\, D^\mu U^+ D_\nu U\, D^\nu U^+) \end{aligned} \quad (19)$$

There are 12 pieces of data determined and 3 free parameters. The results[19] are shown in Tables 2, 3, where both the $O(E^2)$ and $O(E^4)$ results are shown. It can be seen that the corrections to the lowest order results can be significant and hence it is important to carry the expansion to at least $O(E^4)$.

Table 2. The pion scattering lengths and slopes.

	Experimental	Lowest Order	First Two Orders
a_0^0	0.26 ± 0.05	0.16	0.20
b_0^0	0.25 ± 0.03	0.18	0.26
a_0^2	-0.028 ± 0.012	-0.045	-0.041
b_2^2	-0.082 ± 0.008	-0.089	-0.070
a_1^1	0.038 ± 0.002	0.030	0.036
b_1^1	—	0	0.043
a_2^0	$(17 \pm 3) \times 10^{-4}$	0	20×10^{-4}
a_2^2	$(1.3 \pm 3) \times 10^{-4}$	0	3.5×10^{-4}

Table 3. Experiment and theory for K_{l4} decays.

	Data	Lowest Order	First Two Orders
$f_1(0)$	1.47 ± 0.04	1.00	1.45
$f_2(0)$	1.25 ± 0.07	1.00	1.24
λ_1	0.08 ± 0.02	0.00	0.08
λ_2	0.08 ± 0.02	0.00	0.06
$g(0)$	0.96 ± 0.24	0.00	1.00

A more problematic set of reactions are those which are independent of the chiral parameters up to $O(E^6)$. This includes $\gamma\gamma \rightarrow \pi^0\pi^0$, $K_L \rightarrow \pi^0\gamma\gamma$ and $K_S \rightarrow \gamma\gamma$. For these processes, there are no tree level couplings at all in L_2 or L_4, because of the charge neutrality of the mesons involved. The processes arise uniquely at $O(E^4)$ from loop effects, where the photons couple to charge particles and then the charged particles rescatter to neutral ones via the strong interactions. The finite loop contributions are rather small, as can be seen since $\gamma\gamma \rightarrow \pi^0\pi^0$ is an order of magnitude smaller than the tree level process $\gamma\gamma \rightarrow \pi^+\pi^-$. The two isospin amplitudes ($I = 0, 2$), which sum to produce the latter final state, must largely cancel to produce the small $\pi^0\pi^0$ amplitude. The one-loop chiral prediction is not too successful, see Figure 1. It misses some of the low energy structure.[20] There is evidence that this is due to a problem with the one-loop calculation rather than the chiral couplings. This can be seen from a dispersion relation analysis[21] of Morgan and Pennington (MP), who enforced the chiral constraints at threshold. The

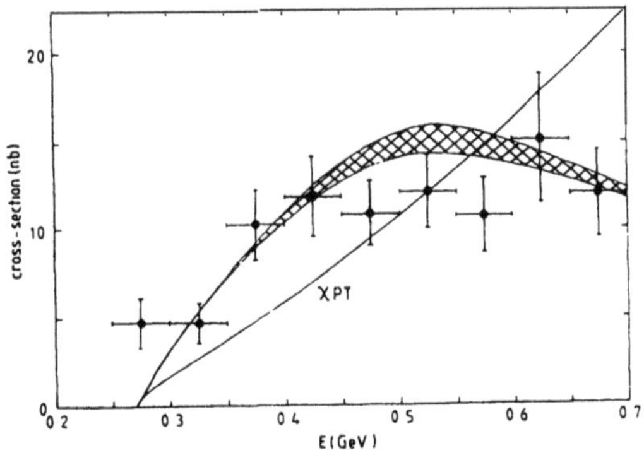

Figure 1. Data for $\gamma\gamma \rightarrow \pi^0\pi^0$ along with the chiral one-loop prediction (χPT) and the MP result. The figure is taken from Ref. 21.

dispersion relation would be satisfied by chiral perturbation theory order by order in the energy expansion. The MP method then involves a summation of a series of diagrams which the chiral calculation treats only at one-loop order. That the results differ at low energies is interesting and suggests that the one-loop calculations may not be adequate for the unitarity effects contained in loops. The energy scale in unitarity effects is $s/4m_\pi^2$ in some cases. Since these unitarity effects are rather small components in most processes, this is not much of a problem. However, in the parameter free sector, which arise primarily from loops, this can play a larger role.

The other reactions which are parameter-free are $K_S \rightarrow \gamma\gamma$ and $K_L \rightarrow \pi^0\gamma\gamma$. For the K_S decay[22] we have

$$BR(K_S \rightarrow \gamma\gamma) = 2.0 \times 10^{-6} \qquad \text{ChPTh}$$
$$(2.4 \pm 1.2) \times 10^{-6} \qquad \text{Data} \tag{20}$$

which agrees within the rather large error bar. In $K_L \rightarrow \pi^0\gamma\gamma$, however, the quoted experimental rate[23] is larger than the one-loop prediction[24] by a factor of 2-3, although the shape of the photon spectrum is in good agreement, see Figure 2. The resolution to this could be related to the same physics seen in $\gamma\gamma \rightarrow \pi^0\pi^0$. The chiral reaction goes through the $K_L \rightarrow \pi^0\pi^+\pi^-$ reaction plus rescattering $\pi^+\pi^- \rightarrow \gamma\gamma$. As a trial, I have added to the EPR calculation modification due to known changes in $K_L \rightarrow \pi^0\pi^+\pi^-$ beyond leading order plus

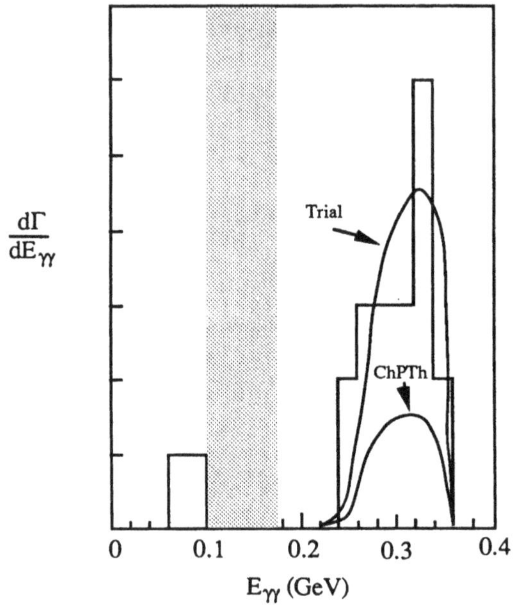

Figure 2. Data on $K_L \to \pi_0\gamma\gamma$ along with the chiral prediction, and a trial estimate of higher-order effects.

the enhancement of the MP two photon calculation. That is, I show

$$\left.\frac{d\Gamma}{dE_{\gamma\gamma}}\right|_{\text{Trial}} = \left.\frac{d\Gamma}{dE_{\gamma\gamma}}\right|_{\text{one loop}} \times \left[\frac{\Gamma(K_L - \pi^+\pi^-\pi^0)}{\Gamma(K_L \to \pi^+\pi^-\pi^0)\Big|_{\text{Lowest order}}}\right]\left[\frac{\Gamma(\gamma\gamma \to \pi^0\pi^0)}{\Gamma(\gamma\gamma \to \pi^0\pi^0)_{\text{one loop}}}\right] \tag{21}$$

which is the other line in Figure 2. While this should not be taken as a true calculation, it does indicate that the resolution may be in higher order corrections.

Overall, the majority of chiral predictions work excellently. The case of the parameter free sector provides an interesting system to study small loop effects, and it seems that some more study should be given to this topic. Even more important than phenomenological success is the fact that chiral perturbation theory provides a systematic approach to low energy physics.

5. Chiral Perturbation Theory vs. Models

In the literature one can sometimes see what appear to be conflicting predictions both claiming to be derived via chiral symmetry. This can never happen (aside from an outright mistake), and what is going on is a confusion between pure chiral perturbation theory and models which incorporate some aspects of chiral symmetry. Practitioners generally have some interest in each, and often mix the usages within the same paper.

Chiral Perturbation Theory is a rigorous use of effective field theory, based on the symmetries and anomalies of QCD. It is a controlled expansion leading to relations between processes. Models, on the other hand, are non-rigorous attempts to mimic QCD. They involve adjustable assumptions based on one's feeling for the physics. They are often very reasonable, but one cannot have complete confidence in them. The two approaches really have different goals. ChPTh categorizes definitively the low energy structure of QCD. If good phenomenology results, so much the better. Models are developed to be more predictive, or to extend the range of predictions outside of the "safe" low-energy limit. A model is not worth anything unless it does better than ChPTh in a phenomenological application.

Models attempt to deal with the two main frustrations experienced by practitioners of ChPTh, i.e., that the theory is limited to low energy and contains many coupling constants which must be measured. These are not flaws of the theory, and are merely symptomatic of the way that all effective field theories work, but in doing phenomenology one would like to transcend these limitations. This can be done by using models, but only at the price of a loss of rigor. Unitarization schemes attempt to extend results to higher energy. Models of the chiral coefficients attempt to predict the couplings constants.

Unitarization is not often thought of as a model, because unitarity is a firm requirement of field theory. Being a field theory, ChPTh satisfies unitarity order by order in the energy expansion. However, attempts to impose exact unitarity go beyond the energy expansion in an uncontrolled way and hence become models. Padé and K-matrix methods are examples of these.[25] The only solid progress that I know of is that recently Gasser and Meissner[26] have extended the pion form factor to two loop order, showing that the Padé method misrepresents some chiral logs. The arguments on this topic that occur in the literature appear excessive when one realizes that unitarization schemes are models.

In attempting to predict the chiral couplings, many models[27-31] exist (see Table 4) but two are worth mentioning further. The most reliable appears to be the influence of the low-lying resonance spectrum,[28] especially vector mesons. It is easy to see how these produce the modifications to low energy scattering and decay which have been parameterized by the chiral coefficients. The other interesting model is the quark loop,[27]

Table 4. Comparison of Models

	Expt	Spectrum (vectors)	Free Quark	Ref. 30	Ref. 31	Nucleon Loop	Linear σ Model
$L_1 + L_{3/2}$	-.9	-2.1	-.8	02.1	01.1	-0.8	-0.5
L_2	1.9	2.1	1.6	1.6	1.8	0.8	1.5
L_9	7.1	7.3	6.3	6.7	6.1	3.3	0.9
L_{10}	-5.6	-5.8	-3.2	-5.8	-5.2	-1.7	-2.0

which is sometimes modified by either single gluon exchange[30] or quark self energy[31] effects. We will address these more fully in the next section.

6. Anatomy of a Chiral Coefficient

This section briefly summarizes some unpublished work of Barry Holstein and myself.[32] The idea is that one can look at the ingredients to one of the chiral coupling constants, L_{10}, in more detail by studying the dispersion relation which predicts its value. The low energy behavior of the vector and axial vector correlation functions

$$i\int d^4x\, e^{iq\cdot x} < 0\,|\,T\left(V_\mu^a(x)\; V_\nu^b(0)\right)|\,0 > = \delta^{ab}\,(q_\mu q_\nu - g_{\mu\nu}q^2)\;\pi_V\,(q^2)$$

$$i\int d^4x\, e^{iq\cdot x} < 0\,|\,T\left(A_\mu^a(x)\; A_\nu^b(0)\right)|\,0 > = \delta^{ab}\left\{(q_\mu q_\nu - g_{\mu\nu}q^2)\;\pi_A\,(q^2) + q_\mu q_\nu\;\pi_A^{(0)}(q^2)\right\} \tag{22}$$

are predicted by chiral symmetry. In particular[9]

$$\pi_V(s) - \pi_A(s) = -\frac{F_\pi^2}{s} - 4L_{10}^{(r)} - \frac{1}{48\pi^2}\left(ln\,\frac{m_\pi^2}{\mu^2} - \frac{1}{3} + H(s)\right) + O(s)$$

$$H(s) \equiv \sigma^2\left[\sigma\, ln\,\frac{\sigma-1}{\sigma+1} + 2\right]$$

$$\sigma \equiv \left(1 - \frac{4m_\pi^2}{s}\right)^{1/2} \tag{23}$$

In the chiral limit this combination obeys an unsubtracted dispersion relation

$$\pi_V(s) - \pi_A(s) = \frac{1}{\pi}\int_{4m_\pi^2}^{\infty} \frac{ds'}{s' - s - i\varepsilon}\left[Im\;\pi_V(s') - Im\;\pi_A(s')\right] \tag{24}$$

This representation leads to the Weinberg sum rules[33]

$$\frac{1}{\pi}\int ds'\, s'[Im\;\pi_V - Im\;\pi_A] = 0 \text{ (only true for } m_q = 0)$$

$$\frac{1}{\pi}\int ds'\,[Im\;\pi_V - Im\;\pi_A] = F_\pi^2 \tag{25}$$

and the prediction for L_{10}

$$\frac{1}{\pi}\int \frac{ds'}{s'}\,[Im\;\pi_V - Im\;\pi_A] = -4L_{10}^r - \frac{1}{48\pi^2}\,(ln\; m_\pi^2/\mu^2 + 1) \tag{26}$$

which is equivalent to the Das Mathus Okubo sum rule.[34]

The interesting feature of these sum rules is that the integrand $Im\ \pi_V$ - $Im\ \pi_A$ is a known function,[35] being measured in $e^+e^- \rightarrow$ hadrons and $\tau \rightarrow \nu$ + hadrons. The shape has a rather simple form. $Im\ \pi_V$ contains a resonance, the ρ, and then approaches a constant ($1/8\pi$), see Figure 3, while $Im\ \pi_A$ has a higher resonance, the a_1, and then approaches the same constant. The difference is then basically entirely described by the ρ and a_1. Thus we know the ingredients to L_{10} in complete detail, as shown in Figure 3, 4. The sum rule accurately reproduces the value of L_{10} previously measured in radiative pion decay.

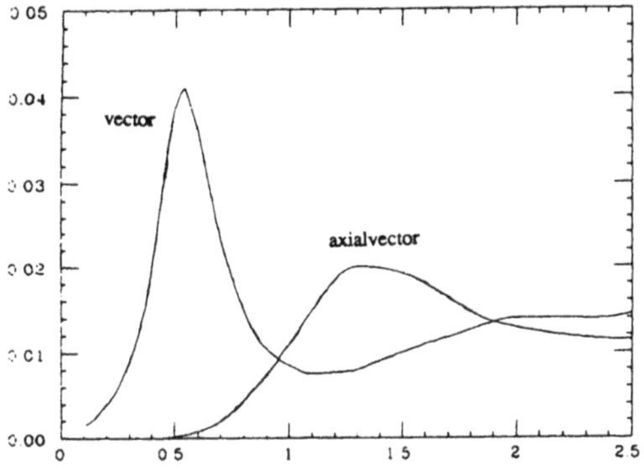

Figure 3. $\frac{1}{\pi} Im\ \pi_V$ and $\frac{1}{\pi} Im\ \pi_A$ as extracted from data in Ref. 35.

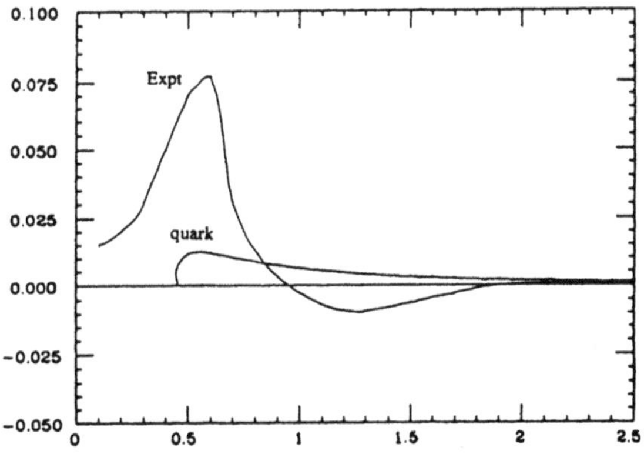

Figure 4. The integrand in the prediction of L_{10}, i.e., $\frac{1}{\pi s}[Im\ \pi_V(s) - Im\ \pi_A(s)]$. Shown are both the experimental result and the quark-loop prediction.

We can ask the models whether they reproduce the physics revealed by the sum rule calculation. The "vector-dominance" model clearly contains exactly the right ingredients. The spin one spectrum is the prime influence in π_V and π_A. One might hope that the quark loop model also would have similar physics, since S-wave $q\bar{q}$ pair form a ρ and P-wave $q\bar{q}$ form an a_1. However, one can calculate $Im\, \pi_V - Im\, \pi_A$ directly in the quark loop model, and no such result emerges. The calculated integrand is structureless, as in Figure 4. This leads one to suspect that it is an accident that the quark-loops model gives a value of L_{10} within a factor of two to the correct one. At the least, one can claim that the physics that binds the $q\bar{q}$ is an important ingredient missing from the free quark loop. The one-gluon effects considered in Ref. 30 may represent a start in this direction, but do not constitute an adequate model of the ρ and a_1

7. Future Directions

During the past decade we have learned how to apply the chiral nonlinear effective field theory in a systematic and phenomenologically useful way. Some sections of the theory are well-developed and mature. Other topics, such as baryons or nonleptonic weak processes, need to be explored more fully. There is the potential for extending the calculational scheme to include dispersion relation techniques. This has been explored in part in Ref. 9, 10, 32, 36. The union of chiral constraints, asymptotic QCD behavior and analyticity properties could be powerful.

There can be application of nonlinear effective field theory techniques in other areas also. While the gravity example that we used in Section 2 is unlikely to be important for phenomenology, it may yet yield some theoretical insights. Extremely interesting at the moment is the application of effective field theory to the symmetry breaking sector of the electroweak theory—i.e., in TeV physics.[37] There are analogues of the QCD reactions which probe the origin of the "Higgs sector", and effective field theory has already proven useful in discussing them.

Acknowledgement

I would like to thank the organizers of this workshop for their great hospitality and for the chance to have such an interesting conference.

References

1. More details about quantum aspects of gravity can be found in N.D. Birrell and P.C.W. Davies, *Quantum Fields in Curved Space* (Cambridge University Press, Cambridge, 1982).

2. M. Veltman in *Methods in Field Theory*, ed. by R. Balian and J. Zinn-Justin (North-Holland, Amsterdam, 1976).
3. K.S. Stelle, *General Relativity and Gravity*, **9**, 353 (1978).
4. A. Starobinsky, *Phys. Lett.* **91B**, 99 (1980).
5. J. Simon, **WISC-MILW-91-TH-11**
6. J. Simon, *Phys. Rev.* **D41**, 3720 (1990); *Phys. Rev.* **D43**, 3308 (1991).
7. A full pedagogical treatment of chiral symmetry is contained in our book: J.F. Donoghue, E. Golowich and B.R. Holstein, *Dynamics of the Standard Model* (Cambridge University Press, Cambridge, 1992). A more experimentally oriented review is: J.F. Donoghue, *Chiral Symmetry as an Experimental Science*, **CERN-TH** to be published in the Proceedings of the Erice School of Antinucleon Physics.
8. S. Weinberg, *Physica* **A96**, 327 (1979).
9. J. Gasser and H. Leutwyler, *Ann. Phys.* **158**, 142 (1984); *Nucl. Phys.* **B250**, 465 (1985).
10. J.F. Donoghue and H. Leutwyler, **UMHEP-321**, to be published in *Z. Phys. C.*
11. J.F. Donoghue and D. Wyler, **UMHEP-350**, to be published in *Phys. Rev. D.*
12. E. Witten, *Nucl. Phys.* **B223**, 422 (1983). J. Wess and B. Zumino, *Phys. Lett.* **B37**, 99 (1971).
13. D. Wyler, these proceedings.
14. J. Bijnens, these proceedings.
15. J. Kambor, these proceedings.
16. A. Manohar, E. Jenkins, U. Meissner and N. Kaiser, these proceedings.
17. T.H.R. Skyrme, *Proc. R. Soc. Lon.* **A260**, 127 (1961). G. Adkins, C. Nappi and E. Witten, *Nucl. Phys.* **B228**, 552 (1983).
18. J.F. Donoghue and B.R. Holstein, *Phys. Rev.* **D40**, 2378 (1989); *Phys. Rev.* **D40**, 3700 (1989).
19. C. Riggenbach, et. al., *Phys. Rev.* **D43**, 127 (1991). J. Bijnens, *Nucl. Phys.* **B337**, 635 (1990).
20. J.F. Donoghue, B.R. Holstein and Y.C. Lin, *Phys. Rev.* **D38**, 2423 (1988).
21. D. Morgan and M. Pennington, *Durham preprint* **DTP-91-44**.
22. J.L. Goity, *Z. Phys.* **C34**, 341 (1986). D. Espriu and G. D'Ambrosio, *Phys. Lett.* **B175**, 237 (1986).
23. G.D. Barr et. al. *Phys. Lett.* **B242**, 523 (1990). V. Papadimitrion et. al. *Phys. Rev.* **D44**, 573 (1991).
24. G. Ecker, A. Pich and E. de Rafael, *Nucl. Phys.* **B303**, 665 (1988).
25. A. Dobado, M.J. Herrero and T.N. Truong, *Phys. Lett.* **B235**, 134 (1990).
26. J. Gasser and U. Meissner, *Nucl. Phys.* **B357**, 90 (1991).
27. J. Balog, *Phys. Lett.* **149B**, 197 (1984). P. Simic, Phys. *Rev. Lett.* **55**, 40 (1985). D. Ebert and H. Reinhardt, *Nucl. Phys.* **B271**, 188 (1986). R.E. Gamboa-Saravi et. al., *Ann. Phys.* **157**, 369 (1984). S.-K. Hu et. al., *Phys. Rev.* **D30**, 836 (1984).

28. J.F. Donoghue, C. Ramirez and G. Valencia, *Phys. Rev.* **D39**, 1949 (1989).
G. Ecker, J. Gasser, A. Pich and E. de Rafael, *Nucl. Phys.* **B34**, 311 (1989).
29. D.Ebert and M.K. Volkov, *Phys. Lett.* **101B**, 252 (1981).
V. Bernard et. al., *Phys. Lett.* **B205**, 16 (1988).
G. Valencia and M. Praszalowicz, *Nucl. Phys.* **B341**, 27 (1990).
30. D. Espriu, E. de Rafael and J. Taron, *Nucl. Phys.* **B345**, 22 (1990), errat. **B355**, 278 (1991).
31. B. Holdom, J. Terning and K. Verbeck, *Phys. Lett.* **B232**, 351 (1989).
32. J.F. Donoghue and B.R. Holstein (in preparation).
33. S. Weinberg, *Phys. Rev. Lett.* **18**, 507 (1967).
34. T. Das, V. Mathus and S. Okubo, *Phys. Rev. Lett.* **19** 859 (1967).
35. R. Peccei and J. Sola, *Nucl. Phys.* **B281**, 1 (1987).
36. J.F. Donoghue, J. Gasser and H. Leutwyler, *Nucl. Phys.* **B343**, 341 (1990).
37. D. Espriu, these proceedings.
D. Espriu, A. Dobado and M.J. Herrero, *Phys. Lett.* **B255**, 405 (1991).
B. Holdom and J. Terning, *Phys. Lett.* **B247**, 88 (1990).
J.F. Donoghue and C. Ramirez, *Phys. Lett.* **B234**, 361 (1990).
A. Dobado and M.J. Herrero, *Phys. Lett.* **B228**, 495 (1989).
B. Holdom, *Phys. Lett.* **B258**, 156 (1991); **B259**, 329 (1991).
C. Ramirez, *Phys. Rev.* **D42**, 1726 (1990).

TOPICS IN CHIRAL PERTURBATION THEORY

NONLOCAL EFFECTIVE FIELD THEORY *†

HOWARD GEORGI
Lyman Laboratory of Physics, Harvard University
Cambridge, MA 02138

ABSTRACT

I give a formal description of the process of matching in effective field theories. I argue that this picture suggests that it should be possible to deal directly with "nonlocal" interaction, so long as they are analytic in momenta in the relevant region. I give an example of the renormalization of a nonlocal interaction.

The general subject of my talk is the question of whether the effective field theory approach to quantum field theory can ever completely replace the "old fashioned" approach in a world where the different important energy scales are not infinitely far apart. That is, can one do calculations in an effective field theory language that are as accurate as those that are done in a "full" theory? There is no question about this in a theory in which the relevant energy scales are well separated from each other. There we understand very well how to set up effective field theory calculations as a perturbation expansion in the ratio of scales. If we know how to do calculations in the full theory, we can then calculate to arbitrary accuracy in the effective theory. Furthermore, even if we cannot do precise calculations in the full theory, the effective theory still contains important information. But what happens as two scales approach one another? I am not sure that this question has much real practical importance. After all, when the two scales get close together, we can simply merge them. However, I am stubborn enough to want to know the answer. Can we find a smooth interpolation between the two well-understood limits?

To this end, I will talk about something a little crazy. I will begin by reviewing the principles by which we calculate in effective field theories. I will then be able to identify the ingredients that are required to calculate efficiently when scales are close together. To answer my question completely, we would need to know how to deal efficiently with "nonlocal" interactions. I will not solve this problem, but I will give some examples that will (I hope) be thought provoking.

Several students have contributed greatly to my understanding of these issues. These include Vineer Bhansali, Peter Cho, Anemarie DeYoung and Rowan Hamilton.

---

*Research supported in part by the National Science Foundation under Grant #PHY-8714654. Harvard preprint #HUTP-91/A051

†Research supported in part by the Texas National Research Laboratory Commission, under Grant #RGFY9106.

I begin with a review of matching.

An effective field theory calculation goes like this. You start at a very large scale, that is with the renormalization scale, μ, very large. In a strongly interacting theory or a theory with unknown physics at high energy, this starting scale should be sufficiently large that nonrenormalizable interactions produced at higher scales are too small to be relevant. In a renormalizable, weakly interacting theory, you start at a scale above the masses of all the particles, where the effective theory is given simply by the renormalizable theory, with no nonrenormalizable terms. You then evolve the theory down to lower scales. As long as no particle masses are encountered, this evolution is described by the renormalization group. However, when μ goes below the mass, Λ, of one of the particles in the theory, we must change the effective theory to a new theory without that particle. In the process, the parameters of the theory change, and new, nonrenormalizable interactions may be introduced. Both the changes in existing parameters, and the coefficients of the new interactions are computed by "matching" the physics just below the boundary in the two theories.[1] It is this process that determines the sizes of the nonrenormalizable terms associated with the heavy particles. Because matching is done for $\mu \approx \Lambda$, the rule for the size of the coefficients of the new operators is simple for $\mu \approx \Lambda$. At this scale, all the new contributions scale with Λ to the appropriate power (set by dimensional analysis) up to factors of coupling constants, group theory or counting factors and loop factors (of $16\pi^2$, *etc.*). Then when the new effective theory is evolved down to smaller μ, the renormalization group introduces additional factors into the coefficients. Thus a heavy particle mass appears in the parameters of an effective field theory in two ways. There is power dependence on the mass that arises from matching conditions. There is also logarithmic dependence that arises from the renormalization group.

Generally, the only way we have of calculating matching corrections is in perturbation theory in some small parameter. In previous discussions of this idea, I have assumed that this is a typical perturbation expansion. Here, I will assume that we are doing a loop expansion instead. This has some technical advantages. The change in the low energy Lagrangian, $\delta\mathcal{L}(\phi)$, can be expanded in number of loops, which if there are no factors of $\hbar$ hidden in the parameters of the Lagrangians is equivalent[2] to an expansion in powers of $\hbar$

$$\Rightarrow \delta\mathcal{L} = \sum_{n=0}^{\infty} \delta\mathcal{L}^n . \tag{1}$$

Now we want to choose $\delta\mathcal{L}(\phi)$ so that the physics of the ϕ particles is the same in the two theories at $\mu = \Lambda$, just at the boundary.

---

[1]In a sense, the renormalization group is simply the "matching" of the theory at the scale μ to the theory at the scale $\mu - d\mu$ without changing the particle content.

[2]There are a number of options here. I will regard the Lagrangian in the high energy theory, including the non-renormalizable terms, as a zero loop effect, with no internal $\hbar$s. However, one could equally well keep track of where the non-renormalizable terms come from, in which case the Lagrangian in the high energy theory would also have a nontrivial loop expansion.

The strongest form of this equivalence that we might impose is that all one-light-particle irreducible ($1LPI$) functions with external light particles are the same. By $1LPI$, I mean a graph that cannot be disconnected by cutting a single **light** particle line. We can describe this in terms of the equality of the light particle effective actions in to two descriptions. The light particle effective action in the low energy theory is just the effective action. In the high energy theory, the light particle effective action is defined as follows. Construct the generating functional for connected Feynman graphs with light particle external lines, and then Legendre transform it. This gives the generating functional for $1LPI$ graphs. Then the matching condition is

$$S_{\mathcal{L}_H+\mathcal{L}}(\Phi) = S_{\mathcal{L}+\delta\mathcal{L}}(\Phi) \,. \tag{2}$$

Formally, let's denote the light particle effective action produced by an arbitrary Lagrangian, $\mathcal{L}$ (in the relevant effective theory), by

$$S_{\mathcal{L}}(\Phi) = \sum_{n=0}^{\infty} S^n_{\mathcal{L}}(\Phi) \,, \tag{3}$$

where n is the number of loops.

At tree level, the light particle effective action in the low energy theory is just the Lagrangian itself, thus

$$S^0_{\mathcal{L}}(\Phi) = \int \mathcal{L}^0(\Phi) \tag{4}$$

$$S^0_{\mathcal{L}+\delta\mathcal{L}}(\Phi) = \int \mathcal{L}^0(\Phi) + \int \delta\mathcal{L}^0(\Phi) \tag{5}$$

In the high energy theory, the light particle effective action is the sum of all $1LPI$ tree graphs. Thus it is a sum of the Lagrangian plus all trees with external light particles and only internal heavy particle lines:

$$S^0_{\mathcal{L}_H+\mathcal{L}}(\Phi) = \int \mathcal{L}^0(\Phi) + \int \left\{ \begin{matrix} \text{virtual heavy} \\ \text{particle trees} \end{matrix} \right\}(\Phi) \tag{6}$$

Thus

$$\begin{aligned} \int \delta\mathcal{L}^0(\Phi) &= S^0_{\mathcal{L}_H+\mathcal{L}}(\Phi) - S^0_{\mathcal{L}}(\Phi) \\ &= \int \left\{ \begin{matrix} \text{virtual heavy} \\ \text{particle trees} \end{matrix} \right\}(\Phi) \end{aligned} \tag{7}$$

This is rather trivial, but you already see the fact that the matching correction is a difference between a calculation in the full theory and one in the low energy effective theory.

The notation here is slightly imprecise, but intentionally so. The change in the "Lagrangian", $\int \delta\mathcal{L}^0(\Phi)$, is still nonlocal, in general because it depends on p/Λ though the virtual heavy particle propagators. However, it is **analytic** in p/Λ everywhere in the region relevant to the low energy theory. Thus it can be expanded in powers with the higher order terms steadily deceasing in importance and dealt with as a local

Lagrangian. However, I will explore the idea that it may be possible to define the low energy theory directly in terms of the nonlocal action $\int(\mathcal{L} + \delta\mathcal{L})$ — that locality is not required, just analyticity. One problem with this speculation is that there is no satisfactory name for this object. It would be natural to call it the effective action, but that name has already been taken for something else. I will call $(\mathcal{L} + \delta\mathcal{L})$ the "nonlocal effective Lagrangian". This mildly oxymoronic name will, I hope, be a reminder that because of its analyticity in momentum, this object can be thought of either as a nonlocal function of momenta, or as a power series in p/Λ.

This situation is much more interesting at one loop. From the general properties of the effective action, we can write

$$S^1_{\mathcal{L}+\delta\mathcal{L}}(\Phi) = S^1_{\mathcal{L}+\delta\mathcal{L}^0}(\Phi) + \int \delta\mathcal{L}^1(\Phi). \tag{8}$$

This is because the one loop term, $\delta\mathcal{L}^1(\Phi)$, can appear only in tree diagrams (otherwise it would be more than one loop) that are not $1LPI$. Thus the matching condition to one loop can be written

$$S^1_{\mathcal{L}_H+\mathcal{L}}(\Phi) - S^1_{\mathcal{L}+\delta\mathcal{L}^0}(\Phi) = \int \delta\mathcal{L}^1(\Phi). \tag{9}$$

We can also write this in terms of the $1LPI$ vertices that $\delta\mathcal{L}^1(\Phi)$ generates, defined by (for any number of loops, n)

$$\delta\mathcal{L}^n(\phi) = \sum_j \phi^j \delta\gamma^{n,j} \tag{10}$$

In terms of the $\delta\gamma$s, (9) becomes

$$\delta\gamma^{1,j} = \Gamma^{1,j}_{\mathcal{L}_H+\mathcal{L}} - \Gamma^{1,j}_{\mathcal{L}+\delta\mathcal{L}^0} \tag{11}$$

where $\Gamma^{n,j}_{\mathcal{L}}$ is the $1LPI$ j-point function in n loops for Lagrangian $\mathcal{L}$

(9) (or equivalently, (11)) is the mathematical statement of the words above about disentangling the long and short distance physics, at least to one loop. Let us pause to consider what it means. Both $S^1_{\mathcal{L}_H+\mathcal{L}}(\Phi)$ and $S^1_{\mathcal{L}}(\Phi + \delta\mathcal{L}^0)$ are nonlocal functions of Φ in which the nonlocality is determined by the long distance physics. If there are massless particles in the theory, the $1PI$ functions may not even be analytic in the momenta, p as $p \to 0$. However, in the difference, all the long distance physics that gives rise to this nonanalyticity cancels. All the remaining nonlocality is due to the propagation of the heavy particles and the result can be expanded in powers of p/Λ with no large coefficients to give the infinite tower of terms in the effective theory. This gives the one loop contribution to the change in the effective Lagrangian.

It is straightforward to go beyond one loop. The only subtlety is that now the lower order changes in $\delta\mathcal{L}$ can also contribute to the long distance physics, and must be subtracted. This can be done using the following:

$$S^n_{\mathcal{L}+\delta\mathcal{L}}(\Phi) = S^n_{\mathcal{L}+\delta\mathcal{L}_{n-1}}(\Phi) + \int \delta\mathcal{L}^n(\Phi) \tag{12}$$

where

$$\delta\mathcal{L}_k \equiv \sum_{n=0}^{k} \delta\mathcal{L}^n . \tag{13}$$

matching condition to n loops can be written

$$\int \delta\mathcal{L}^n(\Phi) = S^n_{\mathcal{L}_H+\mathcal{L}}(\Phi) - S^n_{\mathcal{L}+\delta\mathcal{L}_{n-1}}(\Phi) , \tag{14}$$

or in terms of $1LPI$ functions,

$$\delta\gamma^{n,j} = \Gamma^{n,j}_{\mathcal{L}_H+\mathcal{L}} - \Gamma^{n,j}_{\mathcal{L}+\delta\mathcal{L}_{n-1}} . \tag{15}$$

As usual in perturbation theory, we need the lower order results to compute the higher order results.

After this little formal review, we can return to the main question, which is what to do about the infinite series of terms of higher and higher dimension that are generated by matching. These cause no problem when the scales are well separated, because their effects quickly become negligible. But if there are two or more scales close together (such as m_t and M_Z may be), then we must do something about them. I would like to understand how to interpolate smoothly between the well-understood situation in which the scales are very different and the well-understood situation in which the scales are very close together so that both particles can be integrated out at the same time.

The obvious thing to do is simply to keep track of all the infinite number of higher derivative operators. The question is whether we can do this efficiently. The discussion of matching above suggests a possible approach. The "nonlocal effective Lagrangian" that comes from matching is usually expanded in powers of momenta to produce local interactions. **Why not just deal with it directly without expanding?** The idea is to replace the usual requirement of locality with the assumption of analyticity.

I am going to give you one example of what I mean in a scalar field theory. Suppose that some unknown physics at high energy leaves a single massless scalar field, ϕ, with a discrete reflection symmetry, $\phi \to -\phi$. I will discuss the renormalization of the nonlocal ϕ^4 interaction in this theory. Of course, in addition to the ϕ^4 interaction, there will be ϕ^6 and higher terms. However, as we will see, in one loop, we can ignore them. I will discuss the following diagram:

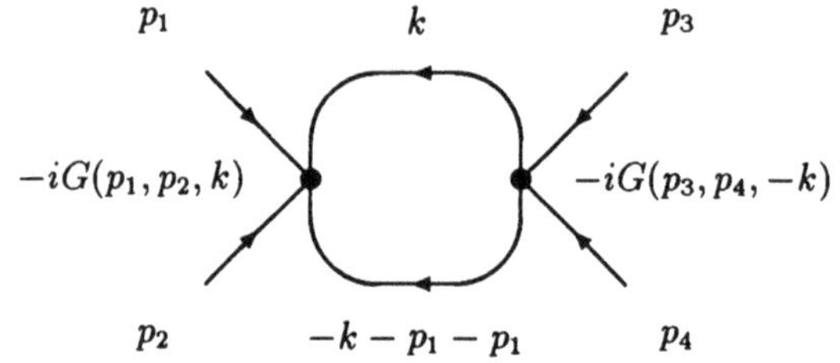

(16)

Fig. 1: Feynman graph contributing to the renormalization of G.

where $p_4 = -(p_1 + p_2 + p_3)$. The nonlocal coupling G is symmetric and satisfies

$$G(p_1, p_2, p_3) = G(p_1, p_2, -p_1 - p_2 - p_3) \tag{17}$$

G is the Fourier transform of the position space function

$$\tilde{G}(x_1 - x_4, x_2 - x_4, x_3 - x_4) \tag{18}$$

The Feynman graph is

$$-\int \frac{d^4k}{(2\pi)^4} \frac{G(p_1, p_2, k)\, G(p_3, p_4, -k)}{[k^2 + i\epsilon][(k + p_1 + p_2)^2 + i\epsilon]} \tag{19}$$

Now combine denominators

$$\begin{aligned} &\frac{1}{[k^2 + i\epsilon][(k + p_1 + p_2)^2 + i\epsilon]} \\ &= \int_0^1 d\alpha \frac{1}{[(1-\alpha)k^2 + \alpha(k + p_1 + p_2)^2 + i\epsilon]^2} \end{aligned} \tag{20}$$

and shift momenta

$$k = \ell - \alpha(p_1 + p_2) \tag{21}$$

to obtain

$$-\int_0^1 d\alpha \int \frac{d^4\ell}{(2\pi)^4} \frac{G(p_1, p_2, \ell - \alpha(p_1 + p_2))\, G(p_3, p_4, -\ell + \alpha(p_1 + p_2))}{[\ell^2 + \alpha(1-\alpha)(p_1 + p_2)^2 + i\epsilon]^2} \tag{22}$$

Now, because the Gs are analytic, we can deal with the ℓ dependence of G by writing a symbolic Taylor expansion

$$\begin{aligned} &G(p_1, p_2, \ell - \alpha(p_1 + p_2))\, G(p_3, p_4, -\ell + \alpha(p_1 + p_2)) \\ &= e^{\ell \frac{\partial}{\partial q}} \left[G(p_1, p_2, q)\, G(p_3, p_4, -q)\right]_{q = -\alpha(p_1 + p_2)} \end{aligned} \tag{23}$$

Then we just have to do the Feynman integral of the exponential

$$\int \frac{d^4\ell}{(2\pi)^4} \frac{e^{\ell \frac{\partial}{\partial q}}}{\left[\ell^2 + \alpha(1-\alpha)(p_1+p_2)^2 + i\epsilon\right]^2} \tag{24}$$

or to be more precise, the dimensionally regularized integral

$$\int \frac{d^{4-\epsilon}\ell}{(2\pi)^{4-\epsilon}\mu^{-\epsilon}} \frac{e^{\ell \frac{\partial}{\partial q}}}{\left[\ell^2 + \alpha(1-\alpha)(p_1+p_2)^2 + i\epsilon\right]^2} \tag{25}$$

We will do this by manipulating the exponential like a power series, because the basic assumption is analyticity in the momenta. Because of symmetry, only even terms in ℓ contribute,

$$\sum_{n=0}^{\infty} \frac{1}{(2n)!} \int \frac{d^{4-\epsilon}\ell}{(2\pi)^{4-\epsilon}\mu^{-\epsilon}} \frac{\left(\ell \frac{\partial}{\partial q}\right)^{2n}}{\left[\ell^2 + \alpha(1-\alpha)(p_1+p_2)^2 + i\epsilon\right]^2} \tag{26}$$

We calculate this as follows

$$\begin{aligned} &\frac{1}{\int d\Omega} \int d\Omega_\ell \, \ell_{\mu_1} \ell_{\mu_2} \cdots \ell_{\mu_{2n}} \\ &= A_n \left(\ell^2\right)^n \overbrace{\left[g_{\mu_1\mu_2} \cdots g_{\mu_{2n-1}\mu_{2n}} + \text{perms}\right]}^{(2n)!/(2^n n!) \text{ terms}} \end{aligned} \tag{27}$$

Hit with $g^{\mu_1\mu_2}$

$$A_{n-1} = (4 + 2(n-1)) \, A_n = 2(n+1) \, A_n \tag{28}$$

So

$$A_n = \frac{1}{2^n (n+1)!} \tag{29}$$

and thus

$$\begin{aligned} &\frac{1}{(2n)!} \frac{1}{\int d\Omega} \int d\Omega_\ell \left(\ell \frac{\partial}{\partial q}\right)^{2n} \\ &= \frac{1}{4^n n!(n+1)!} \left[\ell^2\right]^n \left[\left(\frac{\partial}{\partial q}\right)^2\right]^n \end{aligned} \tag{30}$$

Thus the integral is

$$\sum_{n=0}^{\infty} \frac{1}{4^n n!(n+1)!} \int \frac{d^{4-\epsilon}\ell}{(2\pi)^{4-\epsilon}\mu^{-\epsilon}} \frac{[\ell^2]^n \left[\left(\frac{\partial}{\partial q}\right)^2\right]^n}{\left[\ell^2 + \alpha(1-\alpha)(p_1+p_2)^2 + i\epsilon\right]^2} \tag{31}$$

Now we can calculate this in the usual way. Wick rotate

$$i \sum_{n=0}^{\infty} \frac{(-1)^n}{4^n n!(n+1)!} \int \frac{d^{4-\epsilon}\ell}{(2\pi)^{4-\epsilon}\mu^{-\epsilon}} \frac{[\ell^2]^n \left[\left(\frac{\partial}{\partial q}\right)^2\right]^n}{\left[\ell^2 - \alpha(1-\alpha)(p_1+p_2)^2\right]^2} \tag{32}$$

Now doing the ℓ integral gives

$$\frac{2i}{(4\pi)^2\epsilon\mu^\epsilon}\sum_{n=0}^{\infty}\frac{1}{4^n n!^2}[-\alpha(1-\alpha)(p_1+p_2)^2]^n\left[\left(\frac{\partial}{\partial q}\right)^2\right]^n+\cdots \tag{33}$$

where we have dropped everything but the $1/\epsilon$ pole and the associated μ dependence. If we choose, we can do the sum (it is a Bessel function), however, it is more useful to write it as follows:

$$\frac{i}{(4\pi)^2\epsilon\mu^\epsilon}\left(\frac{\partial}{\partial x}+2\right)\sum_{n=0}^{\infty}\frac{1}{4^n n!(n+1)!}[-x^2\alpha(1-\alpha)(p_1+p_2)^2]^n\left[\left(\frac{\partial}{\partial q}\right)^2\right]^n \tag{34}$$

evaluated at $x=1$.

This can then be turned back into the (Euclidean) angular average of an exponential, just inverting (30).,

$$\frac{i}{(4\pi)^2\epsilon\mu^\epsilon}\left(\frac{\partial}{\partial x}+2\right)\frac{1}{\int d\Omega}\int d\Omega_e\; e^{x\sqrt{-\alpha(1-\alpha)(p_1+p_2)^2}\,e\frac{\partial}{\partial q}} \tag{35}$$

at $x=1$, where e is a Euclidean unit vector. The square root is real for Euclidean momentum p_1+p_2.

Now that we have done the ℓ integral and extracted the $1/\epsilon$ pole, we can now use the Taylor series to put this back into the Gs. The result is

$$\begin{gathered}\frac{i}{(4\pi)^2\epsilon\mu^\epsilon}\left(\frac{\partial}{\partial x}+2\right)\frac{1}{\int d\Omega}\int d\Omega_e\int_0^1 d\alpha \\ \Big[G(p_1,p_2,x\sqrt{\alpha(1-\alpha)(p_1+p_2)^2}\,e-\alpha(p_1+p_2)) \\ G(p_3,-p_1-p_2-p_3,-x\sqrt{\alpha(1-\alpha)(p_1+p_2)^2}\,e+\alpha(p_1+p_2))\Big]_{x=1}\end{gathered} \tag{36}$$

at $x=1$.

Thus we have reduced the divergent part of the original Feynman graph to an integral of the Gs and their derivatives over **finite ranges** (for Euclidean momenta). These last integrals could even be done numerically. This is essentially the one-loop β function for the nonlocal interaction, G.

$$\begin{gathered}\beta_{G(p_1,p_2,p_3)}=\frac{1}{(4\pi)^2}\left(\frac{\partial}{\partial x}+2\right)\frac{1}{\int d\Omega}\int d\Omega_e\int_0^1 d\alpha \\ \Big[G(p_1,p_2,x\sqrt{\alpha(1-\alpha)(p_1+p_2)^2}\,e-\alpha(p_1+p_2)) \\ G(p_3,-p_1-p_2-p_3,-x\sqrt{\alpha(1-\alpha)(p_1+p_2)^2}\,e+\alpha(p_1+p_2))\Big]_{x=1} \\ +\begin{array}{c}\text{cross}\\ \text{terms}\end{array}\end{gathered} \tag{37}$$

The "running" nonlocal coupling satisfies the integrodifferential renormalization group equation

$$\mu \frac{\partial}{\partial \mu} G(p_1, p_2, p_3) = \beta_{G(p_1, p_2, p_3)} \tag{38}$$

In addition, the nonlocal quartic coupling G induces changes in the coupling terms with more fields. The relevant one-loop diagram is

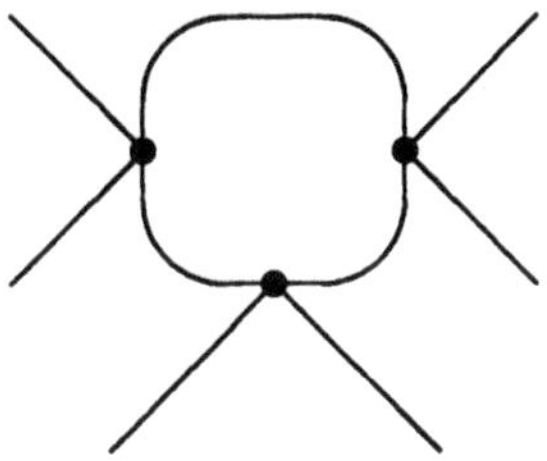

(39)

Fig. 2: Feynman graph contributing to the running of couplings with more than four ϕs.

However, for massless fields, these do not mix back into G in a mass independent renormalization scheme, because the relevant Feynman graph (below) vanishes.

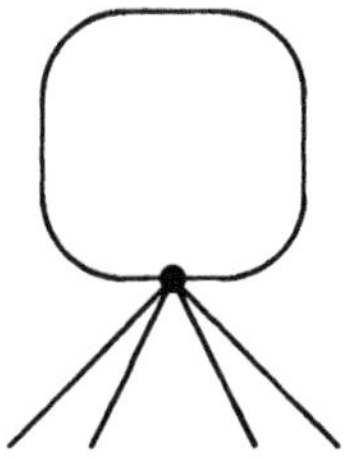

(40)

Fig. 2: Feynman graph that vanishes for massless ϕs.

Light Quark Masses beyond leading order*

D. Wyler
Institut für Theoretische Physik, Universität Zürich,
Schönberggasse 9, CH-8001 Zürich, Schweiz

Abstract

In second order in the chiral expansion there is an ambiguity in the meaning of the quark masses which makes it impossible to determine their ratios. We show that the θ-dependence provides a mean to overcome this problem. Within a multipole expansion in m_ψ, the decays $\psi' \to \psi\pi^0(\eta, 3\pi)$ allow to use the θ-dependence to obtain unique values for the mass ratios. We find $m_u/m_d = 0.3 \pm 0.05 \pm 0.05$ where the first error is experimental and the second our estimate of the remaining theoretical model dependence.

1 Introduction

The masses of the quarks are among the fundamental parameters of the standard model and should be determined as accurately as possible. Because of color confinement, free quarks do not exist and thus no inertial experiment can determine their masses. There are always binding effects of order of the QCD scale (≤ 1 GeV) which overshadow the quark masses, in particular those of the light quarks (u, d and s). Therefore, the light quark masses should rather be viewed as coupling constants which enter the measurable quantities in a well defined way. Usually, the observables are calculated perturbatively from the couplings; however in the case of the light quark masses no perturbative method works. However, one may track down the masses, because they enter observables in a well prescribed fashion, namely by their symmetry properties.

Chiral perturbation[2] theory provides a rigorous, semiphenomenological procedure to extract ratios of quark masses. The method relies on the fact that in the limit

*This paper is based on work done in collaboration with J. Donoghue[1]

of vanishing masses, there is an exact (chiral) symmetry. Deviations from the exact symmetry predictions are functions of the masses themselves and can be expanded in a series in the masses, if these are small compared to hadronic energies.

The knowledge of the masses is particularly important because of the strong CP-problem[3]. The only solution within the standard model is to have $m_u = 0$ (since $m_d > m_u, m_s \gg m_u$, only the up quark mass could possibly vanish), and it is therefore important to know whether $m_u = 0$ can be excluded from low-energy phenomenology.

In lowest order in the chiral expansion (that is the expansion in energy of the particles and the masses) the analysis of Gasser and Leutwyler[4] yields

$$\frac{m_u}{m_s} = \frac{1}{35} \qquad \frac{m_d}{m_s} = \frac{1}{19} \qquad \frac{m_u}{m_d} = \frac{1}{1.8}\,. \tag{1}$$

In next order, there is an ambiguity in the meaning of the masses, pointed out by Kaplan and Manohar[5]. Its origin is the fact that in constructing the chiral expansion, the only rule is that the mass matrix transforms correctly. Clearly then, any function of the mass matrix with the prescribed transformation property is an acceptable mass matrix. Each such "mass" corresponds to a set of low-energy coupling constants; the complete effective Lagrangian is then invariant under a change of the mass and the corresponding redefinitions of the couplings. This ambiguity is, of course, not a property of the original QCD theory, where the mass has—up to renormalization—a well defined meaning, but a consequence of the chiral method, of extracting the masses of the quarks.

In this situation one would like to give a procedure to overcome the ambiguity. That is, one needs quantities which do not share the reparametrization invariance. Whereas all S-matrix elements are invariant, one can pick out certain Green's functions[6] or matrix elements of some operators which are not invariant. The existence of these relies on the fact that the reparametrization transformation depends on the external fields (see below for a detailed argument).

The next question is whether these non-invariant quantities can be directly related to physical observables. So far, the answer is no (with an unrealistic exception). In order to isolate such quantities , additional assumptions have to be provided, usually motivated by physics arguments. We will consider them below. Similarly, calculations of low-energy constants, based on the underlying theory[7], are means to overcome the ambiguity.

Secondly, a condition must be found which selects, among the possible mass matrices, the "one" of QCD. Such a condition must be transparent enough to ensure that the above mentioned physics argument must hold for the QCD mass matrix.

2 Chiral Lagrangians

We will consider an effective Lagrangian of pseudoscalar mesons and work in the framework of Gasser and Leutwyler[4].

Take QCD with three quark flavors u, d, s coupled to external sources l_μ, r_μ, s, p, θ. Since l_μ, r_μ are not of interest to us here, we drop them. We have

$$\mathcal{L}_{QCD} = -\frac{1}{4} F^2_{\mu\nu} + i\bar{\psi}\partial\!\!\!/\psi - \bar{\psi}_L(s+ip)\psi_R - \bar{\psi}_R(s-ip)\psi_L + \frac{\theta\alpha_s}{8\pi} F_{\mu\nu}\tilde{F}^{\mu\nu}, \qquad (2)$$

$$\psi = (u, d, s).$$

The sources s, p are matrices (octets) in flavor space and θ is an SU(3) singlet. For $\theta = \bar{\theta} = \text{const.}$ and

$$s = m = \begin{pmatrix} m_u & 0 & 0 \\ 0 & m_d & 0 \\ 0 & 0 & m_s \end{pmatrix}, \qquad (3)$$

QCD without sources is reproduced. $\mathcal{L}$ is invariant under local $\mathrm{SU}(3)_L \times \mathrm{SU}(3)_R$ transformations ($L \in \mathrm{SU}(3)_L$, $R \in \mathrm{SU}(3)_R$)

$$\psi_L \to L\psi_L \qquad \psi_R \to R\psi_R \qquad (4)$$

if the sources transform according to

$$\begin{aligned} (s+ip) &\to L(s+ip)R^+ \\ \theta &\to \theta\,. \end{aligned} \qquad (5)$$

At low energies, the heavier particles are "integrated out" and $\mathcal{L}$ is replaced by an effective Lagrangian which only depends on the lightest particles, the pseudoscalar mesons. It is convenient to combine them into an unitary matrix U:

$$\begin{aligned} U &= e^{-i\phi} & \det U &= 1 \\ \phi &= \sum_1^8 \frac{\lambda_i}{F}\phi_i & \phi_i &= \{\pi, K, \eta\}\,. \end{aligned} \qquad (6)$$

F is a parameter, $F \approx f_\pi$. λ_i are the eight Gell-Mann matrices and under $\mathrm{SU}(3)_L \times \mathrm{SU}(3)_R$, we have

$$U \to LUR^+\,. \qquad (7)$$

(5) and (7) determine the effective Lagrangian in terms of U and the external field $\chi = s + ip$. The coupling of θ can be inferred from looking at the behaviour of (2) under a $U(1)_A$ transformation $\psi \to e^{-i\alpha\gamma_5}\psi$. ($U$ does not transform under $U(1)_A$, since $U \in \mathrm{SU}(3)$). This chiral rotation will induce a change in θ, according to the usual anomaly equation:

$$\theta \to \theta - 6\alpha\,. \qquad (8)$$

On the other hand, (2) remains invariant, if we let $\chi \to \chi e^{2i\alpha}$. Thus, it is the combination $\chi_\theta = \chi e^{i\theta/3}$ which is invariant under the $U(1)_A$ transformation and which must be used in the chiral Lagrangian. (This conclusion can also be reached by considering all nine pseudoscalar mesons and integrating out the singlet field η'[1].)

The effective Lagrangian now takes the form

$$\mathcal{L} = \mathcal{L}_0 + \mathcal{L}_{mix} + \mathcal{L}_2 + \mathcal{L}_4 + \cdots \tag{9}$$

with

$$\begin{aligned}
\mathcal{L}_0 &= \text{const}\,(\partial\theta)^2 && (10)\\
\mathcal{L}_{mix} &= L_{14}\partial_\mu\theta\,\text{tr}(\partial_\mu\chi_\theta^+ U - U^+\partial_\mu\chi_\theta)\\
&\quad + L_{15}\,\Box\theta\,\text{tr}(\chi_\theta^+ U - U^+\chi_\theta) && (11)\\
\mathcal{L}_2 &= \frac{F^2}{4}\left[\text{tr}(\partial_\mu U\partial^\mu U^+) + \text{tr}(X_\theta^+ U + U^+\chi_\theta)\right] && (12)\\
\mathcal{L}_4 &= L_6\left(\text{tr}(\chi_\theta^+ U + U^+\chi_\theta)\right)^2 + L_7\left(\text{tr}(\chi_\theta^+ U - U^+\chi_\theta)\right)^2\\
&\quad + L_8\,\text{tr}(\chi_\theta U^+\chi_\theta U^+ + \chi_\theta^+ U\chi_\theta^+ U) + \cdots && (13)
\end{aligned}$$

Here, we have omitted pieces which are not relevant for the present investigation. The basic assumption of chiral perturbation theory is that terms with higher powers of momenta and χ are small. The L_i are phenomenological constants (see[4]).

The lowest order Lagrangian $\mathcal{L}_2$ gives, upon expanding U to second power in the field ϕ and setting

$$\chi = 2B_0 m \equiv 2M \tag{14}$$

the well known expressions for the meson masses, such as $m_\pi^2 = B_0(m_u + m_d)$ etc., from which the ratios of the m_q are obtained. However, if $\mathcal{L}_4$ etc. are included, the ambiguity mentioned surfaces: If we replace

$$\begin{aligned}
\chi_\theta &\to \chi_\theta^\lambda = \chi_\theta + \lambda(\det\chi_\theta^+)(\chi_\theta^+)^{-1}e^{-i\theta}\\
L_6 &\to L_6^\lambda = L_6 - \lambda\frac{F^2}{16}\\
L_7 &\to L_7^\lambda = L_7 - \lambda\frac{F^2}{16}\\
L_8 &\to L_8^\lambda = L_8 + \lambda\, 2\frac{F^2}{16}\,,
\end{aligned} \tag{15}$$

then, up to higher powers in χ, the Lagrangian $\mathcal{L}_0 + \mathcal{L}_2 + \mathcal{L}_4$ does not change its form, while $\mathcal{L}_{mix}$ remains invariant.

The proof is a straightforward application of the Caley-Hamilton theorem for 3×3 matrices,

$$0 = -A^3 + A^2\,\mathrm{tr}A + \frac{A}{2}[\mathrm{tr}(A^2) - (\mathrm{tr}A)^2] + \det A\,, \tag{16}$$

which shows that $A^{-1}\det A$ can be written in powers of A. Setting $A = (\chi_\theta^+ U)$ or $(U^+\chi_\theta)$ then yields after a bit of algebra, the rule (15).

This implies that physical quantities derived from (9) can only depend on λ-independent combinations of χ and the L_i. For instance, it is straightforward to show that to second order in χ we have (set $\theta = 0$)

$$f_\pi^2 m_\pi^2 = M_u^* + M_d^* \tag{17}$$

with ($M = B_0 m$)

$$\begin{aligned} M_u^* &= M_u\{1 + \frac{32L_6}{f_\pi^2}(M_u + M_d + M_s) + \frac{32L_8}{f_\pi^2}M_u - 3\mu_\pi - 2\mu K^+ - \frac{1}{3}\mu_\eta \\ &+ \frac{1}{2}\left(\frac{m_d - m_u}{m_s - \hat{m}}\right)(\mu_\eta - \mu_\pi)\} + \frac{32L_7}{f\pi^2}(M_u - M_d)(M_u - M_s) \end{aligned} \tag{18}$$

$$\mu_i = \frac{m_i^2}{32\pi^2 f_\pi^2}\ln\frac{m_i^2}{\mu^2}, \tag{19}$$

etc. Here, μ is the renormalization scale. By applying (15) to M^*, and neglecting higher powers in χ, we easily see that M^* is indeed invariant. The numerical analysis yields

$$\frac{\hat{m}^*}{m_s^*} = \frac{1}{37} \qquad \frac{m_d^* - m_u^*}{m_d^* + m_u^*} = 0.57\,. \tag{20}$$

Therefore, one cannot distinguish among the different χ^λ from purely chiral Lagrangian techniques. In particular, one could have $m_u = 0$ and compensate this by changing the values of the L_i. Of course, this would imply that $\mathcal{L}_4$ changes the predictions of $\mathcal{L}_2$ drastically and therefore the chiral expansion itself would not be too meaningful. However, since m_u is much smaller than all other masses, its relative corrections can still be sizeable anyway, in particular if only some of the higher order corrections are included: We have found that the shift (due to $\mathcal{L}_4$) in the decay constant of the kaon alone implies $\delta m_u/m_u \simeq 0.8$ [1].

3 The θ-dependence of the Lagrangian can overcome the invariance

As emphasized, we must find a non-invariant quantity. It turns out, that the θ-dependence of the transformation rule (15) provides the desired term. In order to see clearly how a non-invariant quantity term arises in an invariant Lagrangian, we expand (9) to first order in θ and write

$$\mathcal{L} \cong \mathcal{L}_C(\chi_0, L_i, U) + \theta \frac{\partial \mathcal{L}}{\partial \theta}(\chi_0, L_i, U) . \tag{21}$$

Under the change (15) we have

$$\delta\mathcal{L} = \left(\frac{\partial \mathcal{L}_C}{\partial \chi_0} \delta\chi_0 + \sum_i \frac{\partial \mathcal{L}_C}{\partial L_i} \delta L_i \right) + \theta\, \delta\left(\frac{\partial \mathcal{L}}{\partial \theta} \right), \tag{22}$$

with

$$\begin{aligned} \delta\chi_0 &= \delta\lambda (\det\chi_0^+)(\chi_0^+)^{-1} e^{-i\theta} . \\ \delta L_6 &= -\delta\lambda \frac{F^2}{16} \qquad \text{etc.} \end{aligned}$$

If we consider only the terms linear in θ in the variation of the first term, we have

$$\delta\mathcal{L} = -i\theta \frac{\partial \mathcal{L}_C}{\partial \chi_0} \det\chi_0^+ (\chi_0^+)^{-1} \delta\lambda + \theta\delta \left(\frac{\partial \mathcal{L}}{\partial \theta} \right) . \tag{23}$$

Since $\mathcal{L}$ is invariant, the non-vanishing variation of the first term must be canceled by $\delta(\partial\mathcal{L}/\partial\theta)$. Thus $\partial\mathcal{L}/\partial\theta$ cannot be invariant and can be used to overcome the ambiguity. This result follows from the presence of θ in the transformation rule (15).

Similarly, Green's functions of non-invariant external fields (such as the pseudoscalar 2-point function) are not reparametrization invariant, and can be used[6].

In the basic QCD-Lagrangian, $\partial\mathcal{L}/\partial\theta$ is given by

$$F\tilde{F} = \frac{\alpha_s}{8\pi} F^a_{\mu\nu} \tilde{F}^{a\mu\nu} . \tag{24}$$

From Eqs. (9)–(13), the effective Lagrangian realization of this operator is ($\chi = \chi_0$)

$$\begin{aligned} F\tilde{F} &= i\,\{ \frac{F^2}{4} \operatorname{tr}(\chi U^+ - U\chi^+) + L_4 \operatorname{tr}(\partial_\mu U \partial^\mu U^+) \operatorname{tr}(\chi U^+ - U\chi^+) + L_5 \operatorname{tr}(\partial_\mu U \partial^\mu U^+) \\ &\quad + 2(L_6 + L_7) \operatorname{tr}(\chi U^+ - U\chi^+) \operatorname{tr}(\chi U^+ - U\chi^+) + 2L_8 \operatorname{tr}(\chi U^+ \chi U^+ - U\chi^+ U\chi^+) \\ &\quad - L_{14} \Box \operatorname{tr}(\chi U^+ - U\chi^+) - L_{15} \partial^\mu \operatorname{tr}(\partial_\mu \chi^+ U - U^+ \partial_\mu \chi) \} + \cdots \end{aligned} \tag{25}$$

When $\chi = 2B_0 m$, the last term can be dropped. The terms with L_4, L_5 are included for completeness [4,1].

It is easily checked that $F\tilde{F}$ is not invariant and thus its matrix elements provide the desired quantity. Furthermore, because of (15), we see that the "masses" M^λ are complex: The expansion in θ allows to pick out that mass which is *real*. Since the $m_u = 0$ solution to the strong CP-Problem only applies to the "real" mass matrix, it is indeed the real mass which is of interest (see discussion in section 5).

4 Determining quark mass ratios from the θ-dependence

As a next step, we must relate $F\tilde{F}$ to observables. If θ has an non-zero vacuum expectation value $\bar{\theta}$, there are CP-violating amplitudes of the form

$$A_{if} = \bar{\theta}\,\langle f|F\tilde{F}|i\rangle \tag{26}$$

(We assume that weak effects are smaller). The matrix element is a combination of the parameters M, L_i, F of the effective Lagrangian which is not reparametrization invariant. By measuring several such amplitudes (like in $\eta \to \pi\pi$), one can eliminate $\bar{\theta}$ and obtain the desired non-invariant information on M and the L_i. Clearly, this is impracticable, as $\bar{\theta}$ is much too small[3].

If $\bar{\theta} = 0$, θ is an external field and we must find appropriate sources which generate it. As pointed out by Voloshin and Zakharov[7], the decays $\psi' \to \psi\pi^0$, $\psi' \to \psi\eta$ are dominated by the operator $F\tilde{F}$. The idea is that these decays are caused by gluons emitted from the $\psi'\psi$-system (which therefore can be viewed as a source for θ) which then couple to the light meson. In particular, these gluons must be in a 0^- state. A corresponding gluon operator is indeed $F\tilde{F}$. Since the c, $\bar{c}$ quarks are quasi static, and ψ', ψ are relatively compact objects, a multipole expansion[8] is expected to work, in which the other 0^- operators should be suppressed by powers of $(m_\psi)^{-1}$. The chiral realization of them—for instance of the operator $f^{abc}F^a_{\mu\nu}F^{b\nu\alpha}\tilde{F}^{c\mu}_{\alpha}$—is similar to that of $F\tilde{F}$, except that the relative weight of the higher order operator with respect to the leading term will change. In particular, the form in the curly bracket of Eq. (25) will remain, except that the constants L_i will be replaced by $L_i + \delta_i$, where δ_i depends on the operator considered.

However, the leading terms is (apart from a constant which drops in the ratios considered) clearly the same for all these operators and their effects will only arise in higher order in the chiral expansion[9]. As a result, we expect their contribution to be $\mathcal{O}(m_\rho/m_\psi)\mathcal{O}(1/3) \sim 0.08$, where $1/3$ is a typical (chiral) higher order effect. But clearly, a more detailed analysis is required. A clean confirmation of the multipole method could be provided by the Υ-system, if the respective decays can be measured.

This discussion illustrates an essential point: In order to isolate non-invariant quantities, physics arguments must be used to suppress additional terms. This often requires uncontrollable "model" assumptions. We feel that the present scheme—which relies on the large mass of m_c—provides a method which works in a physical limit ($m_c \to \infty$) and is—in principle—reliable and testable.

We now write

$$\begin{aligned} A(\psi' \to \psi\pi) &= K\,\langle\pi|F\tilde{F}|0\rangle \\ A(\psi' \to \psi\eta) &= K\,\langle\eta|F\tilde{F}|0\rangle\,, \end{aligned} \tag{27}$$

where K contains the ψ'-ψ matrix element and $F\tilde{F}$ is given by (25). A standard calculation, which also takes into account π^0-η mixing yields[1]

$$r = \frac{\langle\pi^0|F\tilde{F}|0\rangle}{\langle\eta|F\tilde{F}|0\rangle} = \frac{3\sqrt{3}}{4}\frac{m_d - m_u}{m_s - \hat{m}}\frac{f_\eta}{f_\pi}\{1 - \frac{32}{F^2}(M_s - \hat{M})(L_7 - L_8) + \frac{4L_{14}}{F^2}(m_\pi^2 - m_\eta^2) - (3\mu_\pi - 2\mu_K - \mu_\eta)\}, \tag{28}$$

where r is determined from

$$\frac{\Gamma(\psi' \to \psi\pi^0)}{\Gamma(\psi' \to \psi\eta)} = (3.6 \pm 0.9) \cdot 10^{-4} \tag{29}$$

to be

$$r = 0.043 \pm 0.005\ . \tag{30}$$

We can solve (28) for the quark mass ratios and obtain

$$\frac{m_d - m_u}{m_d + m_u}\frac{m_s + \hat{m}}{m_s - \hat{m}} = r\frac{4(f_K^2 m_K^2 - f_\pi^2 m_\pi^2)f_\pi}{3\sqrt{3}\, f_\pi^2 m_\pi^2 f_\eta}(1 - \delta_{GMO}) \times \left[1 + \frac{4L_{14}}{f_\pi^2}(m_\eta^2 - m_\pi^2)\right]\ . \tag{31}$$

Here, δ_{GMO} measures the deviation from the Gell-Mann-Okubo formula[4]; its value is -0.06.

Because of the unknown coupling constant L_{14}, (31) does not determine completely the mass ratios on the left-hand side. L_{14} can, in principle, be determined from a measurement of $\psi' \to \psi\pi\pi\pi$ which also proceeds via the operator $F\tilde{F}$. One obtains

$$\frac{\langle\pi^+\pi^-\pi^0|F\tilde{F}|0\rangle}{\langle\pi^0|F\tilde{F}|0\rangle} = -\frac{2}{9f_\pi^2}\{1 + \frac{24(2L_4 + L_5)}{f_\pi^2}p_+p_- + \frac{32L_7}{f_\pi^2}(m_K^2 - m_\pi^2) + \frac{192(L_6 + L_7)m_\pi^2}{f_\pi^2} + \frac{4L_{14}}{f_\pi^2}(s - m_\pi^2)\}, \tag{32}$$

with $s = (p_+ + p_- + p_0)^2$. Chiral logarithms are not taken into account here. Note, that also the three-pion amplitude is proportional to the isospin violating difference $(m_d - m_u)$ in this order.

Since there is no data on this decay, we can only estimate the range of L_{14}. One expects L_{14} to be dominated by a single resonance (in the spirit of ref. 6), here the η'. This yields

$$L_{14} = \frac{f_\pi^2}{4m_{\eta'}^2} \ . \tag{33}$$

Therefore, we will use the range

$$0 \leq L_{14} \leq \frac{f_\pi^2}{4m_{\eta'}^2} \tag{34}$$

to estimate the effect of L_{14}.

We can now combine all the numerical ingredients and obtain

$$\frac{m_u}{m_d} = 0.30 \pm 0.05 \pm 0.05 \ , \tag{35}$$

where the first uncertainty is experimental and the second comes from (34). This value is below the first order estimate of ~ 0.5, but is safely away from zero.

This lower value of m_u/m_d implies a different value for the constant L_7 that has been previous determined—both from experiment [4] and from a theoretical analysis[6] which used single resonance dominance. However, there the part of the K^0-K^+ mass difference, which originates from the up-down quark mass difference only is required. One obtains[4]

$$\frac{m_K^2}{m_\pi^2}\frac{(m_{K^0}^2 - m_{K^+}^2)_{QM}}{m_K^2 - m_\pi^2} = \frac{m_d - m_u}{m_d + m_u}\frac{m_s + \hat{m}}{m_s - \hat{m}}(1 + 2\Delta_m) \tag{36}$$

with

$$\Delta m = 1 - \frac{f_K^2}{f_\pi^2} - \delta_{GMO} - \frac{32L_7}{f_\pi^2}(m_K^2 - m_\pi^2). \tag{37}$$

(The subscript QM indicates quark mass difference). In order to evaluate $(m_{K^0}^2 - m_{K^+}^2)_{QM}$, a calculation of the electromagnetic effects is needed. In the previous work, use was made of Dashen's theorem, giving 0.29 for the left-hand side of Eq. (36). A direct calculation[10] yields 0.41. When combined with our value for the quark mass ratio in Eq. (35), one obtains $L_7 \simeq -0.3 \cdot 10^{-3}$ very near to the value of $-0.4 \cdot 10^{-3}$ of the refs. 4 and 6. This illustrates that the various approaches to the quark mass ratio are—numerically— not inconsistent, if all physical effects are taken into account.

5 Conclusions and discussion

Exploiting the θ-dependence of QCD, the reparametrization invariance of fourth order chiral perturbation theory can be overcome and the quark mass ratios of the light quarks determined. To implement this procedure, the multipole expansion[7–9] in powers of $(m_\psi)^{-1}$ for the decays $\psi' \to \psi\eta$, $\psi\pi^0$, $\psi\pi^+\pi^-\pi^0$ must be invoked. While higher order terms are expected to yield corrections below the present experimental accuracy (see above), a more quantitative analysis is desirable (A nice check of the multipole expansion would be the measurement of the related processes in the decays of the Υ system, where our approximation is expected to work even better). Furthermore, electromagnetic corrections to these decays are not included so far. We note that our analysis includes η'-η mixing in the decay $\psi' \to \psi\eta$[9]. An advantage of using the θ-dependence is that it probes the real quark mass—the one whose vanishing would solve the strong CP-problem. Our analysis yields $m_u \neq 0$, with a safe distance from zero.

Other reparametrization non-invariant quantities might be considered, such as the pseudoscalar 2-point function[6]. Its successful use requires the hypothesis of single resonance saturation and—to a certain extent—an assumption about the dominance of the lowest order terms. Also, it is not entirely clear to us, whether this method probes the real quark mass matrix which is of interest.

There remains the question whether the reparametrization has any deep meaning, or whether it is just an unwanted property of the effective Lagrangian. In a renormalizable theory, renormalized masses depend on the renormalization scale, and one might wonder whether this ambiguity is related to the one of the effective Lagrangian. (We notice that the form (9)-(13) automatically eliminates the usual arbitrariness of the QCD Lagrangian, where a redefinition of the fermion fields allows to change θ at the expense of adding a phase to the quark masses. Thus, χ_0 is really related to the magnitude of the quark masses.)

In QCD, any calculation is determined by a set of renormalized masses, preferably defined at a large scale $\mu \approx M_W$. In this region, the rescalings of the light masses are mass independent and thus ratios are invariant. At lower energies however, the theory develops more and more structure which influences the masses; for instance, instantons give contributions[11]. Without a reliable calculational tool, they are restrained by the symmetries of the physics which imply[11]

$$\delta m \simeq e^{-i\theta} m m^{-1} , \tag{38}$$

and which—of course—is the same shift as prescribed by (15). This form of the single instanton contribution describes the response to θ. It is useful if m is the renormalized mass at high energies. Then it is plausible to relate M directly to m by $M = \text{const}\, m$, and the mass ratios determined by chiral perturbation theory are indeed the scale independent ones of QCD at high energies.

6 Acknowledgements

It is a great pleasure to thank John Donoghue for a most pleasant collaboration. I thank H. Leutwyler for many important discussions an A. Manohar for comments. I also thank the organizers of this conference for a very stimulating and enjoyable meeting.

7 References

1. J. Donoghue and D. Wyler, *Light Quark Masses beyond leading order;* Univ. of Amherst preprint UMHEP-350
   This paper contains also a more extensive list of references.

2. See the talk by J. Donoghue. Recent papers and overviews include:
   J. Gasser and H. Leutwyler, *Nucl. Phys.* **B 250** (1985) 465; *Phys. Rep.* **87** (1982) 77

   J. Donoghue, E. Golowich and B. Holstein, *Dynamics of the Standard Model*, Cambridge University press 1991.

3. R. D. Peccei in *CP-violation*; Ed. C. Jarlskog; World Scientific, Singapore 1989.

4. See the articles of Gasser and Leutwyler in ref. 2

5. D. Kaplan and A. Manohar, *Phys. Rev. Lett.* **56** (1986) 1994

6. H. Leutwyler, *Nucl. Phys.* **B 337** (1990) 108

7. M. Voloshin and V. Zakharov, *Phys. Rev. Lett.* **45** (1980) 688

8. K. Gottfried, *Phys. Rev. Lett.* **40** (1978) 598. A recent paper which contains further references is Y. P. Kung and T. N. Yan, *Phys. Rev.* **D 41** (1990) 155

9. B. L. Ioffe and M. A. Shifman, *Phys. Lett.* **B 95** (1980) 99

10. R. H. Socolow, *Phys. Rev.* **137** (1965) 1221

11. K. Choi, C. W. Kim and W. K. Sze, *Phys. Rev. Lett.* **61** (1988) 794

$\pi\pi$ and πK SCATTERING IN CHIRAL PERTURBATION THEORY

Norbert Kaiser*

Centre de Recherches Nucléaires et Université Louis Pasteur de Strasbourg
Groupe de Physique Nucléaire Théorique
BP 20, 67037 Strasbourg Cedex, France

ABSTRACT

We investigate $\pi\pi$ and πK scattering in the framework of chiral perturbation theory to one-loop order. The low-energy constants occurring at next-to-leading order in the chiral expansion turn out to be completely saturated by resonance exchange. We introduce explicit (scalar and vector) resonances chirally coupled to the pseudoscalar Goldstone bosons and evaluate the corresponding scattering amplitudes. The momentum dependence of the resonance propagators allows for a partial resummation of yet higher order contributions ($\geq E^6$ in the chiral expansion). The imaginary parts of the resonance propagators can be generated via an unitarizing K–matrix approach together with the proper decay widths of the resonances. We evaluate $S-$ and $P-$waves in $\pi\pi$ and πK scattering and compare them to available data and existing Roy equation fits where a good description is found up to energies of about $\sqrt{s} = 800...900$ MeV. We also point on the importance of a more accurate determination of the πK scattering amplitudes in order to have stricter tests on the quality of the chiral expansion in the strange sector.

I. INTRODUCTION

At low energies the dynamics of the low-lying pseudoscalar (Goldstone) bosons (π, K, η) is accessable to a systematic expansion in powers of external momenta and quark masses – chiral perturbation theory (CHPT). This is due to the fact that in the limit of vanishing quark masses, QCD possesses an exact chiral symmetry $(SU(3)_L \times SU(3)_R)$, which has subtle dynamical consequences. In practice, one makes use of an effective lagrangian of the pseudoscalar fields and expands in the number of meson loops or, equivalently, in the number of derivatives acting on the pseudoscalar fields. To lowest order (E^2), the effective lagrangian is nothing but the well-known non-linear σ-model

$$\mathcal{L}^{(2)} = \frac{F_0^2}{4} Tr(\partial_\mu U \partial^\mu U^\dagger) + \frac{F_0^2}{4} Tr(\chi^\dagger U + \chi U^\dagger) \tag{1}$$

* Address after Oct. 1st 1991: Physik Department T30, TU München, James-Franck-Strasse, W-8046 Garching.

with χ incorporating the quark mass matrix and F_0 is the pion decay constant in the chiral limit, $F_\pi = F_0[1+\mathcal{O}(m_q)]$. The non-linear σ-model contains all the information of the current algebra theorems for various $\pi\pi$ and πK scattering amplitudes.

At next-to-leading order (E^4), the effective theory involves more (higher derivative) terms accompanied by a set of new low-energy constants which are not fixed by symmetry requirements. These constants should be calculated directly from QCD, but since this is not yet possible, they have to be determined phenomenologically. In refs.[1,2] the complete generating functional of QCD at order E^4 has been worked out and all low-energy constants have been determined using empirical information on various mesonic reactions and large N_c arguments. Recently, it has been realized that the effective lagrangian of the pseudoscalars is also constrained by the low-lying (non-Goldstone) resonance spectrum[3,4]. This happens in the sense that the low-energy constants at order E^4 of the pseudoscalar effective theory are completely saturated by resonance exchange contributions. One has, however, to choose the scale μ where the low-energy constants are taken to be close to the resonance mass $\mu \simeq M_R \simeq M_\rho$ in order to have resonance saturation. This feature can be considered as the QCD version of the old vector meson dominance principle (VMD) and has been coined "QCD duality" in ref.[4]. One can therefore calculate Greens functions to order E^4 in two ways. Conventionally, one includes a local action of the pseudoscalars to this order with phenomenologically determined coupling constants or, alternatively, one uses the effective chiral lagrangian to order E^2 and includes explicit resonance fields with appropriate chiral couplings. The second possibility is the one we will explore here and apply to the $\pi\pi$ and πK scattering process. In fact, if one takes the resonance contributions seriously, it becomes obvious that the resonance propagators

$$\frac{1}{M_R^2 - s} = \frac{1}{M_R^2} + \frac{s}{M_R^4} + \frac{s^2}{M_R^6} + \ldots \tag{2}$$

include the dominant terms to the scattering amplitudes at order E^6 or higher (assuming that the resonance saturation goes beyond the local lagrangian of order E^4 considered so far). As an example, let us consider the $I = l = 1$ partial wave in $\pi\pi$ scattering. For the effective theory of pseudoscalars only, the terms of order E^2 and E^4 give essentially a slowly rising phase-shift and are not able to reproduce the rapid rise due to the ρ-resonance starting at energies above 600 MeV. This changes dramatically once the resonance effects are taken into account. However, one only includes some of the higher order terms. As it will turn out, the inclusion of the resonances gives a good description of the existing $\pi\pi$ and πK phase-shifts (either through direct confrontation with the data or with the existing Roy equation fits) at least up to energies of approximately 800...900 MeV.

II. INCLUDING RESONANCE FIELDS

The salient property of QCD at low energies is the spontaneous breakdown of the chiral $SU(3)_L \times SU(3)_R$ symmetry down to its vectorial subgroup $SU(3)_V$. This uniquely specifies the effective lagrangian of the interacting Goldstone bosons (π, K, η) to lowest order (1). At the same order, one can also consider the coupling of the other mesons of scalar (S), pseudoscalar (P)*, vector (V) and axial-vector (A) type. These various mesons do not play the special role as the pseudoscalar Goldstone bosons, but their presence manifests itself in the values of the renormalized coefficients L_i^r appearing at next-to-leading order. One might also add that if one wants to extend the range of validity of the chiral expansion, one is naturally led to explicitely couple in resonance fields since such models are phenomenologically quite successful. One does not have to assign any special character to the resonances but only demands that they carry a non-linear realization of the chiral group. This uniquely specifies their couplings at lowest order. In our case only octet vector as well as singlet and octet scalar resonances can contribute and their most general interactions with the pseudoscalars read[3]:

$$\mathcal{L}^{V\Phi\Phi} = i\frac{G_V}{\sqrt{2}} Tr(V_{\mu\nu}u^\mu u^\nu) \tag{3a}$$

$$\mathcal{L}^{S\Phi\Phi} = c_d Tr(S u_\mu u^\mu) + c_m Tr(S\chi_+) \tag{3b}$$

$$\mathcal{L}^{S_1\Phi\Phi} = \tilde{c}_d S_1 Tr(u_\mu u^\mu) + \tilde{c}_m S_1 Tr(\chi_+) \tag{3c}$$

with the following abbreviations:

$$u(\Phi) = \exp\{i\Phi/2\}, \quad U(\Phi) = u^2(\Phi), \quad u_\mu = iu^\dagger \partial_\mu U u^\dagger . \tag{4}$$

The vector fields are described by an antisymmetric tensor field $V_{\mu\nu}$ with three independent degrees of freedom (for a massive spin-1 particle) having the following particle content:

$$V_{\mu\nu} = \begin{pmatrix} \rho^0 + \sqrt{\frac{1}{3}}\omega_8 & -\sqrt{2}\rho^+ & -\sqrt{2}K^{*+} \\ \sqrt{2}\rho^- & -\rho^0 + \sqrt{\frac{1}{3}}\omega_8 & -\sqrt{2}K^{*0} \\ \sqrt{2}K^{*-} & \sqrt{2}\bar{K}^{*0} & -\sqrt{\frac{4}{3}}\omega_8 \end{pmatrix}_{\mu\nu} \tag{5}$$

The use of an antisymmetric tensor field is convenient (but by no means mandatory) since it avoids the mixing with the scalars present in the usual vector formulation in form of a term $V_\mu \, \partial^\mu S$. For the scalars, there are two types of couplings, one type involving two derivatives of the pseudoscalars and the other one describing the coupling to the vacuum. (Here, χ_+ denotes a scalar source with positive parity as it is also the case for the vacuum). The coupling constants $G_V, c_d, c_m, \tilde{c}_d$ and $\tilde{c}_m$ will be discussed next.

---

* These are of course the non-Goldstone pseudoscalars. From these only the η' is of any practical importance[3].

III. DETERMINATION OF THE RESONANCE PARAMETERS

Next we have to fix the various coupling constants and masses appearing in the resonance lagrangian (3). Following the arguments of ref.[3], which points out the success of the quark counting rules for the vector meson masses, we choose

$$M_V = M_\rho = 769\,\text{MeV}\,. \tag{6}$$

However, when we will later on consider πK-scattering, where $K^*(892)$ exchange occurs in the $I = 1/2$ P-wave, we will choose $M_V = M_{K^*} = 892$ MeV. The singlet and octet scalar masses in the chiral limit can not be determined directly but by using large-N_c arguments and observing that the lowest scalar in the data tables[5] is the a_0, we choose

$$M_S = M_{S_1} = 983\,\text{MeV} \tag{7}$$

Let us now turn to the various coupling constants. The vector-pseudoscalar coupling strength G_V follows directly from the decay width[5]

$$\Gamma(\rho \to 2\pi) = \frac{1}{48\pi}\frac{G_V^2}{F_\pi^4}(M_\rho^2 - 4M_\pi^2)^{3/2} = 153\,\text{MeV} \tag{8}$$

to be

$$|G_V| = 69\,\text{MeV} \tag{9}$$

for $F_\pi = 93.3$ MeV. However, since at this energy both pions are far from being soft, it might be more safe to determine G_V from the vector dominance description of the pion charge form factor[3]. This leads to

$$|G_V| = 53\,\text{MeV}\,. \tag{10}$$

We will use the value given in (10) as our central value for G_V but also quote results using the one given in (9). By similar arguments, one can use the measured decay rate $\Gamma(a_0 \to \eta\pi)$ to determine c_d and c_m assuming that $L_{5,8}^r(\mu = M_\rho)$ are saturated by scalar exchange, i.e.

$$|c_d| = 32\,\text{MeV}, \quad |c_m| = 42\,\text{MeV}, \quad c_d c_m > 0\,. \tag{11}$$

Invoking large-N_c arguments, one can derive the nonet relations

$$\tilde{c}_{d,m} = \pm c_{d,m}/\sqrt{3}\,. \tag{12}$$

Since all the resonance couplings discussed so far are of order E^2, resonance exchange produces contributions of order E^4 (and higher) to the effective pseudoscalar lagrangian. Vector and scalar exchange give rise to the following contributions to the fourth-order coupling constants (taken from ref.[3]):

Vectors :

$$L_1^V = \frac{G_V^2}{8M_V^2}, \quad L_2^V = 2\,L_1^V, \quad L_3^V = -6\,L_1^V \tag{13a}$$

Octet scalars :

$$\begin{aligned} L_1^S &= -\frac{c_d^2}{6M_S^2}, \quad L_3^S = -3\,L_1^S, \quad L_4^S = -\frac{c_d\,c_m}{3M_S^2}, \\ L_5^S &= -3\,L_4^S, \quad L_6^S = -\frac{c_m^2}{6M_S^2}, \quad L_8^S = -3\,L_6^S \end{aligned} \tag{13b}$$

Singlet scalars :

$$L_1^{S_1} = \frac{\tilde{c}_d^2}{2M_S^2}, \quad L_4^{S_1} = \frac{\tilde{c}_d\tilde{c}_m}{M_S^2}, \quad L_6^{S_1} = \frac{\tilde{c}_m^2}{2M_S^2} \tag{13c}$$

The renormalized coupling constants take now the form

$$L_i^r(\mu) = \sum_R L_i^{Res} + \hat{L}_i(\mu)\,. \tag{14}$$

The pertinent result of ref.[3] was that $\hat{L}_i(M_\rho) \simeq 0$ (i = 1,...,10). This is the precise formulation of resonance saturation (like the old vector meson dominance hypothesis) in the context of QCD. Let us now turn to the central subject, namely the $\pi\pi$ and πK scattering process in CHPT with explicit resonances.

IV. THE $\pi\pi$ AND πK SCATTERING AMPLITUDES

We consider the process $\phi_1(p_1) + \phi_2(p_2) \to \phi_3(p_3) + \phi_4(p_4)$ where ϕ_i stands for a pion or a kaon. The lorentz-invariant scattering amplitudes depend on the Mandelstam variables $s = (p_1+p_2)^2 = (p_3+p_4)^2, t = (p_1-p_3)^2 = (p_2-p_4)^2$ and $u = (p_1-p_4)^2 = (p_2-p_3)^2$ subject to the constraint $s+t+u = 4\,M_\pi^2$ or $2\,(M_\pi^2+M_K^2)$ for $\pi\pi$ or πK scattering, respectively. Let us now specialize to these two cases.

$\pi\pi$ scattering :

The $\pi\pi$ scattering amplitude is given in term of one single function $A(s,t,u)$ which is even under $t \leftrightarrow u$ due to Bose symmetry. It can be decomposed into amplitudes with definite total isospin $I = 0, 1, 2$ via

$$\begin{aligned} T^0(s,t,u) &= 3\,A(s,t,u) + A(t,s,u) + A(u,t,s) \\ T^1(s,t,u) &= A(t,s,u) - A(u,t,s) \\ T^2(s,t,u) &= A(t,s,u) + A(u,t,s)\,. \end{aligned} \tag{15}$$

Projection onto partial waves follows via

$$T^I(s,t,u) = 32\pi \sum_{l=0}^{\infty} (2l+1)\, t_l^I(s) P_l(\cos\theta)\,,$$
$$t = 2q^2(\cos\theta - 1)\,, \quad q = \sqrt{s/4 - M_\pi^2} \tag{16}$$

with l the relative angular momentum of the pions and θ the scattering angle in the center-of-mass system. In the elastic region the partial wave amplitudes are parametrized in terms of the real phase-shifts $\delta_l^I(s)$ as:

$$t_l^I(s) = \frac{\sqrt{s}}{2q} \frac{1}{2i} \{ e^{2i\delta_l^I(s)} - 1 \} \tag{17}$$

At small momentum transfer ($q \to 0$), the $\pi\pi$ scattering lengths are defined via

$$Re\ t_l^I(s) = q^{2l}\, [a_l^I + \mathcal{O}(q^2)]\,. \tag{18}$$

The threshold cm energy is $\sqrt{s_{th}} = 2M_\pi \simeq 280\, MeV$.

<u>πK scattering :</u>
Since the kaon has isospin 1/2, one has two independent scattering amplitudes with definite total isospin, $T_{\pi K}^{1/2}(s,t,u)$ and $T_{\pi K}^{3/2}(s,t,u)$, respectively. The latter is directly calculable from the specific process $\pi^+ K^+ \to \pi^+ K^+$ and using crossing symmetry ($s \leftrightarrow u$), one can derive the relation:

$$T_{\pi K}^{1/2}(s,t,u) = \frac{3}{2}\, T_{\pi K}^{3/2}(u,t,s) - \frac{1}{2} T_{\pi K}^{3/2}(s,t,u) \tag{19}$$

The partial wave decomposition and parametrization in terms of phase-shifts are analogous (16,17) with 32π replaced by 16π since the πK system involves non-identical particles and $q = \frac{1}{2\sqrt{s}}\sqrt{[s-(M_K+M_\pi)^2][s-(M_K-M_\pi)^2]}$. Close to threshold, $\sqrt{s_{th}} = M_K + M_\pi \simeq 633\, MeV$, one defines the πK scattering lengths via

$$Re\ t_l^I(s) = \frac{\sqrt{s}}{2}\, q^{2l}\, [a_l^I + \mathcal{O}(q^2)]\,. \tag{20}$$

Notice the different normalization for the πK scattering lengths as compared to the $\pi\pi$ case. This is done for an easier comparison with the existing literature.

Next, we give the on-shell amplitude for the elastic scattering process $\pi\pi \to \pi\pi$. In general one expands the generating functional up to the necessary order in the external sources which allows one to calculate off- and on-shell amplitudes and is necessary to check the corresponding Ward identities. Since we are only interested in the on-shell amplitudes here, one can simplify the calculational procedure and expand the generating functional in powers of the interpolating field Φ and extract the T-matrix from the Φ^4 terms. After some algebra we find for the $\pi\pi$ scattering amplitude $A(s,t,u) = -T(\pi^0(p_1) + \pi^0(p_2) \to \pi^+(p_3) + \pi^-(p_4))$:

$$A(s,t,u) = A^{(2)}(s,t,u) + A_T^{(4)}(s,t,u) + A_U^{(4)}(s,t,u) + A_{Res}(s,t,u) \tag{21}$$

denoting by $A^{(2)}(s,t,u)$ the tree-level (σ-model) part at order E^2, by $A_T^{(4)}(s,t,u)$ the tadpole terms at order E^4, by $A_U^{(4)}(s,t,u)$ the unitarity corrections and by $A_{Res}(s,t,u)$ the resonance contributions, which start out at order E^4. We find for the various terms:

Tree level :

$$A^{(2)}(s,t,u) = \frac{F_0^2}{F_\pi^4}\,(s - \frac{4}{3}M_\pi^2 + \frac{1}{3}\mathring{M}_\pi^2) \tag{22}$$

Tadpole contributions :

$$A_T^{(4)}(s,t,u) = -\frac{F_0^2}{3F_\pi^4}\left[\mu_\pi(12s - 16M_\pi^2 + 3\mathring{M}_\pi^2) + 2\mu_K(3s - 4M_\pi^2 + \mathring{M}_\pi^2) + \frac{1}{3}\mu_\eta\mathring{M}_\pi^2\right] \tag{23}$$

Unitarity corrections :

$$\begin{aligned} A_U^{(4)}(s,t,u) = &\frac{1}{F_\pi^4}\Big\{\frac{\mathring{M}_\pi^4}{18}J_{\eta\eta}^r(s) + \frac{1}{2}(s - 2M_\pi^2 + \mathring{M}_\pi^2)(s - 2M_\pi^2 + 3\mathring{M}_\pi^2)\,J_{\pi\pi}^r(s) \\ &+ \frac{1}{2}(\frac{s}{2} - M_\pi^2 + \mathring{M}_\pi^2)^2\,J_{KK}^r(s) + \Big[\frac{1}{4}(t - 2M_\pi^2)^2 J_{\pi\pi}^r(t) \\ &+ t(s-u)\big[M_{\pi\pi}^r(t) + \frac{1}{2}M_{KK}^r(t)\big] + (t \leftrightarrow u)\Big]\Big\} \end{aligned} \tag{24}$$

Resonance exchanges :

$$\begin{aligned} A_{Res}(s,t,u) = &\frac{G_V^2}{F_\pi^4}\left[\frac{t(s-u)}{M_V^2 - t} + \frac{u(s-t)}{M_V^2 - u}\right] + \frac{4}{F_\pi^4(M_{S_1}^2 - s)}\left[\tilde{c}_d(s - 2M_\pi^2) + 2\tilde{c}_m\mathring{M}_\pi^2\right]^2 \\ &+ \frac{16\tilde{c}_m}{3F_\pi^4 M_{S_1}^2}(\mathring{M}_K^2 + \frac{1}{2}\mathring{M}_\pi^2)\left[\tilde{c}_d(3s - 4M_\pi^2) + \tilde{c}_m\mathring{M}_\pi^2\right] \\ &+ \frac{2}{3F_\pi^4(M_S^2 - s)}\left[c_d(s - 2M_\pi^2) + 2c_m\mathring{M}_\pi^2\right]^2 \\ &+ \frac{16c_m}{9F_\pi^4 M_S^2}(\mathring{M}_K^2 - \mathring{M}_\pi^2)\left[c_d(4M_\pi^2 - 3s) - c_m\mathring{M}_\pi^2\right] \end{aligned} \tag{25}$$

The expression $\mathring{M}_P$ denotes the leading order term in the chiral expansion of the meson masses, $M_P = \mathring{M}_P[1 + \mathcal{O}(m_q)]$ and the quantity μ_P means

$$\mu_P = \frac{\mathring{M}_P^2}{32\pi^2 F_0^2}\,\ln\left(\frac{\mathring{M}_P^2}{\mu^2}\right). \tag{26}$$

The loop functions J^r_{PQ}, K_{PQ} and M^r_{PQ} (with $P,Q = \pi, K, \eta$) are explicitely given in ref.[2]. As a non-trival check on the amplitude, one can analytically show that the scale-dependence appearing in the functions $\mu_P, J^r_{PQ}, M^r_{PQ}$ cancels with the one in the remainder of the low-energy constants $\hat{L}_i(\mu)$ (14) not written down here since we work at the scale $\mu = M_\rho$ where these remainders are close to zero. Obviously, the $\pi\pi$ amplitude (22-25) exhibits (singlet and octet) scalar exchange in the s-channel and vector exchange in the t- and u-channel.

The πK amplitude is of similar structure (21) and it is most convenient to work out first the $I = 3/2$ amplitude $T^{3/2}_{\pi K}(s,t,u)$ from the process $\pi^+(p_1) + K^+(p_2) \to \pi^+(p_3) + K^+(p_4)$ and then determine the $I = 1/2$ amplitude from (19). The corresponding rather lengthy expressions for the tree level amplitude, tadpole contibutions, unitarity corrections and resonance exchange contributions can be found in the research article[6]. We should point out that the πK amplitude $T^{3/2}_{\pi K}$ involves $K^*(892)$-exchange in the u-channel and we will set there $M_V = M_{K^*} = 892$ MeV and leave $M_V = M_\rho = 769$ MeV for the t-channel ρ-meson exchange whenever this is necessary. In general the mass splittings in the various resonance octets or nonets are of higher order in the chiral expansion and a systematic evaluation of these terms is beyond the scope of this work.

V. PHASE-SHIFTS AND THE WIDTH OF THE RESONANCES

Now in the presence of explicit resonances one gets contributions to arbitrary high order E^{2n} ($n = 1, 2, 3, ...$) to the scattering amplitude. We will calculate the phase-shifts from the partial-wave amplitudes via

$$\delta^I_l(s) = \arctan\left(\frac{2q}{\sqrt{s}}\,\mathrm{Re}\,t^I_l(s)\right) \tag{27}$$

which is inspired by the elastic unitarity condition:

$$\mathrm{Im}\,t^I_l(s) = \frac{2q}{\sqrt{s}}\left(\mathrm{Re}\,t^I_l(s)\right)^2 + \mathcal{O}(E^8) \tag{28}$$

In the conventional framework the arctan in (27) can be neglected according the chiral power counting of the phase-shift. The definition (27) which is similar to a K-matrix formulation is very advantageous since it properly unitarizes the resonance exchange and leads to the correct decay widths of the resonances even though we treated them only at tree-level. Let us explain this fact in detail for the ρ-exchange in the $I = 1$ $\pi\pi$ P-wave. The complete elastic partial wave amplitude takes the form:

$$t^I_l(s) = \sqrt{\frac{s}{s - 4M^2_\pi}}\,\frac{1}{\cot\delta^I_l(s) - i} \tag{29}$$

Close to a resonance the phase-shift behaves like:

$$\delta_l^I(s) = \frac{\pi}{2} + \frac{1}{M_R\Gamma_R}(s - M_R^2) + ..., \qquad \tan\delta_l^I(s) = \frac{M_R\Gamma_R}{M_R^2 - s} + ... \tag{30}$$

a paramerization valid for well-separated elastic resonances. Extracting now the ρ-pole term from the $I = 1$ $\pi\pi$ amplitude

$$T^1_{\rho-pole}(s,t,u) = \frac{G_V^2}{F_\pi^4}\frac{2s(t-u)}{M_\rho^2 - s} \tag{31}$$

which leads to the P-wave amplitude

$$t^1_{1,\rho-pole}(s) = \frac{G_V^2}{12\pi F_\pi^4}\frac{sq^2}{M_\rho^2 - s} = \frac{\sqrt{s}}{2q}\tan\delta_1^1(s) \tag{32}$$

we obtain by comparing with (30) the standard result (8) for the $\rho \to \pi\pi$ decay width. These considerations are of course more general and also apply for other resonances in various partial waves. For the K^* one derives the following decay width

$$\begin{aligned} \Gamma_{K^*} &= \frac{G_V^2}{8\pi F_\pi^2 F_K^2}\,|\vec{q}_\pi|^3\,, \\ |\vec{q}_\pi| &= \frac{1}{2M_{K^*}}\sqrt{(M_{K^*}^2 - (M_K + M_\pi)^2)(M_{K^*}^2 - (M_K + M_\pi)^2)}\,. \end{aligned} \tag{33}$$

Even though the arctan-prescription for the phase-shifts works well here we should add that a more thorough investigation loop effects in the presence of resonances to generate imaginary parts in scattering amplitudes (and in particular form factors) would be needed.

VI. RESULTS

Let us first summarize the input parameters. For the pseudoscalars, we use $M_\pi = 139.57$ MeV, $F_\pi = 93.3$ MeV, $M_K = 493.65$ MeV and $F_K = 1.22F_\pi$. As central values for the resonance parameters we use $M_V = M_\rho = 769$ MeV (in the case of K^*–exchange in the πK system we use $M_V = M_{K^*} = 892$ MeV), $G_V = 53$ MeV, $M_S = M_{S_1} = 983$ MeV, $c_d = 32$ MeV, $c_m = 42$ MeV as well as $\tilde{c}_d = c_d/\sqrt{3}, \tilde{c}_m = c_m/\sqrt{3}$. To test the sensitivity of our results against variations in the resonance parameters, we will also use $G_V = 69$ MeV (from the $\rho \to 2\pi$ width) (or $G_V = 76.5$ MeV from the $\Gamma(K^* \to K\pi) = 50$ MeV in the case of πK scattering).

Scattering lengths:
The pertinent scattering lengths (in units of M_π^{-2l}) for the $\pi\pi$ system are given in Table 1, for the standard parameters ($G_V = 53$ MeV) as well as for the case $G_V = 69$

MeV. We find a rather good agreement with the empirical information, which we take from the Roy equation studies of Basdevant *et al.*[7]. For comparison also the results for the conventional approach with pions only (flavor $SU(2)$) are shown. In fact we find $a_0^0 = 0.21$, consistent with the chiral prediction, $a_0^0 = 0.20 \pm 0.01$[1]. The P-wave scattering length a_1^1 is somewhat sensitive to the actual value of G_V, as it is expected. For the higher partial waves, satisfactory agreement with the not too precise Roy equation solutions is found.

	RESONANCES		CHPT	EXP
	G_V =53 MeV	G_V = 69 MeV		
a_0^0	0.21	0.21	0.20	0.26 ± 0.05
a_0^2	-0.043	-0.043	-0.042	-0.028 ± 0.012
$2a_0^0 - 5a_0^2$	0.639	0.639	0.614	0.657 ± 0.052
a_1^1	0.038	0.041	0.037	0.038 ± 0.002
a_2^0	$18.7 \cdot 10^{-4}$	$22.7 \cdot 10^{-4}$	$20 \cdot 10^{-4}$	$(17 \pm 3) \cdot 10^{-4}$
a_2^2	$3.1 \cdot 10^{-4}$	$1.1 \cdot 10^{-4}$	$3.5 \cdot 10^{-4}$	$(1.3 \pm 3.1) \cdot 10^{-4}$
a_3^1	$0.30 \cdot 10^{-4}$	$0.35 \cdot 10^{-4}$	$0.2 \cdot 10^{-4}$	$(0.5 \pm 0.2) \cdot 10^{-4}$

Table 1: $\pi\pi$ scattering lengths. We give the results of the chiral lagrangian with explicit resonance fields for the standard parameters and G_V = 53 MeV and 69 MeV, respectively. The results from the conventional approach ($SU(2)$, pions only) are taken from ref.[1].

	RESONANCES		CHPT	EXP
	G_V =53 MeV	G_V = 76.5 MeV		
$a_0^{1/2}$	0.19	0.19	0.17	0.13 ... 0.24
$a_0^{3/2}$	-0.06	-0.06	-0.05	-0.13 ... -0.05
$a_1^{1/2}$	0.015	0.021	0.014	0.017 ... 0.018
$a_1^{3/2}$	$0.3 \cdot 10^{-4}$	$5.3 \cdot 10^{-4}$	$7.6 \cdot 10^{-4}$	–

Table 2: Some πK scattering lengths in units of M_π^{-2l-1}.

For the πK system, the empirical situation is much worse. The pertinent scattering lengths are only known within a certain error band and no recent Roy equation studies exist. Notice that $R = a_0^{1/2}/a_0^{3/2} \approx -3.2$, close to the cluster of the data $R_{exp} \approx -3$, whereas the lowest order prediction (current algebra) is $R_{CA} = -2.0$. The P–wave scattering lengths are very sensitive to the value of G_V, *i.e.* to the strength of the $K^* K\pi$ vertex. This is particularly visible for the exotic P–wave scattering length $a_1^{3/2}$ and with $G_V = 76.5$ from the empirical $K^* \to K\pi$ width, a good description of the value for $a_1^{1/2}$ is achieved. To summarize the results on the scattering lengths, we can say that the chiral lagrangian including explicit resonances passes the test in the $\pi\pi$–system fairly nicely and gives predictions for the πK system, which should be tested in upcoming experiments at the proposed kaon factories.

Phase-shifts for $\pi\pi$ scattering:
For the central values of the chiral parameters, the resulting phase-shift $\delta_0^0(s)$ (Fig.1a) lies within the Roy equation band spanned by the solution with $a_0^0 = 0.17$ and $a_0^0 = 0.30$, respectively, up to $\sqrt{s} = 640$ MeV. For the solution with $G_V = 69$ MeV this crossing lies at $\sqrt{s} = 600$ MeV. From the figure it becomes obvious that some of the data are in disagreement with the Roy equation fits. In fact, the best data due to the CERN–Munich collaboration[8], only start at $\sqrt{s} = 630$ MeV.

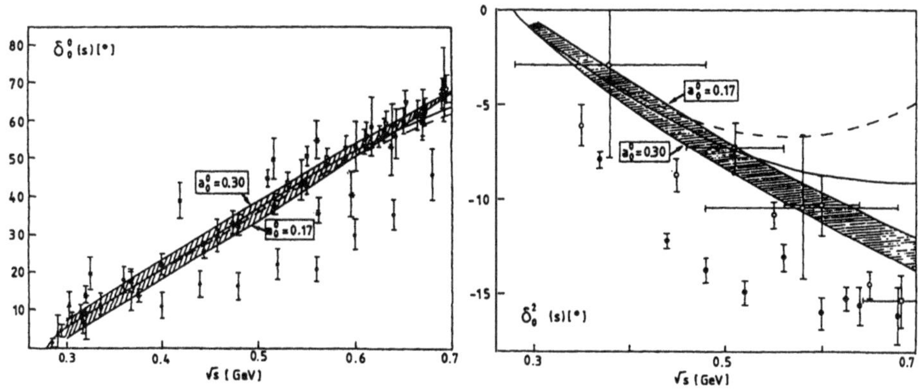

Fig.1: a) The phase $\delta_0^0(s)$ *for* $\sqrt{s} \le 700$ *MeV. The solid line gives the result for our standard values of the chiral parameters, in particular with* $G_V =$ *53 MeV. The dashed line corresponds to* $G_V = 69$ *MeV. The shaded area spanned by the solid lines gives the Roy equation fits*[7] *with* $0.17 \le a_0^0 \le 0.30$.
b) The exotic S-wave $\delta_0^2(s)$.

In Fig.1b, we show results for the $I = 2$ (exotic) S–wave for $\sqrt{s} \le 700$ MeV. Here, for the central set of parameters, our solution leaves the Roy equation band around $\sqrt{s} = 530$ MeV. As was already found in the study in the conventional approach, the actual values of $\delta_0^2(s)$ are rather sensitive to the chiral parameters. The bending of $\delta_0^2(s)$ towards positive values at larger $\sqrt{s}$ can be traced back to the influence of the chiral loops (unitarity corrections). Another quantity of interest related to the S–wave scattering phases is the phase of the parameter ϵ' which measures direct CP–violation in $K \to 2\pi$ decays

$$\phi(\epsilon') = \frac{\pi}{2} - (\delta_0^0 - \delta_0^2)|_{s=M_{K^0}^2} \tag{38}$$

For our standard set of parameters we have $\phi(\epsilon') = 44.5°$ close to the result found in ref.[9], $\phi(\epsilon') = 45 \pm 6°$. Varying the chiral parameters and scalar masses within reasonable bounds we find no change in $\phi(\epsilon')$ that exceeds $\pm 3°$.

In Fig.2a, we show the isovector P–wave together with the available data and the Roy equation fit with $a_0^0 = 0.30$[7]. Up to $\sqrt{s} = 700$ MeV, we achieve a good description of this phase–shift, quite in contrast to the conventional approach, which gives *e.g.* $\delta_1^1(700 \text{ MeV}) = 16.2°$. This improvement is, of course, due to to the presence of the ρ–meson. To illustrate this effect more clearly, we show in Fig.2b the resonance region (720 MeV $\leq \sqrt{s} \leq$ 890 MeV). For the choice $G_V = 69$ MeV (which gives the proper width of the decay $\rho \to 2\pi$) shown by the dashed line, we exactly recover the behaviour of $\delta_1^1(s \approx M_\rho^2)$ as expected.

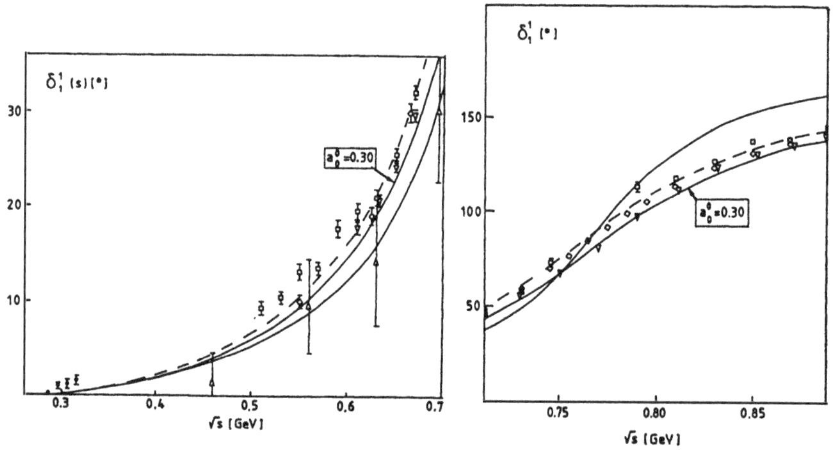

Fig.2: The isovector P-wave $\delta_1^1(s)$. a) $\delta_1^1(s)$ for $\sqrt{s} \leq 700$ MeV. b) $\delta_1^1(s)$ in the resonance region.

Phase-shifts for πK scattering:
As we already mentioned, the situation concerning the data and Roy equation fits is much less satisfactory in the πK system. Also, since the threshold for πK scattering is located at $M_\pi + M_K \approx 633$ MeV, one expects the chiral expansion to work less accurately than for $\pi\pi$. We will therefore put more emphasis onto the comparison with the conventional approach and consider the data mostly indicative. The isospin 1/2 S–wave phase shift, $\delta_0^{1/2}(s)$, is shown in Fig.3a together with the available data and the result of the most recent dispersion–theoretical study of the πK–system performed by Lang and Porod[10]. Up to $\sqrt{s} \approx 750$ MeV, the result obtained from the resonance lagrangian differs from the one obtained in the conventional approach by less than 3°, and the discrepancy increases. In any case, the chiral prediction lies above the bulk of the data in this energy region and above the solution of Lang and Porod[10] (solid line labelled 'LP') (but within the error band). The isospin 3/2 S–wave phase shift, $\delta_0^{3/2}(s)$ is exhibited in Fig.3b for $\sqrt{s} \leq 850$ MeV. The existing data have large error bars in this energy range, but clearly the resonance lagrangian gives an improved description in comparison to the conventional approach. The comparison with the dispersion–theoretical result of Lang and Porod[10] clearly favors the results based on chiral lagrangian with explicit resonance fields.

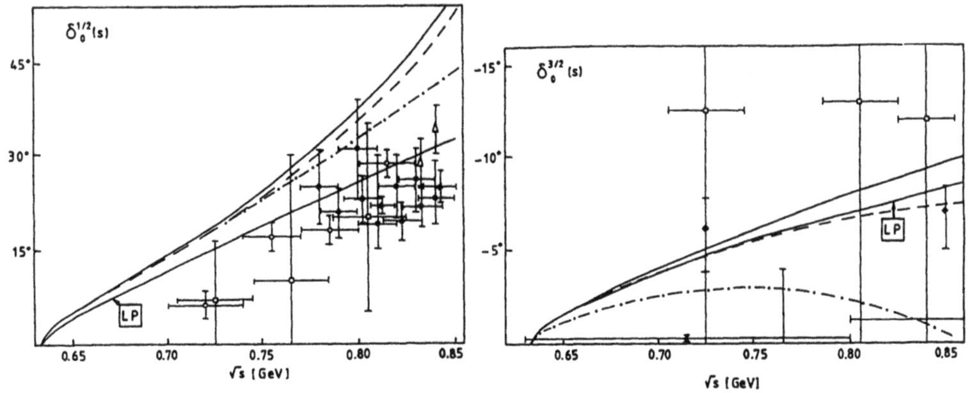

Fig.3: a) The $I = 1/2$ πK S-wave $\delta_0^{1/2}(s)$ for $\sqrt{s} \leq 860$ MeV. The solid line shows the result obtained with the standard set of chiral parameters. The dashed line gives the result with $G_V = 76.5$ MeV obtained from the decay width[5] $\Gamma(K^* \to K\pi) = 50$ MeV. The result of the conventional approach is also shown (dashed-dotted line). The central result of the dispersion-theoretical analysis of Lang and Porod[10] is given by the solid line labelled "LP". b) The $I = 3/2$ S-wave $\delta_0^{3/2}(s)$ for $\sqrt{s} \leq 860$ MeV.

Let us now turn to the πK P-waves. The isospin 1/2 P–wave, $\delta_1^{1/2}(s)$, is of course dominated by the direct $K^*(892)$–exchange. This feature is exhibited in Figs.4. Again, fitting the width of the decay $K^* \to K\pi$ by choosing $G_V = 76.5$ MeV reproduces nicely the slope of the phase-shift in the resonance region as expected from our unitarisation procedure. Not surprisingly, with the explicit K^* resonance present, the $I = 1/2$ P–wave is well described. The $I = 3/2$ is very small up to $\sqrt{s} \approx 1$ GeV and only a few data points with large error bars exist in this energy region. We therefore do not plot $\delta_1^{3/2}(s)$. To summarize these results, it becomes clear that more accurate data are needed and that Roy equation studies should be performed to allow for a more detailed comparison of the chiral prediction with experiment.

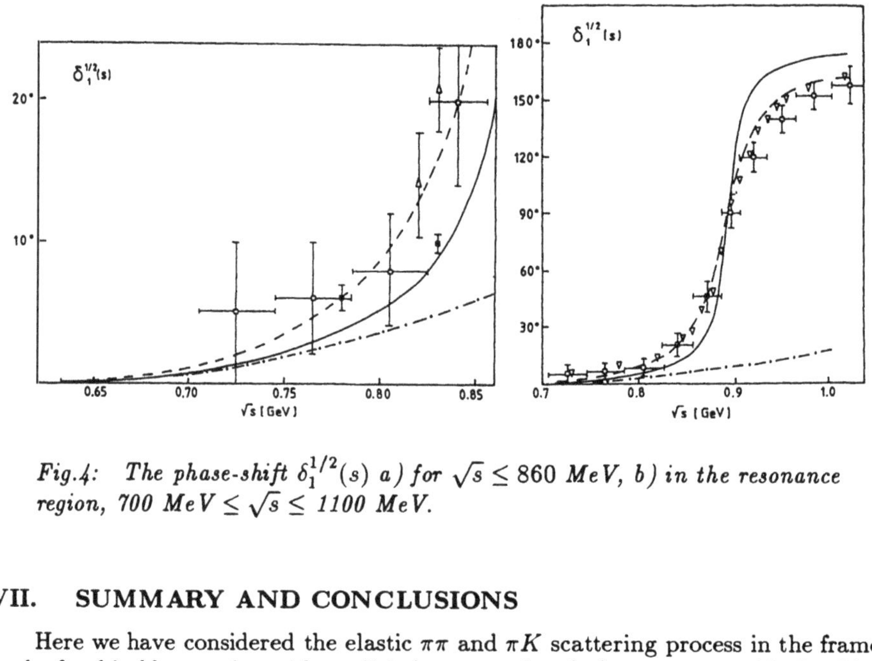

Fig.4: The phase-shift $\delta_1^{1/2}(s)$ a) for $\sqrt{s} \leq 860$ MeV, b) in the resonance region, 700 MeV $\leq \sqrt{s} \leq$ 1100 MeV.

VII. SUMMARY AND CONCLUSIONS

Here we have considered the elastic $\pi\pi$ and πK scattering process in the framework of a chiral lagrangian with explicit (vector and scalar) resonances with the chiral couplings and parameters worked out by Ecker *et al.*[3]. The main assumption in our approach which goes beyond their work is the way we have unitarized the real scattering amplitude by identifying the partial wave amplitude arising from resonance exchange at tree level with $\frac{\sqrt{s}}{2q}\tan\delta_l^I(s)$ to generate the proper imaginary parts and resonance decay widths (K-matrix approach). In fact, such a procedure has to be put onto firmer grounds by a higher loop calculation. The consideration of loop-effects in the presence of resonances is certainly a direction where chiral perturbation theory should be developped further, in particular to get an improved description of the vector form factors of the pion and the kaon (in the ρ-resonance region). Besides these theoretical problems it is more urgent to have improved empirical information on πK scattering, which allows for crucial tests of our understanding of the low-energy structure of QCD in the strange sector.

ACKNOWLEDGEMENTS

The work reported here has been done in collaboration with Véronique Bernard and Ulf-G. Meißner to whom I express my deep gratitude. I thank the CNRS for financial support.

REFERENCES

1. J. Gasser and H. Leutwyler, *Ann. Phys. (N.Y.)* **158** (1984) 142; *Phys. Lett.* **125B** (1983) 321; 325.
2. J. Gasser and H. Leutwyler, *Nucl. Phys.* **B250** (1985) 465; 539.
3. G. Ecker, J. Gasser, A. Pich and E. de Rafael, *Nucl. Phys.* **B321** (1989) 311; G. Ecker, J. Gasser, H. Leutwyler, A. Pich and E. de Rafael, *Phys. Lett.* **B223** (1989) 425.
4. J. F. Donoghue, C. Ramirez and G. Valencia, *Phys. Rev.* **D38** (1988) 2195.
5. Particle Data Group, *Phys. Lett.* **B239** (1990) 1.
6. V. Bernard, N. Kaiser and Ulf-G. Meißner, *Phys. Rev.* **D43** (1991) R2757; *Nucl. Phys.* **B357** (1991) 129; *Nucl. Phys.* **B364** (1991) 283. All further references related to the subject can be traced back from these research articles.
7. J. L. Basdevant, C. D. Froggatt, and J. L. Petersen, *Nucl. Phys.* **B72** (1974) 413; J. L. Basdevant, P. Chapelle, C.Lopez and M. Sigelle, *Nucl. Phys.* **B98** (1975) 285; C. D. Froggatt and J. L. Petersen, *Nucl. Phys.* **B129** (1977) 89; J. L. Petersen, "The $\pi\pi$ Interaction", CERN yellow report 77–04, 1977.
8. B. Hyams *et al.*, *Nucl. Phys.* **B64** (1973) 143; W. Ochs, Thesis, University of Munich (1973).
9. J. Gasser and Ulf-G. Meißner, *Phys. Lett.* **B258** (1991) 129.
10. C. B. Lang and W. Porod, Phys. Rev. **D21** (1980) 1295.

THE ANOMALOUS EFFECTIVE ACTION OF QCD

JOHAN BIJNENS
CERN *Theory Group*
CH 1211 Geneva 23
Switzerland

ABSTRACT

The description of processes involving anomalous vertices in Chiral Perturbation Theory is reviewed shortly. The derivation of the imaginary part of the action in the Euclidean functional integral approach is described and a shorter formula for the higher order contributions derived. These results are then used to derive the order p^6 low energy action in the QCD effective action approach in the chiral limit. For decays of pseudoscalars to photon-(off-shell)-photon the corrections due to non zero current quark masses and gluons are estimated as well.

1 Introduction

The theory for the strong interaction QCD has withstood all experimental tests. It also incorporates naturally all the symmetries which were found to be important in low energy hadronic physics. Low energy hadronic processes have, however, been most succesfully described using an effective Lagrangian that incorporates those symmetries and the relevant degrees of freedom (Chiral Perturbation Theory) as can be seen from most of the contributions in these proceedings. Going beyond the lowest order in these effective Lagrangians can be done consistently but requires the introduction of a large number of free parameters.

In fact if we want to test QCD quantitatively we have to separate out the consequences that follow purely from chiral symmetry and its spontaneous breakdown. These can be separated out most easily using Chiral Perturbation Theory. We write an effective Lagrangian to a certain order in the low energy expansion and leave its parameters free. Loops can then be calculated. For a detailed explanation of the contributions from loops and finite parts at higher order see Refs.[1]. The only difference between different theories with the same spontaneously broken chiral symmetry is in the values of the parameters in the effective Lagrangian. So to test QCD at the quantitative level we have to calculate the values of these parameters. This is what I have done for the anomalous sector within a rather simple model for low energy QCD. Most of the subject of this talk can be found in Ref. [2].

The Lagrangian to next-to-leading order in the normal intrinsic parity sector was determined completely by Gasser and Leutwyler [1]. It has two free parameters at lowest order and 10 more at next-to-leading order. These can all be fixed from experiments and have been used to calculate several processes. In this case it was found that [3][1] several of those free parameters could be well described by direct derivation from a QCD model. A short rederivation of this result will be given later. Their method was extended by Espriu et al. [5] to include gluonic corrections and the effect of chiral symmetry breaking. It is also discussed in the talk by V. Adrianov [6]. The same method has been used to calculate matrix elements of four quark operators [7] and the $\pi^+ - \pi^0$ electromagnetic mass difference [8].

In the abnormal intrinsic parity sector of QCD the low energy effective action was first derived by Wess and Zumino [9] by directly integrating the anomaly consistency conditions. This Lagrangian was later obtained more directly by Witten [10]. This paper also introduced the concept of intrinsic parity. This action only contained one free parameter that has to take integer values, this integer can then be fixed unambiguously from the QCD anomaly. This action was subsequently derived using the global form of the anomaly consistency conditions [11] or by integration of the change in the fermion measure in the functional integral when going from an effective action with quarks incorporating the anomaly or not [12, 13]. The latter method will have some relevance here. The reason for this is that for massive quarks the anomaly is only present if a direct non-derivative pseudoscalar quark coupling is present. If the pseudoscalars couple only derivatively to quarks it can be shown [12] that the anomaly is not present. Therefore it is important to use an effective model that has direct quark pseudoscalar meson couplings. This was essentially the original observation of the anomaly. Two actions that could be related via a symmetry transformation gave a different result.

The effective action in this sector to next-to-leading order has a more patchy recent history. It was known that there were no non-analytic corrections to next-to-leading order to the decays $\pi^0, \eta \rightarrow \gamma\gamma$ [14]. This together with the fixed coefficient of the Wess-Zumino term and the Adler non-renormalization of the anomaly theorem was often used as an argument for the absence of corrections to processes of abnormal intrinsic parity. However, there is no reason why terms not breaking any symmetries cannot contribute to these processes. This was shown by Bijnens et al. [15] and Donoghue and Wyler [16] both by explicitly constructing such terms and calculating the loops of pseudoscalars contributions. The latter were divergent for the decays $\pi^0, \eta \rightarrow \gamma l^+ l^-$ so free parameters to absorb these divergences are needed at the next order.

---

[1]This result was derived in a different context by several other authors, see the references given in [4].

The full infinite part of the next order, i.e., at one loop, was calculated subsequently by Bijnens et al. [17] and Issler [18]. Issler also determined the full set of free parameters at this order. There are 40 free constants here. In the chiral limit there are still 29 left. The relevance of these terms to a number of physical processes was worked out in references [17, 19]. To circumvent the problem of the free parameters the approach of determining the constants by estimating resonance contributions to them was taken there. This had been shown to work well in the normal intrinsic parity sector [1, 20, 21]. A reasonably good agreement with the measurements of the decays π^0 , η , $\eta' \to \gamma\gamma$, γe^+e^- , $\gamma\mu^+\mu^-$ and $\eta \to \pi^+\pi^-\gamma$ and the form factor in $\pi^+\gamma \to \pi^+\pi^0$ was obtained.

In this talk we take the alternative approach of determining the constants via a QCD model. This has been shown to work well in the normal parity sector whenever gluonic corrections are small. We use the same approach here. More care is needed, however, because of the problem of integrating out chiral fermions. In section 2, I give a short overview of Chiral Perturbation Theory and the Wess-Zumino term. This section also serves to define my conventions here. In the next section, I define the QCD model used and rederive the known results in the normal intrinsic parity sector. Section 4 gives the extension to the abnormal intrinsic parity part in general. Then we proceed to calculate the main result, the anomalous effective action at order p^6 in section 5. Section 6 describes the application to $\pi^0, \eta \to \gamma l^+ l^-$ where we have an observable that can be unambiguously compared with my result. This will allow for a determination of a free parameter in the QCD model. In the last section we summarize our results.

2 Chiral Perturbation Theory at lowest order

The strong interaction phenomenology at low energy can be well described by an effective Lagrangian that incorporates the $SU(3)_L \times SU(3)_R$ chiral symmetry of QCD. This symmetry is spontaneously broken to its vector subgroup. There are then 8 Goldstone bosons. These can be identified with the physical π, K and η mesons after the effect of explicit chiral symmetry breaking by the quark masses is taken into account. In this talk we work in the chiral limit and all effects of the explicit quark masses are neglected.

The Goldstone bosons can be described by a 3×3 unitary matrix Σ that transforms under $SU(3)_L \times SU(3)_R$ as

$$\Sigma \to U_L \Sigma U_R^\dagger, \tag{1}$$

where $U_{L(R)} \in SU(3)_{L(R)}$. A parametrization of Σ in terms of the physical pseudoscalar mesons is

$$\Sigma = \exp\frac{2i}{f}M = \exp\frac{2i}{f}\begin{pmatrix} \frac{\pi^0}{\sqrt{2}}+\frac{\eta_8}{\sqrt{6}} & \pi^+ & K^+ \\ \pi^- & -\frac{\pi^0}{\sqrt{2}}+\frac{\eta_8}{\sqrt{6}} & K^0 \\ K^- & \overline{K^0} & -\sqrt{\frac{2}{3}}\eta_8 \end{pmatrix}, \quad (2)$$

where f is a free constant.

The Lagrangian can be expanded in powers of external momenta and fields [1]. Here vector fields and momenta are of $\mathcal{O}(p)$. At lowest order (p^2) there is only one term

$$L_2 = \frac{f^2}{8}\mathrm{tr}D_\mu\Sigma D^\mu\Sigma^\dagger \quad (3)$$

The covariant derivative on Σ is defined in terms of the left- and right-handed external vector fields, L_μ and R_μ, as $D_\mu\Sigma = \partial_\mu\Sigma + iL_\mu\Sigma - i\Sigma R_\mu$. The vector fields transform nonlinearly under the chiral symmetry. We can determine the parameter f from the pion decay constant. Taking the left current from L_2 and the definition of the pion decay constant $< 0|J_\mu^{L1+i2}|\pi^+ >= -if_\pi P_{\pi\mu}/2$, we get $f = f_\pi = 132\ MeV$. We have complete $SU(3)$ symmetry for the decay constants here.

Electromagnetic interactions can be incorporated by the identification

$$L_\mu = R_\mu = eA_\mu Q \quad (4)$$

where e is the electromagnetic coupling constant, A_μ the photon field and $Q = diag(2/3, -1/3, -1/3)$ the quark charge matrix.

This action has not only all symmetries of QCD but also intrinsic parity :

$$\Sigma \leftrightarrow \Sigma^\dagger \qquad L_\mu \leftrightarrow R_\mu \quad (5)$$

So it is not possible to describe processes like $\pi\pi\pi KK$ or $\pi^0 \to \gamma\gamma$. The first terms of this type are $\mathcal{O}(p^4)$ but they cannot be written in a compact form in 4 dimensions. They were first obtained by Wess and Zumino [9] by directly integrating the anomalous Ward identities. A compact form for this action written in 5 dimensions was found by Witten [10]. Here it can be easily seen that the coefficient of these terms had to be quantized. The underlying argument was that the action in 4 dimensions should not depend on the way in which the extrapolation in the fifth dimension was made. One way the fifth dimension comes in is as the integration parameter of the anomalous Ward identities. In section 4 this comes in too in the form of the extra integration over t.

The full lowest order action is given by

$$S = \int d^4x L_2 - N_C S_{WZ}, \quad (6)$$

where N_C is the number of colours and

$$S_{WZ} = \frac{i}{240\pi^2}\int d\Sigma^{ijklm}\mathrm{tr}\left(\Sigma\partial_i\Sigma^\dagger\partial_j\Sigma^\dagger\partial_k\Sigma\partial_l\Sigma^\dagger\partial_m\Sigma^\dagger\right) + \frac{i}{48\pi^2}\int d^4x\epsilon^{\mu\nu\alpha\beta}Z_{\mu\nu\alpha\beta}. \quad (7)$$

Here $Z_{\mu\nu\alpha\beta}$ is zero when all external fields are set to zero. Its explicit expression in my conventions can be found in ref. [17]. The sign of the Wess-Zumino term is the one that reproduces the original calculation of the anomaly.

The action (6) is the full lowest order action in the following sense, for processes of positive intrinsic parity only the term with L_2 should be used, the amplitudes are $\mathcal{O}(p^2)$. For processes of negative (or abnormal) intrinsic parity one vertex from S_{WZ} and the rest from L_2 should be used. These amplitudes are $\mathcal{O}(p^4)$ at lowest nonzero order. Higher orders are similar in the counting.

The amplitude for π^0 , $\eta \to \gamma\gamma^*$ can be derived from the explicit form for $Z_{\mu\nu\alpha\beta}$ and is given by

$$A(M \to \gamma\gamma^*) = \frac{i\sqrt{2}e^2}{4\pi^2 f_M} C_M \epsilon^{\mu\nu\alpha\beta} \epsilon_\mu k_\nu \epsilon^*_\alpha k^*_\beta, \tag{8}$$

where $C_\pi = 1$, $C_{\eta 8} = 1/\sqrt{3}$ and ϵ and ϵ^* are the polarization vectors of the photons with momenta k and k^*, respectively.

3 The normal intrinsic parity Lagrangian

The general method used in this and the next section is the functional integral for fermions regulated by the proper time method. A recent review of this method can be found in the Physics Report by Ball [4]. The relevant formulae will be briefly repeated here.

The action in Euclidean space can be written in the following form :

$$\mathcal{L} = \mathcal{L}_{QCD} + \mathcal{L}_{\mathcal{M}} = \overline{q}\mathcal{D}q \tag{9}$$

where

$$\begin{aligned} \mathcal{D} = \gamma_\mu D_\mu + \mathcal{M} \quad &= \quad \gamma_\mu(\partial_\mu + i g_S G_\mu + i L_\mu P_L + i R_\mu P_R) - m_Q(\Sigma^\dagger P_L + \Sigma P_R) \\ P_L = \frac{1}{2}(1 - \gamma_5) \quad &\text{and} \quad P_R = \frac{1}{2}(1 + \gamma_5) \end{aligned} \tag{10}$$

I use hermitian gamma matrices in Euclidean space. The fields L_μ and R_μ are the same as in the previous section. Gluonic corrections are not included here and I will drop the G_μ in the remainder. The quarks are a 3 vector in the flavour indices. Under the chiral $SU(3)_L \times SU(3)_R$ they transform as

$$q \to (U_L P_L + U_R P_R) q \tag{11}$$

The parameter m_Q can be interpreted as a constituent quark mass. We will assume that the addition of $\mathcal{L}_{\mathcal{M}} = \overline{q}\mathcal{M}q$ provides a good description of QCD at long distances. It can be easily checked using the transformation properties of Σ that the terms containing m_Q are chirally invariant. They do provide a mass m_Q for the quarks.

This is the model as given by Georgi and Manohar [22] with $g_A = 1$ and with the assumption that terms with more derivatives can be neglected. The difference is that we have chosen the form that reproduces the anomaly correctly and have assumed that integrating out the quarks produces the kinetic term for the Goldstone Bosons. In general we can take $g_A \neq 1$, this will change the results presented here. There are arguments using the high energy behaviour of QCD to show that $g_A = 1$, see Ref. [23]. Here taking $g_A \neq 1$ would correspond to adding a one derivative interaction between quarks and Σ.

Hatted quantities are the original ones with P_L and P_R interchanged. The hermitian conjugate of the action term is

$$\mathcal{D}^\dagger = -\gamma_\mu \hat{D}_\mu + \hat{M}. \tag{12}$$

For later convenience we also define

$$\mathcal{D}^\dagger \mathcal{D} = -D^2 + Y + m_Q^2 \quad \text{and} \quad Z_{\mu\nu} = D_\mu D_\nu - D_\nu D_\mu. \tag{13}$$

The effective action at low energies is defined by integrating out the quarks from the functional integral.

$$\exp(\Gamma_E(\Sigma, L_\mu, R_\nu)) = \int [dq][d\bar{q}] \exp(\int d^4x \mathcal{L}_{QCD}) \tag{14}$$

This integral can formally be done resulting in

$$\Gamma_E = \log\det \mathcal{D} \tag{15}$$

and similarly for the complex conjugate of the action, $\Gamma_E^* = \log\det \mathcal{D}^\dagger$. From this we can naively define the real part Γ^+ and the imaginary part Γ^-. Then using the above definitions we get

$$\Gamma^+ = \frac{1}{2} \log\det \mathcal{D}^\dagger \mathcal{D} \tag{16}$$

We now use[2] $\log\det A = \operatorname{tr}\log A$ and the proper time definition of the logarithm to obtain [4]

$$\Gamma^+ = -\frac{1}{2} \int d^4x \operatorname{tr} \int_0^\infty \frac{d\tau}{\tau} \rho(\tau, \mu) \left(e^{-\tau \mathcal{D}^\dagger \mathcal{D}} - e^{-\tau} \right) \tag{17}$$

The function $\rho(\tau, \mu)$ is the regulator. For $\mu = \infty$ it is 1. For μ large, it is 1 for large τ and it must converge to zero for τ near zero fast enough to make the integral above well defined. This can be evaluated as an asymptotic expansion using the Seeley-DeWitt coefficients. These are defined (in the coincidence limit) by

$$< x | e^{-\tau(\mathcal{D}^\dagger \mathcal{D} - m_Q^2)} | x >= \frac{1}{16\pi^2\tau^2} \sum_{n=0}^{\infty} a_n(x, x) \tau^n \tag{18}$$

[2]The det is over all degrees of freedom so the trace is a tr over Dirac, colour and flavour indices and an integral over space.

Putting this into (17) we see that regularization ambiguities only affect the terms with $n = 0, 1, 2$. All other terms are finite. Choosing a renormalization then leads to

$$\Gamma^+ = -\frac{1}{32\pi^2}\int d^4x [\sum_{n\geq 3}^{\infty}\frac{(n-3)!}{(m_Q^2)^{n-2}}a_n(x,x) + \frac{1}{2}m_Q^4\left(\frac{3}{2}+\log\frac{\mu^2}{m_Q^2}\right)a_0(x,x)$$
$$-m_Q^2\left(1+\log\frac{\mu^2}{m_Q^2}\right)a_1(x,x) + \log\frac{\mu^2}{m_Q^2}a_2(x,x)] \quad (19)$$

Here μ is an ultra-violet cutoff parameter.

At this point I would like to make a digression and discuss the concept of intrinsic parity. We can define a parity that only acts on space-time indices by :

$$x_0' = -x_0 \quad L_0'(x') = -L_0(x) \quad R_0'(x') = -R_0(x) \quad (20)$$

with all other quantities unchanged. This is *not* a symmetry of the QCD action. Using the definition of $\mathcal{D}$ we can see that

$$\mathcal{D}(x) \to \gamma_0\mathcal{D}^\dagger(x')\gamma_0 \quad \text{and} \quad \mathcal{D}^\dagger(x) \to \gamma_0\mathcal{D}(x')\gamma_0 \quad (21)$$

From the definitions (15) it can be seen that the real and imaginary parts of the effective action transform as

$$\Gamma^\pm \to \pm\Gamma^\pm. \quad (22)$$

As a consequence Γ^+ must contain an even number of Levi-Civita tensors $\epsilon_{\mu\nu\alpha\beta}$ and Γ^- an odd number. Using the fact that the effective action has to be invariant under the full parity transformation we can then derive that $\Gamma^\pm$ has the same behaviour under the intrinsic parity operation as defined by Witten [10]

$$L_\mu \leftrightarrow R_\mu \quad \text{and} \quad \Sigma \leftrightarrow \Sigma^\dagger. \quad (23)$$

The effective actions also have the same behaviour under intrinsic parity independent of the value of τ and m_Q. So the integral of every a_n contains no $\epsilon_{\mu\nu\alpha\beta}$ independent of the order of m_Q. We will exploit this fact to simplify calculations in the next section.

We can now use the formulas for $a_n(x,x)$ and the values for Y and $Z_{\mu\nu}$ given in Ref. [2] to get the action to $\mathcal{O}(1)$ in an expansion in m_Q^{-2}. The result coincides with that obtained previously ([3] and others see [4]). I only give the terms that depend on Σ. The trace over colours only adds the overall factor of N_C.

$$\Gamma^+ = -\frac{N_C}{32\pi^2}\int d^4x \,\mathrm{tr}\{m_Q^2\log\frac{\mu^2}{m_Q^2}2D_\mu\Sigma^\dagger D_\mu\Sigma - i(D_\mu\Sigma D_\nu\Sigma^\dagger L_{\mu\nu} + D_\mu\Sigma^\dagger D_\nu\Sigma R_{\mu\nu})$$
$$+\frac{1}{3}D_\mu\Sigma^\dagger D_\mu\Sigma D_\nu\Sigma^\dagger D_\nu\Sigma - \frac{1}{6}D_\mu\Sigma^\dagger D_\nu\Sigma D_\mu\Sigma^\dagger D_\nu\Sigma - \frac{1}{3}D_\mu D_\nu\Sigma^\dagger D_\mu D_\nu\Sigma\} \quad (24)$$

This can be rewritten in the notation of ref. [1] using partial integration and the equations of motion and the $SU(3)$ relation quoted in ref. [1].

At this point it is appropriate to mention that the derivative expansion coincides here with an expansion in m_Q^{-2}. The reason is that in the effective action the only dimensional quantities are the derivatives and the external vector fields. This does not coincide with the expansion in Seeley-DeWitt coefficients because Y has a term of $\mathcal{O}(m_Q)$ and one of $\mathcal{O}(m_Q^0)$. E.g., the result for the order 4 Lagrangian above has contributions from a_3 and a_4.

4 The anomalous intrinsic parity effective action.

In this section I extend the methods used in the previous section to the part of the effective action containing $\epsilon_{\mu\nu\alpha\beta}$. A direct definition in terms of the proper time expression for the determinant is not possible here because of ambiguities in the regularization procedure. We will use the definition given by Ball [4]. There it was shown exactly how to define it explicitly. I will review it shortly. It can also be shown that these terms do not depend on the specific regulator chosen. This ambiguity is only present in the normal parity action. Since we are only interested in the terms of $\mathcal{O}(m_Q^{-2})$ I will be sloppy about subtracting the terms with $m_Q \to \infty$. This is important to get the original Wess-Zumino term correctly but does not effect the terms needed here. The subscript ϵ means that only terms containing an odd number of Levi-Civita tensors should be kept. This can always be done by subtracting the same expression with $\mathcal{D}$ and $\mathcal{D}^\dagger$ interchanged.

From (15) we have

$$\Gamma^- = -(\log\det\mathcal{D}^\dagger)_\epsilon. \tag{25}$$

We can always subtract a constant from the effective action. Here we subtract the effective action with $\mathcal{D} = \gamma_\mu\partial_\mu - m_Q$. We define then a family of $\mathcal{D}(t)$ with $\mathcal{D}(1) = \mathcal{D}$ and $\mathcal{D}(0) = \gamma_\mu\partial_\mu - m_Q$. This together with $\delta\log\det A = \mathrm{tr}\delta A A^{-1}$ leads to

$$\Gamma^- = -\int_0^1 dt\int d^4x\,\mathrm{tr}\left(\frac{\partial\mathcal{D}^\dagger(t)}{\partial t}\mathcal{D}^{\dagger-1}\right)_\epsilon. \tag{26}$$

Using $\mathcal{D}^{\dagger-1} = \mathcal{D}(\mathcal{D}^\dagger\mathcal{D})^{-1}$ and $A^{-1} = \int_0^\infty d\tau\exp(-\tau A)$ we obtain

$$\Gamma^- = -\int_0^1 dt\int d^4x\int_0^\infty d\tau\left\{\mathrm{tr}\left(\frac{\partial\mathcal{D}^\dagger}{\partial t}\mathcal{D}e^{-\tau\mathcal{D}^\dagger\mathcal{D}}\right)_\epsilon\right\} \tag{27}$$

We now call the expression in (27) $\Gamma(L_\mu, R_\mu, \mathcal{M})$. To properly reproduce the anomalous behaviour (see ref. [4]) we need to subtract the same expression with $\mathcal{M} = -\infty 1$. So the same expression with L_μ and R_μ unchanged but $\Sigma = 1$ and $m_Q \to \infty$. This can be heuristically seen as a kind of Pauli-Villars regularization. The correct expression for the abnormal intrinsic parity action is

$$\Gamma^- = \Gamma(L_\mu, R_\mu, \mathcal{M}) - \Gamma(L_\mu, R_\mu, -\infty 1) \tag{28}$$

The integral here is over τ and a cylinder in 5 space with 4 space as boundary. Equations (26,27) should be taken as a formal derivation. Eq. (28) is the proper definition of the abnormal intrinsic parity effective action.

We now add and subtract the same integral with external fields L_μ , $R_\mu \neq 0$ and $\Sigma \neq 1$ but with $m_Q \to \infty$.

$$\Gamma^- = \{\Gamma(L_\mu, R_\mu, \mathcal{M}) - \Gamma(L_\mu, R_\mu, \infty\mathcal{M})\} + \{\Gamma(L_\mu, R_\mu, \infty\mathcal{M}) - \Gamma(L_\mu, R_\mu, -\infty 1)\}. \quad (29)$$

Using the fact that we need $\epsilon_{\mu\nu\alpha\beta}$ in Γ^- we can count powers of m_Q and see that there are no positive powers of m_Q anywhere. This was, of course, needed to have sensible $m_Q \to \infty$ limits. So the first pair of brackets only contains terms that are suppressed by powers of m_Q and the second pair of brackets contains terms only of order m_Q^0. This is the Wess-Zumino term [3].

Let us now look closer to the first two terms in (29). The first term is an integral from $\mathcal{D}(0)_1 = \gamma_\mu \partial_\mu - m_Q$ to $\mathcal{D}(1)_1 = \gamma_\mu D_\mu + \mathcal{M}$ while the second term is from $\mathcal{D}(0)_2 = \gamma_\mu \partial_\mu - \infty$ to $\mathcal{D}(1)_2 = \gamma_\mu D_\mu + \infty\mathcal{M}$. The difference between the expression with $\mathcal{D}(0)_1$ and $\mathcal{D}(0)_2$ does not depend on L_μ, R_μ and Σ so it can be dropped and we can integrate from $\mathcal{D}(1)_2$ to $\mathcal{D}(1)_1$. This can be done by defining $\mathcal{D}(t') = \gamma_\mu D_\mu + t'^{-1}\mathcal{M}$. So

$$\Gamma^- = \Gamma_{WZ} - \int d^4x \int_0^1 dt' \int_0^\infty d\tau \mathrm{tr} \left(\frac{\partial \mathcal{D}^\dagger}{\partial t'} \mathcal{D} e^{-\tau \mathcal{D}^\dagger \mathcal{D}} \right)_\epsilon \quad (30)$$

We now replace the exponent by its Seeley-DeWitt expansion and perform the τ integration explicitly. Here it should be kept in mind that it now contains $\exp(-m_Q^2 \tau / t'^2)$. All terms in the t' integral are suppressed by at least one power of m_Q. So the t' integral is well defined and we can do the integration directly. The subscript r in (31) means terms of order m_Q^{-r}.

$$\Gamma^- = \Gamma_{WZ} + \int d^4x \sum_{n=3}^\infty \sum_{r=1}^\infty \frac{(n-2)!}{16\pi^2 r} \left[\frac{1}{m_Q^{2n-2}} \mathrm{tr} \hat{\mathcal{M}} \mathcal{D} a_n \right]_{r\epsilon} \quad (31)$$

This equation is the main result of this section and was previously derived in Ref. [4].

We can now look at some properties of the Seeley-DeWitt coefficients. The coefficients $\mathcal{D}a_n$ are given in terms of more elementary quantities in the coincidence limit by

$$\mathcal{D}a_n = \frac{1}{2}\gamma_\mu (\overline{D_\mu} a_n + D_\mu(a_n)) + \mathcal{M} a_n \quad (32)$$

The last term never gives rise to terms with $\epsilon_{\mu\nu\alpha\beta}$. Since $\hat{\mathcal{M}}\mathcal{M} = m_Q^2$, that only contributes like $\sum_r c_{rn} \mathrm{tr}(a_n)_r$ to the effective action. c_{rn} just indicates that the factor in front depends on n and the order in m_Q. Γ^+ was also a sum over $\mathrm{tr} a_n$ and the

[3]This argument has to be extented for the case of explicit chiral symmetry breaking by current quark masses.

argument why Γ^+ does not contain $\epsilon_{\mu\nu\alpha\beta}$ was independent of τ and m_Q. Therefore this term does not contribute to Γ^-.

A similar but more involved argument can be given for the term containing $D_\mu a_n$. This is given in Ref. [2].

The final result of this section is

$$\Gamma^- = \Gamma_{WZ} + \int d^4x \sum_{n=3}^{\infty}\sum_{r=1}^{\infty} \frac{(n-2)!}{32\pi^2 r} \left[\frac{1}{m_Q^{2n-2}} \mathrm{tr}\gamma_\mu \overline{D_\mu} a_n \hat{\mathcal{M}}\right]_{r\epsilon} \tag{33}$$

Let me repeat that the subscript r means terms of order m_Q^{-r} and the subscript ϵ means that only terms containing an odd number of Levi-Civita tensors should be kept.

One remark is in order here. The abnormal intrinsic parity effective action except for the Wess-Zumino term could be written in 4 dimensions and in a gauge invariant way. This shows that the anomaly is not renormalized by these terms as expected since the Wess Zumino term already saturates the inhomogenous part of the Ward identities.

5 The $\mathcal{O}(p^6)$ abnormal intrinsic parity effective Lagrangian.

In this section I use the main formula from section 4 to obtain the abnormal intrinsic parity effective action at next-to-leading order. There are only contributions from $\overline{D_\mu} a_3$, $\overline{D_\mu} a_4$ and $\overline{D_\mu} a_5$. As we will see $\overline{D_\mu} a_3$ contributes to processes with at least two external vector fields and one pseudoscalar meson. $\overline{D_\mu} a_4$ contains processes with at least one external field and three pseudoscalars and $\overline{D_\mu} a_5$ to processes with at least 5 pseudoscalars.

There is another hidden source of contributions to Γ^-. In rewriting the normal parity action into the standard form we have used the equations of motion. These do contain terms of both normal and intrinsic parity. and can thus produce terms of abnormal intrinsic parity if used in Γ^+. These terms do not contribute in the chiral limit as shown in [2].

The $\mathcal{O}(p^4)$ terms of normal intrinsic parity in the equation of motion could in principle also contribute if the equations of motion are used in S_{WZ}. Because of the presence of $\epsilon_{\mu\nu\alpha\beta}$ in the effective action of abnormal parity all terms of order p^4 besides the standard Wess-Zumino term are total derivatives or depend only on external fields. So there is no need to use the equations of motion and no contributions to Γ^- come from this source.

There are no contributions from $\overline{D_\mu} a_2$. From (33) it can be seen that those terms are suppressed by $m_Q^{-2} m_Q = m_Q$, or at most by m_Q^{-1}. This is however not possible because derivatives have to come in pairs. Similarly the leading term in the term with $\overline{D_\mu} a_6$ is $m_Q^{-10} m_Q^7$ or suppressed by one factor of m_Q more than we consider here.

The contribution from $\overline{D_\mu a_5}$ to the effective action is

$$\Gamma^-(a_5) = \frac{N_C}{32\pi^2}\int d^4x \frac{-\epsilon_{\mu\nu\alpha\beta}}{30} \tag{34}$$
$$\mathrm{tr}\left\{\left(\Sigma D_\gamma D_\mu \Sigma^\dagger - D_\gamma D_\mu \Sigma\Sigma^\dagger\right) D_\nu \Sigma \left(D_\alpha \Sigma^\dagger D_\gamma \Sigma - D_\gamma \Sigma^\dagger D_\alpha \Sigma\right) D_\beta \Sigma^\dagger\right\}$$

Here it can be easily seen that this part of Γ^- describes processes with at least 5 pseudoscalars. The overall factor of N_C comes from the trace over colours. The only trace left now is the one over flavours.

I have not rewritten the result in the set of operators given by [18]. Given the sheer number of terms involved it is easier to calculate directly the different physical processes from the form of the action given here. The next part is the one from $\overline{D_\mu a_4}$.

$$\Gamma^-(a_4) = \frac{N_C}{32\pi^2 m_Q^2}\int d^4x \frac{i}{180}\epsilon_{\mu\nu\alpha\beta}\mathrm{tr}\Big\{2(D_\gamma R_{\gamma\mu} + \Sigma^\dagger D_\gamma L_{\gamma\mu}\Sigma)D_\nu\Sigma^\dagger D_\alpha \Sigma D_\beta \Sigma^\dagger \Sigma$$
$$-3(D_\gamma R_{\mu\nu} + \Sigma^\dagger D_\gamma L_{\mu\nu}\Sigma)D_\alpha\Sigma^\dagger D_\gamma \Sigma D_\beta \Sigma^\dagger \Sigma$$
$$+R_{\mu\gamma}[20 D_\nu\Sigma^\dagger D_\alpha \Sigma\Sigma^\dagger D_\gamma D_\beta \Sigma - 20 D_\gamma D_\nu \Sigma^\dagger \Sigma D_\alpha \Sigma^\dagger D_\beta \Sigma - 2D_\nu \Sigma^\dagger D_\alpha \Sigma D_\gamma D_\beta \Sigma^\dagger \Sigma$$
$$+2\Sigma^\dagger D_\gamma D_\nu \Sigma D_\alpha \Sigma^\dagger D_\beta \Sigma - 8 D_\nu \Sigma^\dagger (\Sigma D_\gamma D_\alpha \Sigma^\dagger - D_\gamma D_\alpha \Sigma\Sigma^\dagger) D_\beta \Sigma]$$
$$-L_{\mu\gamma}[20 D_\nu \Sigma D_\alpha \Sigma^\dagger \Sigma D_\gamma D_\beta \Sigma^\dagger - 20 D_\gamma D_\nu \Sigma\Sigma^\dagger D_\alpha \Sigma D_\beta \Sigma^\dagger - 2 D_\nu \Sigma D_\alpha \Sigma^\dagger D_\gamma D_\beta \Sigma\Sigma^\dagger$$
$$+2\Sigma D_\gamma D_\nu \Sigma^\dagger D_\alpha \Sigma D_\beta \Sigma^\dagger - 8 D_\nu \Sigma(\Sigma^\dagger D_\gamma D_\alpha \Sigma - D_\gamma D_\alpha \Sigma^\dagger \Sigma) D_\beta \Sigma^\dagger]$$
$$+3R_{\mu\nu}[D_\gamma D_\alpha \Sigma^\dagger \Sigma D_\beta \Sigma^\dagger D_\gamma \Sigma + D_\alpha \Sigma^\dagger D_\gamma D_\beta \Sigma\Sigma^\dagger D_\gamma \Sigma$$
$$-D_\gamma \Sigma^\dagger \Sigma D_\gamma D_\alpha \Sigma^\dagger D_\beta \Sigma - D_\gamma \Sigma^\dagger D_\alpha \Sigma\Sigma^\dagger D_\gamma D_\beta \Sigma]$$
$$-3L_{\mu\nu}[D_\gamma D_\alpha \Sigma\Sigma^\dagger D_\beta \Sigma D_\gamma \Sigma^\dagger + D_\alpha \Sigma D_\gamma D_\beta \Sigma^\dagger \Sigma D_\gamma \Sigma^\dagger$$
$$-D_\gamma \Sigma\Sigma^\dagger D_\gamma D_\alpha \Sigma D_\beta \Sigma^\dagger - D_\gamma \Sigma D_\alpha \Sigma^\dagger \Sigma D_\gamma D_\beta \Sigma^\dagger]$$
$$-2D^2 D_\mu \Sigma\{R_{\nu\alpha}\Sigma^\dagger - \Sigma^\dagger L_{\nu\alpha}, \Sigma D_\beta \Sigma^\dagger\} + 2D^2 D_\mu \Sigma^\dagger\{L_{\nu\alpha}\Sigma - \Sigma R_{\nu\alpha}, \Sigma^\dagger D_\beta \Sigma\}$$
$$+i2D_\mu D_\gamma D_\nu \Sigma[2D_\gamma D_\alpha \Sigma^\dagger \Sigma D_\beta \Sigma^\dagger + 2D_\beta \Sigma^\dagger \Sigma D_\gamma D_\alpha \Sigma^\dagger$$
$$+\Sigma^\dagger D_\gamma D_\alpha \Sigma D_\beta \Sigma^\dagger + D_\beta \Sigma^\dagger D_\gamma D_\alpha \Sigma\Sigma^\dagger]$$
$$-i2D_\mu D_\gamma D_\nu \Sigma^\dagger[2D_\gamma D_\alpha \Sigma\Sigma^\dagger D_\beta \Sigma + 2D_\beta \Sigma\Sigma^\dagger D_\gamma D_\alpha \Sigma$$
$$+\Sigma D_\gamma D_\alpha \Sigma^\dagger D_\beta \Sigma + D_\beta \Sigma D_\gamma D_\alpha \Sigma^\dagger \Sigma]\Big\} \tag{35}$$

This part of Γ^- contributes to processes with at least one external field and three pseudoscalars.

The last part of Γ^- is the part coming from the third Seeley-DeWitt coefficient.

$$\Gamma^-(a_3) = \frac{N_C}{32\pi^2 m_Q^2}\int d^4x \frac{-1}{120}\epsilon_{\mu\nu\alpha\beta}\mathrm{tr}\Big\{$$

$$
\begin{aligned}
&7\{D_\gamma R_{\gamma\mu}, R_{\alpha\beta}\}D_\nu\Sigma^\dagger\Sigma - 7\{D_\gamma L_{\gamma\mu}, L_{\alpha\beta}\}D_\nu\Sigma\Sigma^\dagger \\
&+2D_\gamma R_{\gamma\mu}(\Sigma^\dagger L_{\alpha\beta}D_\nu\Sigma - D_\nu\Sigma^\dagger L_{\alpha\beta}\Sigma) - 2D_\gamma L_{\gamma\mu}(\Sigma R_{\alpha\beta}D_\nu\Sigma^\dagger - D_\nu\Sigma R_{\alpha\beta}\Sigma^\dagger) \\
&+3\{R_{\mu\gamma}, R_{\alpha\beta}\}(\Sigma^\dagger D_\gamma D_\nu\Sigma - D_\gamma D_\nu\Sigma^\dagger\Sigma) - 3\{L_{\mu\gamma}, L_{\alpha\beta}\}(\Sigma D_\gamma D_\nu\Sigma^\dagger - D_\gamma D_\nu\Sigma\Sigma^\dagger) \\
&+4L_{\mu\gamma}(D_\gamma D_\nu\Sigma R_{\alpha\beta}\Sigma^\dagger - \Sigma R_{\alpha\beta}D_\gamma D_\nu\Sigma^\dagger) - 4R_{\mu\gamma}(D_\gamma D_\nu\Sigma^\dagger L_{\alpha\beta}\Sigma - \Sigma^\dagger L_{\alpha\beta}D_\gamma D_\nu\Sigma) \\
&+2L_{\mu\gamma}(D_\nu\Sigma D_\gamma R_{\alpha\beta}\Sigma^\dagger - \Sigma D_\gamma R_{\alpha\beta}D_\nu\Sigma^\dagger) - 2R_{\mu\gamma}(D_\nu\Sigma^\dagger D_\gamma L_{\alpha\beta}\Sigma - \Sigma^\dagger D_\gamma L_{\alpha\beta}D_\nu\Sigma) \\
&-13\{R_{\mu\gamma}, D_\gamma R_{\alpha\beta}\}D_\nu\Sigma^\dagger\Sigma + 13\{L_{\mu\gamma}, D_\gamma L_{\alpha\beta}\}D_\nu\Sigma\Sigma^\dagger \\
&\left.+[R_{\alpha\beta}, R_{\mu\gamma}](\Sigma^\dagger D_\gamma D_\nu\Sigma + D_\gamma D_\nu\Sigma^\dagger\Sigma) - [L_{\alpha\beta}, L_{\mu\gamma}](\Sigma D_\gamma D_\nu\Sigma^\dagger + D_\gamma D_\nu\Sigma\Sigma^\dagger)\right\} \quad (36)
\end{aligned}
$$

This part contributes only to processes with at least two external vector fields and one pseudoscalar.

The full order p^6 expression for the abnormal intrinsic parity action is the sum of equations (34,35) and (36). This is the main result of this paper. It is the equivalent of the action derived in [3] for the normal intrinsic parity part in the sense that it is the next-to-leading contribution in the derivative expansion to the relevant physical processes as explained in the paragraph after eq. (7). Here it was necessary to pay more careful attention to the definition of the effective action to get a well-defined expression. It should be kept in mind that the above equations are in Euclidean quantities. Changing to Minkowski quantities can be done simply by multiplying by i.

6 Application to $\pi^0, \eta \to \gamma l^+l^-$.

Now we can look at the decays π^0, $\eta \to \gamma l^+l^-$. I only consider this decay and not the direct two photon decay because we work in the chiral limit here and the decay rate also gets contributions from chiral symmetry breaking terms. The slope at small lepton pair masses is at this order a pure prediction. It already contains six derivatives in its expression so any contribution involving the quark masses is at least $\mathcal{O}(p^8)$.

The singlet component is included in the same way as in Refs. [15, 17] via a nonet approximation and an extra singlet mass. The external currents are the electromagnetic ones and the meson matrix M only contains π^0 and η. This means that all quantities in eq. (36) commute. We have used the equations of motion several times in the previous section. Here they correspond to $\partial^2 M = 0$ so we also drop these terms. We now rewrite it in Minkowski space. This amounts to multiplying by i. Looking at the decay with one photon off-shell ($k^{*2} \neq 0$) and adding the lowest order contribution this leads to

$$
A(M \to \gamma\gamma^*) = \frac{i\sqrt{2}e^2}{4\pi^2 f_M} C_M \epsilon^{\mu\nu\alpha\beta}\epsilon_\mu k_\nu \epsilon^*_\alpha k^*_\beta \left[1 + \frac{k^{*2}}{12m_Q^2}\right] \quad (37)
$$

The most accurate measurements of these formfactors are in photon-photon experiments at electron-positron colliders [24]. These are for large k^{*2}, above $(500\ MeV)^2$ and can only be compared to our result if the pole form for the form factor is explicitly assumed. They do fit the pole form well in the measured regime and lead to

$$\frac{1}{12m_Q^2} = 1.42 \pm 0.21(1.70 \pm 0.14)\ GeV^{-2}. \tag{38}$$

for $\eta(\pi^0)$. These resolve the problem with the sign in the π^0 form factor and are within errors in agreement with the other measurements [25]. Taking a weighted average of the last results we obtain a value for m_Q of

$$m_Q = 223 \pm 9\ MeV. \tag{39}$$

Including the effect of the loops (which are non-leading in N_C) changes this value to

$$m_Q = 243 \pm 9\ MeV. \tag{40}$$

A very similar result result can be obtained directly by comparing our result with the result from vector meson dominance.[4]

$$m_Q = \frac{m_\rho}{\sqrt{12}} \tag{41}$$

This has to be compared with other determinations of m_Q. In ref. [5] an estimate from the vacuum expectation values from QCD sum rules was given, $m_Q \approx 320\ MeV$. This is rather higher than the value we obtain but both estimates have rather large errors. The gluonic corrections to the slope could be sizable. The number obtained is, however, a reasonable value when compared to the other determination.

The corrections due to non zero quark masses and gluonic vacuum expectation values have for this particular process been calculated using Feynman diagrams by A. Pich and myself. For a single quark the relevant Feynman diagrams lead to a combined Feynman integral expression

$$\begin{aligned} F(p^2,q^2) &= \frac{1}{M_Q}\left\{1 + \frac{p^2}{12M_Q^2}\left(1 - \frac{2\pi^2}{45}\frac{<\alpha_S G^2/\pi>}{M_Q^4}\right)\right. \\ &\left. + \frac{q^2}{12M_Q^2}\left(1 - \frac{6\pi^2}{45}\frac{<\alpha_S G^2/\pi>}{M_Q^4}\right)\right\}. \end{aligned} \tag{42}$$

This can also be read off from the QCD sum rule calculations for $J/\psi \to \eta_c\gamma$ in ref. [27]. The quark mass M_Q here is the quark mass propagating in the quark loop. q^2 is the off-shellnes of one photon. The other photon we consider real. p^2 is the meson momentum squared or here the meson mass squared. The quark mass here is the

---

[4] This was noted already a long time ago where this contribution to the slope was calculated using Feynman diagrams[26].

sum of the constituent and the current mass, $M_Q = m_Q + m_q$. We can now put in the relevant numbers for the pion. The pion itself couples to the diagram with the strength m_Q. So the amplitude for the pion (with $m_u = m_d$) is

$$A(\pi^0 \to \gamma\gamma^*) = \frac{i\sqrt{2}\alpha}{\pi f_\pi}\left\{1 - \frac{m_u}{m_Q} + \frac{m_\pi^2}{12m_Q^2}\left(1 - \frac{2\pi^2}{45}\frac{<\alpha_S G^2/\pi>}{m_Q^4}\right) + \frac{q^2}{12m_Q^2}\left(1 - \frac{6\pi^2}{45}\frac{<\alpha_S G^2/\pi>}{m_Q^4}\right)\right\}. \quad (43)$$

The equivalent formula for the η can be written easily. The convergence of the expansion in m_s/m_Q is rather questionable. but it can be seen that the quark mass corrections to the pion decay rate are very small. There are no gluonic corrections to the p^4 term. These are forbidden by the anomaly nonrenormalization theorem. If we use the values for the gluon condensate given in ref. [5] we find very large corrections to the slope. The gluonic corrections are in fact $\mathcal{O}(1)$ and can even change the sign of the slope. The interpretation of that depends on the interpretation of gluonic corrections in this chiral quark model. This is discussed in ref. [5].

7 Conclusions

In this talk I have presented a calculation of the $O(p^6)$ abnormal intrinsic parity low energy effective Lagrangian within the QCD effective action approach. As expected on general grounds it could be written in a manifestly chirally invariant form. As a byproduct we have rederived the formula for the effective action given by Ball [4] and simplified it considerably. This formula can be extended to other dimensions than four in a straightforward way.

We have then used the derived effective action to obtain a value for the parameter m_Q in the QCD model used. The value obtained was in reasonable agreement with other determinations of it though a bit smaller.

References

[1] J. Gasser and H. Leutwyler, Nucl. Phys. B250(1985)465, Ann. Phys. 158(1984)142.

[2] J. Bijnens, The effective action of QCD in the anomalous sector, CERN TH 6023/91, to be published in Nucl. Phys. B.

[3] J. Balog, Phys. Lett. 149B(1984)197,
J. Zuk, Z. Phys. C29(1985)303,
E. D'Hoker and E. Farhi, Nucl. Phys. B248(1984)59.

[4] R.D. Ball, Phys. Rep. 182(1989)1.

[5] D. Espriu, E. de Rafael and J. Taron, Nucl. Phys. B345(1990)22.

[6] V. Adrianov, these proceedings.

[7] A. Pich and E. de Rafael, Nucl. Phys. B358(1991)311,
A. Pich, these proceedings.

[8] J. Bijnens and E. de Rafael, CERN TH 6266/91, CPT / PE2602.

[9] J. Wess and B. Zumino, Phys. Lett. 37B(1971)98.

[10] E. Witten, Nucl. Phys. B223(1983)422.

[11] K. Chou et al., Phys. Lett. B134(1984)67,
H. Kawai and S. Tye, Phys. Lett. B140(1984)403,
J. Mañes, Nucl. Phys. B250(1985)369.

[12] A. Manohar and G. Moore, Nucl. Phys. B243(1984)55.

[13] J. Bijnens, H. Sonoda and M.B. Wise, Can. J. Phys 64(1986)1.

[14] J. Donoghue, B. Holstein and Y. Lin, Phys. Rev. Lett. 55(1985)2766.

[15] J. Bijnens, A. Bramon and F. Cornet, Phys. Rev. Lett. 61(1988)1453.

[16] J. Donoghue and D. Wyler, Nucl. Phys. B316(1989)289.

[17] J. Bijnens, A. Bramon and F. Cornet, Z. Phys. C46(1990)599.

[18] D. Issler, SLAC-PUB 4943 (1989, revised 1990).

[19] J. Bijnens, A. Bramon and F. Cornet, Phys. Lett. B237(1990)488.

[20] G. Ecker et al., Phys. Lett. B223(1989)425.

[21] J. Donoghue, F. Ramirez and G. Valencia, Phys. Rev. D38(1988)2195.

[22] H. Georgi and A. Manohar, Nucl. Phys. B234(1984)189.

[23] S. Weinberg, Phys. Rev. Lett., 65(1990)1181.

[24] CELLO collaboration, Z. Phys. C49(1991)401.

[25] Fonvieille et al., Phys. Lett. 233B(1989)65,
Dzhelyadin et al., Phys. Lett. 94B(1980)548.

[26] A. Pich and J. Bernabéu, Z. Phys. C22(1984)197.

[27] V.A. Beĭlin and A.V. Radyushkin, Sovj. J. Nucl. Phys. 39(1984)800,
A.Y. Khodzmariyan, Sovj. J. Nucl. Phys. 39(1984)614.

NON-LEPTONIC WEAK INTERACTIONS

$K \to 2\pi, 3\pi$ DECAYS AND THE B-PARAMETER IN NEXT-TO-LEADING ORDER CHIRAL PERTURBATION THEORY

J. KAMBOR

Physik Department, Technische Universität München
DW-8046 Garching, FRG

ABSTRACT

We evaluate the next-to-leading corrections to $K \to 2\pi, 3\pi$ decays and the kaon B-parameter in chiral perturbation theory. The lowest order results for the K-decays are substantially affected as loops contribute strongly to $\Delta I = 1/2$ transitions. Fitting the counterterm coupling constants of the next-to-leading order corrections to experimental data, we conclude that the lowest order value of the octet coupling constant is reduced by about 30 % while the corresponding coupling constant for the 27-plet remains unchanged. As a byproduct, the determination of the isospin amplitudes predicts the branching ratio for $K_S \to \pi^+\pi^-\pi^0$ be $(2.4 \pm 0.7) \cdot 10^{-7}$. The kaon B- parameter depends at this order of the chiral expansion on additional coupling constants, which cannot be fixed from the K-decay data.

1. Introduction

Most recent efforts to describe the non-leptonic weak interactions of the light mesons (π, K, η) are based on the chiral langrangian[1,2] as an effective field theory. It's functional form is determined by two requirements: the transformation properties of the interactions under $SU(3)_L \otimes SU(3)_R$ and the identification of the light pseudoscalar mesons as the Goldstone bosons associated with the spontaneous breaking of chiral symmetry to $SU(3)_{L+R}$[3]. In order to give nonvanishing masses to the mesons, soft breaking terms have to be added. These are the only ingredients for the construction of the chiral lagrangian; they constitute the cornerstones of our understanding of the strong and weak interactions of pseudoscalar mesons at low energies.

The theory may be systematically expanded in powers of momenta and masses of the light mesons. This expansion, which is aquivalent to a loop expansion,[4] is the content of chiral perturbation theory (CHPT). Since the chiral lagrangian is nonrenormalizable, each successive loop introduces new counterterms with a priori unknown coupling constants. However, at sufficiently low energies, the higher loops are presumed to yield small corrections. In practice, the corrections are calculated to the one-loop order; physical observables are then expressed in terms of only a few unknown parameters. This program has been carried out for the strong interactions by Gasser and Leutwyler[5]. In a recent paper[6], we have extended their theoretical formalism to the nonleptonic weak interactions.

The coupling constants of the chiral lagrangian parametrize the dynamics of the underlying theory, i.e. the standard model of strong and weak interactions. The ultimate goal of the theory is to calculate these coupling constants from first principles. Attempts in this direction where presented at this conference by A. Pich.[7] A more modest but still useful approach is to eliminate the unknown parameters by relating several physical observables and/or to determine a subset of parameters from experiments. This approach tests the assumptions of softly broken chiral symmetry, independent of the dynamical details of the underlying theory. We have followed this second line of thought and applied the formalism to the CP-conserving $K \to 2\pi$ and $K \to 3\pi$ decays[8] and to the Kaon B-parameter (for applications to radiative K-decays, see Ref. 9). As will be discussed below, the simultaneous analysis of $K \to 2\pi$ and $K \to 3\pi$ decay data implies an enhancement of the $\Delta I = 1/2$ amplitude (relative to $\Delta I = 3/2$) in $K \to 2\pi$ decays by a factor $\simeq 1.5$, coming from chiral loops. This may help in explaining the $\Delta I = 1/2$ rule.

Section 2 contains a brief summary of the chiral lagrangian for strong and weak interactions. In section 3, the standard isospin amplitudes for $K \to 2\pi, 3\pi$ decays are determined from the data. Section 4 presents the results of the next-to-leading order calculation for $K \to 2\pi$ and $K \to 3\pi$ decays and a leeast χ^2 fit of coupling constants to the data. We comment on the question for breakdown of CHPT. In section 5, a complete next-to-leading order calculation of the Kaon B-parameter is given, including all possible counterterms of order p^4. Section 6 summarizes our results.

2. The chiral lagrangian for strong and weak interactions

The effective chiral lagrangian is written in terms of the pseudoscalar meson fields $\phi = (\pi, K, \eta)$. They are collected in a unitary matrix $u(\phi)$ with $u(\phi)^2 \equiv U(\phi) = \exp(i\sqrt{2}\phi/F_0)$ and ϕ is a 3×3 matrix containing the meson fields

$$\phi = \begin{pmatrix} \frac{\pi^0}{\sqrt{2}} + \frac{\eta}{\sqrt{6}} & -\pi^+ & -K^+ \\ \pi^- & \frac{-\pi^0}{\sqrt{2}} + \frac{\eta}{\sqrt{6}} & -K^0 \\ K^- & -\bar{K}^0 & \frac{-2}{\sqrt{6}}\eta \end{pmatrix} . \tag{1}$$

F_0 is the pion decay constant in lowest order CHPT. The action of $G = SU(3)_L \otimes SU(3)_R$ on $G = (SU(3)_L \otimes SU(3)_R)/SU(3)_V$ defines a nonlinear realization of $h(\phi)$[3]

$$u(\phi) \xrightarrow{G} g_R u(\phi) h(\phi)^\dagger = h(\phi) u(\phi) g_L^\dagger \tag{2}$$

where $u(\phi) \in C$, $g_{L,R} \in SU(3)_{L,R}$. Thus

$$U(\phi) \xrightarrow{G} g_R U(\phi) g_L^\dagger \ . \tag{3}$$

The same transformation law, Eq. 3, holds for the covariant derivative $\nabla_\mu U = \partial_\mu U - i(v_\mu + a_\mu)U + iU(v_\mu - a_\mu)$, where v_μ and a_μ are external vector and axialvector sources respectively. The effects of nonvanishing quark masses are introduced by a scalar source χ, transforming like U:[5]

$$\chi \xrightarrow{G} g_R \chi g_L^\dagger \ . \tag{4}$$

The chiral lagrangian is now easily constructed by taking products of U, $\nabla_\mu U$, χ and the hermitean conjugate objects and tracing over apropriate Gell-Mann λ-matrices.[5,6]

For our applications the external fields are not relevant and we may substitute $\nabla_\mu U \to \partial_\mu U$, $\chi \to M$, where in the isospinlimit

$$M = \begin{pmatrix} M_\pi^2 & & \\ & M_\pi^2 & \\ & & 2M_K^2 - M_\pi^2 \end{pmatrix} . \tag{5}$$

In the low energy expansion, $\partial_\mu U$ counts as order p, M as order p^2, where p is a meson momentum.

The chiral lagrangian may then be written as

$$\begin{aligned} \mathcal{L} &= \mathcal{L}_s + \mathcal{L}_{\Delta S=1} \\ &= \mathcal{L}_s^{(2)} + \mathcal{L}_s^{(4)} + \ldots + \mathcal{L}_{\Delta S=1}^{(2)} + \mathcal{L}_{\Delta S=1}^{(4)} + \cdots \end{aligned} \tag{6}$$

where $\mathcal{L}^{(n)}$ means of order p^n. More explicitly

$$\mathcal{L}_s^{(2)} = \frac{F_0^2}{4}\{Tr(L_\mu L^\mu) + Tr(MU + U^\dagger M)\} \tag{7}$$

$$\mathcal{L}_{\Delta S=1}^{(2)} = c_2 Tr(\lambda_6 L\mu L^\mu) + c_3 t_{ik}^{jl} Tr(Q_j^i L_\mu) Tr(Q_l^k L_\mu) \tag{8}$$

with $L_\mu = iU^\dagger \partial_\mu U$. $(Q_j^i)_{kl} = \delta_{il}\delta_{jk}$ are 3×3 matrices in flavor space; $\lambda_6 = Q_3^2 + Q_2^3$ projects the octet and the t_{ik}^{jl} are defined to project the 27-plet part of the interaction. The term quadratic in meson masses, the "weak mass term" can be neglected.[6,10,11] The coupling constants c_2 and c_3 of the $\Delta I = 1/2, 3/2$ operators are fixed by $K \to 2\pi$ decays to be[12]:

$$\begin{aligned} \frac{c_2}{F_\pi^2} &= 0.95 \cdot 10^{-7} \\ \frac{c_3}{F_\pi^2} &= -0.008 \cdot 10^{-7} \ , \end{aligned} \tag{9}$$

where $F_\pi = 93.3 MeV$ is the pion decay constant. The result, $c_2 \gg c_3$ constitutes the mysterious $\Delta I = 1/2$ rule.

In next-to-leading order, the chiral lagrangian has the form

$$\mathcal{L}^{(4)} = \sum_i^{1..12} L_i^r O_i^s + \sum_i^{1..47} E_i^r O_i^8 + \sum_i^{1..33} D_i^r O_i^{27} \ , \tag{10}$$

where L_i^r,E_i^r and D_i^r are renormalized coupling constants of the strong, the weak octet and the weak 27-plet counterterms respectively. [5,6] In the case where external fields are neglected there are 15 independent octet terms and 13 27-plets. To complete the chiral lagrangian to order p^4, one has to add to Eqs. 7,8,10 all one-loop graphs generated from vertices of the lowest order chiral lagrangian, Eqs. 7,8. The WZW-term, [13] which is also of order p^4 does not contribute to the processes considered here. The sum, counterterm plus loops is finite and independent of the renormalization scale appearing in the chiral loops.

3. Determination of isospinamplitudes from the data

The isospin decomposition of the CP-conserving $K \to 2\pi$ amplitudes is

$$\begin{aligned} A(K_S \to \pi^0\pi^0) &= \sqrt{\frac{2}{3}} A_0 e^{i\delta_0} + \frac{2}{\sqrt{3}} A_2 e^{i\delta_2} \\ A(K_S \to \pi^+\pi^-) &= -\sqrt{\frac{2}{3}} A_0 e^{i\delta_0} + \frac{1}{\sqrt{3}} A_2 e^{i\delta_2} \\ A(K^+ \to \pi^+\pi^0) &= \frac{\sqrt{3}}{2} A_2 e^{i\delta_2}. \end{aligned} \tag{11}$$

A_0 and A_2 are real amplitudes corresponding to isospin $I = 0$ and $I = 2$ in the final state, respectively. δ_0 and δ_2 denote the S-wave $\pi\pi$ scattering phases.

The CP-conserving $K \to 3\pi$ amplitudes are conventionally expanded in powers of Dalitz plot variables[14,15]. In our convention, the decomposition reads

$$\begin{aligned} A(K_L \to \pi^+\pi^-\pi^0) =&(\alpha_1+\alpha_3) - (\beta_1+\beta_3)Y \\ &+ (\zeta_1 - 2\zeta_3)(Y^2 + X^2/3) + (\xi_1 - 2\xi_3)(Y^2 - X^2/3) \\ A(K_L \to \pi^0\pi^0\pi^0) =& -3(\alpha_1+\alpha_3) - 3(\zeta_1 - 2\zeta_3)(Y^2 + X^2/3) \\ A(K^+ \to \pi^+\pi^+\pi^-) =&(2\alpha_1 - \alpha_3) + (\beta_1 - \frac{1}{2}\beta_3 + \sqrt{3}\gamma_3)Y \\ &+ 2(\zeta_1+\zeta_3)(Y^2 + X^2/3) - (\xi_1 + \xi_3 - \xi_3')(Y^2 - X^2/3) \\ A(K^+ \to \pi^+\pi^0\pi^0) =& -\frac{1}{2}(2\alpha_1 - \alpha_3) + (\beta_1 - \frac{1}{2}\beta_3 - \sqrt{3}\gamma_3)Y \\ &- (\zeta_1+\zeta_3)(Y^2 + X^2/3) - (\xi_1 + \xi_3 + \xi_3')(Y^2 - X^2/3) \\ A(K_S \to \pi^+\pi^-\pi^0) =& \frac{2}{3}\sqrt{3}\gamma_3 \cdot X - \frac{4}{3}\xi_3' \cdot X \cdot Y. \end{aligned} \tag{12}$$

In Eq. 12,

$$\begin{aligned} Y &= \frac{s_3 - s_0}{M_\pi^2}, \quad X = \frac{s_2 - s_1}{M_\pi^2} \\ s_i &= (k_\mu - p_{i\mu})^2, \quad s_0 = \frac{1}{3}\sum s_i, \end{aligned} \tag{13}$$

where k is the kaon momentum and p^i are those of the three pions π^i. The subscripts 1 and 3 refer to $\Delta I = 1/2, 3/2$, respectively.

The standard analysis of the isospin amplitudes A_0 and A_2 of $K \to 2\pi$ decays and $\alpha_1, \alpha_3, \beta_1, \beta_3, \gamma_3, \zeta_1, \xi_1$ in $K \to 3\pi$ transitions was performed by Devlin and Dickey[14]. Using recent data[16] and including the $\Delta I = 3/2$ quadratic slope parameters ζ_3, ξ_3 and ξ_3' in Eq. 12, we have redone the fit. The results, summarized in the second column of Table 1, are consistent with the values in Ref. 14. The relatively large χ^2 is mainly due to the least well-measured rates: those of $K^+ \to \pi^+\pi^0\pi^0$ and $K_L \to \pi^0\pi^0\pi^0$.

The difference in the magnitudes of $|\delta_2 - \delta_0|$ in Table 1 is that between the value obtained directly from $K \to 2\pi$ and that measured in $\pi - \pi$ scattering[17]. The discrepancy is certainly less than the 3σ effect indicated by the errors, because isospin breaking in the $K \to 2\pi$ amplitudes[18] has been neglected. A 2 % correction in the decay rates would yield the phase measured in $\pi - \pi$ scattering. Following Devlin and Dickey, the phase shifts in the 3π final state have been assumed negligible, since the available phase space is small.

The slope ζ_3 is non-zero and rather large, implying that $\zeta_1 - 2\zeta_3$, which determines the quadratic slope parameter in $K_L \to 3\pi^0$, is compatible with zero. This prediction has recently been confirmed by a new measurement of E736[19]. However, the error bar is large reflecting the big uncertainties in the quadratic PDG slope parameters. New data on the K^+-decays is needed in order to clarify the situation.

The decay rate, $\Gamma(K_S \to \pi^+\pi^-\pi^0)_{CP}$, is determined by the experimentally measured slope parameters, γ_3 and ξ_3'. Performing the phase space integration yields

$$\Gamma(K_S \to \pi^+\pi^-\pi^0)_{CP} \simeq (1.8 \pm 0.5)10^{-21} GeV. \tag{14}$$

The inclusion of the small ξ_3'-term reduces the central value in Eq. 14 by 10 % and increases the error bar by 20 %. The corresponding branching ratio, $B \simeq (2.4 \pm 0.7) \cdot 10^{-7}$, is two orders of magnitude larger than the CP-violating contribution due to $K_L - K_S$ mixing and should be measurable in the near future.

4. $K \to 2\pi$ and $K \to 3\pi$ in next-to-leading order CHPT

In next-to-leading order CHPT the isospin amplitudes $\mathcal{A}^{(i)} = \{A_0, A_2, \alpha_1, \alpha_3, \ldots, \xi_3'\}$ discussed in the previous section are written in the form

$$\mathcal{A}^{(i)} = \mathcal{A}^{(i)}_{\text{tree}} + \mathcal{A}^{(i)}_{\text{C.T.}}(\mu) + \mathcal{A}^{(i)}_{\text{loop}}(\mu). \tag{15}$$

Here, $\mathcal{A}^{(i)}_{\text{tree}}$ is the contribution of the tree-diagrams from the lowest order chiral lagrangian, $\mathcal{A}^{(i)}_{\text{C.T.}}(\mu)$ contains the contributions of strong and weak counterterms and $\mathcal{A}^{(i)}_{\text{loop}}(\mu)$ represents the sum of the one-loop graphs. μ is the renormalization scale; the sum of the last two terms has to be μ-independent, which was a welcome consistency check on our calculations. The explicit expressions for $\mathcal{A}^{(i)}_{tree}$ and $\mathcal{A}^{(i)}_{C.T.}$

TABLE 1: The values of the amplitudes in Eqs. 11 and 12 obtained from fits to experiment are shown in the first two columns. Our value of $\delta_2 - \delta_0$ is obtained from $K \to 2\pi$ decays alone, while some additional constraints were used in Ref. 14. The $K \to 3\pi$ amplitudes $\alpha_1 \ldots \xi_3'$ are in units of 10^{-8}. The results of lowest and next-to-lowest order chiral perturbation theory are displayed in the two columns to the right.

	Devlin and Dickey	Our fit	lowest order	order p^4
$a_{1/2}[keV]$	0.4687 ± 0.0006	0.4699 ± 0.0012	0.4698	0.4698
$a_{3/2}[keV]$	0.0210 ± 0.0001	0.0211 ± 0.0001	0.0211	0.0211
$\delta_2 - \delta_0$	$(-45.6 \pm 5)^\circ$	$(-61.5 \pm 4)^\circ$	0	-29°
α_1	91.46 ± 0.24	91.71 ± 0.32	74.0	91.8
α_3	-7.14 ± 0.36	-7.36 ± 0.47	-4.1	-7.6
β_1	-25.83 ± 0.41	-25.68 ± 0.27	-16.5	-25.6
β_3	-2.48 ± 0.48	-2.43 ± 0.41	-1.0	-2.5
γ_3	2.51 ± 0.36	2.26 ± 0.23	1.8	2.5
ζ_1	-0.37 ± 0.11	-0.47 ± 0.15	---	-0.6
ζ_3	---	-0.21 ± 0.08	---	-0.02
ξ_1	-1.25 ± 0.12	-1.51 ± 0.30	---	-1.5
ξ_3	---	-0.12 ± 0.17	---	-0.05
ξ_3'	---	-0.21 ± 0.51	---	-0.08
χ^2/DOF	12.8/3	10.3/2	4121/5	37/13

as well as the numerical results for $\mathcal{A}^{(i)}_{loop}(\mu = M_\eta)$ are displayed in Ref. 8. I would like to add two comments:

i) The contributions from loop diagrams are substantial, in particular for the $\Delta I = 1/2$ amplitudes. These large loop-contributions stem from the attractiv $I = 0$ S-wave rescattering of pions. The kinematics allows the intermediate states to be on-shell and large imaginary parts are generated. The question arises whether the rescattering diagrams can be summed to all orders in a manner consistent with chiral symmetry.[20] ii) The calculation has been performed in the isospin limit. It cannot be excluded that the $\Delta I = 1/2$ weak Hamiltonian contributes significantly to $\Delta I = 3/2$ processes via isospin breaking effects.[18] Even for $K \to 2\pi$ decays a complete analysis of such effects is still lacking.

4.1 Determination of low-energy parameters

In order to simplify things, we consider the limit $M_\pi^2/M_K^2 \ll 1$. The twelve isospin amplitudes $\mathcal{A}^{(i)}$ depend on 9 parameters, the coupling constants c_2 and c_3 of the lowest order chiral lagrangian and 7 linear combinations $K_1, \ldots, K_7$ of weak counterterm coupling constants. The strong counterterms are taken from Ref. 5. The free parameters may be fitted to the isospin amplitudes by the method of least squares. In order to compare with the lowest order result, we first neglect the one-loop and p^4 counterterm corrections an fit the constants c_2, c_3 to the amplitudes A_0, A_2, α_1, α_3, β_1, β_3 and γ_3 (the quadratic slope parameters $\xi_1, \xi_1, \ldots \xi_3'$ are not included since they vanish identically to lowest order in CHPT). The result of a fit which uses as input data only the $K \to 2\pi$ amplitudes A_0 and A_2 is shown in the third column of Table 1; the predictions for the $K \to 3\pi$ amplitudes $\alpha_1, \ldots$ are systematically too low by about 25–30%. Similarly, fits using as input data only the $K \to 3\pi$ or fits using all the K-decay amplitudes are unsatisfactory. Thus, they demonstrate the necessity of proceeding beyond lowest order in chiral petrurbation theory.

Including the one-loop and counterterm corrections yields the results summarized in the fourth column of Table 1 and in Table 2. In deriving these, we have used as data the set of twelve isospin amplitudes and one phase determined in Section 3 and displayed in column 2 two of Table 1, and required that the nine relative phases of $K \to 3\pi$ be zero with an error of 15°. This requirement is consistent with the assumption used to extract the isospin amplitudes. To this enlarged set of data, we have fitted the coupling constants, c_2, c_3 and the 7 linear combinations of counterterm coupling constants on which the next-to-leading order amplitudes depend. As a check on the possible correlation of the amplitudes induced by the $\Delta I = 1/2$ part of the 27-plet effective interaction, we have fitted the $\Delta I = 1/2$ and $\Delta I = 3/2$ amplitudes separately. As a comparison of the two columns of Table 2 indicates, the correlation is weak; the central values of the well-determined coupling constants are stable.

The most interesting result of including next-to-leading order corrections is the significant reduction of c_2 by 30% to $c_2/F_\pi^2 = 0.66 \cdot 10^{-7}$. The reduction stems from a large loop contribution. The coupling constant c_3 is unaffected by the corrections.

TABLE 2: Values of c_2, c_3 and the counterterms K_i in units of 10^{-7}. Error bars correspond to a change in χ^2 by one unit. No theoretical uncertainties are included.

Fit	8	8 + 27
χ^2/DOF	2.4/4	37/13
$\frac{c_2}{F_\pi^2}$	0.666 ± 0.002	0.662 ± 0.005
$\frac{c_3}{F_\pi^2}$	---	-0.0083 ± 0.0002
K_1	$(-0.7 \pm 7.2)10^{-4}$	$(0.4 \pm 1.2)10^{-4}$
K_2	$(8.4 \pm 0.1)10^{-2}$	$(8.4 \pm 0.1)10^{-2}$
K_3	$(5.8 \pm 0.4)10^{-2}$	$(5.9 \pm 0.4)10^{-2}$
K_4	---	$(0.3 \pm 1.4)10^{-5}$
K_5	---	$(-2.8 \pm 0.5)10^{-4}$
K_6	---	$(-1.4 \pm 0.6)10^{-3}$
K_7	---	$(-6.1 \pm 3.2)10^{-4}$

Thus, the reduction of c_2 implies that a factor, 1.5, of the $\Delta I = 1/2$ enhancement is due to long-distance effects (compare [21]).

A comparison of the two columns of Table 2 shows that the two dominant counterterms, K_2 and K_3 correspond to $\Delta I = 1/2$ operators; agreeing with the conclusions of Donoghue, Golowich and Holstein[22]. They considered only $\Delta I = 1/2$ transitions and found two counterterms which they denoted a_1 and a_2. The values they obtained, corresponding to $K_2 = 5.2 \cdot 10^{-9}$ and $K_3 = 15.5 \cdot 10^{-9}$, differ considerably from ours since they did not include the loops. The well-determined $\Delta I = 3/2$ counterterm coupling constants, K_6 and K_7, are considerably smaller than the $\Delta I = 1/2$ constants, indicating that the pattern of $\Delta I = 1/2$ enhancement observed in the decay amplitudes is preserved at next-to-lowest order.

4.2 Relating $K \to 2\pi$ and $K \to 3\pi$ amplitudes

As described above, in the limit $M_\pi^2/M_K^2 \ll 1$, the twelve isospin amplitudes depend on 9 unknown parameters. Therefore 3 relations between $K \to 2\pi$ and

$K \to 3\pi$ amplitudes valid in nex-to-leading order CHPT follow. As an example, we consider $\Delta I = 1/2$ amplitudes. We define linear combinations which are independent of weak counterterm coupling constants:

$$W_1 \equiv \frac{-1}{\sqrt{6}F_\pi}A_0 + 2\alpha_1 - \frac{M_K^2}{3M_\pi^2}\beta_1 + \frac{2M_K^4}{9M_\pi^4}(4\zeta_1 + \xi_1)$$
$$W_2 \equiv + 2\alpha_1 + \frac{2M_K^2}{3M_\pi^2}\beta_1 + \frac{2M_K^4}{9M_\pi^4}(4\zeta_1 - 2\xi_1) \qquad (16)$$

The contributions from tree diagrams to these linear combinations also vanish, hence W_1, W_2 are pure loop-effects with the form $W_i = c_2/F_\pi^2 \cdot W_i^{loop}$, $i = 1, 2$. From experiments we find (in units 10^{-8}) $W_1^{exp} = (-45 \pm 25)$, $W_2^{exp} = (4 \pm 48)$, whereas the loop calculation yields $W_1^{loop} = -81$, $W_2^{loop} = 5$. This allows a determination of c_2:

$$c_2/F_\pi^2 = 0.56 \pm 0.30 \, . \qquad (17)$$

This shows that the reduction of c_2 in next-to-leading order CHPT is dictated by the structure of the loop amplitude. Alternatively, c_2 may be eliminated which yields a parameterfree relation

$$W_2^{exp} = \frac{W_2^{loop}}{W_1^{loop}} \cdot W_1^{exp} \qquad (18)$$

The large errorbars reflect the uncertainties in the quadratic slope parameters ζ_1 and ξ_1. The combinations W_1 and W_2 have a simple interpretation: they are the discrepancies between the true $K \to 3\pi$ loop amplitude and its quadratic approximation (in dalitz plot variables) evaluated at the soft pion points.

4.3 p^4 versus p^2 contributions

Finally, we consider whether or not CHPT breaks down, as has been claimed. There is a subtlety involved because the physical decay constants, F_P^{Phys}, were used in determining the lowest-order magnitudes of c_2 and c_3. However, the decay constants are parameters of the strong chiral lagrangian which are renormalized at each order of the chiral expansion. In lowest order, there is a single decay constant denoted F_0, and $F_0^{(2)} = F_\pi^0 = F_K^0 \approx 94$MeV. In next-to-leading order[5],

$$F_P = F_0^{(4)}(1 + \Delta F_P) \approx F_P^{\text{Phys}} \quad P = \pi, K, \eta \qquad (19)$$

where the subscript (4) indicates the next-to-leading order value. F_0 is renormalized to an estimated 87 MeV[5], and F_K assumes its physical value of 114 MeV. Thus, using the physical decay constants implies that higher order corrections are hidden in the lowest order determination. Since the decay constants appear as factors,

$1/(F_K F_\pi^2)$ and $1/(F_K F_\pi^3)$, in the amplitudes for $K \to 2\pi$ and 3π decays respectively, the corrections are of order 30% [5]:

$$\frac{1}{F_K F_\pi^2} \sim \frac{1}{F_0^{(4)3}(1 - \Delta F_K - 2\Delta F_\pi)} \tag{20}$$

Moreover, the correction has a sign opposite to the next-to-leading order weak corrections presented in above; the net effect of all corrections is about 20% of the amplitudes, except for β_1, where the corrections are 50%. With this exception, which remains to be explored, there is no evidence for a breakdown of CHPT.

5. The B-parameter in next-to-leading order CHPT

In the standard model, the $K_L - K_S$ mass difference and the CP-violating parameter ϵ are directly proportional to the Kaon B-parameter. Therefore a reliable calculation of the B-parameter is crucial to test the predictions of the standard model in the low energy regime.

The importance of CHPT for the determination of the B-parameter was pointed out by Donoghue, Golowich and Holstein[23] who noticed that the operator $O_{\Delta I=3/2}$ which governs the decay $K^+ \to \pi^+\pi^0$ lies in the same $SU(3)$ multiplet as the operator $O_{\Delta S=2}$ inducing $K^0 - \bar{K}^0$ mixing. In the $SU(3)$ limit, the corresponding matrix elements are then simply related by a Clebsch Gordan coefficient, leading to $B \approx 0.33$. Bijnens, Sonoda and Wise[24] calculated the loop corrections to the lowest order chiral lagrangian. They found that p^4 corrections are large, provided a cutoff of $\mu \approx 1$ GeV is used. However, this result is scale dependent; at the scale $\mu \approx M_\eta$ the corrections are marginal. We have therefore redone the calculation including the weak counterterms. The renormalization scale introduced in the loop calculation drops out in the final result. The loop corrections vanish at $\mu \approx M_\eta$, in agreement with Ref. 24. Thus, the p^4 corrections to the B parameter are given by the counterterm contributions at this scale. Unfortunately, the finite part of the counterterm coupling constants entering the next-to-leading order expression for the B-parameter cannot be fixed in the kaon decays. Therefore, in next-to-leading order CHPT no prediction for the B-parameter emerges unless the counterterm coupling constants of the chiral lagrangian could be determined by some other means.

In the leading-logarithmic approximation, the off diagonal mass matrix element is written as[25]

$$M_{12} = \frac{G_F^2 M_W^2}{4\pi^2}\left\{\lambda_c^2\eta_1\frac{m_c^2}{M_W^2} + \lambda_t^2\eta_2\frac{m_t^2}{M_W^2} + 2\lambda_c\lambda_t\eta_3\frac{m_c^2}{M_W^2}\ln\frac{m_t^2}{m_c^2}\right\}\left(\frac{\alpha_s''(m_c^2)}{\alpha_s'''(\mu^2)}\right)^{2/9}$$
$$\times \frac{1}{2M_K}\langle \bar{K}^0|\mathcal{O}_{\Delta S=2}|K^0\rangle \tag{21}$$

where $\mathcal{O}_{\Delta S=2}$ is the four-quark operator

$$\mathcal{O}_{\Delta S=2}(\mu) = (\bar{s}_L^\alpha\gamma_\mu d_L^\alpha)(\bar{s}_L^\beta\gamma^\mu d_L^\beta) \tag{22}$$

and $q_L^\alpha \equiv \frac{1}{2}(1-\gamma_5)q^\alpha$. α, β are colour indices. In Eq. 21 G_F is the Fermi constant, λ_i are products of two elements of the CKM-mixing matrix $\lambda_i = V_{id}V_{is}^*$, $i = u, c, t$ and η_1, η_2, η_3 are Wilson coefficients, where the overall renormalization scale dependent factor $[\alpha_s''(\mu^2)]^{-2/9}$ has been factored out explicitly.

The scale independent parameter B is defined by

$$B = B(\mu^2)[\alpha_s(\mu^2)]^{-2/9} \tag{23}$$

where α_s is the QCD running coupling contant, μ is the renormalization scale and $B(\mu^2)$ is defined as

$$\langle \bar{K}^0 | \mathcal{O}_{\Delta S=2}(\mu) | K^0 \rangle = B(\mu^2)\frac{4}{3}F_K^2 M_K^2 \,. \tag{24}$$

The chiral realization of the operator $\mathcal{O}_{\Delta S=2}(\mu)$ is

$$\begin{aligned} \mathcal{O}^{chiral}_{\Delta S=2} = g_{\Delta S=2}\Big\{ & Tr(Q_3^2 L_\mu) Tr(Q_3^2 L^\mu) \\ & + \sum_j^{1\ldots 33} \frac{D^{(j)}}{c_3} Tr(Q_3^2 A^{(j)}) Tr(Q_3^2 B^{(j)}) \Big\} \end{aligned} \tag{25}$$

Here, Q_3^2 are the 3×3 matrices introduced in Eq. 8 and the sum runs over all 27-plet counterterms, section 2.

As mentioned above, the four quark operator governing the decay $K^+ \to \pi^+\pi^0$,

$$\mathcal{O}_{\Delta I=3/2}(\mu) = \left(\bar{s}_L^\alpha \gamma_\mu d_L^\alpha\right)\left(\bar{u}_L^\beta \gamma^\mu u_L^\beta - \bar{d}_L^\beta \gamma^\mu d_L^\beta\right) + \left(\bar{s}_L^\alpha \gamma_\mu u_L^\alpha\right)\left(\bar{u}_L^\beta \gamma^\mu d_L^\beta\right) \tag{26}$$

lies in the same multiplet as $O_{\Delta S=2}$. Its chiral realization is

$$\begin{aligned} \mathcal{O}^{chiral}_{\Delta I=3/2} = g_{\Delta I=3/2}\Big\{ & \frac{3}{10} t^{jl}_{ik} (Tr(Q^i_j L_\mu) Tr(Q^k_l L^\mu) \\ & + \sum_r^{1\ldots 33} \frac{D^{(r)}}{c_3} Tr(Q^i_j A^{(r)}) Tr(Q^k_l B^{(r)})) \Big\} \end{aligned} \tag{27}$$

t^{jl}_{ik} is the tensor projecting out the 27-plet component of the interaction. The normalization in Eq. 27 has been chosen such that in the $SU(3)$ limit $g_{\Delta S=2} = g_{\Delta I=3/2}$.

To relate $g_{\Delta S=2}$ to the coupling constant of the effective chiral lagrangian, c_3, we need also the short distance OPE of the effective $\Delta I = 3/2$ Lagrangian,

$$\mathcal{L}^{eff}_{\Delta I=3/2} = \sqrt{2} G_F \sin_1 \cos_1 \cos_3 \eta_{3/2} \left[\alpha_s(\mu^2)\right]^{-2/9} \mathcal{O}_{\Delta I=3/2} \,. \tag{28}$$

where $\sin_i$ and $\cos_i$ are sines and cosines of the CKM-angles and $\eta_{3/2} \approx 0.4$ is the Wilson coefficient associated with $\mathcal{O}_{\Delta I=3/2}$.[25] Combining Eqs. 25-28 we find

$$g_{\Delta S=2} = \left[\alpha_s(\mu^2)\right]^{2/9} \frac{1}{\sqrt{2}G_F \sin_1 \cos_1 \cos_3 \eta_{3/2}} \frac{10}{3} c_3 \tag{29}$$

The B-parameter is now calculable in terms of c_3 and weak counterterm coupling constants D_i^r. The explicit one-loop calculation yields

$$M \equiv \langle \bar{K}^0 | \mathcal{O}^{chiral}_{\Delta S=2}(\mu) | K^0 \rangle = \frac{g_{\Delta S=2}}{F_K^2} \{ M_{tree} + M_{loop} + M_{C.T.} \} \tag{30}$$

with

$$\begin{aligned} M_{\rm tree} &= -4M_K^2 \\ M_{C.T.} &= -\frac{1}{c_3}\left\{16M_K^4(D_1^r - D_2^r - D_4^r - D_7^r) - 8M_K^2M_\pi^2(4D_1^r + D_7^r) + 16M_\pi^4 D_1^r\right\} \\ M_{\rm loop} &= \frac{M_K^4}{4\pi^2 F_K^2} + 2M_K^2(7\mu_\pi + 12\mu_K + 9\mu_\eta) + 2M_\pi^2(\mu_\pi + \mu_\eta) \end{aligned} \tag{31}$$

with

$$\mu_p = \frac{1}{32\pi^2} \frac{M_p^2}{F_p^2} \ln \frac{M_p^2}{\mu^2} \tag{32}$$

and the D_i are defined in Ref. 6 Using Eqs. 24,29 the scale invariant B-parameter becomes

$$B = \frac{-c_3}{F_K^4} \frac{10}{\sqrt{2}G_F \sin_1 \cos_1 \cos_3 \eta_{3/2}} \left\{1 - \frac{M_{loop} + M_{C.T.}}{4M_K^2}\right\}. \tag{33}$$

In lowest order CHPT, $F_K = F_0^{(2)} \simeq 94\text{GeV}$. However, a factor $1/F_K^2$ in Eq. 33 comes from the definition of B, Eq. 24, and has to be fixed at the physical value. With $c_3 = -0.008F_\pi^2 10^{-7}$ (see Table 2) one obtains in lowest order CHPT $B = 0.34$, the result of Ref. 23.

At next-to-leading order, we first observe that the sum, $M_{loop} + M_{C.T.}$, is independent of the renormalization scale ν of the chiral logs. The variation with ν is strong: $M_{loop}(\nu = M_\eta) \approx 0$ but $M_{loop}(\nu = 1GeV) > 4M_K^2$! Neglecting the counterterms leads to the misleading conclusion of breakdown of CHPT. The B-parameter may be written as

$$B = B_{tree} + B_{C.T.}(\nu = M_\eta). \tag{34}$$

Unfortunately, the second term in Eq. 34 is unknown, as it corresponds to counterterms not fixed by K-decays. The first term in Eq. 34 contains also a p^4-effect, the renormalization of F_K (see previous discussion). It reduces B_{tree} considerably, to about 0.25. Therefore, larger values for B would require a sizeable positive counterterm contribution.

6. Conclusions

The nonleptonic kaon decays, $K \to 2\pi$ and $K \to 3\pi$ and the B-parameter have been analyzed in next-to-leading order CHPT. First, the standard analysis of isospin amplitudes performed by Devlin and Dickey has been revised including the $\Delta I = 3/2$ quadratic sope parameters. The limits obtained yield predictions for the quadratic PDG slope parameter in the decay $K_L \to 3\pi^0$ as well as a refined estimate of the CP-conserving contribution to the decay $K_S \to \pi^+\pi^-\pi^0$.

Second, the $K \to 2\pi$ and $K \to 3\pi$ amplitudes have been calculated in next-to-leading order CHPT and the relevant counterterm coupling constants have been determined from fits to experiments. We conclude that the lowest order octet coupling constant is reduced by 30 % while the coupling constant of the 27-plet remains unchanged. This implies a factor, 1.5, of the $\Delta I = 1/2$ enhancement is due to long distance dffects. Once the coupling constants have been fixed, the order p^2 and p^4 contributions may be disentangled. With one exception, the amplitudes show no evidence for breakdown of CHPT. We observe a partial cancelation between order p^4 corrections in the renormalization of F_π, F_K and the next-to-leading order weak corrections.

The kaon B-parameter depends in next-to-leading order CHPT on an additional weak counterterm coupling constant which cannot be fixed from the K-decays. The large variation of the loop amplitude with the renormalization scale is éxactly cancelled by these counterterms, as it must. No breakdown of CHPT can be inferred from the loop amplitude alone. It remains to be shown how large the finite part of the relevant counterterm is.

Acknowledgements

The results discussed in this talk have been obtained together with J. Missimer and D .Wyler, to whom I thank for the enjoyable collaboration. I am also grateful to J. Donoghue, G. Ecker, J. Gasser, H. Leutwyler and A. Pich for helpful discussions.

References

[1] S. Weinberg, *Phys. Rev.* **166** (1968) 1568.
[2] L.-F. Li and H. Pagels, *Phys. Rev. Lett.* **26** (1971) 1204; *Phys. Rev. Lett.* **27** (1971) 1089.
[3] S. Coleman, J. Wess and B. Zumino, *Phys. Rev.* **177** (1969). 2239
C.G. Callan, S. Coleman, J. Wess and B. Zumino, *Phys. Rev.* **177** (1969) 2247.

[4] S. Weinberg, *Physica* **96A** (1979) 327.
[5] J. Gasser and H. Leutwyler, *Ann. of Phys.* **158** (1984) 142; *Nucl. Phys.* **B 250** (1985) 465.
[6] J. Kambor, J. Missimer and D. Wyler, *Nucl. Phys.* **B 346** (1990) 17.
[7] A. Pich, these proceedings.
[8] J. Kambor, J. Missimer and D. Wyler, *Phys. Lett.* **B 261** (1991) 496.
[9] G. Ecker, A. Pich and E. de Rafael, *Nucl. Phys.* **B 303** (1988) 665.
[10] C. Bernard, T. Draper, A. Soni, H. Politzer and M. Wise, *Phys. Rev.* **D 32** (1985) 2343.
[11] R.J. Crewther, *Nucl. Phys.* **B 264** (1986) 277.
[12] A. Pich, B. Guberina and E. de Rafael, *Nucl. Phys.* **B 277** (1986) 197.
[13] J. Wess and B. Zumino, *Phys. Lett.* **B 37** (1971) 95.
[14] T.J. Devlin and J.O. Dickey, *Rev. Mod. Phys.* **51** (1979) 237.
[15] L.-F. Li and L. Wolfenstein, *Phys. Rev.* **D 21** (1980) 178;
H.-Y. Cheng, *Phys. Lett.* **B 238** (1990) 399.
[16] Particle Data Group, *Phys. Lett.* **B 204** (1988) 1.
[17] K.L. Au, D. Morgan and R. Pennington, *Phys. Rev.* **D 35** (1987) 1633.
W. Hoogland et al., *Nucl. Phys.* **B 126** (1977) 109.
[18] J.F. Donoghue, *Nucl. Phys.* **B** (Proc. Suppl.) **7A** (1989) 59.
[19] B. Winstein and S. Samolvar, private communication.
[20] J. Gasser, H. Leutwyler and J. Kambor, work in progress.
[21] T.N. Truong, *Phys. Lett.* **B 207** (1988) 495.
N. Isgur, K. Maltman, J. Weinstein and T. Barnes, *Phys. Rev. Lett.* **64** (1990) 161.
M. Neubert and B. Stech, *Phys. Lett.* **B 231** (1989) 477.
[22] J.F. Donoghue, E. Golowich and B.R. Holstein, *Phys. Rev.* **D 30** (1984) 587.
[23] J.F. Donoghue, E. Golowich and B.R. Holstein, *Phys. Lett.* **B 119** (1982) 412.
[24] J. Bijnens, H. Sonoda and M. Wise, *Phys. Rev. Lett.* **53** (1984) 2367.
[25] F.J. Gilman and M.B. Wise, *Phys. Rev.* **D 20** (1979) 2392; Phys. Rev. **D 27** (1982) 1128.

LONG-DISTANCE EFFECTS IN NONLEPTONIC K DECAYS AND THE $\Delta I = 1/2$ RULE

T. N. PHAM

Centre de Physique Théorique, CNRS-UPR-A0014, Ecole Polytechnique, 91128 Palaiseau Cedex, France.

ABSTRACT

I discuss the long-distance contributions to the nonleptonic K decay amplitude and argue that the contributions from the low-lying pseudoscalar meson intermediate states to the Cottingham-like formula for the $K - \pi$ matrix elements, could enhance the $\Delta I = 1/2$ amplitude. This together with possible enhancement of the four-quark operator matrix elements by other effects, could explain a large part of the $\Delta I = 1/2$ amplitude.

In the original *current* × *current* form of the weak interactions of leptons and hadrons given by Gell – Mann[1] as a generalisation of the Fermi interaction, the $\Delta I = 1/2$ and $\Delta I = 3/2$ nonleptonic weak interactions of strange hadrons would be of comparable strength. This is however not observed in Kaon and Hyperon nonleptonic decays which show a strong suppression of the $\Delta I = 3/2$ amplitude. Experimentally, the $\Delta I = 1/2$ rule is obeyed to a few percent in amplitude and remains to this day a puzzle in particle physics. The short-distance effects as suggested by Wilson[2] and analysed by Gaillard and Lee[3] and by Altarelli and Maiani[4] in the framework of perturbative QCD, produce a large enhancement of the $\Delta I = 1/2$ interactions, by a factor of 4 to 5 relative to the $\Delta I = 3/2$ part. This enhancement, though important, is still far below the observed enhancement by a factor of about 20 in Kaon and Hyperon decays.

For some time, it has been argued that the Penguin interactions with its left-right structure with respect to chiral symmetry as given by the one gluon exchange in perturbative QCD could produce a large amplitude[5] , however recent analyses[6] indicate that this contribution is small, at most amounts to 10–20% of the decay amplitude. Thus approaches that go beyond the standard perturbative QCD must be explored to explain the $\Delta I = 1/2$ rule. One approach is based on a sum rule derived by Wilson[2] for the nonleptonic decay amplitudes. Recent analyses by Nardulli, Preparata and by Rotondi[7] and by Sutherland and myself[8] indicate that additional contributions to the decay amplitudes from the low-lying states in the T-product of the current correlation function could significantly increase the $\Delta I = 1/2$ amplitude from the value obtaind with the vacuum insertion approximation(VIA). In this talk, I would like to discuss the work of Sutherland and myself on this contribution. Details have appeared in the published works[8] so I will only present here the main arguments in favor of the long-distance contributions.

Our starting point is the nonleptonic decay Hamiltonian in the standard model

$$H_{NL} = -(\frac{ig^2}{8})\sin\theta_C\cos\theta_C \int d^4x\Delta_{\mu\nu}(x,m_W^2) \times T\left[J_\mu^{1+i2}(x)J_\nu^{4-i5}(0)\right] \\ +\text{other terms} + h.c \qquad (1)$$

where $\Delta_{\mu\nu}(x,m_W^2)$ is the W boson propagator and $g^2/8m_W^2 = G_F/\sqrt{2}$. By Wilson's argument the $x-$integral is important only where $|x| \sim 1/m_W^2$ so that H_{NL} is a product of local operators. The leading operators in QCD are of dimension 6 since the off-diagonal $\bar{s}_L d_R$ operator with singularity $1/x^2$ can be rotated away. QCD gives us the expansion

$$T\left[J_\mu^{1+i2}(x)J_\mu^{4-i5}(0)\right] \sim c_1O_1(0) + c_4O_4(0) \tag{2}$$

where we neglect other operators with small coefficients for simplicity. Here[7]

$$c_1(1/m_W) = \left[1 - b\frac{g^2(\mu)}{24\pi^2}\ln(\frac{\mu}{m_W})\right]^{0.48} \tag{3}$$

$$c_4(1/m_W) = \frac{2}{3}\left[1 - b\frac{g^2(\mu)}{24\pi^2}\ln(\frac{\mu}{m_W})\right]^{-0.24} \tag{4}$$

with $b = 33 - 2n_F$, n_F being the number of flavour and

$$g^2(\mu) = \frac{4\pi}{(1 + (b/6\pi)\ln(\mu/\Lambda))} \tag{5}$$

the QCD coupling constant. μ is a renormalisation scale parameter for the operators O_1 and O_4 . The short-distance enhancement of the $\Delta I = 1/2$ O_1 operator over the $\Delta I = 3/2$ O_4 operator is due to the large value of c_1 relative to c_4 as given in the above expressions obtained in perturbative QCD. To compute the K decay amplitude, we use Current Algebra to express the $K \to 2\pi$ and $K \to 3\pi$ amplitudes in terms of the $K \to \pi$ transition via the Callan-Treiman relation. This can be more conveniently done using an effective chiral Lagrangian for nonleptonic decay of Kaons first written down by Cronin[9]

$$\mathcal{L}_W(\Delta I = 1/2) = \frac{cG_F}{(4\sqrt{2}f^4)}\mathrm{Tr}\lambda_6\partial_\mu M\partial_\mu M^\dagger \tag{6}$$

which are of the *current* $\times$ *current* form with the left-handed quark currents replaced by the $V - A$ hadronic currents. A chiral Lagrangian of the above form can also be written for the $\Delta I = 3/2$ amplitude. In the standard non-linear realisation of chiral symmetry and for $SU(3) \times SU(3)$, we have

$$M = exp(2if\phi), \qquad \phi = \frac{1}{\sqrt{2}}\sum_i \lambda_i\phi_i \tag{7}$$

where ϕ is the octet pseudoscalar meson field operators, f is the inverse of the usual pion decay constant f_π . The $(V - A)$ octet currents take the following form for the pseudoscalar meson:

$$(V - A)_\mu = \frac{i}{2f^2}M\partial_\mu M^\dagger \tag{8}$$

The chiral Lagrangian Eq. (6) then predicts the $\Delta I = 1/2$ part of the $K \to 3\pi$ decay amplitude in terms of the $K \to 2\pi$ amplitude. The predicted value agrees with experiment to within 20 % . This shows that the lowest order chiral Lagrangian given above describes correctly the K decays and that the dominant contribution comes from the p^2−terms. The computation of the $K \to 2\pi$ and $K \to 3\pi$ amplitudes is reduced to computing the $K\pi$ matrix element. For this purpose, we use a sum rule obtained by Wilson[2] . Starting from the operator product expansion

$$T\left[J_{\mu N}(x)J_{\mu S}(0)\right] = B_1'(x)\sigma_m'(0) \tag{9}$$

Ignoring terms which do not contribute to the $K\pi$ transitions, we have in momentum space

$$\int^{q_{max}} \left[M_{AB}(q) - B_1'(q)\sigma_{AB}'\right] d^4q = 0 \tag{10}$$

where

$$M_{AB} = \int d^4x \, exp(iq\cdot x) \left\langle A \left| J_{\mu N}(x)J_{\mu S}(0)\right| B\right\rangle \tag{11}$$

and

$$B_1'(q) = \int d^4x \, exp(iq\cdot x) B_1'(x) \tag{12}$$

If $B_1'(x)$ scales as $(x^2)^0$ as in QCD, and for q_{max} not too large, we obtain

$$\int^{q_{max}} M_{AB} d^4q = \sigma_{AB}' \tag{13}$$

Comparing Eq. (13) with the following Cottingham formula[10]

$$A_{K\pi} = \int \frac{d^4q}{(2\pi)^4} \frac{T(q)}{(m_W^2 - q^2 - i\epsilon)} \tag{14}$$

where

$$T(q) = -i \int d^4x \, exp(iq\cdot x) \times \left\langle \pi \left| T[J_{\mu N}(x)J_{\mu S}(0)]\right| K\right\rangle \tag{15}$$

we see that the $K\pi$ transition can be obtained from $c_1(1/m_W)$ and $c_4(1/m_W)$ and an evaluation of the Cottingham formula assuming only moderate q dominate the integral. Since $A_{K\pi}$ is quadratic in momenta because of the $(V - A) \times (V - A)$ structure of the weak interactions, $T(q)$ must be proportional to p^2 and $p\cdot q$ and the effective contributions to the Cottingham integral come from terms in $T(q)$ of the form

$p^2T_0(q)$ which is suppressed at high energy by a power of q^2 or $p\cdot q$. This suppression factor at high energy makes the Cottingham integral for $A_{K\pi}$ looks similar to the Cottingham formula for the $\pi^+ - \pi^0$ mass difference where the suppression factor $1/q^2$ comes from the photon propagator. We can see this by looking at the low-lying pseudoscalar intermediate state contributions to $A_{K\pi}$

$$A_{K\pi}(p^2) = i\frac{G_F}{\sqrt{2}}C^i_{AB}\int^{q_{max}} \frac{d^4q}{(2\pi)^4}\frac{(2p+q)^2F^2(q^2)}{((p+q)^2-m^2)} \tag{16}$$

where $F(q^2)$ is the pseudoscalar electromagnetic form factor, C^i_{AB} , for $i = 1,4$, are sum of products of Clebsch-Gordan coefficients of vector current coupling to pseudoscalar meson for O_1 and O_4 respectively.

On the other hand, the $\pi^+ - \pi^0$ mass difference due to the pseudoscalar meson intermediate states(Born term) is given by

$$\delta m^2 = ie^2\int \frac{d^4q}{(2\pi)^4}\frac{F^2(q^2)}{q^2}\left[\frac{(2p+q)^2}{((p+q)^2-m_\pi^2-i\epsilon)} - 4\right] \tag{17}$$

As can be seen from Eq. (16) and Eq. (17), the p^2−term in $A_{K\pi}$ and the term independent of the pion momenta in δm^2 come from integrals of the form

$$I = \int \frac{d^4q}{(2\pi)^4}\frac{F^2(q^2)}{q^2} \tag{18}$$

This shows that the Born term which makes the bulk of δm^2 must also makes an important contribution to the $K\pi$ matrix elements. To obtain the long-distance contribution, we integrate the Cottingham integral in Eq. (16) up to a momentum q_{max} which separates the long-distance part of the Cottingham integral from the short-distance part factorised by the short-distance Wilson coefficients $c_1(1/m_W)$ and $c_4(1/m_W)$ for the $\Delta I = 1/2$ and $\Delta I = 3/2$ operators respectively. The $K\pi$ matrix elements are thus given by

$$\begin{aligned} A_{1/2} &= c_1(1/m_W)M_{1/2}(q_{max}) \\ A_{3/2} &= c_4(1/m_W)M_{3/2}(q_{max}) \end{aligned} \tag{19}$$

(this is equivalent to setting $q_{max} = \mu$ in the sumrule (Eq. (13))). Using vector meson dominance for the form factor $F(q^2)$, we find

$$M_{1/2}(q_{max}) = \frac{G_F}{\sqrt{2}}\frac{4C^1_{AB}p^2m_\rho^2}{16\pi^2}\left(1 - \frac{m_\rho^2}{q^2_{max}+m_\rho^2}\right) \tag{20}$$

which is large and is comparable to the vacuum insertion approximation value. The important fact is that this Born term contribution increases the $\Delta I = 1/2$ amplitude because of the large Clebsch-Gordan coefficients associated with the O_1 operator. This result is also not very sensitive to the variation of q_{max}. It is reasonable to choose q_{max} between 1 and 2 GeV as a compromise in the separation of long-distance and short-distance effects. For $q_{max} = 2\,\text{GeV}$, terms depending on q_{max} in $M_{1/2}$ can be neglected. Including other contributions to $M_{1/2}$ from the contact term as well as the disconnected piece of $T(q)$ which can be also obtained easily by the vacuum insertion approximation, we obtain

$$A_{1/2} = \frac{G_F}{\sqrt{2}}\frac{c_1}{3}\sin\theta_C\cos\theta_C f_\pi^2 p^2\left(1+\frac{18m_\rho^2}{16\pi^2 f_\pi^2}\right) \tag{21}$$

Similarly, we find that for the $\Delta I = 3/2$, the Born term and the vacuum insertion approximation are of opposite sign and tend to cancel out leaving a suppression of the $\Delta I = 3/2$ amplitude. Our numerical values for the $\Delta I = 1/2$ and $\Delta I = 3/2$ amplitudes as given in the published paper[8] are

$$c = 0.42\,, \qquad c' = -0.004 \tag{22}$$

to be compared with the experimental values

$$c_{exp} = 1.1 \pm 0.2\,, \qquad c'_{exp} = 0.032 \pm 0.002 \tag{23}$$

where we have taken as positive sign the experimental amplitudes as indicated by the positive sign of the relative 1/2 and 3/2 amplitudes obtained from experiment.

The result shows that the Born term and the vacuum insertion approximation gives too small a value for the $\Delta I = 3/2$ amplitude. It is then expected that contributions from higher mass intermediate states such as the vector mesons or axial vector mesons , e.g ρ, A_1 etc. must be evaluated to get a more reliable prediction. This involves more theoretical uncertainties such as our knowledge on the transition form factors involved as discussed in details in our work. For the $\Delta I = 1/2$ amplitude, this result is encouraging since it tells us that it is possible to increase the $\Delta I = 1/2$ amplitude by various contributions in a more or less reliable manner. In a recent work on the $\Delta I = 1/2$ rule, Pich and de Rafael[11] have shown that the gluon condensate could increase by an appreciable amount the $\Delta I = 1/2$ amplitude. This would bring the total computed amplitude to be about 70 to 80 % of the experimental value. The remaining term may be obtained by looking at the higher mass intermediate states such as vector and axial vector mesons mentioned above.

We have emphasised in this work the importance of the long-distance contributions to the K decay amplitudes with respect to the vacuum insertion approximation. These contributions can be considered as some sort of multi-soft gluon exchange Penguin interactions in the $q\bar{q}$ intermediate states of the Cottingham formula. These soft-gluon exchange processes then give rise to bound states such as the low-lying pseudoscalar intermediate states which we show here to be important and is comparable to the vaccuum insertion approximation value. A way to see this is to look at the expression for the pion decay constant in a bound state model of the pion. It is given by

$$f_\pi = \frac{3m}{2\pi} \tag{24}$$

where m is the dynamical quark mass[12] .

Putting this in Eq. (21) , we find that the Born term and the vacuum insertion value are comparable. This also tells us that it is not surprising to find a large deviation from the vacuum insertion approximation in the nonleptonic K decay amplitudes.

I would like to thank the organisers of the workshop for the warm hospitality extended to me at Dobogoko.

References

1. M. Gell-Mann, *Phys.Rev.* **125** (1962) 1067.
2. K. G. Wilson, *Phys.Rev.* **179** (1969) 1499.
3. M. K. Gaillard and B. W. Lee, *Phys. Rev. Lett.* **33** (1974) 108.
4. G. Altarelli and L. Maiani, *Phys. Lett.* **B52** (1974) 351.
5. M. A. Shifman, A. I. Vainshtein and V. I. Zakharov, *Nucl. Phys.* **B120** (1977) 315.
6. T. N. Pham, *Phys. Lett.* **B145** (1984) 115; *Phys. Lett.* **B182** (1986) 378.
   J. F. Donoghue, *Phys. Rev.* **D30** (1984) 1499.
   M. B. Gavèla et al., *Phys. Lett.* **B148** (1984) 225.
7. G. Nardulli and Preparata, *Phys. Lett.* **B104** (1981) 399;
   G. Nardulli, G. Preparata and D. Rotondi, *Phys. Rev.* **D27** 91983) 557.

8. T. N. Pham and D. G. Sutherland, *Phys. Lett.* **B135** (1984) 209; *Z. Phys. C* **41** (1988) 327.

9. J. A. Cronin, *Phys.Rev.* **161** (1967) 1483.

10. W. N. Cottingham, *Ann. Phys.* (NY) **25** (1963) 424.

11. A. Pich and E. de Rafael, *Four-quark operators and nonleptonic weak transitions* Marseille and CERN preprint, CPT-90/P.2393, CERN-TH.5906; A. Pich, *Talk at this workshop.*

12. T. N. Pham, *Phys. Rev.* **D28** (1983) 472.
H. Pagels and S. Stokar, *Phys. Rev.* **D20** (1979) 2947.

BOSONIZATION OF THE SHORT-DISTANCE STRANGENESS-CHANGING EFFECTIVE HAMILTONIAN

A. Pich
Theory Division, CERN, CH-1211 Geneva 23, Switzerland

ABSTRACT

The low-energy chiral realization of the $\Delta S = 1, 2$ short-distance effective Hamiltonians is calculated by integrating out the quark fields in a gluonic background and in the presence of a source term which triggers spontaneous chiral symmetry breaking.

1. Introduction

The origin of the empirically observed enhancement of strangeness-changing non-leptonic weak amplitudes with isospin transfer $|\Delta I| = 1/2$ is a long-standing question in particle physics. The short-distance analysis of the product of weak hadronic currents results in an effective $\Delta S = 1$ Hamiltonian,

$$H^{\Delta S=1} = \frac{G_F}{\sqrt{2}} V_{ud} V_{us}^* \sum_i C_i(\mu^2)\, Q_i \tag{1.1}$$

(G_F is the Fermi coupling and V_{ud}, V_{us} are matrix elements of the Cabibbo - Kobayashi - Maskawa (CKM) flavour mixing matrix), which is a sum of local four-quark operators, constructed with the light (u, d, s) quark fields only,

$$\begin{aligned}
Q_1 &\equiv 4(\bar{s}_L \gamma^\mu d_L)(\bar{u}_L \gamma_\mu u_L),\\
Q_2 &\equiv 4(\bar{s}_L \gamma^\mu u_L)(\bar{u}_L \gamma_\mu d_L),\\
Q_3 &\equiv 4(\bar{s}_L \gamma^\mu d_L) \sum_{q=u,d,s} (\bar{q}_L \gamma_\mu q_L),\\
Q_4 &\equiv 4 \sum_{q=u,d,s} (\bar{s}_L \gamma^\mu q_L)(\bar{q}_L \gamma_\mu d_L),\\
Q_5 &\equiv 4(\bar{s}_L \gamma^\mu d_L) \sum_{q=u,d,s} (\bar{q}_R \gamma_\mu q_R),\\
Q_6 &\equiv -8 \sum_{q=u,d,s} (\bar{s}_L q_R)(\bar{q}_R d_L),
\end{aligned} \tag{1.2}$$

modulated by Wilson coefficients $C_i(\mu^2)$, which are functions of the heavy (W, t, b, c) masses and an overall renormalization scale μ. Only five of these operators are independent, since $Q_4 = -Q_1 + Q_2 + Q_3$.

From the point of view of chiral $SU(3)_L \otimes SU(3)_R$ and isospin quantum numbers, $Q_- \equiv Q_2 - Q_1$ and Q_i $(i = 3,4,5,6)$ transform as $(8_L, 1_R)$ and induce $|\Delta I| = 1/2$ transitions, while $Q_1 + 2/3Q_2 - 1/3Q_3$ transforms like $(27_L, 1_R)$ and induces both $|\Delta I| = 1/2$ and $|\Delta I| = 3/2$ transitions. In the absence of strong interactions, $C_2(\mu^2) = 1$ and all other Wilson coefficients vanish. The standard electroweak model gives rise then to $|\Delta I| = 1/2$ and $|\Delta I| = 3/2$ amplitudes of nearly equal size, while experimentally the ratio between both amplitudes is a factor of twenty! To solve this big discrepancy, QCD effects should be enormous.

The leading α_s corrections indeed give, for μ-values around 1 GeV, an enhancement by a factor two to three of the Q_- Wilson coefficient with respect to the $Q_+ \equiv Q_2 + Q_1$ one $[C_\pm(\mu^2) \equiv (C_2(\mu^2) \pm C_1(\mu^2))/2]$. Moreover, the gluonic exchanges generate the additional $|\Delta I| = 1/2$ operators $Q_i\,(i = 3,4,5,6)$. Nevertheless, this by itself is not enough to explain the experimentally observed rates, without simultaneously appealing to a further enhancement in the hadronic matrix elements of at least some of the isospin 1/2 four-quark operators.

The evaluation of hadronic matrix elements of local four-quark operators is a difficult problem due to the fact that they are governed by the long-distance behaviour of the strong interactions, i.e. the confinement regime of QCD. Moreover, these operators have non-zero anomalous dimensions and, therefore, their matrix elements depend on the renormalization scale μ. Since physical amplitudes are renormalization-scale independent quantities, this μ-dependence should exactly cancel the one appearing in the Wilson coefficients. In order to keep control of the renormalization-scale dependence, and therefore to get a meaningful result, a full QCD calculation of the relevant matrix elements is required. This is a highly non-trivial task.

The Standard Model also predicts strangeness-changing transitions with $\Delta S = 2$ via two W-exchanges, the so-called box diagrams. At long distances this results in an effective Hamiltonian,

$$H^{\Delta S=2} = \frac{G_F^2 M_W^2}{4\pi^2} \left[\lambda_c^2 F_1 + \lambda_t^2 F_2 + 2\lambda_c\lambda_t F_3\right] C_{\Delta S=2}(\mu^2)\, Q_{\Delta S=2}, \qquad (1.3)$$

which is also proportional to a local four-quark operator,

$$Q_{\Delta S=2} \equiv (\bar{s}_L \gamma^\mu d_L)(\bar{s}_L \gamma_\mu d_L), \qquad (1.4)$$

modulated by products of CKM-matrix elements, $\lambda_q = V_{qd}^* V_{qs}\,(q = u,c,t)$, times functions $F_i\,(i = 1,2,3)$ of the heavy masses $m_t^2, M_W^2, m_b^2, m_c^2$ of the fields which have been integrated out. The operator $Q_{\Delta S=2}$ is multiplicatively renormalizable and has a non-zero anomalous dimension; hence the μ^2 dependence via $C_{\Delta S=2}(\mu^2) \approx \alpha_s(\mu^2)^{-2/9}$ in eq. (1.3), which has to be cancelled by the corresponding μ^2-dependence of the hadronic matrix element. Under the chiral flavour group $SU(3)_L \otimes SU(3)_R$, $Q_{\Delta S=2}$ transforms as a $(27_L, 1_R)$ operator.

It has become usual to parametrize the $K^0 - \overline{K}^0$ matrix element as

$$\langle \overline{K}^0 | Q_{\Delta S=2} | K^0 \rangle \equiv \frac{4}{3} f_K^2 M_K^2 B(\mu^2) \equiv \frac{4}{3} f_K^2 M_K^2 \alpha_s(\mu^2)^{2/9} \widehat{B}, \qquad (1.5)$$

where $B(\mu^2) = 1$ corresponds to the so-called vacuum saturation approximation. Theoretical knowledge of this parameter is urgently required for a useful confrontation of the observed CP-violation in the $K^0 - \overline{K}^0$ system with the Standard Model predictions.

2. Two-Point Functions of Four-Quark Operators at Short Distances

Some insight into the dynamics responsible for the octet enhancement of $\Delta S = 1$ transitions can be obtained by looking at this problem in an inclusive way. Technically, this means the study of two-point functions of the type

$$\psi_{ij}(q^2) \equiv i \int d^4x\, e^{iqx} \langle \emptyset | T(Q_i(x) Q_j^\dagger(0)) | \emptyset \rangle, \qquad (2.1)$$

with Q_i one of the four-quark operators listed in the previous section. The spectral functions associated to these two-point functions, when modulated with the appropriate Wilson coefficients, are quantities with definite physical information: they describe in an inclusive (and averaged) way, how the weak operators couple the vacuum to physical states of a given invariant mass. If the $|\Delta I| = 1/2$ enhancement is an intrinsic property of the $\Delta S = 1$ Hamiltonian, it should also show up at the inclusive level; i.e., the strength of the $|\Delta I| = 1/2$ spectral functions should be much bigger than the $|\Delta I| = 3/2$ one.

At large q^2, one can use perturbative QCD to study these two-point functions. A complete (μ-independent) calculation at the leading logarithm approximation was done in refs. 1 and 2, showing that the only enhancement present at this order is the one coming from the Wilson coefficients (i.e. a factor of about two). The non-logarithmic α_s-correction to the 27-correlators was also computed in ref. 1, where it was found to be moderate; for instance, for the operator $Q_{\Delta S=2}$ one has[1,3,4]

$$\begin{aligned} C_{\Delta S=2}(\mu^2)^2 \frac{1}{\pi} \mathrm{Im}\psi_{\Delta S=2}(t) &= \theta(t) \frac{2}{15} \frac{t^4}{(16\pi^2)^3} \alpha_s(\mu^2)^{-4/9} \\ &\times \left\{ 1 + \frac{\alpha_s(\mu^2)}{\pi} \left[\log(t/\mu^2) - \frac{1217}{810} \right] + \dots \right\}. \end{aligned} \qquad (2.2)$$

The corresponding calculation for the octet case is much more involved due to the fact that there are several operators (Q_-, Q_3, Q_5, Q_6) which mix under renormalization. One needs to compute, at the four-loop level, all possible two-point functions (2.1) built with the different octet operators, i.e. a 4×4 matrix correlator which must be renormalized in matrix form. The complete results of this calculation can be found in ref. 4. Here, I prefer to show some simplified results, obtained by using two approximations which eliminate the mixing among operators, while keeping at the same time the important physical effects:

i) If "Penguin" diagrams are neglected, the operators $Q_\pm$ are multiplicatively renormalizable.

ii) The interesting "Penguin" operator Q_6 can be isolated by noting that in the large-N_c limit (N_c = number of colours) the anomalous dimension matrix γ_{ij} of the set of operators Q_i becomes zero, but for γ_{66}; i.e. in this limit there is no mixing among operators and only Q_6 gets renormalized. [Note that the large-N_c limit is not appropriate for studying the operators $Q_\pm$, since $\gamma_{++} = \gamma_{--} = 0$ implies $C_+(\mu) = C_-(\mu) = 1/2$, and the short-distance enhancement is therefore lost.]

With these approximations, one obtains[3,4]

$$C_+(\mu^2)^2\,\frac{1}{\pi}\mathrm{Im}\psi_{++}(t)\big|_{\text{no Penguin}} = \theta(t)\,\frac{32}{15}\,\frac{t^4}{(16\pi^2)^3}\left(\frac{\alpha_s(M_W^2)}{\alpha_s(\mu^2)}\right)^{4/9} \times\left\{1+\frac{\alpha_s(\mu^2)}{\pi}\left[\log(t/\mu^2)-\frac{1217}{810}\right]+\ldots\right\}, \tag{2.3a}$$

$$C_-(\mu^2)^2\,\frac{1}{\pi}\mathrm{Im}\psi_{--}(t)\big|_{\text{no Penguin}} = \theta(t)\,\frac{16}{15}\,\frac{t^4}{(16\pi^2)^3}\left(\frac{\alpha_s(M_W^2)}{\alpha_s(\mu^2)}\right)^{-8/9} \times\left\{1+\frac{\alpha_s(\mu^2)}{\pi}\left[-2\log(t/\mu^2)+\frac{6709}{810}\right]+\ldots\right\}, \tag{2.3b}$$

$$C_6(\mu^2)^2\,\frac{1}{\pi}\mathrm{Im}\psi_{66}(t)\big|_{1/N_c} \sim \theta(t)\,\frac{12}{5}\,\frac{t^4}{(16\pi^2)^3}\left(\frac{\alpha_s(M_W^2)}{\alpha_s(\mu^2)}\right)^{-18/11} \times\left\{1+\frac{\alpha_s(\mu^2)}{\pi}\left[-\frac{9}{2}\log(t/\mu^2)+\frac{117501}{4840}\right]+\left(\frac{\alpha_s(\mu^2)}{\pi}\right)^2\left[470.72+\ldots\right]+\ldots\right\}. \tag{2.3c}$$

The coefficients of the logarithmic terms are the known anomalous dimension factors, as they should be. The non-logarithmic α_s-corrections to the octet correlators turn out to be very big and positive, while for ψ_{++} this correction is moderate and negative. Eq. (2.3c) also includes the $O(\alpha_s^2)$ correction to the ψ_{66} correlator, which is easily computed in the large N_c limit, thanks to the factorization property of the quark currents; this shows that, at least in this case, the big octet enhancement also appears at higher orders. If one does the usual $\mu^2 = t$ rescaling to eliminate all logarithms in the spectral functions, summing them into the Wilson coefficients, the remaining α_s-terms produce corrections to the octet correlators which are bigger than 100%, even at values of momentum transfer as high as 10 GeV^2. The perturbative calculation has therefore blown up! The short-distance behaviour of the $\Delta S = 1$ correlators clearly shows a dynamical enhancement in the octet channel as a consequence of the interplay of gluonic corrections.

3. Chiral Symmetry

The QCD Lagrangian with three light flavours $q \equiv \begin{pmatrix} u \\ d \\ s \end{pmatrix}$ and in the presence of external vector $v_\mu(x)$, axial-vector $a_\mu(x)$, scalar $s(x)$ and pseudoscalar $p(x)$ fields is defined by

$$\mathcal{L}_{\rm QCD}(q,\bar q,G;v,a,s,p) = \mathcal{L}^0_{\rm QCD} + \bar q\gamma^\mu(v_\mu+\gamma_5 a_\mu)q - \bar q(s-i\gamma_5 p)q, \tag{3.1}$$

where

$$\mathcal{L}^0_{\rm QCD} = -\frac{1}{4}\sum_{a=1}^{8} G^{(a)}_{\mu\nu}G^{(a)\mu\nu} + i\bar q\gamma^\mu(\partial_\mu + iG_\mu)q. \tag{3.2}$$

The external fields v_μ, a_μ, s and p are Hermitian 3×3 matrices in flavour space, with ${\rm tr} v_\mu = {\rm tr} a_\mu = 0$. The field $s(x)$ contains in particular the quark mass matrix $\mathcal{M} = {\rm diag}(m_u, m_d, m_s)$.

The Lagrangian $\mathcal{L}_{\rm QCD}(x)$ is invariant under local chiral $SU(3)_L \otimes SU(3)_R$ transformations:

$$q_L \equiv \frac{1}{2}(1-\gamma_5)q(x) \to V_L(x)q_L, \tag{3.3a}$$

$$q_R \equiv \frac{1}{2}(1+\gamma_5)q(x) \to V_R(x)q_R, \tag{3.3b}$$

$$l_\mu \equiv v_\mu - a_\mu \to V_L l_\mu V_L^+ + iV_L\partial_\mu V_L^+, \tag{3.3c}$$

$$r_\mu \equiv v_\mu + a_\mu \to V_R r_\mu V_R^+ + iV_R\partial_\mu V_R^+, \tag{3.3d}$$

$$s + ip \to V_R\,(s+ip)\,V_L^+, \tag{3.3e}$$

where V_L and V_R are unitary unimodular matrices of $SU(3)_L$ and $SU(3)_R$. At the level of the hadronic physical states, this symmetry is spontaneously broken down to $SU(3)_{V=L+R}$. The pattern of spontaneously broken symmetry implies specific constraints on the dynamics of the strong interactions between the low-lying octet of pseudoscalar states (π, K, η), which are the Goldstone bosons associated to the broken axial symmetry generators. The best way to formulate the chiral symmetry constraints is by means of an effective chiral Lagrangian involving the pseudoscalar degrees of freedom only. The meson fields

$$\Phi \equiv \frac{\vec\lambda}{\sqrt2}\cdot\vec\phi = \begin{pmatrix} \pi^o/\sqrt2 + \eta/\sqrt6 & \pi^+ & K^+ \\ \pi^- & -\pi^o/\sqrt2+\eta/\sqrt6 & K^o \\ K^- & \overline{K}^o & -2\eta/\sqrt6 \end{pmatrix} \tag{3.4}$$

are collected in a 3×3 special unitary matrix, $U \equiv \exp\left(i\sqrt2\Phi/f\right)$, which transforms as

$$U \to V_R\, U\, V_L^\dagger. \tag{3.5}$$

The form of the effective chiral Lagrangian is determined by symmetry requirements. It consists of a string of terms

$$\mathcal{L}_{\text{eff}}^{\text{strong}} = \mathcal{L}^{(2)} + \mathcal{L}^{(4)} + \mathcal{L}^{(6)} + \ldots \tag{3.6}$$

ordered in increasing number of derivatives and external fields. The first term $\mathcal{L}^{(2)}$ in eq. (3.6) is the non-linear sigma model Lagrangian ($f \approx f_\pi = 93.3$ MeV)

$$\mathcal{L}^{(2)} = \frac{f^2}{4}\text{tr}\Big(D_\mu U D^\mu U^\dagger + \chi U^\dagger + U\chi^\dagger\Big). \tag{3.7}$$

The external vector and axial-vector fields are incorporated in the covariant derivative

$$D_\mu U = \partial_\mu U - i(v_\mu + a_\mu)U + iU(v_\mu - a_\mu), \tag{3.8}$$

and $\chi \equiv 2B_0(s + ip)$. The parameter B_0 ($\approx -\langle\bar{u}u\rangle/f^2$) relates the squares of the pseudoscalar meson masses to the quark masses,

$$B_0 = \frac{m_{\pi^+}^2}{m_u + m_d} = \frac{m_{K^+}^2}{m_u + m_s} = \frac{m_{K^0}^2}{m_d + m_s}. \tag{3.9}$$

To next-to-leading order there are two types of contributions: one-loop graphs generated by $\mathcal{L}^{(2)}$ and tree graphs involving one vertex from $\mathcal{L}^{(4)}$. In fact $\mathcal{L}^{(4)}$ is the sum of the Wess-Zumino[5] effective Lagrangian $\mathcal{L}_{WZ}$ which accounts for the chiral anomaly and a chiral invariant piece[6] which in full generality involves ten dimensionless real constants L_i (i = 1,2,...,10):

$$\begin{aligned}
\mathcal{L}^{(4)}(x) = {} & \mathcal{L}_{WZ} + L_1\,[\text{tr}(D_\mu U^\dagger D^\mu U)]^2 + L_2\,\text{tr}(D_\mu U^\dagger D_\nu U)\,\text{tr}(D^\mu U^\dagger D^\nu U) \\
& + L_3\,\text{tr}(D_\mu U^\dagger D^\mu U D_\nu U^\dagger D^\nu U) + L_4\,\text{tr}(D_\mu U^\dagger D^\mu U)\,\text{tr}(\chi^\dagger U + U^\dagger \chi) \\
& + L_5\,\text{tr}(D_\mu U^\dagger D^\mu U(\chi^\dagger U + U^\dagger\chi)) + L_6\,[\text{tr}(\chi^\dagger U + U^\dagger\chi)]^2 \\
& + L_7\,[\text{tr}(\chi U - U^\dagger\chi)]^2 + L_8\,\text{tr}(\chi^\dagger U\chi^\dagger U + \chi U^\dagger\chi U^\dagger) \\
& - i\,L_9\,\text{tr}(F_R^{\mu\nu}D_\mu U D_\nu U^\dagger + F_L^{\mu\nu}D_\mu U^\dagger D_\nu U) + L_{10}\,\text{tr}(U^\dagger F_R^{\mu\nu} U F_{L\mu\nu}) \\
& + \text{ two contact terms in external fields.}
\end{aligned} \tag{3.10}$$

The tensors $F_R^{\mu\nu}$ and $F_L^{\mu\nu}$ stand for the field strengths associated to the non-Abelian external fields r_μ and l_μ. The phenomenological determination of the L_i's has been discussed in detail in ref. 6.

The effect of strangeness-changing non-leptonic weak transitions can be incorporated as a perturbation to the strong Lagrangian (3.6),

$$\mathcal{L}_{\text{eff}} = \mathcal{L}_{\text{eff}}^{\text{strong}} + \mathcal{L}_{\text{eff}}^{\Delta S=1} + \mathcal{L}_{\text{eff}}^{\Delta S=2}. \tag{3.11}$$

The most general form of the effective bosonic Lagrangians which, to lowest order in the number of derivatives, have the same $SU(3)_L \otimes SU(3)_R$ transformation properties as the four-quark Hamiltonians in eqs. (1.1) and (1.3), can be written as

$$\mathcal{L}_{\rm eff}^{\Delta S=1} = \frac{-G_F}{\sqrt{2}} V_{ud} V_{us}^* \, 4 \left\{ g_{\underline{8}}^{(1/2)} \mathcal{L}_{\underline{8}}^{(1/2)} + g_{\underline{27}}^{(1/2)} \mathcal{L}_{\underline{27}}^{(1/2)} + g_{\underline{27}}^{(3/2)} \mathcal{L}_{\underline{27}}^{(3/2)} \right\}, \qquad (3.12a)$$

$$\mathcal{L}_{\rm eff}^{\Delta S=2} = -\frac{G_F^2 M_W^2}{4\pi^2} \left[\lambda_c^2 F_1 + \lambda_t^2 F_2 + 2\lambda_c\lambda_t F_3\right] \frac{4}{3} \frac{f_K^2}{f_\pi^2} \widehat{B} \, \mathcal{L}_{\underline{27}}^{\Delta S=2}, \qquad (3.12b)$$

where

$$\mathcal{L}_{\underline{8}}^{(1/2)} = \sum\nolimits_i (\mathcal{L}_\mu)_{2i} (\mathcal{L}^\mu)_{i3}, \qquad (3.13a)$$

$$\mathcal{L}_{\underline{27}}^{(1/2)} = (\mathcal{L}_\mu)_{21} (\mathcal{L}^\mu)_{13} + (\mathcal{L}_\mu)_{23} \Big[4(\mathcal{L}^\mu)_{11} + 5(\mathcal{L}^\mu)_{22} \Big], \qquad (3.13b)$$

$$\mathcal{L}_{\underline{27}}^{(3/2)} = (\mathcal{L}_\mu)_{21} (\mathcal{L}^\mu)_{13} + (\mathcal{L}_\mu)_{23} \Big[(\mathcal{L}^\mu)_{11} - (\mathcal{L}^\mu)_{22} \Big], \qquad (3.13c)$$

$$\mathcal{L}_{\underline{27}}^{\Delta S=2} = \mathcal{L}_\mu(x)_{23} \mathcal{L}^\mu(x)_{23}. \qquad (3.13d)$$

$\mathcal{L}_\mu$ is shorthand for the 3×3 matrix

$$\mathcal{L}_\mu \equiv -\frac{i f^2}{2} U^\dagger D_\mu U, \qquad (3.14)$$

representing the left-handed currents at lowest order in the momentum expansion. Note that, at order p^2, only one chiral structure is possible for each set of chiral and flavour quantum numbers. $g_{\underline{8}}^{(1/2)}, g_{\underline{27}}^{(1/2)}, g_{\underline{27}}^{(3/2)}$ and $\widehat{B}$ are dimensionless constants, real constants when CP-violation effects are neglected, which cannot be determined from symmetry arguments alone. In the $SU(3)_V$-limit, $g_{\underline{27}}^{(3/2)} = 5 g_{\underline{27}}^{(1/2)} = (8/9) C_+(\mu^2) \alpha_s(\mu^2)^{2/9} \widehat{B} (f_K/f_\pi)^2$. From the observed $K \to \pi\pi$ decay rates, one gets

$$|g^{(1/2)}|_{\rm exp} \equiv |g_{\underline{8}}^{(1/2)} + g_{\underline{27}}^{(1/2)}|_{\rm exp} = 5.1, \qquad |g_{\underline{27}}^{(3/2)}|_{\rm exp} = 0.16, \qquad (3.15)$$

clearly indicating the enhancement of the octet $|\Delta I| = 1/2$ transitions. The $SU(3)_V$ relation among the different $\underline{27}$ couplings implies then $\widehat{B} \approx 0.35$.

4. The QCD Effective Action at Long Distances

Chiral symmetry implies that the generating functional for the Green's functions of vector, axial-vector, scalar and pseudoscalar quark currents,

$$e^{i\Gamma(v,a,s,p)} = \frac{1}{Z} \int \mathcal{D}G_\mu \mathcal{D}\bar{q} \mathcal{D}q \, \exp\left(i \int d^4x \mathcal{L}_{\rm QCD}(q, \bar{q}, G; v, a, s, p) \right) \qquad (4.1)$$

(the normalization factor Z is fixed so that $\Gamma(0,0,0,0) = 1$), admits a low-energy representation

$$e^{i\Gamma(v,a,s,p)} = \int \mathcal{D}U \exp\left(i \int d^4x \mathcal{L}_{\text{eff}}^{\text{strong}}(U; v, a, s, p)\right), \tag{4.2}$$

where $\mathcal{L}_{\text{eff}}^{\text{strong}}$ is the effective chiral Lagrangian (3.6).

Once $\Gamma(v, a, s, p)$ is known, the functional dependence on U of the low energy effective action can in principle be obtained from the chiral transformation properties of Γ. However, although the integral over the quark fields in eq. (4.1) can be explicitly done, we do not know at present how to perform analytically the remaining integration over the gluon fields. A perturbative evaluation of the gluonic contributions would obviously fail in reproducing the correct dynamics of spontaneous chiral symmetry breaking (SCSB). A possible way out is to parametrize phenomenologically the SCSB and make a weak gluon-field expansion around the resulting physical vacuum. The simplest parametrization[7] is obtained by adding to the QCD Lagrangian (3.1) the following term

$$\Delta\mathcal{L}_{QCD} \equiv -M_Q(\bar{q}_R U q_L + \bar{q}_L U^{\dagger} q_R), \tag{4.3}$$

which serves to introduce the U field in a way non invariant under $U \to -U$, and a mass parameter M_Q which regulates the infra-red behaviour of the low energy effective action. In the presence of this term the operator $\bar{q}q$ acquires a vacuum expectation value:

$$\langle\bar{q}q\rangle \equiv \langle\bar{q}q\rangle_{G_\mu} - \langle\bar{q}q\rangle_{G_\mu=0} = -\frac{1}{12}\frac{\alpha_S}{\pi}\frac{\langle GG\rangle}{M_Q} - \frac{1}{360}\frac{\alpha_S}{\pi}g_s f_{abc}\frac{\langle G^a G^b G^c\rangle}{M_Q^3} + \ldots \tag{4.4}$$

(M_Q is assumed to be sufficiently large so that an expansion in inverse powers of M_Q of the low energy effective action can be accepted). The coupling (4.3) is therefore an effective way to generate the order parameter due to SCSB.

It is convenient to make a chiral rotation of the quark fields,

$$Q_L = \xi q_L, \qquad Q_R = \xi^{\dagger} q_R, \tag{4.5}$$

with ξ chosen such that $U = \xi^2$. Under chiral $SU(3)_L \otimes SU(3)_R$ transformations (V_L, V_R)

$$\xi(x) \to V_R \xi(x) h^{\dagger}(x) = h(x)\xi(x) V_L^{\dagger}, \tag{4.6}$$

which defines the compensating $SU(3)_V$ transformation $h(\phi(x))$, the wanted ingredient for a non-linear representation of the chiral group[8]. The quark fields of the rotated basis $Q_{L,R}$ transform like

$$Q_L \to h(x) Q_L, \qquad Q_R \to h(x) Q_R, \tag{4.7}$$

and the new term (4.3),

$$\Delta\mathcal{L}_{QCD} = -M_Q\left(\bar{Q}_R Q_L + \bar{Q}_L Q_R\right), \tag{4.8}$$

is invariant. The quark fields $Q_{L,R}$ can be interpreted as "constituent chiral quarks" and M_Q as a "constituent quark mass".

The QCD Lagrangian in the rotated basis and in Euclidean space has then the following form:

$$\mathcal{L}^{(E)}_{\rm QCD} = \frac{-1}{4} G^{(a)}_{\mu\nu} G^{(a)}_{\mu\nu} + \bar{Q} D_E Q, \tag{4.9}$$

with D_E the Euclidean Dirac operator ($\widetilde{\gamma}_\mu \equiv -i\gamma_\mu$ are Hermitian gamma matrices with positive metric)

$$D_E = \widetilde{\gamma}_\mu \nabla_\mu + \mathbf{M} = \widetilde{\gamma}_\mu(\partial_\mu + \mathcal{A}_\mu) + \mathbf{M}, \tag{4.10}$$

where

$$\mathcal{A}_\mu = iG_\mu + \Gamma_\mu - \frac{i}{2}\gamma_5 \xi_\mu \quad \text{and} \quad \mathbf{M} = \frac{-1}{2}(\Sigma - \gamma_5 \Delta) - M_Q. \tag{4.11}$$

The external vector v_μ and axial-vector a_μ fields appear now in Γ_μ and ξ_μ:

$$\Gamma_\mu = \frac{1}{2}\left[\xi^\dagger(\partial_\mu - ir_\mu)\xi + \xi(\partial_\mu - il_\mu)\xi^\dagger\right], \tag{4.12a}$$

$$\xi_\mu = i\left[\xi^\dagger(\partial_\mu - ir_\mu)\xi - \xi(\partial_\mu - il_\mu)\xi^\dagger\right]; \tag{4.12b}$$

while the external s and p fields have been frozen to the quark mass matrix $\mathcal{M}$, i.e.

$$\Sigma = \xi^\dagger \mathcal{M} \xi^\dagger + \xi \mathcal{M} \xi \quad \text{and} \quad \Delta = \xi^\dagger \mathcal{M} \xi^\dagger - \xi \mathcal{M} \xi. \tag{4.13}$$

The Σ and Δ terms break explicitly the chiral symmetry.

The Euclidean effective action $W_E(U; v, a, \mathcal{M}, M_Q)$ associated to $\mathcal{L}^{(E)}_{\rm QCD}$ is given by

$$\begin{aligned} e^{W_E} &= \frac{1}{Z}\int \mathcal{D}G_\mu \exp\left(-\int d^4z \frac{1}{4} G^{(a)}_{\mu\nu} G^{(a)}_{\mu\nu}\right) \exp\Gamma_E(\mathcal{A}, \mathbf{M}) \\ &\equiv \frac{1}{Z}\int [\mathcal{D}G_\mu] \exp\Gamma_E(\mathcal{A}, \mathbf{M}), \end{aligned} \tag{4.14}$$

with

$$\exp\Gamma_E(\mathcal{A}, \mathbf{M}) = \int \mathcal{D}\bar{Q}\mathcal{D}Q \exp\left(\int d^4z\, \bar{Q} D_E Q\right) = \det D_E. \tag{4.15}$$

Here we are concerned with the non-anomalous part of the effective action. It is then sufficient to consider the modulus of the determinant; i.e.

$$\Gamma_E(\mathcal{A}, \mathbf{M}) = \frac{1}{2} \log\left[\det\left(D_E^\dagger D_E\right)\right]. \tag{4.16}$$

Since $D_E^\dagger D_E$ is a second-order elliptic operator, det $D_E^\dagger D_E$ can be defined using, e.g., the ξ-function regularization method (for a review, see ref. 9),

$$\det\left(D^\dagger D\right) = \exp\left(\frac{-d}{ds}\zeta_{D^\dagger D}(s)\big|_{s=o}\right), \tag{4.17}$$

with

$$\zeta_{D^\dagger D}(s) = \frac{\nu^{2s}}{\Gamma(s)} \int_0^\infty d\tau\, \tau^{s-1} \mathrm{Tr}\left[\exp\left(-\tau D^\dagger D\right)\right], \tag{4.18}$$

where Tr stands for trace over Dirac γ-matrices, colour $SU(3)$ matrices, flavour $SU(3)$ matrices and Euclidean space. ν is an arbitrary mass scale, so that $\zeta(s)$ remains dimensionless, which serves to regulate ultra-violet divergences (one should distinguish between this UV-regulator mass scale ν^2 and the perturbative QCD scale μ^2 which appears, via $\overline{MS}$-renormalization, in the Wilson coefficients of the effective weak Hamiltonian). Explicit calculations can now be made using the powerful technique of the heat kernel expansion[9] applied to the operator $D_E^\dagger D_E - M_Q^2$, i.e.

$$(x|e^{-\tau D^\dagger D}|y) = \frac{1}{16\pi^2\tau^2}\, e^{-(x-y)^2/4\tau} \sum_{n=0}^{\infty} H_n(x,y)\, \tau^n\, e^{-\tau M_Q^2}. \tag{4.19}$$

The infra-red (large τ) behaviour is regulated by the $\exp(-\tau M_Q^2)$ factor. The functions $H_n(x,x)$ are the so-called Seeley-DeWitt coefficients[9].

The derivation of the low energy effective chiral Lagrangian within this framework has been extensively discussed in ref. 7. In the chiral limit, and including the leading gluonic contributions, one gets:

$$f_\pi^2 = \frac{N_c}{16\pi^2} 4M_Q^2 \left[\log\frac{\nu^2}{M_Q^2} + \frac{\pi^2}{6N_c}\frac{\langle\frac{\alpha_s}{\pi}GG\rangle}{M_Q^4} + \frac{1}{360N_c}\frac{\langle g^3GGG\rangle}{M_Q^6} + \ldots\right], \tag{4.20a}$$

$$8L_1 = 4L_2 = L_9 = \frac{N_c}{48\pi^2}\left[1 + O\left(\frac{1}{M_Q^6}\right)\right], \tag{4.20b}$$

$$L_3 = L_{10} = -\frac{N_c}{96\pi^2}\left[1 + \frac{\pi^2}{5N_c}\frac{\langle\frac{\alpha_s}{\pi}GG\rangle}{M_Q^4} + O\left(\frac{1}{M_Q^6}\right)\right], \tag{4.20c}$$

The relation (4.20a) allows us to trade the ultra-violet cut-off ν^2 by the physical pion decay constant. Due to dimensional reasons, the leading contributions to the O(p^4) couplings only depend on N_c and geometrical factors. It is remarkable that the couplings L_1, L_2 and L_9 do not get any gluonic corrections at this order; this result

Table 1
Leading-order ($\alpha_s = 0$) predictions for the L_i's, compared with their phenomenological determination[6] at the M_ρ scale. All numbers are given in units of 10^{-3}.

	L_1	L_2	L_3	L_9	L_{10}
$L_i^{th}(\alpha_s = 0)$	0.79	1.58	−3.17	6.33	−3.17
$L_i^{exp}(M_\rho)$	0.7 ± 0.3	1.3 ± 0.7	-4.4 ± 2.5	6.9 ± 0.7	-5.2 ± 0.3

is independent of the way the SCSB has been parametrized (M_Q can be taken to be infinite). Table 1 compares the predictions obtained with only the leading term in eqs. (4.20b, c) (i.e. neglecting the gluonic corrections) with the phenomenological determination of the L_i-couplings done by Gasser and Leutwyler[6]. The numerical agreement is quite impressive; both the order of magnitude and the sign are correctly reproduced. Moreover, the gluonic corrections shift the values of L_3 and L_{10} in the right direction, making them more negative.

5. The Effective Action of Four-Quark Operators

The effective action of a given operator $O(x)$ is formally defined by the functional average

$$\langle O \rangle = \frac{\int \mathcal{D}G_\mu \mathcal{D}\bar{q}\mathcal{D}q \, i \int d^4x \, O(x) \exp\left(i \int d^4z \, \mathcal{L}_{\rm QCD}(q, \bar{q}, G; v, a, s, p)\right)}{\int \mathcal{D}G_\mu \mathcal{D}\bar{q}\mathcal{D}q \, \exp\left(i \int d^4z \, \mathcal{L}_{\rm QCD}(z)\right)}. \tag{5.1}$$

With $\mathcal{L}_{\rm QCD}^{(E)}$ defined in eq. (4.9), this corresponds to an effective bosonic operator

$$\langle O(x) \rangle = \frac{\int [\mathcal{D}G_\mu] \mathcal{D}\bar{Q}\mathcal{D}Q \; O^{(E)}(x) \exp\left(\int d^4z \, \bar{Q} D_E Q\right)}{\int [\mathcal{D}G_\mu] \mathcal{D}\bar{Q}\mathcal{D}Q \, \exp\left(\int d^4z \, \bar{Q} D_E Q\right)}. \tag{5.2}$$

In the simpler case of a bilinear quark operator $\bar{q}_{jL}(x)\gamma^\mu q_{iL}(x)$ the corresponding effective bosonic current can be easily obtained by differentiation with respect to the external left-handed field

$$\langle \bar{q}_{jL} \tilde{\gamma}^\mu q_{iL} \rangle (x) = i \frac{\delta W_E}{\delta l_\mu(x)_{ji}} = i \frac{\int [\mathcal{D}G_\mu] \frac{\delta \Gamma_E}{\delta (l_\mu)_{ji}} \exp \Gamma_E}{\int [\mathcal{D}G_\mu] \exp \Gamma_E}, \tag{5.3a}$$

$$\frac{\delta \Gamma_E}{\delta l_\mu} = \frac{1}{2} \mathrm{Tr} \left(\frac{\delta D_E}{\delta l_\mu} D_E^{-1} + \frac{\delta D_E^\dagger}{\delta l_\mu} (D_E^\dagger)^{-1} \right). \tag{5.3b}$$

Equations (5.3a,b) provide an explicit recipe to construct the effective action, in an external gluonic field background, of the quark current operator associated to l_μ. We have then two ways to compute the bosonized current operator. Either we already know the effective action W_E in which case we simply have to take the variation with respect to an external source; or otherwise, in general, we will have to compute the r.h.s. in eq. (5.3b) and integrate over gluons as indicated in eq. (5.3a).

The heat kernel expansion technique can be applied to compute eq. (5.3b). Using the proper time representation of the Dirac propagator,

$$(x|D^{-1}|x) = \frac{d}{ds}\left(\frac{\nu^{2s}}{\Gamma(s)}\int_0^\infty d\tau\,\tau^s \frac{e^{-\tau M_Q^2}}{16\pi^2\tau^2}\sum_{n=0}^{\infty}\mathcal{H}_n(x,x)\tau^n\right)\Bigg|_{s=0}, \tag{5.4}$$

the problem is reduced to the calculation of the derivative coefficient functions[9]

$$\mathcal{H}_n(x,x) \equiv H_n(x,y)\overleftarrow{D}_y^\dagger|_{y=x}, \tag{5.5}$$

where $H_n(x,y)$ are the coefficient functions of the heat kernel expansion in eq. (4.19).

The effective action of four-quark operators can be obtained in a similar way. Let us consider as a specific example the $Q_{\Delta S=2}$ operator of eq. (1.4). We can make a point-splitting of the two currents and rewrite $\langle Q_{\Delta S=2}\rangle$ as $[(\lambda_{ab})_{ij} = \delta_{ai}\delta_{bj}]$

$$\langle Q_{\Delta S=2}(x)\rangle = \frac{1}{Z\,\exp W_E}\int d^4y\int\frac{d^4k}{(2\pi)^4}\,e^{-ik.(x-y)}\int[\mathcal{D}G_\mu]\int\mathcal{D}\bar{Q}\mathcal{D}Q$$
$$\left(\bar{Q}_L(x)\xi(x)\lambda_{32}\xi^\dagger(x)\tilde{\gamma}_\mu Q_L(x)\right)\left(\bar{Q}_L(y)\xi(y)\lambda_{32}\xi^\dagger(y)\tilde{\gamma}_\mu Q_L(y)\right)\exp\left(\int d^4z\,\bar{Q}D_E Q\right). \tag{5.6}$$

The integral over the four-vector k is a delta function which, upon integration over the y-variable, brings all the fields of the $Q^{(E)}_{\Delta S=2}$-operator to the same point. The technical advantage of writing $\langle Q_{\Delta S=2}(x)\rangle$ in such a form is that now we can introduce effective bosonic operators in an external gluonic background as variations of the Euclidean action $\Gamma_E(\mathcal{A},\mathbf{M})$ in eq. (4.15) with respect to appropriate external sources:

$$\langle Q_{\Delta S=2}(x)\rangle = \int d^4y\int\frac{d^4k}{(2\pi)^4}\,e^{-ik.(x-y)}\,\frac{\int[\mathcal{D}G_\mu]\,\frac{(-1)\delta^2}{\delta l_\mu(x)_{32}\delta l_\mu(y)_{32}}\exp\Gamma_E(\mathcal{A},\mathbf{M})}{\int[\mathcal{D}G_\mu]\exp\Gamma_E(\mathcal{A},\mathbf{M})}$$
$$= \frac{\int[\mathcal{D}G_\mu]\,\langle Q_{\Delta S=2}(x)\rangle_{\mathcal{A},\mathbf{M}}\exp\Gamma_E(\mathcal{A},\mathbf{M})}{\int[\mathcal{D}G_\mu]\exp\Gamma_E(\mathcal{A},\mathbf{M})}, \tag{5.7}$$

where

$$\langle Q_{\Delta S=2}(x)\rangle_{A,\mathrm{M}} = -\,\mathrm{Tr}\Big(D_E^{-1}\frac{\delta D_E}{\delta l_\mu(x)_{32}}\Big)\,\mathrm{Tr}\Big(D_E^{-1}\frac{\delta D_E}{\delta l_\mu(x)_{32}}\Big) + \mathrm{Tr}\Big(D_E^{-1}\frac{\delta D_E}{\delta l_\mu(x)_{32}}D_E^{-1}\frac{\delta D_E}{\delta l_\mu(x)_{32}}\Big). \tag{5.8}$$

The first term on the r.h.s. of eq. (5.8) corresponds to the factorization of the two currents, while the second term contains the non-factorizable contributions.

The bosonization of the $\Delta S = 1,2$ Hamiltonians in eqs. (1.1) and (1.3) generates the effective Lagrangians (3.12a,b), with explicit values for the different couplings. The calculation has been done in ref. 4 including, in addition to the well-known $O(N_c^2)$ contributions, the subleading $O(N_c)$ and $O(\alpha_s N_c)$ terms. The resulting couplings are:

$$\widehat{B} = \alpha_s(\mu^2)^{-2/9}\,\frac{3}{4}\left[1+\frac{1}{N_c}\left(1-\frac{N_c}{2}\frac{\langle\frac{\alpha_s}{\pi}GG\rangle}{16\pi^2 f_\pi^4}+0(\alpha_s N_c)^2\right)\right], \tag{5.9a}$$

$$g_{\underline{27}}^{(3/2)} = 5g_{\underline{27}}^{(1/2)} = C_+(\mu^2)\,\frac{2}{3}\left[1+\frac{1}{N_c}\left(1-\frac{N_c}{2}\frac{\langle\frac{\alpha_s}{\pi}GG\rangle}{16\pi^2 f_\pi^4}+0(\alpha_s N_c)^2\right)\right], \tag{5.9b}$$

$$\begin{aligned} g_{\underline{8}}^{(1/2)} = {} & C_-(\mu^2)\left[1-\frac{1}{N_c}\left(1-\frac{N_c}{2}\frac{\langle\frac{\alpha_s}{\pi}GG\rangle}{16\pi^2 f_\pi^4}+0(\alpha_s N_c)^2\right)\right] \\ & + C_+(\mu^2)\,\frac{1}{5}\left[1+\frac{1}{N_c}\left(1-\frac{N_c}{2}\frac{\langle\frac{\alpha_s}{\pi}GG\rangle}{16\pi^2 f_\pi^4}+0(\alpha_s N_c)^2\right)\right] \\ & + C_4(\mu^2) - C_6(\mu^2)\,16L_5\left(\frac{\langle\bar\psi\psi\rangle}{f_\pi^3}\right)^2 + 0(\alpha_s N_c). \end{aligned} \tag{5.9c}$$

6. Discussion

The leading terms in eqs. (5.9) are the well-known large-N_c predictions, which originate from the two factorized traces on the r.h.s. of eq. (5.8). To the extent that gluons communicating the two fermionic traces are neglected (this would be an $O(\alpha_s^2)$ effect, since two gluons at least are needed because of colour conservation), QCD (and quark-mass) corrections are already reabsorbed in the bosonization of each separate current, i.e. in the values of the physical decay constants $f_{\pi,K}$.

The non-factorizable contribution in eq. (5.8) gives rise to the $1/N_c$ corrections in eqs. (5.9). The results of the vacuum saturation approximation (VSA), often used in the literature, are reproduced in the limit $\alpha_s = 0$; in fact, the VSA results include also those gluonic effects which can be reabsorbed in $f_{\pi,K}$. However, there are additional gluonic contributions, not included in the VSA, which generate $\langle\frac{\alpha_s}{\pi}GG\rangle$ corrections to the $1/N_c$ terms. These gluonic corrections turn out to be quite big. Using for the gluon condensate the range $(360\mathrm{MeV})^4 \le \langle\frac{\alpha_s}{\pi}GG\rangle \le (390\mathrm{MeV})^4$, one has

$$-0.63 \le \frac{1}{N_c}\left(1-\frac{N_c}{2}\frac{\langle\frac{\alpha_s}{\pi}GG\rangle}{16\pi^2 f_\pi^4}\right) \le -0.37. \tag{6.1}$$

Note that the gluonic contributions flip the sign of the $1/N_c$ corrections, with respect to the VSA result; this is precisely what is needed in order to explain the $|\Delta I| = 1/2$ enhancement.

As shown in eqs. (5.9),there is a correlation between the $1/N_c$ corrections to the different chiral couplings. The same gluon condensate correction which decreases $\widehat{B}$, $g_{\underline{27}}^{(3/2)}$ and $g_{\underline{27}}^{(1/2)}$, increases $g_{\underline{8}}^{(1/2)}$. The measured $K \to \pi\pi$ rates require a big and positive $1/N_c$ correction to the octet coupling; this necessarily implies a value for the $\widehat{B}$ factor smaller than the large-N_c prediction $\widehat{B} = 3/4$.

The Wilson coefficients $C_i(\mu^2)$ (i=3,4,5,6) are already $O(\alpha_s)$; therefore, at the level of approximation used in eqs. (5.9), one only needs the effective bosonic realization of the corresponding operators to leading order in $1/N_c$. In this limit, $\langle Q_3 \rangle = \langle Q_5 \rangle = 0$, $\langle Q_4 \rangle = \langle Q_- \rangle$, and $\langle Q_6 \rangle$ reduces to a product of factorized (scalar-pseudoscalar) currents, which is related to the L_5 term in the $O(p^4)$ strong Lagrangian. The contribution of these operators to the octet coupling is given in the last two terms in eq. (5.9c). $C_4(\mu^2)$ and $C_6(\mu^2)$ are both negative; therefore, the $\langle Q_4 \rangle$ contribution gives a (very) small suppression, while $\langle Q_6 \rangle$ produces an additional enhancement of $g_{\underline{8}}^{(1/2)}$.

At leading non-trivial order in $1/N_c$, the renormalization-scale dependence of the $C_6(\mu^2)$ Wilson coefficient is exactly cancelled by the μ^2 dependence of $\langle \bar{\psi}\psi \rangle^2$. The cancellation of the μ^2 dependence of the other Wilson coefficients is a little bit more involved, because it is a next-to-leading order (in $1/N_c$) effect. In order to see such a cancellation it is necessary to compute the perturbative gluonic corrections to the non-factorizable part of the bosonization of the corresponding four-quark operators. It is easy to check[4] that the logarithmic α_s-corrections to the bosonized operators are weighted by the same anomalous dimension factors as the original four-quark operators, as they should be in order to get a μ^2-independent result. The scale of these logarithms is given by M_Q, i.e. $\langle Q_i \rangle \sim -\frac{1}{2}\gamma_{ij}^{(1)}\frac{\alpha_s}{\pi}\log\left(\frac{\mu^2}{M_Q^2}\right)\langle Q_j \rangle$. Therefore, taking $\mu = M_Q$, all these logarithmic corrections are reabsorbed into the Wilson coefficients.

The size of the M_Q scale can be phenomenologically obtained in different ways[4,7]. For instance, inserting the values of the quark and gluon condensates in eq. (4.4), one gets[7] $M_Q \approx 320 MeV$, i.e. the typical value of a constituent quark mass in phenomenological models[10]. Similar values are obtained from other sources. The scaling $\mu^2 = M_Q^2$ generates a sizable enhancement of $g_{\underline{8}}^{(1/2)}$ and a mild suppression of the remaining couplings in eqs. (5.9). Table 2 shows a compilation of numerical results for various values of the input parameters.

The numbers in Table 2 show a clear improvement with respect to the leading $N_c \to \infty$ predictions. However, they should be taken more as a qualitative indication that the correct dynamics is reproduced than as true quantitative predictions. M_Q is certainly a very low scale, and therefore one could worry about the meaning of the perturbatively calculated Wilson coefficients at such small value of μ. Perhaps it would be better to explicitly incorporate all $\alpha_s \log(\mu/M_Q)$ corrections to

Table 2

Numerical values of the coupling constants in eqs. (5.9) for different input values of the gluon condensate and $\Lambda_{\overline{MS}}$. The scaling $\mu^2 = M_Q^2$ has been taken.

$\langle \frac{\alpha_s}{\pi} GG \rangle$ (MeV^4)	$\Lambda_{\overline{MS}}$ (MeV)	$g^{1/2}$	$g_{\underline{27}}^{(3/2)}$	$\widehat{B}$
$(360)^4$	150	2.1	0.13	0.48
$(360)^4$	200	2.6	0.12	0.43
$(390)^4$	150	2.4	0.08	0.28
$(390)^4$	200	3.0	0.07	0.25
Leading $N_c \to \infty$ limit		2/3	1/3	3/4
Experiment		5.1	0.16	?

the bosonized operators, and take a bigger value for the renormalization scale; to do that in a sensible way, however, it is necessary to perform a detailed perturbative calculation including also the non-logarithmic corrections. One should also check the size of the $O(\alpha_s N_c)^2$ corrections and chiral breaking effects. Progress in these directions is expected in the near future.

Acknowledgements

I would like to thank the organizers for the charming atmosphere of this workshop. The results discussed here have been obtained in collaboration with E.de Rafael. This work has been supported in part by CICYT, Spain, under grant No. AEN90-0040.

References

1. A.Pich and E.de Rafael, *Phys. Lett.* **158B** (1985) 477.
2. B.Guberina, A.Pich and E.de Rafael, *Phys. Lett.* **163B** (1985) 198; A.Pich, B.Guberina and E.de Rafael, *Nucl. Phys.* **B277** (1986) 197.
3. A.Pich, *Proc. of the Ringberg Workshop on "Hadronic Matrix Elements and Weak Decays"*, eds. A.J.Buras, J.-M.Gerard and W.Huber, *Nucl. Phys. B (Proc. Suppl.)* **7A** (1989) 194.
4. A.Pich and E.de Rafael, *Nucl. Phys.* **B358** (1991) 311.
5. J.Wess and B.Zumino, *Phys. Lett.* **37B** (1971) 95; E.Witten, *Nucl. Phys.* **B223** (1983) 422.
6. J.Gasser and H.Leutwyler, *Nucl. Phys.* **B250** (1985) 465.
7. D.Espriu, E.de Rafael and J.Taron, *Nucl. Phys.* **B345** (1990) 22.
8. S.Coleman, J.Wess and B.Zumino, *Phys. Rev.* **177** (1969) 2239.
9. R.D.Ball, *Phys. Reports* **182** (1989) 1, and references therein.
10. A.Manohar and H.Georgi, *Nucl. Phys.* **B234** (1984) 189.

BARYONS

Baryon Chiral Perturbation Theory

Elizabeth Jenkins and Aneesh V. Manohar

Department of Physics 0319, University of California, San Diego
9500 Gilman Drive, La Jolla, CA 92093-0319, U.S.A.

ABSTRACT

A new formulation of baryon chiral perturbation theory is developed using an effective Lagrangian for broken $SU(3)_L \times SU(3)_R$ chiral symmetry, in which the baryons appear as heavy static fermions. This formulation of baryon chiral perturbation theory has a consistent expansion in powers of momentum and light quark masses because the baryon mass does not appear in the effective Lagrangian. The chiral expansion in powers of the physical strange quark mass requires the inclusion of both the spin-1/2 octet and the spin-3/2 decuplet in the effective theory. The chiral Lagrangian for octet and decuplet baryons is used to compute the leading non-analytic in m_s corrections to baryon axial vector currents, masses, and non-leptonic decays. The results are compared with experiment. There is a large non-analytic strange quark contribution to the baryon masses which changes the value of $\langle p| m_s \bar{s} s |p\rangle$ from 411 MeV to approximately zero. The decoupling of the non-analytic contributions due to intermediate decuplet states in the chiral limit is studied in some detail. It is possible to define an $SU(6)$ symmetry of the effective Lagrangian which relates octet and decuplet states. The best fit values for the octet and decuplet pion couplings respect approximate $SU(6)$ symmetry, but the non-leptonic weak decay parameters do not.

1. Introduction

The QCD Lagrangian for the three light quarks has an approximate $SU(3)_L \times SU(3)_R$ chiral symmetry, which is spontaneously broken by the strong interactions to the diagonal $SU(3)_V$ subgroup. There is a pseudoscalar octet of approximate Goldstone bosons, the π's, K's and η. Low momentum processes involving Goldstone bosons can be related to each other using chiral symmetry. The implications of chiral symmetry for low-momentum processes are best derived using an effective Lagrangian which implements the spontaneous breaking of chiral symmetry.[1–6] The Goldstone boson effective Lagrangian is the most general possible Lagrangian with broken $SU(3)_L \times SU(3)_R$ symmetry, including terms with arbitrary numbers of derivatives. Higher dimension operators in the effective Lagrangian are suppressed by inverse powers of the chiral symmetry breaking scale $\Lambda_\chi \sim 1$ GeV. Thus the pion scattering amplitudes in the effective Lagrangian are determined as a power series in k/Λ_χ, where k is the pion momentum. The explicit $SU(3)_L \times SU(3)_R$ symmetry breaking by the light quark masses can be incorporated by including factors of the quark mass matrix M in the effective Lagrangian, where M transforms as $(\mathbf{3},\overline{\mathbf{3}}) \oplus (\overline{\mathbf{3}},\mathbf{3})$ under $SU(3)_L \times SU(3)_R$. The power counting scheme of dividing factors of ∂_μ and M by Λ_χ can be shown to be a consistent scheme to all orders in the loop expansion if $\Lambda_\chi \leq 4\pi f$. Even though the effective Lagrangian is non-renormalizable, and has an infinite number of coupling constants, one only needs to know a finite number of couplings to determine scattering amplitudes to any fixed order in k/Λ_χ.

Octet baryon fields can be included in the effective Lagrangian as matter fields, following the prescription of Callan *et al.*[5] The higher dimension terms in the baryon Lagrangian are also suppressed by inverse powers of Λ_χ but there is no longer a consistent derivative expansion for processes involving baryons. Consider, for example,

$$L = \overline{B}\left(i\partial\!\!\!/ - m_B\right) B + a\overline{B}\left(i\partial\!\!\!/ - m_B\right)\frac{\partial^2}{\Lambda_\chi^2} B$$

which is the baryon kinetic term plus a higher dimension term with two additional derivatives. The higher dimension term is important even for low-momentum transfer processes, because the time derivatives in $\partial^2/\Lambda_\chi^2$ produce a factor of m_B^2/Λ_χ^2 which is not small. Thus a computation of low-momentum baryon-pion scattering amplitudes requires that one first sum all the time derivatives in the effective Lagrangian; it is not consistent to just use the lowest order term. A similar problem occurs in the loop expansion. Weinberg's power counting argument fails because m_B is a dimensionful number of order Λ_χ. Higher order loop graphs can produce amplitudes which are only "suppressed" by factors of $m_B/\Lambda_\chi \sim 1$, and are therefore as important as lower order amplitudes. A new formalism is developed in this article which circumvents these problems. The baryon is included in the chiral Lagrangian as a heavy static fermion. This theory is shown to have a consistent derivative and loop expansion. The leading non-analytic corrections to the baryon axial currents, masses, and non-leptonic decays are computed in this formalism, and the results are compared with experiment. We show that the experimental data requires that the baryon chiral Lagrangian include both the octet and decuplet baryons.

2. The Heavy Baryon Theory and Power Counting

Low-energy theorems of current algebra only require that the pion momentum be small and that the baryons are nearly on- shell; the large baryon mass is irrelevant. The heavy baryon formalism described in this section is useful under precisely the same conditions. The heavy baryon Lagrangian describes the interactions of a heavy baryon with low- momentum pions. The velocity of the baryon is nearly unchanged when it exchanges some small momentum with the pion. Thus, it is convenient to use a variant of the formalism developed recently for the study of the heavy quark limit in QCD.[8,9] A nearly on-shell baryon with velocity v^μ has momentum

$$p^\mu = m_B v^\mu + k^\mu, \tag{1}$$

where m_B is the baryon mass, and $k \cdot v \ll \Lambda_\chi$ is proportional to the amount by which the baryon is off-shell. The effective theory is written in terms of baryon fields B_v with definite velocity v^μ, which are related to the original baryon fields by[9]

$$B_v(x) = e^{i m_B \not{v}\, v_\mu x^\mu} B(x). \tag{2}$$

The new baryon fields obey a modified Dirac equation,

$$i\not{\partial} B_v = 0 \tag{3}$$

which no longer contains a baryon mass term. Derivatives acting on the field B_v produce factors of k, rather than p, so that higher derivative terms in the Lagrangian are suppressed by powers of k/Λ_χ, which is small. Thus, the heavy baryon Lagrangian has a consistent derivative expansion. The effective theory can be used provided the pion momentum and the off-shellness of the baryon are small compared with Λ_χ.

The heavy baryon Lagrangian also has a $1/m_B$ expansion. The $1/m_B$ effects in the original Dirac theory can be reproduced in the effective theory by including higher dimension operators suppressed by inverse powers of m_B.[10] The relative magnitude of m_B and Λ_χ is irrelevant for the validity of the chiral expansion. In general, one might want to separate terms which are formally $1/m_B$ from terms which are formally $1/\Lambda_\chi$. In this article, we will restrict our discussion to the lightest baryon multiplets, the octet and decuplet, which have masses of order Λ_χ. Thus the $1/m_B$ and $1/\Lambda_\chi$ expansions can be combined into a single expansion in $1/\Lambda_\chi$.

The heavy baryon chiral theory has a consistent power counting expansion. The proof is rather elementary, and is essentially the same as the proof of the power counting expansion for the chiral quark model.[11] The terms in the effective theory are of the form

$$f^2\Lambda_\chi^2\left(\frac{\overline{B}_v}{f\sqrt{\Lambda_\chi}}\right)^{n_1}\left(\frac{B_v}{f\sqrt{\Lambda_\chi}}\right)^{n_2}\left(\frac{\partial}{\Lambda_\chi}\right)^{n_3}\left(\frac{M}{\Lambda_\chi}\right)^{n_4}(v)^{n_5}(\Sigma)^{n_6}\left(\Sigma^\dagger\right)^{n_7} \tag{4}$$

with coefficients of order one, provided $\Lambda_\chi \leq 4\pi f$. Note that the Lagrangian includes explicit factors of the baryon velocity v^μ. This power counting scheme is consistent because the effective heavy baryon theory does not have any terms which contain the baryon mass m_B. Factors of m_B therefore cannot occur in any loop graph, and the dimensional analysis arguments used in the proof of (4) are valid.

The heavy baryon chiral theory is similar to the non-relativistic formulation of baryon chiral perturbation theory discussed recently by Weinberg.[12] The non-relativistic baryon theory is the heavy baryon theory in the Lorentz frame in which $v^\mu = (1,0,0,0)$. There are some advantages to using the heavy baryon theory. Lorentz invariance is manifest, and quantum corrections can be computed in a straightforward manner using ordinary Feynman graphs, rather than time ordered perturbation theory.

3. The Heavy Baryon Chiral Lagrangian

The heavy baryon chiral Lagrangian is written in terms of the baryon fields

$$B_v = \begin{pmatrix} \frac{1}{\sqrt{2}}\Sigma_v^0 + \frac{1}{\sqrt{6}}\Lambda_v & \Sigma_v^+ & p_v \\ \Sigma_v^- & -\frac{1}{\sqrt{2}}\Sigma_v^0 + \frac{1}{\sqrt{6}}\Lambda_v & n_v \\ \Xi_v^- & \Xi_v^0 & -\frac{2}{\sqrt{6}}\Lambda_v \end{pmatrix}, \tag{5}$$

and the Goldstone boson fields

$$\pi = \frac{1}{\sqrt{2}} \begin{pmatrix} \frac{1}{\sqrt{2}}\pi^0 + \frac{1}{\sqrt{6}}\eta & \pi^+ & K^+ \\ \pi^- & -\frac{1}{\sqrt{2}}\pi^0 + \frac{1}{\sqrt{6}}\eta & K^0 \\ K^- & \overline{K}^0 & -\frac{2}{\sqrt{6}}\eta \end{pmatrix}. \tag{6}$$

The fields ξ and Σ are related to the pion fields by

$$\xi = e^{i\pi/f}, \quad \Sigma = \xi^2 = e^{2i\pi/f}. \tag{7}$$

With this normalization, the pion decay constant is $f \approx 93$ MeV. Under a $SU(3)_L \times SU(3)_R$ transformation, these fields transform as

$$\Sigma \to L\Sigma R^\dagger, \quad B_v \to UB_vU^\dagger, \tag{8}$$

$$\xi \to L\xi U^\dagger = U\xi R^\dagger, \tag{9}$$

where U is defined implicitly by Eq. (9).

The Dirac structure of the heavy baryon field simplifies considerably, because the B_v field is a two-component spinor.[9] The heavy baryon field B_v is projected onto the particle portion of the spinor using the projection operator

$$P_v = \left(\frac{1+\not{v}}{2}\right), \quad B_v = P_vB_v. \tag{10}$$

The effects of antibaryon spinor components are represented in the effective theory by including higher dimension operators suppressed by $1/m_B$. One can also define spin operators S_v^μ that act on the baryon fields,[7] with the properties

$$v \cdot S_v = 0, \quad S_v^2\, B_v = -\tfrac{3}{4}\, B_v, \tag{11}$$

$$\left\{S_v^\lambda, S_v^\sigma\right\} = \tfrac{1}{2}\left(v^\lambda v^\sigma - g^{\lambda\sigma}\right), \quad [S_v^\lambda, S_v^\sigma] = i\epsilon^{\lambda\sigma\alpha\beta} v_\alpha S_{v\beta}, \tag{12}$$

where $\epsilon_{0123} = +1$. The spin operators reduce to the usual spin operators $\vec{\sigma}/2$ for a non-relativistic spin-1/2 particle in the rest frame where $v^\mu = (1,0,0,0)$. Spin operators are defined in an arbitrary Lorentz frame by boosting from the rest frame. The commutation relations (12) can be derived by taking the usual relations which are valid in the rest frame, such as $[S^i, S^j] = i\epsilon^{ijk}S^k$, and writing them in four-vector notation using v^μ. These relations are then valid in an arbitrary Lorentz frame. The Dirac structure of the theory can now be eliminated; all Lorentz tensors made from spinors can be written in terms of v^μ and S_v^μ, using the identities

$$\begin{aligned}&\overline{B}_v\gamma_5 B_v = 0, \quad \overline{B}_v\gamma^\mu B_v = v^\mu \overline{B}_v B_v, \quad \overline{B}_v\gamma^\mu\gamma_5 B_v = 2\overline{B}_v S_v^\mu B_v, \\ &\overline{B}_v\sigma^{\mu\nu}B_v = 2\epsilon^{\mu\nu\alpha\beta}v_\alpha \overline{B}_v S_{v\beta} B_v, \quad \overline{B}_v\sigma^{\mu\nu}\gamma_5 B_v = 2i\left(v^\mu \overline{B}_v S_v^\nu B_v - v^\nu \overline{B}_v S_v^\mu B_v\right).\end{aligned} \tag{13}$$

$\overline{B}_v\gamma_5 B_v$ vanishes because it is order $1/m_B$. The operator $\overline{B}\gamma_5 B$ matches onto an operator involving B_v fields which is suppressed by $1/m_B$ in the heavy baryon theory,

$$\overline{B}\gamma_5 B \to \frac{1}{m_B}\overline{B}_v\ S_v \cdot \overleftrightarrow{\partial}\, B_v. \tag{14}$$

The most general Lagrangian at lowest order (one derivative or one M) is

$$L = \int \frac{d^3v}{2v^0} L_v \tag{15}$$

with

$$\begin{aligned}L_v = &\ i\,\mathrm{Tr}\,\overline{B}_v\,(v\cdot\mathcal{D})\,B_v + 2D\ \mathrm{Tr}\,\overline{B}_v\,S_v^\mu\,\{A_\mu, B_v\} + 2F\ \mathrm{Tr}\,\overline{B}_v\,S_v^\mu\,[A_\mu, B_v] \\ &+ b_D\ \mathrm{Tr}\,\overline{B}_v\left\{\xi^\dagger M\xi^\dagger + \xi M\xi, B_v\right\} + b_F\ \mathrm{Tr}\,\overline{B}_v\left[\xi^\dagger M\xi^\dagger + \xi M\xi, B_v\right] \\ &+ \sigma\ \mathrm{Tr}\ M(\Sigma+\Sigma^\dagger)\ \mathrm{Tr}\,\overline{B}_v B_v + \frac{f^2}{4}\ \mathrm{Tr}\,\partial_\mu\Sigma\partial^\mu\Sigma^\dagger + a\ \mathrm{Tr}\ M(\Sigma+\Sigma^\dagger),\end{aligned} \tag{16}$$

where

$$\mathcal{D}^\mu B_v = \partial^\mu B_v + [V^\mu, B_v], \tag{17}$$

$$V^\mu = \frac{1}{2}\left(\xi\partial^\mu\xi^\dagger + \xi^\dagger\partial^\mu\xi\right), \quad A^\mu = \frac{i}{2}\left(\xi\partial^\mu\xi^\dagger - \xi^\dagger\partial^\mu\xi\right), \tag{18}$$

and

$$M = \begin{pmatrix} m_u & 0 & 0 \\ 0 & m_d & 0 \\ 0 & 0 & m_s \end{pmatrix}, \tag{19}$$

is the quark mass matrix. The integration of L_v over all v in Eq. (15) ensures that the theory is Lorentz invariant. It also allows one to consider several baryons with different velocities simultaneously. From now on, we will write down velocity dependent Lagrangians L_v without indicating the integral over v explicitly. All the terms in Eq. (16) have either one factor of the symmetry breaking mass matrix M, or at least one derivative. Note that the mass term $m_B\overline{B}B$ in the usual chiral Lagrangian, which has no powers of ∂ or M, is absent because of the redefinition Eq. (2).

4. The Spin-3/2 Decuplet

The spin-3/2 decuplet can also be included in the effective chiral theory. The decuplet can be described by a Rarita- Schwinger field $(T^\mu)_{abc}$, which contains both spin-1/2 and spin-3/2 pieces. The spin-1/2 pieces are projected out using the constraint $\gamma^\mu T_\mu = 0$. Under $SU(3)_L \times SU(3)_R$, T^μ transforms as

$$T^\mu_{abc} \to U_a^d\, U_b^e\, U_c^f\, T^\mu_{def}, \tag{20}$$

where U is defined in Eq. (9). The field T_v^μ is defined in terms of T^μ by

$$T_v^\mu(x) = e^{im_T \not v v_\mu x^\mu} T^\mu(x). \tag{21}$$

Spin operators S_v^ν can be defined which act on the spinor indices of T_v^μ, and satisfy the same spin algebra as the baryon spin operators Eq. (12). The spin operator S_v^ν is not the total angular momentum operator J_v^ν of the spin-3/2 particle. The total angular momentum operator J_v^ν, which is defined by

$$(J_v^\nu\, T_v)^\mu = S_v^\nu\, T_v^\mu + i\epsilon^{\mu\nu\alpha\beta} v_\alpha\, T_{v\beta}, \tag{22}$$

acts on both the spinor and vector indices, and satisfies

$$J_v^2\, T_v^\mu = -\tfrac{15}{4}\, T_v^\mu. \tag{23}$$

The constraint $\gamma_\mu T^\mu = 0$ implies that

$$v^\mu T_\mu = 0, \qquad S_v^\mu T_\mu = 0. \tag{24}$$

The propagator for the Rarita-Schwinger field contains a polarization projector which projects out the four physical positive energy spinor solutions to the equation of motion $\mathcal{U}_i^\mu$, $i = 1-4$. The polarization sum is

$$P_v^{\mu\nu} = \sum_i \mathcal{U}_i^\mu \overline{\mathcal{U}}_i^\nu = (v^\mu v^\nu - g^{\mu\nu}) - \tfrac{4}{3} S_v^\mu S_v^\nu. \tag{25}$$

The identities

$$\begin{aligned} &P_v^{\mu\nu} P_{v\nu}{}^{\lambda} = -P_v^{\mu\lambda}, \quad P_v^{\mu\nu} v_\nu = P_v^{\mu\nu} v_\mu = 0, \quad P_v^{\mu\nu} g_{\mu\nu} = -2, \\ &P_v^{\mu\nu} S_{v\,\nu} = S_{v\,\mu} P_v^{\mu\nu} = 0, \quad P_v^{\mu\nu} S_{v\,\mu} = -\tfrac{4}{3} S_v^\nu, \quad S_{v\nu} P^{\mu\nu} = -\tfrac{4}{3} S_v^\mu, \end{aligned} \tag{26}$$

are useful for the computation of Feynman diagrams.

The lowest order decuplet Lagrangian is

$$\begin{aligned} L_v^{10} = &-i\overline{T}_v^\mu (v\cdot\mathcal{D})\, T_{v\,\mu} + \Delta m\, \overline{T}_v^\mu\, T_{v\,\mu} + \mathcal{C}\left(\overline{T}_v^\mu A_\mu B_v + \overline{B}_v A_\mu T_v^\mu\right) \\ &+ 2\mathcal{H}\, \overline{T}_v^\mu S_{v\,\nu} A^\nu T_{v\,\mu} + c\,\overline{T}_v^\mu \left(\xi^\dagger M \xi^\dagger + \xi M \xi\right) T_{v\,\mu} - \bar\sigma \operatorname{Tr} M(\Sigma + \Sigma^\dagger) \overline{T}_v^\mu T_{v\,\mu}. \end{aligned} \tag{27}$$

The kinetic term for the decuplet has the opposite sign from that for the octet because the spinor solutions $\mathcal{U}^\mu$ are spacelike, so that $\mathcal{U}^2 < 0$. Eq. (27) uses a modified definition of T_v^μ, in which a factor of $\exp(im_B \not v\, v_\mu x^\mu)$ multiplies T^μ in Eq. (21). This definition avoids the introduction of factors of $\exp(i(m_T - m_B)\not v\, v_\mu x^\mu)$ into the Lagrangian in terms which contain both decuplet and octet fields. The decuplet Lagrangian thus contains an explicit decuplet mass term proportional to $\Delta m = m_T - m_B$ since only part of the decuplet mass was removed by the transformation to a velocity dependent field.

5. A Sample Calculation: $T \to B\pi$

The heavy baryon Lagrangian can now be used to compute Green's functions in the effective theory. The Feynman rules can be read off from Eqs. (16) and (27). The octet propagator is $i/(k \cdot v)$, the decuplet propagator is $iP_v^{\mu\nu}/(k \cdot v - \Delta m)$, the pion nucleon vertex is $k \cdot S_v$ times a Clebsch-Gordan coefficient, *etc.* Graphs are much easier to compute in the heavy baryon theory than in the traditional baryon chiral perturbation theory using four-component spinors, because the Dirac matrix structure has disappeared.

We compute the decay width $T \to B\pi$ using the heavy baryon Lagrangian. Instead of computing the decay width directly, we will instead compute the imaginary part of the Feynman diagram shown in fig. 1. The optical theorem implies that the desired decay rate is twice the imaginary part of fig. 1. The calculation illustrates several techniques which are useful in computing Feynman graphs in the heavy baryon theory. From Eq. (27) and Eq. (18), the $TB\pi$ vertex equals $Ck^\mu/\sqrt{2}f$ times a Clebsch-Gordan coefficient. The Feynman integral is

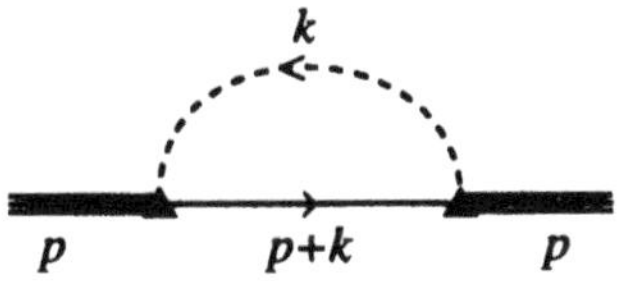

FIGURE 1.
The imaginary part of this graph is $\Gamma(T \to B\pi)/2$.

$$I = \frac{C^2}{2f^2} (\text{Clebsch})^2 \int \frac{d^D k}{(2\pi)^D} \left(\frac{i}{(k+p)\cdot v + i\bar{\epsilon}}\right) \left(\frac{i}{k^2 - M^2 + i\bar{\epsilon}}\right) \overline{T}_\mu (-k^\mu)(k^\nu) T_\nu, \tag{28}$$

where $\bar{\epsilon}$ has been used in the propagators, to avoid confusion with $\epsilon = 2 - D/2$, which is used in dimensional regularization. The two denominators can be combined using the identity

$$\frac{1}{ab} = 2\int_0^\infty \frac{d\lambda}{(a + 2b\lambda)^2}, \tag{29}$$

so that*

$$I = \frac{C^2}{f^2} (\text{Clebsch})^2 \int_0^\infty d\lambda \int \frac{d^D k}{(2\pi)^D} \frac{\overline{T}_\mu k^\mu k^\nu T_\nu}{(k^2 - M^2 + 2\lambda (k+p)\cdot v + i\bar{\epsilon})^2}. \tag{30}$$

---

* Since $\lambda \geq 0$, $\bar{\epsilon} + 2\lambda\bar{\epsilon}$ can be replaced by $\bar{\epsilon}$.

Shifting the momentum integral, using $v_\mu T^\mu = 0$ and $D = 4 - 2\epsilon$, gives

$$
\begin{aligned}
I &= \frac{\mathcal{C}^2}{f^2} (\text{Clebsch})^2 \overline{T}_\mu T^\mu \int_0^\infty d\lambda \int \frac{d^D k}{(2\pi)^D} \frac{k^2/D}{\left(k^2 - M^2 - \lambda^2 + 2\lambda p\cdot v + i\bar{\epsilon}\right)^2} \\
&= -\frac{i}{32\pi^2} \frac{\mathcal{C}^2}{f^2} (\text{Clebsch})^2 \; \overline{T}_\mu T^\mu \; \mu^{2\epsilon} \; \Gamma(-1+\epsilon) \int_0^\infty d\lambda \left(\lambda^2 - 2\lambda p\cdot v + M^2 - i\bar{\epsilon}\right)^{1-\epsilon} .
\end{aligned}
\tag{31}
$$

Let $\mathcal{I}(\alpha)$ be defined by

$$
\mathcal{I}(\alpha; b, c) \equiv \int_0^\infty d\lambda \left(\lambda^2 + 2\lambda b + c\right)^\alpha . \tag{32}
$$

The recurrence relation

$$
(1+2\alpha)\,\mathcal{I}(\alpha; b, c) = \left(\lambda^2 + 2\lambda b + c\right)^\alpha (\lambda + b) \Big|_0^\infty + 2\alpha\left(c - b^2\right)\mathcal{I}(\alpha - 1; b, c) , \tag{33}
$$

can be derived by integration by parts. In dimensional regularization,

$$
\lim_{\lambda\to\infty} \lambda^{1-\epsilon}, \lambda^{-\epsilon} = 0, \tag{34}
$$

so Eq. (33) relates $\mathcal{I}(1-\epsilon; -p\cdot v, M^2)$ in Eq. (31) to $\mathcal{I}(-1-\epsilon; -p\cdot v, M^2)$ which is finite as $\epsilon \to 0$,

$$
\begin{aligned}
\mathcal{I}(1-\epsilon; -p\cdot v, M^2) &= \frac{1}{3-2\epsilon}\left[p\cdot v\left(M^2\right)^{1-\epsilon} + 2(1-\epsilon)\left(M^2 - (p\cdot v)^2\right)\mathcal{I}(-\epsilon; -p\cdot v, M^2)\right] \\
\mathcal{I}(-\epsilon; -p\cdot v, M^2) &= \frac{1}{1-2\epsilon}\left[p\cdot v\left(M^2\right)^{-\epsilon} + 2(-\epsilon)\left(M^2 - (p\cdot v)^2\right)\mathcal{I}(-1-\epsilon; -p\cdot v, M^2)\right] .
\end{aligned}
\tag{35}
$$

We need the imaginary part of the integral Eq. (31) to compute the decay width $T \to B\pi$. The only term with an imaginary part is $\mathcal{I}(-1-\epsilon; -p\cdot v, M^2)$. Since this integral is finite, we can set ϵ to zero, and evaluate it for an on-shell decuplet baryon T in the rest frame, so $p = (\Delta m, 0, 0, 0)$, $p\cdot v = \Delta m$, and

$$
\begin{aligned}
\operatorname{Im}\mathcal{I}(-1; -p\cdot v, M^2) &= \operatorname{Im}\int_0^\infty \frac{d\lambda}{\left(\lambda^2 - 2\lambda\Delta m + M^2 - i\bar{\epsilon}\right)} \\
&= \operatorname{Im}\frac{1}{2\sqrt{(\Delta m)^2 - M^2}} \ln\left(\frac{\lambda - \Delta m - \sqrt{(\Delta m)^2 - M^2 - i\bar{\epsilon}}}{\lambda - \Delta m + \sqrt{(\Delta m)^2 - M^2 + i\bar{\epsilon}}}\right)\Bigg|_0^\infty \\
&= \frac{\pi}{\sqrt{(\Delta m)^2 - M^2}} .
\end{aligned}
\tag{36}
$$

Combining Eq. (36), Eq. (35), and Eq. (31) gives

$$\operatorname{Im} I = \tfrac{1}{2}\Gamma(T \to B\pi) = (\text{Clebsch})^2 \, \frac{\mathcal{C}^2 \left((\Delta m)^2 - M^2\right)^{3/2}}{24\pi f^2}. \tag{37}$$

The experimentally measured $T \to B\pi$ decays are $\Delta \to N\pi$, $\Sigma^* \to \Lambda\pi$, $\Sigma^* \to \Sigma\pi$, and $\Xi^* \to \Xi\pi$, with Clebsch-Gordan coefficients 1, $1/\sqrt{2}$, $1/\sqrt{3}$, and $1/\sqrt{2}$ respectively. The coefficients $\mathcal{C}$ determined from these decay modes[7] using Eq. (37) are 1.8, 1.5, 1.5, and 1.3 respectively, with an average value of

$$|\mathcal{C}| = 1.53. \tag{38}$$

The differences in the four numbers reflect $SU(3)$ breaking effects.

6. Why the Decuplet is Important

The effective chiral theory with octet and decuplet baryons has three scales, the pseudo-Goldstone mass M, the chiral symmetry breaking scale Λ_χ, and the decuplet-octet mass difference Δm. In the chiral limit, $M \ll \Delta m$, so one can construct an effective theory with only octet baryons and pions by integrating out the decuplet fields. Decuplet effects in the original theory are reproduced in the new theory by higher dimension operators involving only octet baryons and pions. Higher dimension operators obtained by integrating out the decuplet are suppressed by powers of $\mathcal{C}^2/\Delta m$ (where $\mathcal{C}$ is the $TB\pi$ coupling), whereas higher dimension operators due to other QCD effects are suppressed by $1/\Lambda_\chi$. In the real world, the $\Delta - N$ mass difference is approximately 300 MeV, and the $TB\pi$ coupling is approximately 1.5, as determined in Eq. (38). Thus decuplet contributions are expected to be 6–7 times more important than other higher dimension operators in the chiral theory. This enhancement of decuplet effects occurs because the $T - B$ mass difference is much smaller than the intrinsic hadronic scale of 1 GeV. Since the decuplet effects are so important, it is best to retain explicit decuplet fields in the effective theory, rather than integrate them out.

There are higher baryon resonances such as the $N(1440)$, which is only 500 MeV above the $N(939)$. These higher resonances are not as important as the decuplet. The $N(1440) \to N\pi$ width determines the $N(1440)N\pi$ coupling constant. One can estimate that the effect of the $N(1440)$ on octet baryon amplitudes is only about 10% that of the $\Delta(1232)$. The large suppression factor is due to the larger mass of the $N(1440)$, and the much smaller $N(1440)N\pi$ coupling. The size of the $N(1440)$ contribution is comparable to the estimate of higher dimension effects. Thus all higher resonances other than the decuplet do not have to be included explicitly; their effects can be mimicked by higher dimension operator which satisfy naive dimensional analysis.[11]

The above conclusions are not surprising if one believes that the quark model provides a reasonable qualitative picture of baryons. The octet and decuplet baryons have essentially identical wave functions, differing only in the arrangement of their spins. The higher resonances, on the other hand, differ in their spatial wavefunctions. A spin-flip does not cost very much energy, because the hyperfine spin-spin interaction is rather weak. Thus it is relatively easy for an octet baryon to be converted into a decuplet baryon, but it is more difficult to convert it into other excited states.

In the calculations described here, the decuplet-octet mass difference is treated as small compared with the kaon mass M_K. All computation are performed treating Δm as a perturbation. The chiral limit in which M_K is small compared with Δm is discussed in more detail in Section 11.

7. SU(6) Symmetry and Heavy Baryons

The heavy baryon chiral theory has octet and decuplet fields labeled by a velocity vector v. One can define an $SU(6)_v$ symmetry transformation which rotates an octet baryon B_v into a decuplet baryon T_v with the same velocity. There is an independent $SU(6)$ transformation for each v, and B_v and T_v together form a **56** of $SU(6)_v$. The **56** can be described by a tensor $Q_v^{abc;\alpha\beta\gamma}$ which has three flavor indices and three spinor indices, and is symmetric under the simultaneous permutation of spinor and flavor indices. T_v is the part of Q_v which is completely symmetric in flavor and spin, and B_v has mixed symmetry under flavor and spin. The tensor Q_v only has two non-zero components for each spinor index, because it satisfies the projection relation Eq. (10) on each spinor index. What we have constructed is the decomposition of the **56** of $SU(6)$ into $(\mathbf{8},\mathbf{2})+(\mathbf{10},\mathbf{4})$ of $SU(3)\times SU(2)$ under the embedding $\mathbf{6}\to(\mathbf{3},\mathbf{2})$. The symmetry can be written in a relativistic theory because there are velocity dependent fields. The non-relativistic $SU(6)$ symmetry for velocity dependent fields has also been discussed in a recent paper by Carone and Georgi.[13]

There does not seem to be any way to implement the $SU(6)_v$ transformation on the pion fields. The field A_μ defined in Eq. (18) is an $SU(3)$ octet, and has spin one. If the field is treated as part of the **35** of $SU(6)$, then one obtains the $SU(6)_v$ symmetry relations[14]

$$F=\tfrac{2}{3}D,\quad \mathcal{C}=-2D,\quad \mathcal{H}=-3D, \tag{39}$$

for the various pion-baryon coupling constants. One could also obtain these relations by using non-relativistic quark model wavefunctions for the baryons. As we will see, the best fit values for these constants is very close to this $SU(6)$ prediction. Thus there appears to be some evidence that, in some sense, the real world is close to the $SU(6)$ point of Eq. (39).

The field A_μ couples to baryon fields with different velocities, so that the $SU(6)_v$ transformation must be the same for all values of v. It is not clear how to extend the $SU(6)$ symmetry further, or how to study the symmetry breaking in a systematic manner.

8. The Axial Currents

The baryon axial vector currents in the $SU(3)$ symmetry limit to lowest order in the derivative expansion are

$$\begin{aligned} J_\mu^A &= D \operatorname{Tr} \overline{B}_v S_v^\mu \{\xi T^A \xi^\dagger + \xi^\dagger T^A \xi, B_v\} + F \operatorname{Tr} \overline{B}_v S_v^\mu [\xi T^A \xi^\dagger + \xi^\dagger T^A \xi, B_v] \\ &+ \tfrac{1}{2} v^\mu \operatorname{Tr} \overline{B}_v [\xi T^A \xi^\dagger - \xi^\dagger T^A \xi, B_v] + i \tfrac{f^2}{2} \operatorname{Tr} T^A \left(\partial^\mu \Sigma^\dagger \Sigma - \partial^\mu \Sigma \Sigma^\dagger\right) \\ &+ \tfrac{1}{2} v^\mu \, \overline{T}_v^\nu (\xi T^A \xi^\dagger - \xi^\dagger T^A \xi) T_{v\nu} + \mathcal{H} \, \overline{T}_v^\nu S_v^\mu (\xi T^A \xi^\dagger + \xi^\dagger T^A \xi) T_{v\nu} \\ &+ \tfrac{1}{2} C \, \overline{T}_{v\mu} (\xi T^A \xi^\dagger + \xi^\dagger T^A \xi) B_v + \tfrac{1}{2} C \, \overline{B}_v (\xi T^A \xi^\dagger + \xi^\dagger T^A \xi) T_{v\mu}. \end{aligned} \tag{40}$$

The leading non-analytic correction to the axial currents is a chiral logarithmic correction of the form $m_s \ln m_s$ produced by one-loop Feynman graphs which can be found in Refs. 7 and 14. The correction to the axial currents can be written in the form

$$\langle B_i | \, J_\mu^A \, | B_j \rangle = \left(\alpha_{ij}^A + \left(\overline{\beta}_{ij}^A - \overline{\lambda}_{ij} \alpha_{ij}^A \right) \frac{M_K^2}{16\pi^2 f^2} \ln \left(M_K^2/\mu^2 \right) \right) \overline{u}_{B_i} \, \gamma_\mu \gamma_5 \, u_{B_j}, \tag{41}$$

where α_{ij}^A is the lowest order result, $\overline{\lambda}_{ij} = \lambda_{ij} + \lambda'_{ij}$ is the one-loop correction due to wavefunction renormalization,

$$\sqrt{Z_i Z_j} = 1 + \overline{\lambda}_{ij} \frac{M_K^2}{16\pi^2 f^2} \ln \left(M_K^2/\mu^2 \right), \quad \overline{\lambda}_{ij} = \tfrac{1}{2} \left(\overline{\lambda}_i + \overline{\lambda}_j \right),$$

$\overline{\beta}_{ij}^A = \beta_{ij}^A + \beta'^A_{ij}$ is the correction due to all other graphs, and u is a spinor. All coefficients are written in the form $\bar{c} = c + c'$. The total correction $\bar{c}$ equals the sum of the correction from graphs with intermediate octet lines (c) and the correction from graphs with intermediate decuplet lines (c'). Thus one can determine the correction in a theory without any decuplet fields by dropping all c' type corrections. Corrections are computed for the currents that occur in hyperon semileptonic decay, as well as for the proton matrix element of the T^8 current, which is needed for an analysis of the proton spin problem.[15] The matrix element of the current (41) depends on μ because Eq. (41) contains only the leading non-analytic term in m_s. There are also M_K^2 terms with arbitrary coefficients which come from higher dimension operators in the chiral Lagrangian. The μ dependence of Eq. (41) is equal to the anomalous dimensions of these coefficients, so that the total μ dependence cancels.

The lowest order coefficients α_{ij}^A are:

$$\begin{aligned} \alpha_{pn}^{1+i2} &= (D+F), & \alpha_{\Lambda\Xi^-}^{4+i5} &= -\tfrac{1}{\sqrt{6}}(D-3F), \\ \alpha_{\Lambda\Sigma^-}^{1+i2} &= \tfrac{2}{\sqrt{6}} D, & \alpha_{n\Sigma^-}^{4+i5} &= (D-F), \\ \alpha_{\Xi^0\Xi^-}^{1+i2} &= (D-F), & \alpha_{\Sigma^0\Xi^-}^{4+i5} &= \tfrac{1}{\sqrt{2}}(D+F) = \tfrac{1}{\sqrt{2}} \alpha_{\Sigma^+\Xi^0}^{4+i5}, \\ \alpha_{p\Lambda}^{4+i5} &= -\tfrac{1}{\sqrt{6}}(D+3F), & \alpha_{pp}^{8} &= \tfrac{1}{\sqrt{12}}(3F-D). \end{aligned} \tag{42}$$

The coefficients β^A_{ij} are:

$$
\begin{aligned}
\beta^{1+i2}_{pn} &= \tfrac{2}{9}D^3 + \tfrac{2}{9}D^2F + \tfrac{2}{3}DF^2 - 2F^3 - \tfrac{1}{2}D - \tfrac{1}{2}F, \\
\beta^{1+i2}_{\Lambda\Sigma^-} &= \tfrac{1}{\sqrt{6}}\left(\tfrac{17}{9}D^3 - DF^2 - D\right), \\
\beta^{1+i2}_{\Xi^0\Xi^-} &= \tfrac{2}{9}D^3 - \tfrac{2}{9}D^2F + \tfrac{2}{3}DF^2 + 2F^3 - \tfrac{1}{2}D + \tfrac{1}{2}F, \\
\beta^{4+i5}_{p\Lambda} &= \tfrac{1}{\sqrt{6}}\left(\tfrac{19}{18}D^3 - \tfrac{5}{2}D^2F - \tfrac{7}{2}DF^2 + \tfrac{9}{2}F^3 + \tfrac{5}{4}D + \tfrac{15}{4}F\right), \\
\beta^{4+i5}_{\Lambda\Xi^-} &= \tfrac{1}{\sqrt{6}}\left(\tfrac{19}{18}D^3 + \tfrac{5}{2}D^2F - \tfrac{7}{2}DF^2 - \tfrac{9}{2}F^3 + \tfrac{5}{4}D - \tfrac{15}{4}F\right), \\
\beta^{4+i5}_{n\Sigma^-} &= \tfrac{7}{18}D^3 - \tfrac{13}{18}D^2F + \tfrac{7}{6}DF^2 + \tfrac{1}{2}F^3 - \tfrac{5}{4}D + \tfrac{5}{4}F, \\
\beta^{4+i5}_{\Sigma^0\Xi^-} &= \tfrac{1}{\sqrt{2}}\left(\tfrac{7}{18}D^3 + \tfrac{13}{18}D^2F + \tfrac{7}{6}DF^2 - \tfrac{1}{2}F^3 - \tfrac{5}{4}D - \tfrac{5}{4}F\right) = \frac{1}{\sqrt{2}}\beta^{4+i5}_{\Sigma^+\Xi^0}, \\
\beta^{8}_{pp} &= \tfrac{1}{\sqrt{12}}\left(-\tfrac{11}{9}D^3 + 3D^2F + 3DF^2 - 3F^3 - \tfrac{9}{2}F + \tfrac{3}{2}D\right),
\end{aligned}
\tag{43}
$$

and the coefficients β'^A_{ij} are:

$$
\begin{aligned}
\beta'^{1+i2}_{pn} &= -\tfrac{10}{81}\mathcal{H}\mathcal{C}^2 + \left(\tfrac{2}{3}D + \tfrac{2}{9}F\right)\mathcal{C}^2, \\
\beta'^{1+i2}_{\Lambda\Sigma^-} &= -\tfrac{5}{27\sqrt{6}}\mathcal{H}\mathcal{C}^2 + \tfrac{8}{3\sqrt{6}}(D+F)\mathcal{C}^2, \\
\beta'^{1+i2}_{\Xi^0\Xi^-} &= \tfrac{20}{81}\mathcal{H}\mathcal{C}^2 + \left(\tfrac{14}{27}D + 2F\right)\mathcal{C}^2, \\
\beta'^{4+i5}_{p\Lambda} &= \tfrac{5}{9\sqrt{6}}\mathcal{H}\mathcal{C}^2 - \tfrac{1}{\sqrt{6}}(D+F)\mathcal{C}^2, \\
\beta'^{4+i5}_{\Lambda\Xi^-} &= -\tfrac{5}{9\sqrt{6}}\mathcal{H}\mathcal{C}^2 + \tfrac{1}{\sqrt{6}}\left(\tfrac{34}{18}D - F\right)\mathcal{C}^2, \\
\beta'^{4+i5}_{n\Sigma^-} &= \tfrac{5}{81}\mathcal{H}\mathcal{C}^2 + \left(-\tfrac{1}{27}D + F\right)\mathcal{C}^2, \\
\beta'^{4+i5}_{\Sigma^0\Xi^-} &= -\tfrac{55}{81\sqrt{2}}\mathcal{H}\mathcal{C}^2 + \tfrac{1}{\sqrt{2}}\left(\tfrac{19}{9}D + \tfrac{17}{9}F\right)\mathcal{C}^2 = \tfrac{1}{\sqrt{2}}\beta'^{4+i5}_{\Sigma^+\Xi^0}. \\
\beta'^{8}_{pp} &= \tfrac{1}{\sqrt{3}}(D-F)\mathcal{C}^2.
\end{aligned}
\tag{44}
$$

The wavefunction renormalization coefficients $\overline{\lambda}_i$ are given in Section 9.

The chiral corrections to the axial octet currents were computed by Bijnens, Sonoda and Wise (BSW).[16] The above results differ from BSW because decuplet contributions have been included, and also because BSW inadvertently omitted wavefunction renormalization in their calculations. Wavefunction renormalization is the biggest part of the chiral correction, and hence makes a significant difference in a fit to the experimental results. A fit to the experimental data neglecting all chiral corrections gives[15] $D = 0.80$ and $F = 0.50$. The chiral corrections are large ($\gtrsim$ 100%) if one only includes contributions from loop graphs involving internal octet field propagators. The experimental results have very small $SU(3)$ breaking effects. Thus chiral corrections including only octet chiral logarithms are in conflict with experiment. The situation is rather different if one includes both octet and decuplet graphs. There is a significant cancellation between octet and decuplet chiral logarithms, so that the $SU(3)$ breaking effects are much smaller. There are two new parameters when decuplets are included, $\mathcal{C}$ and $\mathcal{H}$. $\mathcal{C}$ was determined from $T \to B\pi$ decay, Eq. (38). There is therefore only one additional parameter, $\mathcal{H}$, for the decuplet fields. The best fit to the axial currents gives[14]

$$F = 0.40 \pm 0.03, \quad D = 0.61 \pm 0.04, \quad \mathcal{H} = -1.91 \pm 0.7. \tag{45}$$

$\mathcal{H}$ is not determined very well since it only occurs in the loop correction. The chiral corrections including the decuplet are all less than 30%, with the exception of $\Sigma \to n$ (67%) and $\Lambda \to p$ (46%). The theoretical and experimental results agree to within errors. Note that the values Eq. (45) and Eq. (38) approximately satisfy the $SU(6)$ relations Eq. (39).

9. Baryon Masses

Leading non-analytic corrections to baryon octet and decuplet masses[17] are calculated using Lagrangians L_v and L_v^{10}. The one-loop formulæ for baryon masses can be written in the form

$$\begin{aligned} M_i =& \overline{m} - 2\overline{\sigma}m_s + \overline{\alpha}_i m_s - \overline{\beta}_i \frac{M_K^3}{16\pi f^2} + \left(\overline{\gamma}_i - \overline{\lambda}_i \overline{\alpha}_i\right) m_s \frac{M_K^2}{16\pi^2 f^2} \ln\left(M_K^2/\mu^2\right) \\ & \pm 2\lambda'_i(\sigma - \tilde{\sigma}) m_s \frac{M_K^2}{16\pi^2 f^2} \ln\left(M_K^2/\mu^2\right) \pm \lambda'_i \Delta m \frac{M_K^2}{16\pi^2 f^2} \ln\left(M_K^2/\mu^2\right), \end{aligned} \tag{46}$$

where plus and minus signs apply for octet and decuplet masses, respectively. The term $\overline{m}$ is the $SU(3)$ invariant mass of the baryon multiplet; $\overline{m} = m_B$ for octet baryons and $\overline{m} = m_T = m_B + \Delta m$ for decuplet baryons. The quantity $-2\overline{\sigma}m_s$ is the tree-level contribution to the baryon masses from sigma terms; $\overline{\sigma} = \sigma$ for octet baryons and $\overline{\sigma} = \tilde{\sigma}$ for decuplet baryons. The tree-level coefficients $\overline{\alpha}_i = \alpha_i$

$$\begin{aligned} \alpha_N &= -2(b_D - b_F), & \alpha_\Delta &= 0, \\ \alpha_\Sigma &= 0, & \alpha_{\Sigma^*} &= \tfrac{2}{3}c, \\ \alpha_\Lambda &= -\tfrac{8}{3}b_D, & \alpha_{\Xi^*} &= \tfrac{4}{3}c, \\ \alpha_\Xi &= -2(b_D + b_F), & \alpha_{\Omega^-} &= 2c, \end{aligned} \tag{47}$$

describe baryon mass splittings which are linear in m_s. Octet and decuplet masses depend on distinct parameters at tree level. The four octet masses are given in terms of the three parameters $(m_B - 2\sigma m_s)$, $b_D m_s$, and $b_F m_s$, whereas the four decuplet masses are written in terms of the two parameters $m_T - 2\tilde{\sigma} m_s$ and $c m_s$. The tree-level masses satisfy the baryon mass relations,

$$\tfrac{3}{4}M_\Lambda + \tfrac{1}{4}M_\Sigma - \tfrac{1}{2}\left(M_N + M_\Xi\right) = 0, \tag{48}$$

$$M_{\Sigma^*} - M_\Delta = M_{\Xi^*} - M_{\Sigma^*} = M_{\Omega^-} - M_{\Xi^*}. \tag{49}$$

These two mass relations, the Gell-Mann–Okubo formula for octet masses and the equal spacing rule for decuplet masses, are a direct consequence of the fact that $SU(3)$ symmetry breaking is purely octet.

Three non-analytic terms appear at one loop — contributions to the baryon masses which vary as $m_s^{3/2}$, $m_s^2 \ln m_s$, and $(\Delta m) m_s \ln m_s$. The $(\Delta m) m_s \ln m_s$ non-analytic mass contribution is a new term which arises only if octet and decuplet baryons are treated together in the chiral Lagrangian. There is no non-analytic correction proportional to $m_B m_s \ln m_s$ since m_B does not appear in the chiral Lagrangian for heavy baryon fields.

The coefficients $\overline{\beta}_i = \beta_i + \beta_i'$ parametrize the non-analytic $m_s^{3/2}$ contribution:

$$\begin{aligned}
\beta_N &= (\tfrac{5}{3}D^2 - 2DF + 3F^2) + \tfrac{4}{9\sqrt{3}}(D^2 - 6DF + 9F^2), & \beta_N' &= \tfrac{1}{3}\mathcal{C}^2,\\
\beta_\Sigma &= 2(D^2 + F^2) + \tfrac{16}{9\sqrt{3}}D^2, & \beta_\Sigma' &= (\tfrac{10}{9} + \tfrac{8}{9\sqrt{3}})\mathcal{C}^2,\\
\beta_\Lambda &= \tfrac{2}{3}D^2 + 6F^2 + \tfrac{16}{9\sqrt{3}}D^2, & \beta_\Lambda' &= \tfrac{2}{3}\mathcal{C}^2,\\
\beta_\Xi &= (\tfrac{5}{3}D^2 + 2DF + 3F^2) + \tfrac{4}{9\sqrt{3}}(D^2 + 6DF + 9F^2), & \beta_\Xi' &= (1 + \tfrac{8}{9\sqrt{3}})\mathcal{C}^2,
\end{aligned} \tag{50}$$

$$\begin{aligned}
\beta_\Delta &= (\tfrac{5}{27} + \tfrac{20}{81\sqrt{3}})\mathcal{H}^2, & \beta_\Delta' &= \tfrac{1}{3}\mathcal{C}^2,\\
\beta_{\Sigma^*} &= \tfrac{40}{81}\mathcal{H}^2, & \beta_{\Sigma^*}' &= (\tfrac{2}{9} + \tfrac{4}{9\sqrt{3}})\mathcal{C}^2,\\
\beta_{\Xi^*} &= (\tfrac{5}{9} + \tfrac{20}{81\sqrt{3}})\mathcal{H}^2, & \beta_{\Xi^*}' &= (\tfrac{1}{3} + \tfrac{4}{9\sqrt{3}})\mathcal{C}^2,\\
\beta_{\Omega^-} &= (\tfrac{10}{27} + \tfrac{80}{81\sqrt{3}})\mathcal{H}^2, & \beta_{\Omega^-}' &= \tfrac{2}{3}\mathcal{C}^2.
\end{aligned} \tag{51}$$

Wavefunction renormalization

$$Z_i = 1 + \overline{\lambda}_i \frac{M_K^2}{16\pi^2 f^2} \ln\left(M_K^2/\mu^2\right)$$

determines the coefficients $\overline{\lambda}_i = \lambda_i + \lambda_i'$,

$$\begin{aligned}
\lambda_N &= \tfrac{17}{6}D^2 - 5DF + \tfrac{15}{2}F^2, & \lambda_N' &= \tfrac{1}{2}\mathcal{C}^2,\\
\lambda_\Sigma &= \tfrac{13}{3}D^2 + 3F^2, & \lambda_\Sigma' &= \tfrac{7}{3}\mathcal{C}^2,\\
\lambda_\Lambda &= \tfrac{7}{3}D^2 + 9F^2, & \lambda_\Lambda' &= \mathcal{C}^2,\\
\lambda_\Xi &= \tfrac{17}{6}D^2 + 5DF + \tfrac{15}{2}F^2, & \lambda_\Xi' &= \tfrac{13}{6}\mathcal{C}^2,
\end{aligned} \tag{52}$$

$$\begin{aligned}
\lambda_\Delta &= \tfrac{25}{54}\mathcal{H}^2, & \lambda_\Delta' &= \tfrac{1}{2}\mathcal{C}^2,\\
\lambda_{\Sigma^*} &= \tfrac{20}{27}\mathcal{H}^2, & \lambda_{\Sigma^*}' &= \tfrac{2}{3}\mathcal{C}^2,\\
\lambda_{\Xi^*} &= \tfrac{55}{54}\mathcal{H}^2, & \lambda_{\Xi^*}' &= \tfrac{5}{6}\mathcal{C}^2,\\
\lambda_{\Omega^-} &= \tfrac{35}{27}\mathcal{H}^2, & \lambda_{\Omega^-}' &= \mathcal{C}^2.
\end{aligned} \tag{53}$$

The coefficients $\overline{\beta}$ are determined by the same diagrams as wavefunction renormalization coefficients $\overline{\lambda}$. These coefficients are equal to Clebsch-Gordan coefficients times a power of the Goldstone boson mass. The wavefunction renormalization coefficient is multiplied by M^2, whereas the $m_s^{3/2}$ contribution is multiplied by M^3. Thus the η and K contributions to $\overline{\beta}$ and $\overline{\lambda}$ are proportional to each other, except that the η contribution to $\overline{\beta}$ has an additional factor of $\sqrt{4/3}$, because $M_\eta^2 = 4M_K^2/3$. The constant of proportionality is determined by explicit computation to be 2/3. Thus

$$\overline{\beta} = \tfrac{2}{3}\left(\overline{\lambda}(K) + \tfrac{2}{\sqrt{3}}\overline{\lambda}(\eta)\right) = \tfrac{2}{3}\left(\overline{\lambda} + (\tfrac{2}{\sqrt{3}} - 1)\overline{\lambda}(\eta)\right),$$

where $\overline{\lambda} = \overline{\lambda}(K) + \overline{\lambda}(\eta)$ equals the sum of contributions from diagrams with K and η exchange. The coefficients $\overline{\gamma}_i = \gamma_i + \gamma_i'$ are:

$$\begin{aligned}
\gamma_N &= \tfrac{43}{9}b_D - \tfrac{25}{9}b_F - b_D(\tfrac{4}{3}D^2 + 12F^2) + b_F(\tfrac{2}{3}D^2 - 4DF + 6F^2) + \tfrac{52}{9}\sigma, \\
\gamma_\Sigma &= 2b_D - b_D(6D^2 + 6F^2) - b_F(12DF) + \tfrac{52}{9}\sigma, \\
\gamma_\Lambda &= \tfrac{154}{27}b_D - b_D(\tfrac{50}{9}D^2 + 18F^2) + b_F(12DF) + \tfrac{52}{9}\sigma, \\
\gamma_\Xi &= \tfrac{43}{9}b_D + \tfrac{25}{9}b_F - b_D(\tfrac{4}{3}D^2 + 12F^2) - b_F(\tfrac{2}{3}D^2 + 4DF + 6F^2) + \tfrac{52}{9}\sigma,
\end{aligned} \tag{54}$$

$$\gamma_N' = \tfrac{1}{3}c\mathcal{C}^2, \quad \gamma_\Sigma' = \tfrac{8}{9}c\mathcal{C}^2, \quad \gamma_\Lambda' = \tfrac{4}{3}c\mathcal{C}^2, \quad \gamma_\Xi' = \tfrac{29}{9}c\mathcal{C}^2, \tag{55}$$

$$\begin{aligned}
\gamma_\Delta &= -c + \tfrac{5}{27}c\mathcal{H}^2 + \tfrac{52}{9}\tilde{\sigma}, & \gamma_\Delta' &= 0, \\
\gamma_{\Sigma^*} &= -\tfrac{52}{27}c + \tfrac{40}{81}c\mathcal{H}^2 + \tfrac{52}{9}\tilde{\sigma}, & \gamma_{\Sigma^*}' &= -\tfrac{2}{3}b_D\mathcal{C}^2, \\
\gamma_{\Xi^*} &= -\tfrac{77}{27}c + \tfrac{95}{81}c\mathcal{H}^2 + \tfrac{52}{9}\tilde{\sigma}, & \gamma_{\Xi^*}' &= -(\tfrac{4}{3}b_D + \tfrac{2}{3}b_F)\mathcal{C}^2, \\
\gamma_{\Omega^-} &= -\tfrac{34}{9}c + \tfrac{20}{9}c\mathcal{H}^2 + \tfrac{52}{9}\tilde{\sigma}, & \gamma_{\Omega^-}' &= -2(b_D + b_F)\mathcal{C}^2.
\end{aligned} \tag{56}$$

The one-loop formulæ for the baryon masses depend on five parameters: $m_B - 2\sigma m_s + \frac{52}{9}\sigma m_s x$, $m_T - 2\tilde{\sigma} m_s + \frac{52}{9}\tilde{\sigma} m_s x$, $b_D m_s$, $b_F m_s$, and $c m_s$, where $x = (M_K^2/16\pi f^2)\ln M_K^2/\mu^2$. A sixth parameter $\Delta m + 2(\sigma - \tilde{\sigma})m_s$ is equal to the difference of the first two parameters up to terms which are higher order in the strange quark mass expansion.

The corrections to the mass relations Eqs. (48), (49) are

$$\begin{aligned}
\tfrac{3}{4}M_\Lambda + \tfrac{1}{4}M_\Sigma - \tfrac{1}{2}(M_N + M_\Xi) &= \left(\tfrac{2}{3}(D^2 - 3F^2) - \tfrac{1}{9}\mathcal{C}^2\right)\left(1 - \tfrac{2}{\sqrt{3}}\right)\frac{M_K^3}{16\pi f^2} \\
&\quad - \left(\tfrac{16}{3}b_D D^2 + \left(\tfrac{2}{3}b_D + \tfrac{5}{3}b_F\right)\mathcal{C}^2 + \tfrac{5}{9}c\mathcal{C}^2\right) m_s \frac{M_K^2}{16\pi^2 f^2}\ln\left(M_K^2/\mu^2\right)
\end{aligned} \tag{57}$$

$$\begin{aligned}
&(M_{\Sigma^*} - M_\Delta) - (M_{\Xi^*} - M_{\Sigma^*}) = (M_{\Xi^*} - M_{\Sigma^*}) - (M_{\Omega^-} - M_{\Xi^*}) \\
&= \left(\tfrac{2}{9}\mathcal{C}^2 - \tfrac{20}{81}\mathcal{H}^2\right)\left(1 - \tfrac{2}{\sqrt{3}}\right)\frac{M_K^3}{16\pi f^2} + \left(\tfrac{2}{9}\mathcal{C}^2 c + \tfrac{2}{3}\mathcal{C}^2 b_F\right) m_s \frac{M_K^2}{16\pi^2 f^2}\ln\left(M_K^2/\mu^2\right).
\end{aligned} \tag{58}$$

The correction to the GMO formula (with only octet diagrams) was computed previously by BSW.[16] Corrections to the GMO formula and the equal spacing rule depend only on the non-singlet and non-octet portions of the chiral corrections. Wavefunction coefficients λ_i and λ_i' which are purely singlet and octet automatically satisfy the mass relations, and hence all terms proportional to σ and $\tilde{\sigma}$ and the logarithmic mass contribution proportional to Δm cancel out of the mass relations. In addition, to $O(m_s^2)$, differences of decuplet mass splittings are simply related, so that there is a new mass relation

$$\tfrac{1}{2}\left(M_{\Sigma^*} - M_{\Delta}\right) - \left(M_{\Xi^*} - M_{\Sigma^*}\right) + \tfrac{1}{2}\left(M_{\Omega^-} - M_{\Xi^*}\right) = 0, \tag{59}$$

which is valid upto second order in symmetry breaking, including all leading non-analytic corrections.

The quantity $M_K^3/16\pi f^2 \approx 279$ MeV is numerically quite large. This term, however, always appears with the suppression factor $(2/\sqrt{3} - 1)$ in the baryon mass relations, since the coefficients $\overline{\beta}_i$ would be proportional to the wave-function coefficients $\overline{\lambda}_i$ if the contribution of η exchange graphs did not receive an extra factor of $2/\sqrt{3}$ from the M^3 behavior of the graph. The quantity $(2/\sqrt{3} - 1)(M_K^3/16\pi f^2) \approx 43$ MeV is the generic correction expected from the M_K^3 piece. The chiral logarithm $(M_K^2/16\pi^2 f^2)\ln\left(M_K^2/\mu^2\right)$ multiplies baryon mass differences which are typically of order (100 MeV), and hence this term should give corrections of order 25 MeV. Finally, the calculation of chiral corrections to the mass relations ignores corrections of order m_s^2, whose expected magnitude is $M_K^2/16\pi^2 f^2$ times a baryon mass difference of $O(100$ MeV), or ≈ 20 MeV. Thus, our calculation can only be trusted at the 20 MeV level. Experimentally, the GMO formula works to a precision of 6.5 MeV. Using the best fit values of the parameters $D = 0.61$, $F = 0.40$, obtained from a fit to baryon octet semileptonic weak decays with $|\mathcal{C}| = 1.6$ as determined previously, the $M_K^3/16\pi f^2$ correction is numerically ≈ 15.4 MeV and the logarithmic correction is numerically $\approx$ 3.8 MeV. The total theoretical correction is less than 20 MeV and agrees with experiment to within the accuracy of the theoretical calculation. The relation (59) is satisfied to -3.6 MeV experimentally, while the three differences of mass differences are measured to be 2.6 MeV, 9.8 MeV, and 12.4 MeV, respectively. The calculated M_K^3 and logarithmic corrections to the equal spacing rule are 13.9 MeV and 9.3 MeV for the parameter values $D = 0.61$, $F = 0.40$, $|\mathcal{C}| = 1.6$, and $\mathcal{H} = -1.9$. Agreement with experiment to within 20 MeV is again obtained.

The M_K^3 contribution to the baryon masses is only suppressed by the factor $(2/\sqrt{3} - 1)$ in the mass relations Eqs. (48) and (49). There is no such suppression factor in the contribution to the individual mass terms. The contribution of the M_K^3 term is ≈ 279 MeV times a Clebsch-Gordan coefficient. The Clebsch-Gordan coefficients can be quite large, so that the M_K^3 term contributes as much as -1 GeV to some of the baryon masses. However, we have already seen that the dominant contribution of the M_K^3 term is $SU(3)$ singlet and octet. The **27** piece, which contributes to the GMO formula and the equal spacing rule, is suppressed by $(2/\sqrt{3}-1)$, and is small. Thus there is a large non-linear m_s dependence in the baryon masses which is purely $SU(3)$ singlet or octet, and therefore does not affect the mass relations. The large $m_s^{3/2}$ term does not invalidate the chiral expansion in m_s. The proton matrix element $\langle p|\, m_s \bar{s}s\, |p\rangle$, obtained by differentiating the proton mass with respect to m_s, gets a large contribution from the $m_s^{3/2}$ term so that

$$\langle p|\, m_s \bar{s}s\, |p\rangle = m_s \frac{\partial m_p}{\partial m_s} \approx 0 \pm 150 \text{ MeV} \tag{60}$$

instead of the usual value quoted of around 411 MeV. A more detailed discussion can be found in Ref. 18.

10. Hyperon Non-leptonic Decays

According to the $\Delta I = \frac{1}{2}$ rule, $\Delta S = 1$ nonleptonic weak decays are dominated by an interaction which transforms as $(8,1)$ under $SU(3)_L \times SU(3)_R$. This component is represented in the chiral Lagrangian by

$$\begin{aligned} L_v^{\Delta S=1} &= h_D \operatorname{Tr} \overline{B}_v \left\{ \xi^\dagger h \xi, B_v \right\} + h_F \operatorname{Tr} \overline{B}_v \left[\xi^\dagger h \xi, B_v \right] \\ &\quad + h_C \overline{T}_v^\mu \left(\xi^\dagger h \xi \right) T_{v\,\mu} + h_\pi \frac{f^2}{4} \operatorname{Tr} h \partial_\mu \Sigma \partial^\mu \Sigma^\dagger \end{aligned} \tag{61}$$

to leading order in the derivative and quark mass expansions. The $s \to d$ transition is specified by the matrix

$$h = \begin{pmatrix} 0 & 0 & 0 \\ 0 & 0 & 1 \\ 0 & 0 & 0 \end{pmatrix}. \tag{62}$$

The Lagrangian $L_v^{\Delta S=1}$ contains three undetermined parameters h_D, h_F, and h_C which have dimensions of mass. The parameter h_π is determined from $\Delta S = 1$ K decays, $h_\pi = 3.2 \times 10^{-7}$. Although formally higher order in the derivative expansion, the term proportional to h_π is included because it produces corrections to the hyperon decay amplitudes which are order one compared to the effects of h_D, h_F and h_C, since h_π is over an order of magnitude larger than its expected value based on dimensional analysis.

The heavy baryon calculation of hyperon nonleptonic decay produces decay amplitudes of the form

$$\mathcal{A}(B_i \to B_j \pi) = G_F m_{\pi^+}^2 \, \overline{u}_{B_j} \left\{ \mathcal{A}_{ij}^{(S)} + 2(k \cdot S_v) \mathcal{A}_{ij}^{(P)} \right\} u_{B_i}, \tag{63}$$

where k is the outgoing momentum of the pion. The decay amplitude reduces to the non-relativistic amplitude

$$\mathcal{A}(B_i \to B_j \pi) = G_F m_{\pi^+}^2 \, \overline{u}_{B_j} \left\{ \mathcal{A}_{ij}^{(S)} + |\vec{k}|(\hat{k} \cdot \vec{\sigma}) \mathcal{A}_{ij}^{(P)} \right\} u_{B_i}, \tag{64}$$

in the rest frame of the heavy baryon where $v^\mu = (1,0,0,0)$ and $S^\mu_{v=0} = (0,\frac{1}{2}\vec{\sigma})$. The dimensionless amplitudes $s = \mathcal{A}^{(S)}_{ij}$ and $p = -|\vec{k}|\mathcal{A}^{(P)}_{ij}$ are order one in both the m_B and m_s expansions and are the same amplitudes s and p which are used to define the decay asymmetry parameters α, β, and γ. The heavy baryon P-wave amplitude p is related to the conventional amplitude by

$$p = \frac{|\vec{k}|\mathrm{A}^{(P)}_{ij}}{(E_j + M_j)} = \left\{\frac{(M_i - M_j)^2 - m_\pi^2}{(M_i + M_j)^2 - m_\pi^2}\right\}^{1/2} \mathrm{A}^{(P)}_{ij} \tag{65}$$

where E_j is the energy of the final baryon and

$$|\vec{k}| = \frac{1}{2M_i}\left\{\left[(M_i + M_j)^2 - m_\pi^2\right]\left[(M_i - M_j)^2 - m_\pi^2\right]\right\}^{1/2}. \tag{66}$$

The conventional definition of the P-wave amplitude $\mathrm{A}^{(P)}$ involves multiplication of p by an unnatural factor which is $O(m_B/m_s)$. Multiplication by this factor obscures the comparison of S- and P-wave hyperon decay amplitudes since P-wave amplitudes appear to be enhanced by an order of magnitude.

The S- and P-wave hyperon nonleptonic decay amplitudes are calculated to one loop using the weak Lagrangian $L_v^{\Delta S=1}$ and the baryon-pion Lagrangians L_v and L_v^{10}. The one-loop S- and P-wave decay amplitudes can be written in the form

$$\mathcal{A}_{ij\phi} = \frac{i}{\sqrt{2}f}\left(\overline{\alpha}_{ij} + \left(\overline{\beta}_{ij} - \overline{\lambda}_{ij\phi}\,\overline{\alpha}_{ij}\right)\frac{M_K^2}{16\pi^2 f^2}\ln\left(M_K^2/\mu^2\right)\right), \tag{67}$$

where f is related to the physical pion decay constant $f_\pi \approx 93$ MeV by

$$f_\pi = f\left(1 - \frac{1}{2}\frac{M_K^2}{16\pi^2 f^2}\ln\left(M_K^2/\mu^2\right)\right).$$

The coefficients $\overline{\alpha}_{ij} = \alpha_{ij}$ are tree-level amplitudes. The wavefunction coefficients $\overline{\lambda}_{ij\phi} = \overline{\lambda}_{ij} + \lambda_\phi$ include pion wavefunction renormalization,

$$\sqrt{Z_\phi} = 1 + \lambda_\phi \frac{M_K^2}{16\pi^2 f^2}\ln\left(M_K^2/\mu^2\right),$$

where the pion wavefunction renormalization coefficient

$$\lambda_\pi = -\tfrac{1}{6}. \tag{68}$$

Four independent hyperon decay amplitudes remain after imposition of isospin. These amplitudes are taken to be $\Sigma^+ \to n\pi^+$, $\Sigma^- \to n\pi^-$, $\Lambda \to p\pi^-$, and $\Xi^- \to \Lambda\pi^-$. The other amplitudes are determined by the isospin relations

$$\begin{aligned}
&\sqrt{2}\,\mathcal{A}(\Sigma^+ \to p\pi^0) - \mathcal{A}(\Sigma^+ \to n\pi^+) + \mathcal{A}(\Sigma^- \to n\pi^-) = 0,\\
&\mathcal{A}(\Lambda \to p\pi^-) + \sqrt{2}\,\mathcal{A}(\Lambda \to n\pi^0) = 0,\\
&\mathcal{A}(\Xi^- \to \Lambda\pi^-) + \sqrt{2}\,\mathcal{A}(\Xi^0 \to \Lambda\pi^0) = 0,
\end{aligned} \tag{69}$$

where eqs. (69) apply to S- and P-wave amplitudes separately.

The tree-level $SU(3)$ symmetry predictions are

$$\begin{aligned}
\alpha^{(S)}_{\Sigma^+ n} &= 0, \\
\alpha^{(S)}_{\Sigma^- n} &= -h_D + h_F, \\
\alpha^{(S)}_{\Lambda p} &= \tfrac{1}{\sqrt{6}}(h_D + 3h_F), \\
\alpha^{(S)}_{\Xi^- \Lambda} &= \tfrac{1}{\sqrt{6}}(h_D - 3h_F),
\end{aligned} \tag{70}$$

for the S-wave amplitudes and

$$\begin{aligned}
\alpha^{(P)}_{\Sigma^+ n} &= -D(h_D - h_F)/(M_\Sigma - M_N) - \tfrac{1}{3}D(h_D + 3h_F)/(M_\Lambda - M_N), \\
\alpha^{(P)}_{\Sigma^- n} &= -F(h_D - h_F)/(M_\Sigma - M_N) - \tfrac{1}{3}D(h_D + 3h_F)/(M_\Lambda - M_N), \\
\alpha^{(P)}_{\Lambda p} &= \tfrac{2}{\sqrt{6}}D(h_D - h_F)/(M_\Sigma - M_N) + \tfrac{1}{\sqrt{6}}(D + F)(h_D + 3h_F)/(M_\Lambda - M_N), \\
\alpha^{(P)}_{\Xi^- \Lambda} &= -\tfrac{2}{\sqrt{6}}D(h_D + h_F)/(M_\Xi - M_\Sigma) - \tfrac{1}{\sqrt{6}}(D - F)(h_D - 3h_F)/(M_\Xi - M_\Lambda).
\end{aligned} \tag{71}$$

for the P-wave amplitudes. Elimination of the two parameters h_D and h_F amongst the three non-vanishing S-wave amplitudes yields one $SU(3)$ symmetry relation, the Lee-Sugawara relation,

$$\tfrac{3}{\sqrt{6}}\,\mathcal{A}^{(S)}(\Sigma^- \to n\pi^-) + \mathcal{A}^{(S)}(\Lambda \to p\pi^-) + 2\,\mathcal{A}^{(S)}(\Xi^- \to \Lambda\pi^-) = 0. \tag{72}$$

There is no $SU(3)$ relation for the P-wave amplitudes. The formulæ for S- and P-wave coefficients β_{ij} are quite lengthy and will not be given here. The coefficients are listed in Ref. 19.

Experimental S- and P-wave hyperon nonleptonic decay amplitudes are compared with theory in Table 1. The theoretical formulæ have been evaluated using parameter values determined from a best fit to one-loop S-wave formulæ. The theoretical S-wave amplitudes are in good agreement with experimental values. Chiral logarithmic corrections to tree-level S-wave amplitudes are larger than anticipated. The correction from octet graphs is small for all amplitudes. The large correction from decuplet graphs is dominated by the contribution of wavefunction renormalization coefficients times tree-level amplitudes. Since the wavefunction coefficients λ' contain a large singlet piece, most of the chiral logarithmic correction is not $SU(3)$ violating. The correction to the Lee-Sugawara relation -0.31 is in fairly good quantitative agreement with the experimental value of -0.24. The correction to the Lee-Sugawara relation can be naturally smaller than typical corrections to individual S-wave amplitudes since only the $SU(3)$-violating portion of the chiral logarithmic corrections contributes.

The tree-level $SU(3)$ symmetric P-wave amplitudes are suppressed compared to S-wave amplitudes. The two terms in the tree-level formulæ Eq. (71) cancel

TABLE 1

Decay	s_{expt}	s_{theory}	s_{tree}	Δs_{loop}	Δs_{octet}	Δs_{decup}
$\Sigma^+ \to n\pi^+$	0.06	−0.09	0.00	−0.09	0.13	−0.22
$\Sigma^+ \to p\pi^0$	−1.43	−1.41	−0.85	−0.55	−0.04	−0.51
$\Sigma^- \to n\pi^-$	1.88	1.90	1.21	0.69	0.18	0.51
$\Lambda \to p\pi^-$	1.42	1.44	0.91	0.53	0.16	0.37
$\Lambda \to n\pi^0$	−1.04	−1.02	−0.64	−0.37	−0.11	−0.27
$\Xi^- \to \Lambda\pi^-$	−1.98	−2.04	−1.19	−0.84	−0.14	−0.71
$\Xi^0 \to \Lambda\pi^0$	1.52	1.44	0.84	0.60	0.10	0.50
Decay	p_{expt}	p_{theory}	p_{tree}	Δp_{loop}	Δp_{octet}	Δp_{decup}
$\Sigma^+ \to n\pi^+$	1.81	0.82	−0.06	0.89	0.16	0.72
$\Sigma^+ \to p\pi^0$	1.17	0.36	−0.13	0.49	−0.06	0.55
$\Sigma^- \to n\pi^-$	−0.06	0.34	0.13	0.21	0.26	−0.05
$\Lambda \to p\pi^-$	0.52	−0.52	−0.28	−0.24	−0.31	0.08
$\Lambda \to n\pi^0$	−0.39	0.38	0.21	0.17	0.23	−0.05
$\Xi^- \to \Lambda\pi^-$	0.48	0.35	0.11	0.24	0.30	−0.07
$\Xi^0 \to \Lambda\pi^0$	−0.33	−0.24	−0.07	−0.16	−0.21	0.05

S- and P-wave amplitudes for parameter values determined from a fit to one-loop S-wave formulæ. The fit was performed setting $D = 0.61$, $F = 0.40$, $\mathcal{C} = 1.6$, and $\mathcal{H} = -1.9$. The best fit parameters are $h_D = -0.35 \pm 0.09$, $h_F = 0.86 \pm 0.05$, and $h_C = -0.36 \pm 0.65$ in units of $G_F m_{\pi^+}^2 \sqrt{2} f_\pi$. Quoted errors reflect the 20% theoretical error of the one-loop calculation. Experimental amplitudes are given in column 2. The total theoretical one-loop amplitudes (column 3) are the sum of tree-level amplitudes (column 4) and chiral logarithmic corrections (column 5). The chiral logarithmic correction is the sum of the chiral correction produced by octet and decuplet graphs, respectively. The chiral correction which results from renormalization of the pion decay constant is contained in the octet chiral correction. The theoretical S-wave amplitudes s_{theory} should agree with measured values to within a nominal theoretical error of 0.30 in amplitude. Agreement with experiment for P-wave amplitudes is only expected to within corrections which are comparable to Δp_{loop}.

against each other for the parameter values determined by the S-wave fit. Because the tree-level contribution is suppressed, chiral logarithmic corrections are order one compared to tree-level amplitudes. Thus, it is not surprising that $SU(3)$ symmetry predictions fail for P-wave amplitudes. Because the neglected $O(m_s)$ correction is similar in magnitude to the computed $O(m_s \ln m_s)$ correction (and the chiral logarithmic correction is order one compared to the tree-level contribution), one expects corrections to the one-loop theoretical calculation of order 100%. Despite such a large uncertainty in the theoretical calculation, some of the features of the experimental P-wave amplitudes are reproduced by the one-loop theoretical calculation.

Nonleptonic $\Delta S = 1$ decays of the Ω^- can also be computed to one-loop in terms of the parameters of the octet weak Lagrangian $L_v^{\Delta S=1}$. Tree-level formulæ are given in Ref. 19.

11. Decuplet Decoupling

The calculations performed in Sections 8–10 exploited the near degeneracy of octet and decuplet baryons in the real world. These calculations were done in the parameter regime $\Delta m < M_K$. The decuplet-octet mass difference Δm was dropped from the decuplet propagator and treated as a explicit vertex in the chiral Lagrangian. This approximation keeps only terms linear in the small parameter $\Delta m / M_K$; higher order terms are neglected. In the limit of exact degeneracy, the decuplet contributes to chiral logarithmic corrections to octet processes and to the leading non-analytic M_K^3 correction to baryon octet masses. For nearly degenerate octet and decuplet baryons, an additional non-analytic baryon mass correction arises, the $\Delta m M_K^2 \ln M_K^2$ correction. Although the above approximation scheme is sufficient for comparison with experiment, the approximation breaks down in the chiral limit. As the chiral limit is approached, the inequality $\Delta m < M_K$ is replaced by the opposite inequality $M_K \ll \Delta m$. In the chiral limit, the decuplet cannot contribute to the non-analytic corrections for octet processes since these corrections come from infrared divergences. The crossover from the degeneracy regime to the chiral limit is studied in detail in this section. The precise manner in which the decuplet decouples in the chiral limit is explained.

The decuplet M_K^3 and $\Delta m M_K^2 \ln M_K^2$ corrections to octet masses are produced by the Feynman diagram shown in fig. 2(a). For momenta small compared with Δm, the decuplet can be integrated out explicitly. The higher dimension operators represented in fig. 2(b) are generated by expanding the decuplet propagator

$$\frac{1}{k \cdot v - \Delta m} = -\frac{1}{\Delta m}\left(1 + \frac{k \cdot v}{\Delta m} + \ldots\right). \tag{73}$$

Since all terms are suppressed by $1/\Delta m$, these graphs cannot produce the M_K^3 and $\Delta m M_K^2 \ln M_K^2$ terms. The counterterm fig. 2(c) can be proportional to positive powers of Δm. This counterterm cannot produce the above terms, however, since

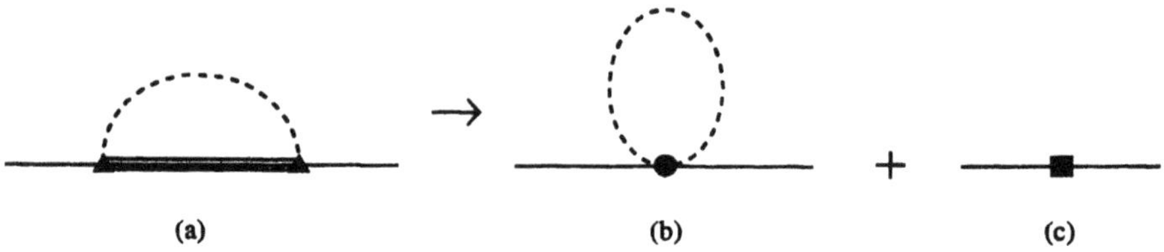

FIGURE 2.

Decoupling of the decuplet: Diagram (a) produces M_K^3 and $\Delta m M_K^2 \ln M_K^2$ corrections to octet masses when $\Delta m < M_K$. In the opposite limit, $M_K \ll \Delta m$, the intermediate decuplet baryon can be integrated out of the theory. Contracting the decuplet line produces graphs (b), which are suppressed by powers of $1/\Delta m$, where the dot in (b) represents higher dimension operators. Matching between the two theories at $\mu = \Delta m$ also produces a counterterm (c).

an $SU(3)$ symmetric mass term cannot generate any chiral logarithms because it can be removed from the Lagrangian,* and any $SU(3)$ violating counterterm with factors of the mass matrix M produces corrections of order $M_K^4 \ln M_K^2$. Hence there is no term in the effective theory without decuplets which can produce the terms required. The M_K^3 and $\Delta m M_K^2 \ln M_K^2$ terms must therefore disappear as one crosses over to the parameter regime $M_K \ll \Delta m$. Similar reasoning leads to the conclusion that other chiral logarithmic corrections due to the decuplet also must disappear in the chiral limit.

The decoupling of the decuplet can be seen explicitly by computing fig. 2 for arbitrary values of Δm and M_K. The Feynman integral for this diagram is proportional to

$$\begin{aligned} I &= \int \frac{d^D k}{(2\pi)^D} \frac{i}{(k\cdot v - \Delta m)} \frac{i}{(k^2 - M^2)} k_\mu(-k_\nu) \quad \overline{B}_v \, P_v^{\mu\nu} B_v \\ &= 2\int_0^\infty d\lambda \int \frac{d^D k}{(2\pi)^D} \frac{k^\mu k^\nu}{\left[k^2 - M^2 + 2\lambda(k\cdot v) - 2\lambda\Delta m\right]^2} \quad \overline{B}_v \, P_v^{\mu\nu} B_v \\ &= -\frac{4}{D}\int_0^\infty d\lambda \int \frac{d^D \ell}{(2\pi)^D} \frac{\ell^2}{\left[\ell^2 - \lambda^2 - 2\lambda\Delta m - M^2\right]^2} \quad \overline{B}_v B_v \\ &= \frac{4}{D}\frac{i}{16\pi^2}\frac{\Gamma(1+D/2)\Gamma(1-D/2)}{\Gamma(D/2)\Gamma(2)} \int_0^\infty d\lambda \left[\lambda^2 + 2\lambda\Delta m + M^2\right]^{(D/2)-1} \overline{B}_v B_v, \end{aligned} \tag{74}$$

where the momentum integral is shifted to $\ell^\mu = k^\mu + \lambda v^\mu$ in the third line, and the identities $v_\mu P_v^{\mu\nu} = 0$ and $g_{\mu\nu} P_v^{\mu\nu} = -2$ are used. Setting $D = 4 - 2\epsilon$ and using the

* This observation is equivalent to the result that no $m_B M_K^2 \ln M_K^2$ term appears in Eq. (46).

identity Eq. (33) gives

$$
\begin{aligned}
I &= \frac{2i}{16\pi^2}\Gamma(-1+\epsilon)\int_0^\infty d\lambda\,\left[\lambda^2+2\lambda\Delta m+M^2\right]^{1-\epsilon} \\
&= \frac{i}{24\pi^2}\left[4\left(M^2-(\Delta m)^2\right)^2\int_0^\infty d\lambda\,\frac{1}{[\lambda^2+2\lambda\Delta m+M^2]}\right. \\
&\quad \left. +\left(2(\Delta m)^3-3M^2\Delta m\right)\ln M^2/\mu^2+4M^2\Delta m-\frac{10}{3}(\Delta m)^3\right]
\end{aligned}
\tag{75}
$$

The remaining integral can be evaluated explicitly,

$$
\begin{aligned}
&\int_0^\infty d\lambda\,\frac{1}{[\lambda^2+2\lambda\Delta m+M^2]} = \int_0^\infty d\lambda\,\frac{1}{\left[(\lambda+\Delta m)^2+M^2-(\Delta m)^2\right]} \\
&= \frac{\pi}{2\sqrt{M^2-(\Delta m)^2}}-\frac{1}{\sqrt{M^2-(\Delta m)^2}}\tan^{-1}\left(\frac{\Delta m}{\sqrt{M^2-(\Delta m)^2}}\right); \quad (M>\Delta m) \\
&= -\frac{1}{2\sqrt{(\Delta m)^2-M^2}}\ln\left(\frac{\Delta m-\sqrt{(\Delta m)^2-M^2}}{\Delta m+\sqrt{(\Delta m)^2-M^2}}\right); \quad (\Delta m>M).
\end{aligned}
\tag{76}
$$

The exact expression (75) and (76) can be studied in the two limits $\Delta m \ll M$ and $M \ll \Delta m$. Expanding the integral for $\Delta m \ll M$, and keeping only the leading terms gives

$$
I = \frac{i}{24\pi^2}\left[2\pi M^3-3M^2\Delta m\ln(M^2/\mu^2)\right]. \tag{77}
$$

The non-analytic terms are of the form included in Eq. (46). Note that the M_K^3 term comes from a $\tan^{-1}\infty=\pi/2$, and thus has an extra factor of π. This is a general result; all odd powers of M_K arise from integrals similar to Eq. (76), and have an additional factor of π. The opposite limit, $M \ll \Delta m$ yields the expression

$$
\begin{aligned}
I &= \frac{i}{24\pi^2}\left[-2\left((\Delta m)^2-M^2\right)^{3/2}\ln\left(\frac{M^2}{4(\Delta m)^2}\right)\right. \\
&\quad \left. +\left(2(\Delta m)^3-3M^2\Delta m\right)\ln M^2/\mu^2+\text{analytic in } M\right].
\end{aligned}
\tag{78}
$$

Expanding $\left((\Delta m)^2 - M^2\right)^{3/2}$,

$$\begin{aligned} I &= \frac{i}{24\pi^2}\left[\left(-2\,(\Delta m)^3 + 3M^2\Delta m + \mathcal{O}\left(\frac{1}{\Delta m}\right)\right)\ln\frac{M^2}{4(\Delta m)^2}\right. \\ &\qquad \left. + \left(2(\Delta m)^3 - 3M^2\Delta m\right)\ln M^2/\mu^2 + \text{ analytic in } M\right] \\ &= \frac{i}{24\pi^2}\left[\left(-2\,(\Delta m)^3 + 3M^2\Delta m\right)\ln\frac{\mu^2}{4(\Delta m)^2}\right. \\ &\qquad \left. + \mathcal{O}\left(\frac{1}{\Delta m}\right)\ln\frac{M^2}{4(\Delta m)^2} + \text{ analytic in } M\right]. \end{aligned} \tag{79}$$

Note that there is no longer any decuplet contribution to the M_K^3 term, nor any $\Delta m M_K^2 \ln M_K^2$ term. The $\ln\left(\mu^2/4(\Delta m)^2\right)$ term can be reabsorbed into local counterterms, as can the analytic terms. The only infrared non-analytic pieces left are of order $1/\Delta m$, which are precisely the pieces that can be generated by the higher dimension operators produced on integrating out the decuplet.

The decoupling of the decuplet non-analytic corrections in the limit that $M_K \ll \Delta m$ agrees with theorems on the behavior of non-analytic corrections in the chiral limit.[20] The precise nature of the decoupling is rather subtle, and requires the complete expression for the Feynman integrals as a function of $\Delta m/M_K$. We have explicitly computed the decoupling in this section for the baryon masses. Similar results also hold for the axial currents and the non-leptonic decays. In the real world, $\Delta m < M_K$, so we have simplified our computations by only retaining the terms linear in Δm. The expressions given in Sections 8–10 cannot be used to study the chiral limit, as can be seen in the example above. However, correct formulæ in the extreme chiral limit can be readily obtained from these expressions by dropping all non-analytic corrections produced by graphs with decuplet intermediate states. For computations of chiral $SU(2)$ breaking effects due to the pion mass, it is important to retain the complete form of the integrals, because Δm is not small compared with M_π.

Acknowledgements

We would like to thank the organizers of the Dobogókő workshop, especially András Patkós. We greatly benefitted from discussions with participants at the workshop—N. Kaiser, U.-G. Meißner, and especially H. Georgi and A. Pich. We also thank M.B. Wise for helpful discussions over the last year. This work was supported in part by DOE grant #DE-FG03-90ER40546. A.M. was also supported in part by a grant from the Alfred P. Sloan Foundation, and a PYI award (PHY-8958081) from the National Science Foundation.

References

1. S. Weinberg, *Phys. Rev. Lett.* **18** (1967) 188.
2. R. Dashen and M. Weinstein, *Phys. Rev.* **183** (1969) 1261.
3. L.- F. Li and H. Pagels, *Phys. Rev. Lett.* **26** (1971) 1204, *Phys. Rev. Lett.* **27** (1971) 1089;
   P. Langacker and H. Pagels, *Phys. Rev.* **D8** (1973) 4595.
4. S. Weinberg, *Physica* **96A** (1979) 327.
5. S. Coleman, J. Wess and B. Zumino, *Phys. Rev.* **177** (1969) 2239;
   C. Callan, S. Coleman, J. Wess and B. Zumino, *Phys. Rev.* **177** (1969) 2247.
6. J. Gasser and H. Leutwyler, *Ann. Phys.* **158** 1984 142.
7. E. Jenkins and A. Manohar, *Phys. Lett.* **255B** (1991) 558.
8. W.E. Caswell and G.P. Lepage, *Phys. Lett.* **167B** (1986) 437;
   M.B. Voloshin and M.A. Shifman, *Sov. J. Nucl. Phys.* **45** (1987) 463, *Sov. J. Nucl. Phys.* **47** (1988) 511;
   N. Isgur and M.B. Wise, *Phys. Lett.* **232B** (1989) 113, *Phys. Lett.* **237B** (1990) 527;
   E. Eichten and B. Hill *Phys. Lett.* **234B** (1990) 511.
9. H. Georgi, *Phys. Lett.* **240B** (1990) 447.
10. E. Eichten and B. Hill, *Phys. Lett.* **243B** (1990) 427;
    M.E. Luke, *Phys. Lett.* **252B** (1990) 447;
    A.F. Falk, B. Grinstein, and M.E. Luke, *Nucl. Phys.* **B357** (1991) 185.
11. A. Manohar and H. Georgi, *Nucl. Phys.* **B234** (1984) 189.
12. S. Weinberg, *Phys. Lett.* **251B** (1990) 288, *Nucl. Phys.* **B363** (1991) 3.
13. C. Carone and H. Georgi, HUTP- 91/A053.
14. E. Jenkins and A. Manohar, *Phys. Lett.* **259B** (1991) 353.
15. See for example, A.V. Manohar, The g_1 Problem: Much Ado About Nothing, in *Polarized Collider Workshop*, edited by J. Collins, S.F. Heppelman, and R.W. Robinett, A.I.P Conference Proceedings No. 223 (American Institute of Physics, New York, 1991) and references therein;
    R.L. Jaffe and A.V. Manohar, *Nucl. Phys.* **B337** (1990) 509.
16. J. Bijnens, H. Sonoda, and M.B. Wise, *Nucl. Phys.* **B261** (1985) 185.
17. E. Jenkins, UCSD/PTH 91-12, to be published in *Nuclear Physics B.*
18. E. Jenkins and A.V. Manohar, UCSD/PTH 91-31.
19. E. Jenkins, UCSD-PTH 91-24.
20. J. Gasser and A. Zepeda, *Nucl. Phys.* **B174** (1980) 445.

PHOTO–NUCLEON PROCESSES IN CHIRAL PERTURBATION THEORY

Ulf-G. Meißner[†]

Universität Bern
Institut für Theoretische Physik
Sidlerstr. 5
CH-3012 Bern, Switzerland

ABSTRACT

I consider photo–nucleon processes in the framework of chiral perturbation theory to one loop order. First, the low energy behaviour of neutral pion photoproduction is worked out. Virtual pions generate infrared singularities which modify the conventional low–energy theorem for the electric dipole amplitude E_{0+}. The one-loop calculation is confronted with recent accurate measurements of the total cross section in the threshold region. A related topic is the chiral expansion of the Compton scattering amplitude in forward direction. The non–trivial information about the nucleon structure is embodied in the electric (α) and magnetic (β) polarizabilities of the proton and the neutron. The one–loop calculation sheds light on the two most prominent empirical features that a) the sum $\alpha + \beta$ is approximatively the same for the proton and the neutron and that b) the proton and the neutron behave essentially as electric dipoles.

1. INTRODUCTION

With the advent of CW machine, the electromagnetic structure and interactions of hadrons at low energies can be investigated with a superbe accuracy. Such an improvement on the experimental side is an enormously fruitful trigger for improved theoretical investigations. Although hadrons are built from quarks and gluons, at low energy they show a different face. In fact, the spontaneous breakdown of the chiral symmetry of QCD leads to massless pseudoscalar particles, the Goldstone bosons (pions). These govern the dynamics at low energies. The theoretical precision tool to take this into account together with the pertinent Ward–Takahashi identities of QCD is called chiral perturbation theory (CHPT) – a systematic and simultaneous expansion of the QCD Greens functions in external momenta and quark masses. In the following I will discuss the calculation of two presently very much debated photo–nucleon processes (threshold pion photoproduction and the electromagnetic polarizabilities) in the framework of CHPT.

[†] Heisenberg Fellow

2. CHPT WITH BARYONS

At low energies, the QCD Greens functions can be systematically determined. For that, one replaces the fundamental QCD lagrangian $\mathcal{L}_{QCD}$ by an effective lagrangian $\mathcal{L}_{eff}$ expressed in terms of the asymptotically observed nucleon (Ψ) and pion (π) fields.[1,2,3,4] This effective lagrangian is constructed in harmony with the underlying discrete and continuous symmetries of the fundamental theory and it abides to the pertinent Ward-Takahashi identities. Furthermore, the effective theory admits a systematic expansion in external momenta and quark masses. This framework is called chiral perturbation theory (CHPT). In what follows, we will work in the one-loop approximation. The two main reasons to consider loops are unitarity and the possible infrared singular structure of certain Greens functions. In the presence of baryon fields (here, the nucleons), complications arise from the fact that the nucleon mass in the chiral limit is non-vanishing, $\mathring{m} \simeq 800$ MeV.[5] This is in sharp contrast to the meson sector, where the pseudoscalars related to the spontaneous chiral symmetry breaking become massless in the chiral limit of vanishing quark masses. This means that in the baryon sector there is no exact one-to-one mapping of the loop expansion onto the expansion in external momenta and quark masses. Stated differently, there is no strict proof (as can be done in the meson sector) that the leading infrared singularities in the Greens functions are completely given by the one-loop graphs.* Nevertheless, such a behaviour is plausible and no explicit counter example has been worked out so far. A detailed account of the low-energy structure in the baryon sector for the case of πN scattering can be found in Ref.[3].

The effective lagrangian relevant to one-loop order reads (we work in SU(2) and the isospin limit $m_u = m_d = \hat{m}$):

$$\begin{aligned} \mathcal{L}_{eff} &= \mathcal{L}_1 + \mathcal{L}_2 \\ \mathcal{L}_1 &= \mathcal{L}_{\pi N}^{(1)} + \mathcal{L}_{\pi\pi}^{(2)} \end{aligned} \tag{2.1}$$

where the superscript (i) denotes terms of order E^i in the low-energy expansion.** In the one-loop approximation, the structure of the effective theory is rather simple: One has to evaluate tree and one-loop diagrams involving the vertices from $\mathcal{L}_1$ and tree graphs involving $\mathcal{L}_2$ vertices. Explicitely, $\mathcal{L}_1$ can be written as:[3,4]

$$\begin{aligned} \mathcal{L}_{\pi N}^{(1)} &= \bar{\Psi}(i\gamma^\mu D_\mu - \mathring{m})\Psi + \frac{\mathring{g}_A}{2}\bar{\Psi}\gamma_\mu\gamma_5 u^\mu \Psi \\ \mathcal{L}_{\pi\pi}^{(2)} &= \frac{F^2}{4}\,\mathrm{Tr}\left(\nabla_\mu U^\dagger \nabla^\mu U + M^2(U + U^\dagger)\right) \end{aligned} \tag{2.2}$$

---

* Here, one-loop graphs means meson loops, *i.e.* closed fermion lines are not considered.

** E^i stands for any term of the type $q^n \hat{m}^{(i-n)/2}$, with q a generic baryon 3-momentum or a generic meson 4-momentum.

where U embodies the pion fields

$$U = u^2 = \sigma + i\phi/F, \qquad \sigma^2 + \phi^2/F^2 = 1$$
$$\phi = \begin{pmatrix} \pi^0 & \sqrt{2}\,\pi^+ \\ \sqrt{2}\,\pi^- & -\pi^0 \end{pmatrix} \tag{2.3}$$

The covariant derivatives $\nabla_\mu U$ and $D_\mu \Psi$ are given by

$$\begin{aligned} \nabla_\mu U &= \partial_\mu U - ieA_\mu [Q, U] \\ D_\mu \Psi &= \partial_\mu \Psi + \Gamma_\mu \Psi \\ \Gamma_\mu &= \frac{1}{2}\{u^\dagger(\partial_\mu - ieA_\mu Q)u + u(\partial_\mu - ieA_\mu Q)u^\dagger\} \end{aligned} \tag{2.4}$$

where $Q = \mathrm{diag}(1, 0)$ is the (nucleon) charge matrix. Furthermore,

$$u_\mu = i u^\dagger \nabla_\mu U u^\dagger \tag{2.5}$$

and A_μ denotes the photon field (I restrict myself here to those terms which are explicitely needed.). $F, \overset{\circ}{g}_A$ and $\overset{\circ}{m}$ denote the pion decay constant, the axial vector coupling constant and the nucleon mass in the chiral limit $\hat{m} = 0$, respectively,

$$(F, \overset{\circ}{g}_A, \overset{\circ}{m}) = (F_\pi, g_A, m)_{|\hat{m}=0} \tag{2.6}$$

and M^2 stands for the leading term in the expansion of the pion mass squared in the current quark mass

$$m_\pi^2 = M^2 \{1 + \mathcal{O}(\hat{m})\} \,. \tag{2.7}$$

To leading order, the Goldberger-Treiman relation is exact, $\overset{\circ}{g}_{\pi N} = \overset{\circ}{g}_A\, \overset{\circ}{m}/F$, with $\overset{\circ}{g}_{\pi N}$ the pion-nucleon coupling constant in the chiral limit. It is instructive to expand $\mathcal{L}_1$ in powers of the pion and the photon fields. One finds

$$\begin{aligned} \mathcal{L}_{\pi N}^{(1)} =& \bar{\Psi}\{i\partial\!\!\!/ - \overset{\circ}{m} + \frac{e}{2}(1 + \tau_3)A\!\!\!/ + \frac{i}{8F^2}[\phi, \partial\!\!\!/\phi] \\ & - \frac{\overset{\circ}{g}_A}{2F}\partial\!\!\!/\phi\, \gamma_5 + \frac{ie\overset{\circ}{g}_A}{4F}[\tau_3, \phi]\, A\!\!\!/\, \gamma_5\}\Psi + \ldots \end{aligned} \tag{2.8}$$

which embodies the conventional pseudovector pion-nucleon coupling as well as the celebrated $NN\gamma\pi$ and $NN\pi\pi$ contact terms (Kroll-Ruderman and Weinberg terms).

At next-to-leading order in CHPT, one has two kinds of contributions. First, there are the pion loop graphs with all vertices from $\mathcal{L}_1$. These account for unitarity corrections as well as vertex and mass renormalization and secondly, there are polynomial counter terms up to and including order $\mathcal{O}(q^3)$. In the meson sector, the two lowest order constants F and M are not changed by the loops, which is different for the baryon sector – the values of $\overset{\circ}{m}$ and $\overset{\circ}{g}_A$ are in fact influenced by one-loop contributions. Therefore one has to add appropriate counter terms in $\mathcal{L}_2$. The full expression for $\mathcal{L}_2$ reads:

$$\mathcal{L}_2 = \Delta\mathcal{L}^{(0)}_{\pi N} + \Delta\mathcal{L}^{(1)}_{\pi N} + \mathcal{L}^{(2)}_{\pi N} + \mathcal{L}^{(3)}_{\pi N} + \mathcal{L}^{(4)}_{\pi\pi} \tag{2.9}$$

Here, $\Delta\mathcal{L}^{(0,1)}_{\pi N}$ are the terms necessary for the renormalization of the nucleon mass and the axial vector coupling constant in the chiral limit. I do not give the explicit form of the polynomial terms $\mathcal{L}^{(2,3)}_{\pi N}$ and $\mathcal{L}^{(4)}_{\pi\pi}$. A complete list of these terms can be found in Ref.[4].

3. NEUTRAL PION PHOTOPRODUCTION AT THRESHOLD

Let me consider the reactions $\gamma(k) + p(p_1) \rightarrow \pi^0(q) + p(p_2)$ and $\gamma(k) + n(p_1) \rightarrow \pi^0(q) + n(p_2)$. First, I concentrate on the proton case. The threshold is defined when the photon energy is just sufficient to produce a pion at rest. In the conventional Mandelstam variables, this means for the cm energy $\sqrt{s} = m + M_\pi$, or in the photon lab frame

$$E_{\gamma,\text{thr}} = M_{\pi^0}\left(1 + \frac{M_{\pi^0}}{2m_p}\right) = 144.7\,\text{MeV} \tag{3.1}$$

In the threshold region, the produced pions will only be in relative $S-$ or $P-$waves and one can analyse the data in terms of electric dipole and magnetic and electric quadrupole amplitudes. However, I want to issue a warning here which both experimenters and theorists should take seriously. Close to threshold, the total cross section dies out very fast, and one has to use also measurements of the differential X-section at energies not too far above threshold to eventually unravel the relative importance of these multipoles. This is a very non–trivial undertaking, and 30 years of analyzing πN scattering data (phase shifts) should have taught us that only by strictly adhering to analyticity constraints one gets proper information in the threshold region. At present, such a machinery exists for πN scattering but has not been developed in full detail for the photoproduction case. I will pick up this theme later on. Now what is all the fuzz concerning this threshold region about? In fact, as the pion momentum $\vec{q}$ tends to zero, one has

$$\left.\frac{|\vec{k}|}{|\vec{q}|}\frac{d\sigma}{d\Omega}\right|_{\vec{q}\rightarrow 0} = (E_{0+})^2 \tag{3.2}$$

This means that at threshold the differential cross section is entirely given in terms of the electric dipole amplitude (the subscript '+' refers to the total pion angular momentum $j = l + 1/2$). Furthermore, a so–called "low–energy theorem" (LET) has been derived 20 years ago. It states that to second order in the mass ratio $\mu = M_\pi/m$, the amplitude E_{0+} is given by[6]

$$\begin{Bmatrix} E^{\pi^0 p}_{0+} \\ E^{\pi^0 n}_{0+} \end{Bmatrix} = -\frac{e g_{\pi N}}{8\pi m}\mu \begin{Bmatrix} 1 - \frac{1}{2}(\kappa_p + 3)\mu \\ \frac{1}{2}\kappa_n \mu \end{Bmatrix} + \mathcal{O}(\mu^3) \tag{3.3}$$

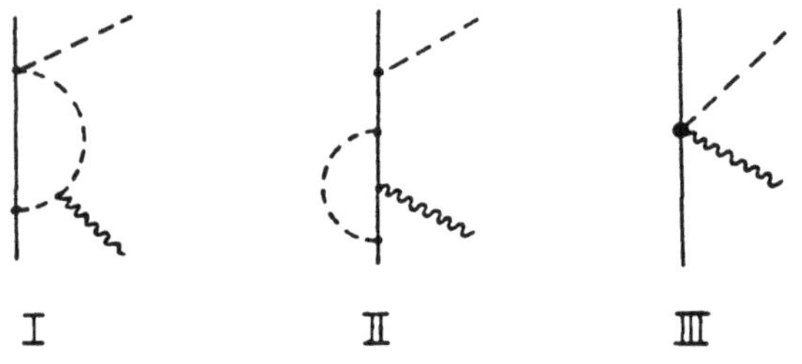

Fig. 1: Representative diagrams for neutral pion photoproduction. Shown are one diagram of classes I, II and III, in order. The class I diagram is the triangle graph discussed in the text and the class II graph obviously contributes to the anomalous magnetic moment.

The first term proportional to μ is safely based on PCAC, the only assumption being that the non–Born parts are zero in the chiral limit. The μ^2–term is more interesting, there exist various proofs in the literature that this term is also entirely a consequence of PCAC (combined with some seemingly harmless extra assumptions, see below). In the units used in photo–nuclear reactions, one finds

$$\left\{ \begin{array}{c} E_{0+}^{\pi^0 p} \\ E_{0+}^{\pi^0 n} \end{array} \right\} = \left\{ \begin{array}{c} -2.26 \\ 0.48 \end{array} \right\} \cdot 10^{-3}/M_{\pi^+} \tag{3.4}$$

Before 1986, the best experimental determination gave (-1.8 ± 0.6)* for $\gamma p \to \pi^0 p$, no big deal. Then, however, came the measurements at Saclay and Mainz.[7] The value of E_{0+} published by the Saclay people was (-0.5 ± 0.3). The Mainz group found two solutions, one giving $\mathrm{Re}E_{0+}(146.8\mathrm{MeV}) = -0.25 \pm 0.19$ and the other -1.61 ± 0.16. These results seem to indicate a substantial deviation from the LET, and hectic activity set in trying to explain them. It was then, however, realized that the Saclay group had subtracted a particular rescattering correction of approximately -1.0 and the Mainz group had not taken into account the constraints from the total X–section.[8] If one does that, one finds that the "LET" is ok at π^0 threshold, but that there is a rapid variation in $\mathrm{Im}E_{0+}$ due to the coupling to the channel $\gamma p \to \pi^+ n \to \pi^0 p$ which opens at $E_\gamma = 149.9$ MeV, *i.e.* approximately 5 MeV above the $\pi^0 p$ threshold. In view of this somewhat confused situation, it seems worthwile to perform a complete one–loop calculation and to see whether the "LET" really is a direct consequence of chiral symmetry. Before I discuss this calculation in detail, I should point out that it is done in the isospin limit $m_u = m_d = \hat{m}$ and one therefore has only *one* threshold. Thus one has to be careful in comparing the results of the calculation directly with the data. Despite this (and it remains an open problem how to include isospin breaking) one can make some firm statements on the form of the "LET", as I will describe now.

* I will drop the factor $10^{-3}/M_{\pi^+}$ as long as no confusion can occur

If one draws all the diagrams to one–loop order, one finds that there are 66. Life is made easier by the fact that these diagrams fall into three classes which are separately gauge invariant, $k \cdot T(\gamma p \to \pi^0 p) = 0$. Some representative diagrams are shown in Fig.1. One can easily convince oneself that the diagrams of class I carry a common factor $g_A/F_\pi^3 \sim g_{\pi N}$, whereas class II scales as $g_A^3/F_\pi^3 \sim g_{\pi N}^3$ * and class III are the counterterms from $\mathcal{L}_{\pi N}^{(2,3)}$ – there is only one and it contributes at order $\mathcal{O}(\mu^3)$ to E_{0+}. Class II clearly contains the diagrams which generate the anomalous magnetic moment. Let me first concentrate on class I. One can express the electric dipole amplitude E_{0+} in terms of 4 Lorentz–invariant functions à la CGLN,[9]

$$E_{0+} = -i\frac{em}{8\pi}\frac{\mu(2+\mu)}{(1+\mu)^{3/2}}\left\{A_1 + \frac{\mu(1+\mu/2)}{1+\mu}mA_3 + \frac{\mu^2}{2}\frac{m}{1+\mu}A_4\right\} \tag{3.5}$$

so we see that A_1 is potentially most dangerous since it carries the lowest power of μ. An explicit calculation of the diagrams of class I gives[10]

$$\delta A_1(s,u) = \frac{2ig_{\pi N}}{F_\pi^2}\left\{(s-m^2)[\gamma_6(s) - \gamma_4(s)] + (s \leftrightarrow u)\right\} \tag{3.6}$$

and analogous expressions for δA_3, δA_4 (they are not needed for the further discussion). Expanding δA_1 around threshold, one finds

$$\begin{aligned}\delta A_1^{\pi^0 p}(s,u)|_{\rm th} = \delta A_1^{\pi^0 n}(s,u)|_{\rm th} &= i\frac{eg_{\pi N}}{32F_\pi^2}\mu + \mathcal{O}(\mu^2)\\ &= c\cdot\sqrt{\hat{m}} + \mathcal{O}(\hat{m})\end{aligned} \tag{3.7}$$

There we have it – a non–analytic term in $\hat{m}$. Now it is obvious that the good–old "LET" is spoiled, in fact, it just is not **correct**. It receives a correction at order μ^2 from the graphs of class I (only from the triangle diagram and its crossed partner). The modified expression for E_{0+} reads

$$\begin{aligned}&\left\{\begin{matrix} E_{0+}^{\pi^0 p} \\ E_{0+}^{\pi^0 n}\end{matrix}\right\} = -\frac{eg_{\pi N}}{8\pi m}\mu\left\{F_1 + \mu F_2 + \mathcal{O}(\mu^2)\right\}\\ &F_1 = \begin{pmatrix}1\\0\end{pmatrix}\ ,\ F_2 = -\frac{1}{2}\begin{pmatrix}3+\kappa_p\\ \kappa_n\end{pmatrix} - \Delta\begin{pmatrix}1\\1\end{pmatrix} + \begin{pmatrix}\tilde{\Delta}_p\\ \tilde{\Delta}_n\end{pmatrix}\end{aligned} \tag{3.8}$$

with $\Delta = (m/4F_\pi)^2$ and $\tilde{\Delta}_{p,n}$ in (3.8) stands for terms from class II which are discussed below. It is important to notice that the PCAC prediction for F_1 is correct – this should lay at rest all models in which the leading term has been modified. Where does this non–analytic term come from? It can be traced back to the triangle diagram

* In fact, the graphs $\sim g_{\pi N}^3$ fall into two separately gauge invariant subsets.

in Fig.1 and the fact that the loop–function γ_6 contains a logarithmic singularity in the chiral limit. In fact, as M_π tends to zero, one has:

$$\begin{aligned}\gamma_6(s_{\rm th}) &= \frac{1}{64\pi^2 m^2}\left\{-\ln\mu + \frac{3\pi^2}{8} - \frac{1}{2} + \mathcal{O}(\mu^2)\right\}\\ \gamma_6(u_{\rm th}) &= \frac{1}{64\pi^2 m^2}\left\{-\ln\mu - \frac{\pi^2}{8} - \frac{1}{2} + \mathcal{O}(\mu^2)\right\}\end{aligned} \tag{3.9}$$

where $u_{th} = m^2[1-(\mu^2+2\mu)/(1+\mu)]$. This invalidates the usual assumption that δA_1 is of order μ^2, based on the assumption that $\delta A_1(\nu)$ is a differentiable function under crossing $\nu \to -\nu$ (*i.e.* interchange of s and u since $\nu = (s-u)/4m$). Clearly, this is not the case for the function γ_6! Stated differently, when one expands in powers of ν and $\nu_B = (t - M_\pi^2)/(4m)$, $\delta A_1 = a_{20}\nu^2 + a_{02}\nu_B + \ldots$. Now ν^2 and ν_B are of order μ^2 and if the Taylor coefficients would be non–singular, the contribution of δA_1 to E_{0+} would start out at order μ^3. However, due to the singularity in the function γ_6, some of the Taylor coefficients are singular. In particular, one has $a_{20} = \mathcal{O}(\mu^{-1})$ so that one ends up with a contribution of order μ^2. Let me now consider the diagrams from class II (proportional to $g_{\pi N}^3$). The subset of diagrams, which gives rise to the anomalous magnetic moment $\kappa_{p,n}$, contains no further contibutions at order μ^2. From the remaining graphs of this class (like e.g. the boxes) one finds $\tilde\Delta_p = \tilde\Delta_n = 0$. The calculation of $\tilde\Delta_{p,n}$ is much more involved than the one for Δ (class I) since it stems from many different diagrams and loop functions. This result is also born out by the numerical calculations. It is not completely surprising since one knows that in πN–scattering the box diagrams only contribute marginally.[3]

A few remarks on the numerology are in order. If you plug in the numbers for the corrections displayed in (3.8), you will find that at order μ^2, the term proportional to $\Delta = (m/4F_\pi)^2$ is dominant and the prediction for E_{0+} would even change sign if taken face value. This is, however, an artefact of the expansion in μ. If one does the full one–loop calculation for all diagrams and the counterterm from class III (with its strength estimated from $\Delta(1232)$ exchange) (plus the tree level terms), one finds $E_{0+}^{\pi^0 p} = -1.3$ – a sizeable reduction from the "old LET" value but also different from the one given by the first two terms in the expansion (3.8). This means that the expansion in μ is slowly converging. To my opinion, the moral to be drawn from this calculation is that if one wants to make statements about the constraints from chiral symmetry at next–to–leading order, one has to consistently evaluate all one–loop graphs. We know that in CHPT the anomalous magnetic moment is generated by loops – why should that be the only such contribution? In fact, it would be rather mysterious if it were – but it is not! I should repeat that (3.3) is not a low–energy theorem, the terms of order μ^2 are not complete. There are numerous proofs in the literature that this old form is indeed correct. All these so–called proofs can be shown

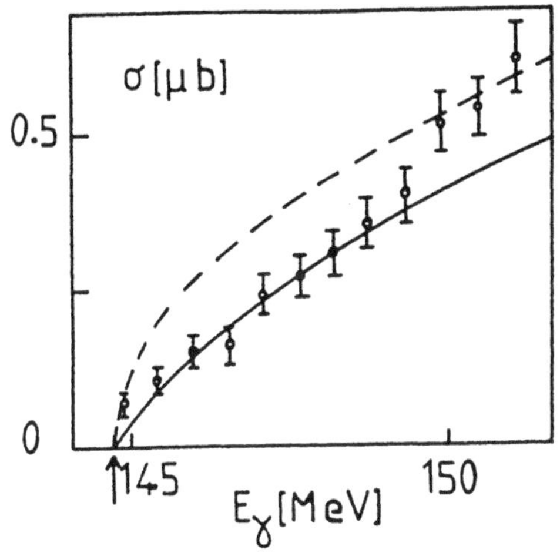

Fig. 2: Total cross section for $\gamma p \to \pi^0 p$ close to threshold, $E_\gamma \leq 150$ MeV. The solid line gives the one-loop result, the tree-level prediction is depicted by the dashed line. The data are taken from the Mainz measurement.[7]

to be incorrect at various places, and people should (have to!) forget to talk about (3.3) as a LET!

Finally, I show the total cross section as computed from the one-loop graphs together with the (class III) counterterm (with the strength estimated from Δ, ρ and ω exchange) in comparison to the accurate Mainz data[7] for $E_\gamma \leq 150$ MeV. The calculation compares favorably with these data and shows that a smaller value of $E_{0+,\mathrm{thr}}$ is compatible with the total cross section. It is important to notice that the imaginary part of E_{0+} increases strongly very fast. A more detailed investigation of this topics will be presented soon.[11]

4. ELECTROMAGNETIC POLARIZABILITIES

The electric charge radii of the neutron and the proton as well as their magnetic moments are known to a high accuracy. Low-energy Compton scattering reveals further important information about the internal structure of the nucleon, parametrized by the so-called electric and magnetic polarizabilities. These are the first non-trivial structure constants in two-photon observables. Let us place a neutron/proton into an external electric (magnetic) field $\vec{E}$ ($\vec{B}$). The particle will acquire an induced electric (magnetic) dipole moment,

$$\vec{d}_E = \alpha \vec{E} \quad ; \quad \vec{d}_M = \beta \vec{B} \tag{4.1}$$

The electric (α) and magnetic (β) polarizabilities therefore characterize the ease with which a dipole moment can be induced in a composite system. This information is as fundamental as the one related to the one-photon processes (charge radii, magnetic moments).

Before discussing the CHPT calculation of the polarizabilities, let me give a brief overview about the experimental situation:

- The sum of the electric and the magnetic polarizabilities of the proton and the neutron can be determined rather precisely by use of a forward dispersion relation sum rule,

$$\bar{\alpha} + \bar{\beta} = \frac{1}{2\pi^2} \int_{E_{\gamma,\mathrm{thr}}}^{\infty} d\omega \frac{\sigma_\gamma(\omega)}{\omega^2} \tag{4.2}$$

  with $\sigma_\gamma(\omega)$ the total photoabsorption cross section of the proton or neutron, and $\omega = E_\gamma$ the photon energy in the lab frame. The threshold value $E_{\gamma,\mathrm{thr}}$ is given in (3.1). The most commonly quoted values are $\bar{\alpha}_p + \bar{\beta}_p = (14.3 \pm 0.3) \cdot 10^{-4}\,\mathrm{fm}^3$ and $\bar{\alpha}_n + \bar{\beta}_n = (15.8 \pm 0.5) \cdot 10^{-4}\,\mathrm{fm}^3$.

- The separation into the individual contributions ($\bar{\alpha}_{p,n}$, $\bar{\beta}_{p,n}$) is afflicted with much larger uncertainties. However, there exists clear evidence that the proton and the neutron behave as electric dipoles, $\bar{\alpha}_{p,n} >> \bar{\beta}_{p,n}$. Typical values quoted in the literature span the ranges $\alpha_n \approx \bar{\alpha}_p \approx 8 \ldots 12 \cdot 10^{-4}\,\mathrm{fm}^3$ and $\bar{\beta}_n \approx \bar{\beta}_p \approx -1 \ldots 4 \cdot 10^{-4}\,\mathrm{fm}^3$. There is also some evidence that the intrinsic electric polarizability is larger for the neutron than for the proton.*

To perform a systematic one-loop calculation, we need an operational definition of the polarizabilities. Consider the low-energy expansion of the spin-independent Compton scattering amplitude. In the rest frame it takes the form

$$T(\gamma N \to \gamma N) = T_0 + \bar{\alpha}\,\omega'\omega\,\vec{\epsilon}\,' \cdot \vec{\epsilon} + \bar{\beta}\,(\vec{\epsilon}\,' \times \vec{k}') \cdot (\vec{\epsilon} \times \vec{k}) + \mathcal{O}(\omega^4) \tag{4.3}$$

with $(\omega, \vec{k}, \vec{\epsilon})$ and $(\omega', \vec{k}', \vec{\epsilon}\,')$ the frequencies, momenta and polarization vectors of the incoming and outgoing photon, respectively. The constant term T_0 is nothing but the Thomsom scattering amplitude which carries only information about the charge and mass of the particle the photon scatters off. The internal structure is, therefore, embodied in the two structure constants $\bar{\alpha}$ and $\bar{\beta}$ which parametrize the corrections to T_0 at order ω^2. In quantum field theory, the calculation of these structure constants

* The bar on α and β refers to the fact that we are dealing with the so-called Compton polarizabilities. In what follows, I will only consider these. They differ from the static polarizabilities (4.1) by recoil effects and alike. In the literature, one can find a variety of prescriptions how to relate these two definitions (mostly in terms of non-relativistic physics). Although it would be worthwhile to clarify this connection in more detail, lack of space forbids to do so.

(the polarizabilities) proceeds as follows. Obviously, we need the Fourier transformed matrix-element of two electromagnetic currents in a nucleon state,

$$T_{\mu\nu}(p,k) = \int d^4x \, e^{ik\cdot x} < N(p)|T \, J_\mu^{em}(x) J_\nu^{em}(0)|N(p) > \tag{4.4}$$

with p and k the nucleon and photon four-momentum, respectively. From this, we can form the spin-averaged Compton tensor in forward direction,[12,13]

$$\begin{aligned}\theta_{\mu\nu} &= \frac{1}{4} \mathrm{Tr}\,\{(\not p + m)\, T_{\mu\nu}(p,k)\} = e^2 \{g_{\mu\nu} \, A(s) \\ &+ k_\mu k_\nu \, B(s) + (p_\mu k_\nu + p_\nu k_\mu)\, C(s) + p_\mu p_\nu \, D(s)\}\end{aligned} \tag{4.5}$$

with $e^2/4\pi = 1/137$ the fine structure constant. The kinematics is rather simple here. The Mandelstam variables are $s = (p+k)^2$, $t = 0$ and $u = (p-k)^2 = 2m^2 - s$ (in forward direction, $t = 0$ since $\cos\theta = 1$ and $t \sim (1 - \cos\theta)$). Gauge invariance ($k_\mu \theta^{\mu\nu} = 0$) and crossing symmetry reduce the number of independent scalar functions in (4.5) to two. If we choose to work with $A(s)$ and $B(s)$, the comparison with (4.3) leads to the following expressions for the polarizabilities

$$\bar{\alpha} + \bar{\beta} = -\frac{e^2 m}{2\pi} A''(m^2) \; , \; \bar{\beta} = -\frac{e^2}{4\pi m} B(m^2) \tag{4.6}$$

where the prime denotes differentiation with respect to s. The non-renormalization theorem of the electric charge leads to the additional constraint $A(m^2) = Z^2$.

What is most amazing about the CHPT calculation of $\bar{\alpha}_{p,n}$ and $\bar{\beta}_{p,n}$ to one-loop order is that there are no contributions from polynomial counterterms. This is essentially due to the fact that the kinematics is rather special and that there are no external pions. The complete discussion of the vanishing of possible counterterms from $\mathcal{L}_{\pi N}^{(0,1,2,3)}$ is given in Ref.13, here I just want to stress the similarity to the mesonic reactions $\gamma\gamma \to \pi^0\pi^0$,[14] $K_L^0 \to \pi^0\gamma\gamma$[15] or $K_S^0 \to \gamma\gamma$.[16] They are also given purely by loops (to leading order) and therefore serve as a clean probe concerning the loop contributions. To my knowledge, the electric and magnetic polarizabilities are the first observables including baryons in which this intriguing phenomenon appears.

For the proton/neutron one has to evaluate 52/22 one-loop diagrams. This task is, however, tremendously simplified by the special kinematics discussed before. In fact, one can perform all Feynman parameter integrations. The full expressions are given in Ref.12, here let me only discuss the chiral expansion of the polarizabilities. Since we are dealing with a forward amplitude, it does not come as a surprise that they diverge in the chiral limit. The chiral expansion reads

$$\begin{aligned}
\bar{\alpha}_p &= \frac{e^2 g_{\pi N}^2}{192\pi^3 m^3} \left\{ \frac{5\pi}{2\mu} + 18 \ln\mu + \frac{33}{2} + \mathcal{O}(\mu) \right\} \\
\bar{\alpha}_n &= \frac{e^2 g_{\pi N}^2}{192\pi^3 m^3} \left\{ \frac{5\pi}{2\mu} + 6 \ln\mu - \frac{3}{2} + \mathcal{O}(\mu) \right\} \\
\bar{\beta}_p &= \frac{e^2 g_{\pi N}^2}{192\pi^3 m^3} \left\{ \frac{\pi}{4\mu} + 18 \ln\mu + \frac{63}{2} + \mathcal{O}(\mu) \right\} \\
\bar{\beta}_n &= \frac{e^2 g_{\pi N}^2}{192\pi^3 m^3} \left\{ \frac{\pi}{4\mu} + 6 \ln\mu + \frac{5}{2} + \mathcal{O}(\mu) \right\}
\end{aligned} \tag{4.7}$$

The first interesting observation from (4.7) is that the empirical result $\bar{\alpha}_p + \bar{\beta}_p \approx \bar{\alpha}_n + \bar{\beta}_n$ follows naturally from the leading term, in fact, we find $\bar{\alpha}_p + \bar{\beta}_p = \bar{\alpha}_n + \bar{\beta}_n = 15 \cdot 10^{-4} \mathrm{fm}^3$, almost too close to the data. Furthermore, the coefficient of the term proportional to M_π^{-1} is ten times bigger for $\bar{\alpha}_{p,n}$ than for $\bar{\beta}_{p,n}$, in agreement with the empirical trends. This can be traced back to the fact that there are 8 diagrams which contribute to the leading singularity – and they have the same isospin factors for the proton and the neutron.

The full one-loop prediction has also been worked out with the following results

$$\begin{aligned} \bar{\alpha}_p &= 7.4 \cdot 10^{-4}\ \mathrm{fm}^3 \qquad & \bar{\beta}_p &= -2.0 \cdot 10^{-4}\ \mathrm{fm}^3 \\ \bar{\alpha}_n &= 10.1 \cdot 10^{-4}\ \mathrm{fm}^3 \qquad & \bar{\beta}_n &= -1.2 \cdot 10^{-4}\ \mathrm{fm}^3 \end{aligned} \tag{4.8}$$

The values for the proton and neutron electric polarizabilities compare favorably with the data. The empirical trend $\bar{\alpha}_n > \bar{\alpha}_p$ is supported by this result. The values for $\bar{\beta}_p$ and $\bar{\beta}_n$ are considerably smaller than the corresponding electric ones. The negative sign for $\bar{\beta}_{p,n}$ leads, of course, to too small values of the sums $(\bar{\alpha} + \bar{\beta})_{p,n}$. There are, however, important higher loop effects which eventually have to be considered like e.g. the contribution from the $\Delta(1232)$ resonance which in many hadron models gives a large positive magnetic polarizability. While a two–loop calculation by itself is a formidable task, there is a "cheap way" of getting a handle on the two-loop effects for the sum $\bar{\alpha} + \bar{\beta}$ based on the dispersion sum rule (4.2). Once one has performed the one–loop calculation of the pion photopoduction amplitudes (neutral and charged pions) and tree level processes like $\gamma N \to \pi\pi N$, the dispersive integral can be evaluated to give $\bar{\alpha} + \bar{\beta}$ to two-loop order (in the isospin limit). This can be most easily understood if one considers a typical two-loop Feynman diagram for Compton scattering and cuts it in the interaction region (after/before the absorption/emission of the incoming/outgoing photon). Since these questions are presently under investigation, I will finish the discussion here. I only want to stress one more time that for the baryon sector the polarizabilities are a clean probe of loop effects and that the leading singular terms proportional to M_π^{-1} are generated by four Feynman diagrams (and their crossed partners). At present we do not know how to fish out these most singular diagrams without explicit calculations.

5. OUTLOOK

For the case of pion photoproduction, it remains to also work out the channels involving charged pions ($\gamma p \to \pi^+ n$, $\gamma n \to \pi^- p$). Here, the one loop corrections are much more tame since there is the Kroll-Ruderman term of order $\mathcal{O}(1)$, i.e. the neutral channels are much more sensitive to the chiral loop corrections. It would be particularly interesting to have a measurement of the reaction $\gamma n \to \pi^0 n$ since here the loop corrections show up at leading order ! It is also interesting to consider the electroproduction of pions (virtual photons) to achieve a better understanding of the longitudinal multipoles (and the validity of the LET for L_{0+}). Ultimately, one wants to learn to which degree of accuracy the axial form factor of the nucleon can be extracted from the electroproduction process. Having performed the one–loop calculation for all the pion photoproduction amplitudes, one can attack the problem of calculating the sum of electric and magnetic polarizabilities of the nucleon to two-loop order via the dispersion sum rule. Obviously, these are interesting processes and improved theoretical calculations as well as empirical determinations are very much needed.

6. ACKNOWLEDGEMENTS

The work reported here has been done in collaborations with Véronique Bernard, Jürg Gasser and Norbert Kaiser to whom I express my gratitude. This work was supported in part by Deutsche Forschungsgemeinschaft and by Schweizerischer Nationalfonds.

7. REFERENCES

1. for a review, see: Ulf-G. Meißner, "Chiral Perturbation Theory with Nucleons", preprint BUTP–91/32, 1991, to appear in the proceedings of the Summer School on Nucleons and Nucleon Structure, Institute for Nuclear Theory, University of Washington, Seattle, July 1991.
2. R. Dashen and M. Weinstein, *Phys. Rev.* **183** (1969) 1261; *Phys. Rev.* **188** (1969) 2330;
   L.-F. Li and H. Pagels, *Phys. Rev. Lett.* **26** (1971) 1204;
   P. Langacker and H. Pagels, *Phys. Rev.* **D8** (1973) 4595;
   H. Pagels, *Phys. Rep.* **16** (1975) 219;
   S. Weinberg, *Physica* **96A** (1979) 327;
   J. Gasser and H. Leutwyler, *Ann. Phys. (N.Y.)* **158** (1984) 142.
3. J. Gasser, M.E. Sainio and A. Švarc, *Nucl. Phys.* **B307** (1988) 779.
4. A. Krause, *Helv. Phys. Acta.* **63** (1990) 3.
5. J. Gasser, H. Leutwyler and M.E. Sainio, *Phys. Lett.* **253B** (1991) 252; 260.
6. A. I. Vainshtein and V. I. Zakharov, *Nucl. Phys.* **B36** (1972) 589;
   P. de Baenst, *Nucl. Phys.* **B24** (1970) 633.
7. E. Mazzucato *et al.*, *Phys. Rev. Lett.* **57B** (1986) 3144;
   R. Beck *et al.*, *Phys. Rev. Lett.* **65B** (1990) 1841.
8. A.M. Bernstein and B.R. Holstein, *Comments Nucl. Part. Phys.* **20** (1991) 197.
9. G.F. Chew, M.L. Goldberger, F.E. Low and Y. Nambu, *Phys. Rev.* **106** (1957) 1345.
10. V. Bernard, J. Gasser, N. Kaiser and Ulf-G. Meißner, *Phys. Lett.* **B268** (1991) 291.

11. V. Bernard, N. Kaiser and Ulf-G. Meißner, in preparation.

12. V. Bernard, N. Kaiser and Ulf-G. Meißner, *Phys. Rev. Lett.* **67** (1991) 1515.

13. V. Bernard, N. Kaiser and Ulf-G. Meißner, "Nucleons with Chiral Loops: Electromagnetic Polarizabilities", preprint BUTP-91/30, 1991, to appear in *Nucl. Phys. B.*

14. J. Bijnens and F. Cornet, *Nucl. Phys.* **B296** (1988) 557.

15. G. Ecker, A. Pich and E. de Rafael, *Phys. Lett.* **189B** (1987) 363.

16. G. D'Ambrosio and D. Espriu, *Phys. Lett.* **175B** (1986) 237;
J. Goity, *Z. Phys.* **C34** (1987) 341.

EFFECTIVE LAGRANGIANS AND THE PROTON MATRIX ELEMENT OF THE AXIAL SINGLET CURRENT

J. SCHECHTER
Department of Physics
Syracuse University, Syracuse, NY 13244–1130

V. SONI
National Physical Laboratory, Dr. K.S. Krishnan Road
New Delhi 110012, India

A. SUBBARAMAN
Department of Physics
Syracuse University, Syracuse, NY 13244–1130

and

H. WEIGEL+
Division de Physique Théorique, Institut de Physique Nucléaire
91406 Orsay Cedex, France

ABSTRACT

The "proton spin puzzle" is examined from the effective Lagrangian point of view. A generalized (to include "short distance" information) Skyrme model is shown to give a satisfactory picture. Then we examine the question of the breakup of the axial singlet matrix element into "matter" and "glue" pieces in this framework.

1. Introduction

The polarized muon–polarized proton scattering experiment[1] performed by the European Muon Collaboration has been interpreted as indicating that the expectation value of the quark spin operator in the proton state $<S_i>_{prot.}$ approximately vanishes. Of course, the quark spin operator is $(\frac{1}{2})$ times the space integral of the U(1) axial vector current, J_i^5:

+Address after October 1, 1991: Institut für Theoretische Physik, Universität Tübingen, 7400 Tübingen, Germany.

$$J_i^5 = i \sum_{a=1}^{3} \bar{q}_a \gamma_i \gamma_5 q_a . \tag{1}$$

This result seems counterintuitive in that it flatly contradicts the naive non-relativistic quark model. Another unusual result follows if SU(3) flavor symmetry is assumed: the matrix element of the strange axial vector current seems to be relatively large in magnitude.

These features have justifiably attracted a lot of theoretical attention[2] and even some controversy.

In this talk we will look at the situation from the effective Lagrangian and Skyrme model point of view. Unlike the naive quark model, the naive Skyrme model gives[3] about the experimental value (zero) for the axial singlet proton matrix element. Hence it would seem to be a very reasonable starting point for further investigation.

We should say a word about the relation with chiral perturbation theory (CPT) which is of great interest at this conference. CPT is concerned with the systematic extension of the (already quite good) tree level chiral Lagrangian results away from zero pion momentum. Here we will be probing the interactions of the $\eta'(960)$ meson which, though of great importance for QCD, can hardly be considered soft. We will be working at tree level. However if we succeed in explaining the relevant η' interactions at tree level we may consider the resulting effective Lagrangian as a good candidate for a starting "U(1)" Lagrangian on which to use CPT techniques.

2. Form Factors

For the axial current matrix element of the ath quark between proton states of momenta p and p′ we write:

$$\frac{\sqrt{p_0 p_0'}}{M_N} \langle p' | i\bar{q}_a \gamma_\mu \gamma_5 q_a | p\rangle = i\bar{u}(\underline{p}')\{\gamma_\mu \gamma_5 H_a(q^2) + \frac{i q_\mu}{2M_N} \gamma_5 \tilde{H}_a(q^2)\}\, u(\underline{p}), \tag{2}$$

where $q_\mu = p_\mu - p'_\mu$. The singlet form factors are defined by

$$H(q) \equiv \sum_{a=1}^{3} H_a(q^2) \; ; \; \tilde{H}(q^2) \equiv \sum_{a=1}^{3} \tilde{H}_a(q^2). \tag{3}$$

The EMC experiment measured the linear combination $4H_1(0)+H_2(0) + H_3(0)$. This can be supplemented by the neutron decay axial constant: $1.25 \approx g_A = H_1(0)-H_2(0)$. A third linearly independent combination $R \equiv H_1(0)+H_2(0)-2H_3(0)$ can be estimated (to be $3F-D \approx 0.68$) by SU(3) rotation of the hyperon semi-leptonic decay amplitudes. Although SU(3) works reasonably well for relating the various baryon weak amplitudes it should be recognized that these amplitudes are all of flavor changing type. The extension to the flavor conserving matrix element R has not been previously tested and indeed some models[4] suggest that R could differ non-negligibly from its SU(3) symmetric value. It seems safer to consider R as a parameter. Then one finds (see the discussion in sect.2 of ref. 5) that i) H(0) still remains small but with fairly sizable error bars ii) the "strange" component $H_3(0)$ may be quite a bit closer to zero than its SU(3) value (about −0.23).

3. Results in generalized Skyrme models – a brief review

The by now well-known result[3] that H(0)=0 in the basic (Skyrme term plus quadratic term) SU(3) Skyrme model can be seen immediately.[6] First note the decomposition of the nonet matrix U:

$$U = \exp\left(\frac{2i\eta'}{\sqrt{3}F_\pi}\right) \tilde{U} \ , \ \det \tilde{U} = 1. \tag{4}$$

Using Noether's theorem, the U(1) axial current turns out to be the gradient of the η' field:

$$J^5_\mu = \sqrt{3}F_\pi \partial_\mu \eta'. \tag{5}$$

The structure of this current does not change when one evaluates its matrix elements in the soliton sector of the theory. Hence, going over to momentum space, it must be proportional to q_μ in Eq. 2 and can only contribute to $\tilde{H}(q^2)$. Thus H(0)=0 in this model.

Before going further we should remark that the predictions of the basic SU(3) Skyrme model are actually fairly reasonable when one diagonalizes the collective Hamiltonian exactly[7] rather than by using lowest order perturbation theory (as was originally done).

Nevertheless, a great deal of effort has been made by many workers to construct "generalized" Skyrme models by including additional terms or degrees of freedom which would probe more deeply the short distance structure of the nucleon. Examples are higher derivative terms, explicit quarks and vector mesons. A critical discussion of the relevant literature is contained, among other things, in ref. 5. It seems to us that a particularly neat choice is to construct an effective Lagrangian using the low lying s-wave $q\bar{q}$ states – the pseudoscalars and vectors – and keeping a minimal number of derivatives. Such a model[8] has been studied in detail and the predictions are rather encouraging. Interestingly, both the non-electromagnetic part of the n-p mass difference (which cannot be explained in the basic model containing pseudoscalars only) and the possibility of a non-zero value of H(0) require terms proportional to the Levi-Civita symbol $\epsilon_{\mu\nu\alpha\beta}$ in this model. The singlet axial current now takes the form

$$J^5_\mu = \sqrt{3}F_\pi \partial_\mu \eta' + s\mathfrak{J}^5_\mu, \tag{6}$$

where $\mathfrak{J}^5_\mu$ is a complicated expression given in eq. (4.5) of ref. 5. A typical term of $\mathfrak{J}^5_\mu$ is proportional to $\epsilon_{\mu\nu\alpha\beta}\mathrm{Tr}(\rho_\nu\rho_\alpha\rho_\beta)$, where ρ_ν is the vector meson nonet matrix, which clearly is not a pure gradient.

We have included in Eq. 6 for our present purpose a factor s which would be unity if, in the relevant term of $\mathscr{L}_{\text{eff}}$, $-(s/\sqrt{3}F_\pi)\ \partial_\mu \eta' J^5_\mu$, η' only appears via U. This expresses the possibility of additional nonet rule violation for the η' interactions. The actual calculation yields[5,9]

$$H(0) = (0.30 \pm 0.03)\ s, \tag{7}$$

wherein an estimate of the "theoretical uncertainty" was made. The choice s=1 agrees with present experiment but a reduction in the experimental errors might necessitate a different choice.

The strange form factor $H_3(0)$ is much harder to calculate since details of SU(3) symmetry breaking are very important. Making the attempt, one typically finds $H_3(0) \sim -0.02$ [including vectors, ref. 10] to -0.08 [pseudoscalars only, first of ref. 4]. This should be compared with the central value of the SU(3) symmetric estimate $H_3(0) \sim -0.23$.

Evidently the overall picture in the Skyrme model generalized to include vectors is a satisfactory one. From a practical point of view one might want to stop here. However, our present goal is to try to make some contact with the 2-component schemes which are suggested by intuitive consideration of the QCD parton model.

4. Attempts at a Two-Component Decomposition

To make intuitive sense of $H(0) \approx 0$ in the QCD parton model it has been suggested[11] to make a decomposition based on the anomalous divergence equation:

$$\partial_\mu J^5_\mu = 2i \sum_{a=1}^{3} m_a \bar{q}_a \gamma_5 q_a + \partial_\mu K_\mu, \tag{8}$$

where $K_\mu = \frac{-3ig^2}{4\pi^2} \epsilon_{\mu\beta\gamma\delta} \operatorname{Tr}(A_\beta \partial_\gamma A_\delta - \frac{2ig}{3} A_\beta A_\gamma A_\delta)$ is the Chern–Simons current. Writing, with the neglect of quark mass terms,

$$\langle J^5_\mu \rangle_{\text{prot}} = \langle J^5_\mu - K_\mu \rangle_{\text{prot}} + \langle K_\mu \rangle_{\text{prot}} \ , \tag{9}$$

it is perhaps natural to associate the conserved (but not gauge invariant) quantity $(J^5_\mu - K_\mu)$ with the ordinary "matter" contribution to the spin current and to thereby expect $\langle J^5_\mu - K_\mu \rangle_{\text{prot}} \sim 1$. This is supposed to be cancelled by a "gluon spin" contribution $\langle K_\mu \rangle_{\text{prot}} \sim -1$. Although this decomposition for the relevant special case of the "forward" matrix element has been claimed to be gauge invariant, a recent calculation[12] in the Schwinger model puts this in doubt. We must conclude that the procedure is suspicious. Nevertheless it is intriguing so one may wish to search for alternate approaches to a matter vs. glue decomposition.

One suggestion[13] is to simply take matrix elements of Eq. 8 and use pole saturation to evaluate the right hand side. The picture is:

$$\begin{aligned} <\partial_\mu J^5_\mu>_p &= <2i\Sigma m_a \bar{q}_a \gamma_5 q_a>_p + <\partial_\mu K_\mu>_p \\ 2m_N H(0) &= \quad \text{"matter"} \quad + \quad \text{"glue"}. \end{aligned} \tag{10}$$

However there are two problems with this. The first is that, unlike Eq. 9, it doesn't work for $m_a \to 0$. Secondly,[14] because of the large isospin violation in the current quark masses ($m_d \approx 2m_u$) a large isospin violation is induced for the matrix elements of interest; for examples $<\bar{s}\gamma_5 s>_n \neq <\bar{s}\gamma_5 s>_p$ and $<\partial_\mu K_\mu>_n \neq <\partial_\mu K_\mu>_p$. This may be easily understood from the equation [see section 3 of ref. 15 for more details]:

$$\sum_j (M^2)_{ij} <\varphi_j>_{\substack{p\\n}} = g_{i\substack{p\,p\\n\,n}}\,, \tag{11}$$

where $\varphi_i \sim \bar{q}_i \gamma_5 q_i$, $(M^2)_{ij}$ is proportional to the "bare" quark mass matrix including instanton effects and g_{ipp} and g_{inn} are the Yukawa coupling constants of the pseudoscalar fields to the proton and to the neutron respectively.. The right hand side of Eq. 11 is isospin invariant (e.g. $g_{3pp} \approx g_{3nn}$). Thus the large isospin violation in $(M^2)_{ij}$ induces a large isospin violation in $<\varphi_j>$.

5. Two Components for the Yukawa Couplings

The previous remark contains a hint about a consistent way to proceed. Since the Yukawa couplings are isospin conserving it is reasonable to compare baryon couplings to η' and to the gauge invariant quantity $\partial_\mu K_\mu$ as a measure of "matter" vs. "glue". This gives a generalization[14,16] of the old fashioned Goldberger–Treiman relation.

In carrying out such an approach it is necessary to take account of the "U(1) problem" of QCD. The following effective Lagrangian[17] generates a mass for the η' while still maintaining the (massless) anomaly Eq. 8:

$$\mathcal{L}_{\text{eff}} = \frac{-F_\pi^2}{8} \text{Tr}(\partial_\mu U \partial_\mu U^\dagger) + \frac{G^2}{6F_\pi^2 m_{\eta'}^2} + \frac{iG}{12} \ln \left(\frac{\det U}{\det U^\dagger}\right), \tag{12}$$

where $G = \partial_\mu K_\mu$. It is crucial that G is constrained by its equation of motion to be:

$$G = \sqrt{3}F_\pi m^2_{\eta'}\eta'. \tag{13}$$

Now let us construct an old fashioned chiral Lagrangian in which the nucleon fields are explicitly included (It is more conceptually elegant to generate the nucleons as solitons of the meson Lagrangian but we will come to this later). Of course, a Yukawa term for the η' must be added to $\mathscr{L}_{eff}$:

$$-\frac{g^{(2)}_{\eta' NN}}{2m_N}\,\mathrm{Tr}(\bar{N}\gamma_\mu\gamma_5 N)\,\partial_\mu\eta'. \tag{14}$$

[For more details and care see section 2 of ref. 15]. This results in an addition to J^5_μ (corresponding to the second term in Eq. 6) so that now

$$J^5_\mu = \sqrt{3}F_\pi\partial_\mu\eta' + iH(0)\,\mathrm{Tr}(\bar{N}\gamma_\mu\gamma_5 N), \tag{15}$$

where $H(0) = \sqrt{3}F_\pi g^{(2)}_{\eta' NN}/2m_N$ [Goldberger–Treiman formula]. Eqs. 14 and 15 represent pre QCD physics. Taking QCD into account suggests the possibility of an additional effective coupling of the gauge invariant gluonic quantity $G=\partial_\mu K_\mu$ to the nucleon, namely

$$-\frac{ig_{GNN}}{2m_N}\,\mathrm{Tr}(\bar{N}\gamma_\mu\gamma_5 N)\,\partial_\mu G\,. \tag{16}$$

Notice that G is a chiral singlet so the addition of Eq. 16 modifies neither the axial anomaly Eq. 8 nor the U(1) axial current Eq. 15. What does change is the formula for eliminating G; instead of Eq. 13 we now have

$$G = \sqrt{3}F_\pi m^2_{\eta'}\eta' - \frac{3ig_{GNN}}{2m_N}F_\pi m^2_{\eta'}\partial_\mu \mathrm{Tr}(\bar{N}\gamma_\mu\gamma_5 N). \tag{13a}$$

Substituting this back into $\mathscr{L}_{eff}$ yields an extra contribution to the η' Yukawa interaction so that the relevant coupling constant is no longer $g^{(2)}_{\eta' NN}$ [which is still however related to H(0) by the Goldberger–Treiman formula]. Expressed in terms of the <u>net</u> coupling constant $g_{\eta' NN}$ the Goldberger–Treiman formula now[16,15,18] takes the "two–component" form:

$$H(0) = \frac{\sqrt{3}F_\pi}{2m_N}\left(g_{\eta' NN} - \sqrt{3}F_\pi m^2_{\eta'}g_{GNN}\right). \tag{17}$$

This formula was first derived by Shore and Veneziano[16] in a somewhat different way. The first term [which, due to Eq. 13a , contains a "hidden" glue part] is analogous to $\langle J^5_\mu - K_\mu \rangle$ in Eq. 9 and is the "matter" piece while the second term corresponds to the "glue" piece. Regardless of the status of the decomposition (Eq. 9), we may note that the two terms in Eq. 17 are, by construction, each gauge invariant.

While Eq. 17 is apparently well defined it is not immediately helpful since $g_{\eta' NN}$ and g_{GNN} are not well known. Thus we cannot conclude that the first term in Eq. 17 is around 1 as expected intuitively in the quark picture. One might find $g^2_{\eta' NN}$ in principle from analysis[19] of the SU(3) singlet exchange contribution to low energy baryon baryon scattering:

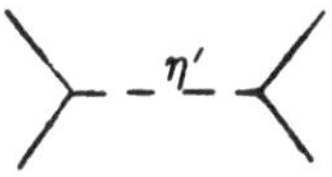

However there is an interesting new wrinkle in our model. Substituting Eq. 13a back in $\mathcal{L}_{eff}$ also yields a contact term:

The net singlet exchange amplitude, $A(q^2)$ is the sum of these two contributions. We find,[15] in fact,

$$A(0) = cH(0), \tag{18}$$

where $c = 2m_N \left(g_{GNN} + g_{\eta' NN}/\sqrt{3}F_\pi m^2_{\eta'}\right)$. Eq. 18 shows that A(0) is a direct physical "measure" of the singlet axial matrix element H(0).

6. "Matter" vs. "Glue" in a Soliton Model

Now let us evaluate $g_{\eta' NN}$ in Eq. 17 using the effective Lagrangian[8] of pseudoscalars and vectors mentioned before and where the nucleon is treated as a soliton excitation. The model actually does a reasonable job of accounting for many baryon properties without introducing any parameters other than those known experimentally from mesonic interactions. The crucial modification of this model required is to allow the gluonic field G to couple to the "short distance" part of the axial current as we did in Eq. 16 for the model with explicit baryons. We thus add[20] to the effective Lagrangian the term:

$$\frac{t}{3F_\pi m^2_{\eta'}} \, \partial_\mu G \, \mathcal{J}^5_\mu \,, \tag{19}$$

where $\mathcal{J}^5_\mu$ has been discussed after Eq. 6 and where t is a new dimensionless constant (proportional to g_{GNN}).

We observe that the equation of motion for the η' field in this model is

$$(-\Box + m^2_{\eta'})\eta' = \frac{s-t}{\sqrt{3}F_\pi}\,\partial_\mu J^5_\mu\,, \tag{20}$$

where s is the "nonet fudge factor" introduced in Eq. 6. Taking the expectation value of both sides of Eq. 20 in the proton state immediately gives the desired formula:

$$g_{\eta' NN} = \left(\frac{s-t}{s}\right)\frac{2m_N H(0)}{\sqrt{3}\ F_\pi}\,. \tag{21}$$

When $t \to 0$ this reduces to the old "one–component" Goldberger Treiman relation. We already found H(0)/s in this model (see Eq. 7). It remains to find (s–t) which is seen from Eq. 20 to measure the strength of the source of the η' field. This may be obtained in an indirect fashion. It was explained in detail in ref. 21 that the non–electromagnetic part of the n–p mass difference arises in the Skyrme model in large part due to the excitation of the non–strange components of the η' field. In turn this excitation is proportional to the source of the η' field. Adapting[20] an earlier calculation[21] of the n–p mass difference gives the estimate: $s-t=1.3\pm0.5$. Taking this together with Eq. 7 in Eq. 21 yields $g_{\eta' NN}=3.2\pm1.6$. Writing Eq. 17 as

$$H(0) = \text{"matter} + \text{"glue"}, \tag{22}$$

then gives

$$\text{"matter"} = 0.4 \pm 0.2. \tag{23}$$

Choosing a value for H(0) within the allowed experimental range then fixes "glue" by subtraction. It also fixes the value of the fudge factor s. Specifically we have the chart

H(0)	s	"glue"
–0.1	–0.33	–0.5 ± 0.2
0	0	–0.4 ± 0.2
0.1	0.33	–0.3 ± 0.2
0.2	0.67	–0.2 ± 0.2
0.3	1	–0.1 ± 0.2

From this chart and Eq. 23 we first conclude that there is a tendency for "glue" to cancel "matter" as desired from intuitive considerations. However, rather than a situation like $0 \approx 1-1$ we get small=small′–small". Of course, taking the uncertainties into account one might argue that an intermediate situation is consistent. We do expect, however, that s should be close to 1, which would imply that the "glue" piece is small[22].

7. Summary

The soliton approach to calculating the proton axial singlet current matrix element has the nice feature that the simplest model (SU(3) Skyrme model with 2 basic terms) already gives the desired result of zero. However, experimentally H(0) may not be precisely zero. In the soliton approach, deviations from zero arise from the "short distance" part of the effective Lagrangian. It is therefore a probe of some very interesting physics. We have stressed and made use of the fact that the same physics is probed by the non–electromagnetic part of the n–p mass difference. The model in which short distance physics is described by vector mesons was found to result in a reasonable picture. Of course other descriptions of short distance effects, like explicit quarks, could also be used.[23]

We have also seen that a gauge invariant decomposition of H(0) into "matter" and "glue" pieces can be made in the present framework. This provides a further probe of the part of the effective Lagrangian which describes the U(1) anomaly. A simple derivation of the Shore–Veneziano formula was given and the proportionality between H(0) and the SU(3) singlet exchange amplitude for baryon baryon scattering was noted. An estimate revealed that the "matter" and "glue" contributions to H(0) do tend to cancel each other but each is most likely small compared to unity. In progress is an attempt to improve this estimate by considering the n–p mass difference calculation in a "fully cranked"[10] SU(3) vector model. The ultimate goal is a good understanding of the connection between the QCD parton model and the soliton approach.

8. References

1. E.M.C., J. Ashman, et.al., *Phys. Lett.* **B206**, (1988) 364; *Nucl. Phys.* **B238** (1989) 1.
2. Reviews are provided by R. Jaffe and A. Manohar, *Nucl. Phys.* **B337** (1990) 509; G. Altarelli and W.J. Stirling, CERN report No. Th5249/88; G. Altarelli, CERN report No. Th5675/90; H.–Y. Cheng. Institute of Physics, Academia Sinica, Taipei report IP–ASTP–01–91.
3. S. Brodsky, J. Ellis and M. Karliner, *Phys. Lett.* **B206** (1988) 309.
4. See N.W. Park, J. Schechter and H. Weigel, *Phys. Rev.* **D43** (1991) 869 and references therein; see also H.J. Lipkin in *Future Polarization Physics at Fermilab*, edited by E. Berger et.al. (1988).
5. R. Johnson, N.W. Park, J. Schechter, V. Soni and H. Weigel, *Phys. Rev.* **D42** (1990) 2998.
6. N.W. Park, J. Schechter and H. Weigel, *Phys. Lett.* **B228** (1989) 420; T.D. Cohen and M.K. Banerjee, *Phys. Lett.* **B230** (1989) 129.
7. H. Yabu and K. Ando, *Nucl. Phys.* **B301** (1988) 601. See also the first of ref. 4.
8. Ö. Kaymakcalan and J. Schechter, *Phys. Rev.* **D31** (1985) 1109; T. Fujiwara, T. Kugo, H. Terao, S. Uehara and K. Yamawaki, *Prog. Theor. Phys.* **73** (1985) 926; P. Jain, R. Johnson, U.–G. Meissner, N.W. Park and J. Schechter, *Phys. Rev.* **D37** (1988) 3252; U.–G. Meissner, N. Kaiser, H. Weigel and J. Schechter, *Phys. Rev.* **D39** (1989) 1956; B. Schwesinger, H. Weigel, G. Holzwarth and A. Hayashi, *Phys. Rep.* **173** (1989) 173; H. Weigel, *Phys. Lett.* **B215** (1988) 24.
9. V. Bernard, N. Kaiser and Ulf–G. Meissner, *Phys. Lett.* **B237** (1990) 545.
10. N.W. Park and H. Weigel, Orsay report IPNO (TH91)–57.
11. A.V. Efremov and O.J. Teryaev, Dubna report JINR E2–88–297; G.

Altarelli and G.G. Ross, *Phys. Lett.* **B212** (1989) 391; R.D. Carlitz, J.C. Collins and A.H. Mueller, *Phys. Lett.* **B214** (1988) 229.
12. A. Manohar, *Phys. Rev. Lett.* **66** (1991) 1663 and references therein.
13. See for example, T.P. Cheng and L.–F. Li, *Phys. Rev. Lett.* **62** (1989) 1441.
14. G. Veneziano, *Mod. Phys. Lett.* **A4** (1989) 1605; T. Hatsuda, *Nucl. Phys.* **B329** (1990) 376.
15. J. Schechter, V. Soni, A. Subbaraman and H. Weigel, *Mod. Phys. Lett.* **A5** (1990) 2543.
16. G.M. Shore and G. Veneziano, *Phys. Lett.* **B244** (1990) 75.
17. C. Rosenzweig, J. Schechter and G. Trahern, *Phys. Rev.* **D21** (1980) 3388; P. DiVecchia and G. Veneziano, *Nucl. Phys.* **B171** (1980) 253; E. Witten, *Ann. Phys.* **128** (1981) 1789; P. Nath and R. Arnowitt, *Phys. Rev.* **D23** (1981) 473.
18. T. Hatsuda, Univ. of Washington report INT–5–90.
19. See for example A.V. Efremov, J. Soffer and N.A. Tornqvist, *Phys. Rev. Lett.* **64** (1990) 1495.
20. J. Schechter, V. Soni, A. Subbaraman and H. Weigel, *Phys. Rev. Lett.* **65** (1990) 2955.
21. P. Jain, R. Johnson, N.W. Park, J. Schechter and H. Weigel, *Phys. Rev.* **D40** (1989) 855.
22. This is also claimed in a lattice QCD approach: J. Mandula, *Phys. Rev. Lett.* **65** (1990) 1403.
23. Still another approach involves QCD sum rules. See J. Pasupathy, *Proceedings of the International Seminar on Direct Nuclear Reactions*, Jan. 1989.

Proton Spin and Quark Spin *

Peter Minkowski
Institute for Theoretical Physics
University of Bern
Sidlerstr. 5, CH-3012 Bern, Switzerland

December 1991

Abstract

The spin and angular momentum structure of the proton is discussed, within the framework of QCD involving the three light quark flavors u, d , s. It is argued, that unless large angular (sub)momenta - all adding up to $J_{tot} = 1/2$ - compose a considerable fraction of the proton, contrary to the simple ordering pattern seen in the spectroscopy of baryonic resonances, the asymptotic sum rule for the flavor singlet spin dependent structure function satisfies the inequality :

$$\lim_{Q^2 \to \infty} 2 \int g_1^{0p}(x, Q^2)\, dx = \sum_{q=u,d,s} g_q \gtrsim 0.4 \pm 0.1$$

The quantities $g_{u,d,s}$ in the equation above denote the axial current proton coupling constants for the respective quark flavor. The superscript ^{0p} denotes the singlet (vector)currrent correlation function of the proton, involving the spin asymmetry.

*Work supported in part by Schweizerischer Nationalfonds

We consider the spin dependent proton structure function [1] - [7]

$$g_1^{p\,jj}(x, Q^2)$$

corresponding to the (electromagnetic) current correlation function

$$W_{\mu\nu}^{jj}(q, p; h', h) =$$

$$\frac{1}{4\pi} \langle \mathcal{N}, p; h' | \int d^4 z\, e^{iqz} (j_\mu(z)\, j_\nu(0)) | \mathcal{N}, p; h \rangle$$

$$W_{\mu\nu}^{jj}(h', h) = \begin{array}{l} \left\{ \left(-g_{\mu\nu} + \dfrac{q_\mu q_\nu}{q^2} \right) F_1^{p\,jj} + (p_\mu p_\nu)_\perp F_2^{p\,jj} \right\} \delta_{h'h} + \\ + i\varepsilon_{\mu\nu\sigma\tau} q^\sigma\, \bar{u}(p, h') \gamma^\tau \gamma_5 u(p, h) \dfrac{1}{2\nu} g_1^{p\,jj} + \cdots \end{array} \tag{1}$$

In eq. (1) standard notation for deep inelastic scattering is used :

$$f^{p\,jj} = f^{p\,jj}(x, Q^2) \quad \text{for} \quad f = F_1, F_2, g_1, \cdots$$

$$\nu = pq \quad , \quad Q^2 = -q^2 \quad , \quad x = \frac{Q^2}{2\nu} \tag{2}$$

$$(p_\mu p_\nu)_\perp = \left(p_\mu - \frac{\nu}{q^2} q_\mu \right) \left(p_\nu - \frac{\nu}{q^2} q_\nu \right)$$

whereas $h', h = \pm 1/2$ denote the helicities of the nucleons. For $|\vec{p}| \to \infty$ the (covariant) spin vector of the nucleon is given by the axial bilinear :

$$\text{for} \quad |\vec{p}| \to \infty \quad : \quad \left\{ \begin{array}{l} \bar{u}(p, h) \gamma^\tau \gamma_5 u(p, h) \approx 2 m_{\mathcal{N}} s^\tau(h) \\ s(h) \approx (2h)\, p / m_{\mathcal{N}} \end{array} \right.$$

The currents in the correlation function in eq. (1) are considered as neutral vector currents within QCD involving the three light quark flavors u, d, s. In flavor space the current j_μ is determined by a diagonal flavor matrix Q_j (Q^{em} for the electromagnetic current) :

$$j_\mu = \bar{q}_A Q^j_{AB} \gamma_\mu q \quad ; \quad A, B = u, d, s$$

$$Q^j_{AB} = Q^j_A \delta_{AB} \quad ; \quad Q^{em}_A = (2/3, -1/3, -1/3)$$

We are interested here only in the mean value of $g_1^{p\,jj}(x, Q^2)$ over x :

$$G_1^{p\,jj}(Q^2) = \int_0^1 dx\, g_1^{p\,jj}(x, Q^2)$$

$$G_1^{p\,jj} = G_1^{p\,em} \quad ; \quad \text{for} \quad j = j^{em} \tag{3}$$

We shall call $G_1^{p\,jj}$ in eq. (3) the spin moment or more precisely the quark spin moment of the structure function $g_1^{p\,jj}$.
Spin moments are decomposed according to the flavor matrices $(Q^j)^2$

$$(Q^j)^2 = q^{jj}_{(0)} Q^{(0)} + q^{jj}_{(8)} Q^{(8)} + q^{jj}_{(3)} Q^{(3)}$$

$$Q^{(0)} = \begin{pmatrix} 1 & & \\ & 1 & \\ & & 1 \end{pmatrix} \quad Q^{(8)} = \begin{pmatrix} 1 & & \\ & 1 & \\ & & -2 \end{pmatrix} \quad Q^{(3)} = \begin{pmatrix} 1 & & \\ & -1 & \\ & & 0 \end{pmatrix}$$

$$G_1^{p\,jj} = \frac{1}{2} \left(q^{jj}_{(0)}\, G_1^{p\,(0)} + q^{jj}_{(8)}\, G_1^{p\,(8)} + q^{jj}_{(3)}\, G_1^{p\,(3)} \right) \tag{4}$$

The basis spin moments

$$G_1^{p\,(X)} \quad \text{for} \quad X = 0, 8, 3$$

are in one to one correspondence with the associated axial vector current operators $j_\mu^{5\,(X)}$:

$$G_1^{p\,(X)} \quad \leftrightarrow \quad j_{\tau\,5}^{(X)} = \bar{q}\, Q^{(X)} \gamma_\tau \gamma_5\, q \tag{5}$$

The current operators $j_\tau^{5\,(X)}$ in eq. (5) generate the nucleon forward matrix elements :

$$\langle \mathcal{N}, p; h' | j_\tau^{5\,(X)}(0) | \mathcal{N}, p; h \rangle = \bar{u}(p, h') \gamma^\tau \gamma_5 u(p, h)\, g_{\mathcal{N}}^{(X)}$$

$$g_{(\mathcal{N}=p)}^{(X)} = g^{(X)} \quad , \quad g_{(\mathcal{N}=n)}^{(0),(8)} = g^{(0),(8)} \quad , \quad g_{(\mathcal{N}=n)}^{(3)} = -g^{(3)}$$

$$g^{(0)} = g_u + g_d + g_s \quad , \quad g^{(8)} = g_u + g_d - 2g_s \quad , \quad g^{(3)} = g_u - g_d \tag{6}$$

The relations in eq. (6) are a consequence of isospin ($SU2_{u,d}$) invariance. A multiplicative renormalization of the axial currents $j_\tau^{5\,(X)}$ entails the identical renormalization of the (proton) matrix elements $g^{(X)}$. We denote this property by the formal relation :

$$g^{(X)} = g^{(X)}\left[j_\tau^{5\,(X)} \right] \quad \propto \quad j_\tau^{5\,(X)}$$

The vector currents in the correlation function in eq. (1) as well as the traceless axial currents generating together the charges of $SU3_{fl} \times SU3_{fl}$ have identically vanishing anomalous dimensions. The spin moments $G_1^{p\,(X)}$ in eq. (5) are the product of a Wilson coefficient function and the associated axial current matrix element $g^{(X)}$ defined in eq. (6) :

$$G_1^{p\,(X)}\left[Q^2 / \Lambda^2 \right] = C^{(X)}\left[Q^2/\mu^2, \kappa_\mu \right] g^{(X)}(\mu^2) + O(1/Q^2)$$

$$g^{(X)}(\mu^2) = g^{(X)}\left[j_\tau^{5\,(X)} \Big|_\mu \right] \quad ; \quad \kappa_\mu = \frac{g_\mu^2}{16\pi^2} \tag{7}$$

The coefficient functions $C^{(X)}$ satisfy a renormalization group equation :

$$-\partial_\tau C^{(X)} = \left[\gamma^{(X)}\left(j_\tau^{5\,(X)} \right) + b(\kappa)\partial_\kappa \right] C^{(X)}$$

$$\tau = \log(Q^2/\mu^2) \quad , \quad b(\kappa) = -(\beta/g)\kappa = 9\kappa^2 + 64\kappa^3 + \cdots$$

$$\gamma^{(X)} = \gamma^{(X)}(\kappa) = \mu^2 \frac{\partial}{\partial \mu^2} \log\left[Z\left(j_\tau^{5\,(X)} \right) \right]$$

$$\gamma^{(3)} = \gamma^{(8)} = 0 \quad , \quad \gamma^{(0)}(\kappa) = 24\kappa^2 + O(\kappa^3) \tag{8}$$

In eq. (8) $b(\kappa)$ denotes the coupling constant rescaling function for three color triplet quark flavors, whereas

$$\gamma^{(X)}(\kappa) = \mu^2 \frac{\partial}{\partial \mu^2} \log \left[Z \left(j_\tau^{5\,(X)} \right) \right]$$

stands for the anomalous dimension of the axial current $j_\tau^{5\,(X)}$ respectively [8] . We define the axial coupling constants $g_{u,d,s}$ as referring to the axial currents normalized at infinite scale :

$$g^{(X)} = g^{(X)}(\infty) = g^{(X)} \left[j_\tau^{5\,(X)} \Big|_{\mu=\infty} \right] = \frac{g^{(X)}(\mu'^2)}{E^{(X)}(\kappa_{\mu'})} \tag{9}$$

$$j_\tau^{5\,(X)} = j_\tau^{5\,(X)} \Big|_{\mu=\infty}$$

In eq. (9) the functions $E^{(X)}(\kappa)$ denote the expoinential factors :

$$E^{(X)}(\kappa) = \exp \int_0^\kappa \frac{\gamma^{(X)}(\kappa')}{b(\kappa')} d\kappa' \quad ; \quad E^{(X)}(0) = 1 \tag{10}$$

Eq. (8) yields the solution (to first order in the running coupling constant $\bar{\kappa}(Q^2/\Lambda^2)$) [8] , [9] :

$$G_1^{p\,(X)} [Q^2/\Lambda^2] = C^{(X)} [1, \bar{\kappa}] \; E^{(X)}(\bar{\kappa}) \; g^{(X)}$$

$$E^{(3)\,,\,(8)} \equiv 1 \quad , \quad E^{(0)}(\bar{\kappa}) = 1 + \frac{8}{3} \bar{\kappa} + O(\bar{\kappa}^2)$$

$$G_1^{p\,(3)} \approx (1 - 4\bar{\kappa})(g_u - g_d) \quad , \quad G_1^{p\,(8)} \approx (1 - 4\bar{\kappa})(g_u + g_d - 2g_s)$$

$$G_1^{p\,(0)} \approx (1 - 4\bar{\kappa}/3)(g_u + g_d + g_s)$$

(11)

The relations in eq. (11) have been widely discussed in terms of quark and gluon spin asymmetry distributions in connection with the anomalous Ward identity of the singlet axial current [9] , [10] , [11] , [12] .
The singlet axial current normalized at infinite scale (eq. (9)) defines the (uniquely normalized) quark spin density operator

$$\mathcal{F}_{(q)\ \mu\nu}{}^{\sigma} = (1/2)\, \varepsilon_{\mu\nu}{}^{\sigma\tau}\, j_{\tau}^{5\,(0)}$$

contributing to the angular momentum density (in QCD) :

$$\begin{aligned}
&\mathcal{J}_{\mu\nu}{}^{\sigma} = \mathcal{F}_{(q)\ \mu\nu}{}^{\sigma} + T_{\mu\nu}{}^{\sigma} \quad , \quad \partial_{\sigma}\, \mathcal{J}_{\mu\nu}{}^{\sigma} = 0 \\
&\mathcal{F}_{(q)\ \mu\nu}{}^{\sigma} = (1/2)\, \varepsilon_{\mu\nu}{}^{\sigma\tau} \textstyle\sum_q \bar{q}\, \gamma_{\tau}\, \gamma_5\, q \\
&T_{\mu\nu}{}^{\sigma} = x_{\mu}\, \Theta_{\nu}{}^{\sigma} - x_{\nu}\, \Theta_{\mu}{}^{\sigma} \\
&\Theta_{\nu}{}^{\sigma} = \Theta_{(V)\ \nu}{}^{\sigma} + \Theta_{(q)\ \nu}{}^{\sigma} \\
&\Theta_{(V)\ \nu}{}^{\sigma} = F_{\varrho\ \nu}^{\ a}\, F^{\sigma\varrho\ a} + (1/4)\, \delta_{\nu}{}^{\sigma}\, F_{\alpha\ \beta}^{\ a}\, F^{\alpha\beta\ a} \\
&\Theta_{(q)\ \nu}{}^{\sigma} = \textstyle\sum_q \bar{q} \left[(i/2)\, \gamma^{\sigma}\, \overleftrightarrow{D}_{\nu} \right] q - \delta_{\nu}{}^{\sigma}\, \mathcal{L}_q
\end{aligned} \tag{12}$$

In eq. (12) $F_{\alpha\ \beta}^{\ a}$ denotes the (eight) field strength tensors of the color group and $D_{\nu}\, q$ stands for the covariant derivative of the color triplet quark fields.
The equal time commutators of spacelike components of (conserved) vector currents involve specifically the singlet axial current $j_{\tau}^{5\,(0)} \Big|_{\mu=\infty}$:

$$j_{\mu} = \bar{q}\; Q^{j}\, \gamma_{\mu}\, q$$

$$[\, j_1(t, \vec{x}),\, j_2(t, \vec{y})\,] = 2\, i\, \varepsilon_{123}\, \delta^3(\vec{x} - \vec{y})\; q_{(X)}^{jj}\; j^{\tau=3,\,5\,(X)}(t, \vec{x}) \tag{13}$$

The normalization of the singlet axial current (eq. (9)) is fixed by the commutation rules of the spin operators :

$$\vec{S}_q(t) = \frac{1}{2} \int d^3x \, \vec{j}^{5\,(0)}(t, \vec{x}) \tag{14}$$

$$\left[S_q^m(t), S_q^n(t) \right] = i \varepsilon_{mnr} \, S_q^r(t) \quad , \quad m, n, r = 1, 2, 3$$

Eqs. (9) and (13) lead to the asymptotic sum rules [1] , [2] .

$$G_\infty^{\mathcal{N}\,em} = \lim_{Q^2 \to \infty} G_1^{\mathcal{N}\,em}(Q^2 / \Lambda^2)$$

$$G_\infty^{p\,em} = \frac{1}{9} g^{(0)} + \frac{1}{36} g^{(8)} + \frac{1}{12} g^{(3)} \quad , \quad G_\infty^{p\,em} - G_\infty^{n\,em} = \frac{1}{6} g^{(3)} \tag{15}$$

$SU3_{fl}$ analysis of $g^{(8)}$

In the limit of exact $SU3_{fl}$ symmetry the following relations hold between the constants $g^{(3)\,,\,(8)}$ in eq. (15) and the axial current matrix elements obtained from hyperon decays :

decay	g_A / g_V	experiment	$SU3_{fl}$ *fit*
$n \to p e^- \bar{\nu}_e$	$F + D$	1.260 ± 0.004	1.259
$\Lambda \to p e^- \bar{\nu}_e$	$F + D/3$	0.718 ± 0.015	0.726
$\Xi^- \to \Lambda e^- \bar{\nu}_e$	$F - D/3$	0.250 ± 0.050	0.192
$\Sigma^- \to n e^- \bar{\nu}_e$	$F - D$	− 0.340 ± 0.017	− 0.341

Table 1

The values in the rightmost column of Table 1 correspond to the following parameters, which we compare to the corresponding quantities inferred by Bourquin et al. [13], given in parentheses :

$$F = 0.46 \pm 0.025 \quad , \quad D = 0.80 \pm 0.025$$
$$(0.458 \pm 0.016) \qquad (0.734 \pm 0.017)$$

$$F / D = 0.58 \pm 0.06$$
$$(0.62)$$

$$g^{(3)} = F + D = 1.260 \pm 0.004 \quad , \quad g^{(8)} = 3F - D = 0.58 \pm 0.10$$
$$(1.192) \qquad (0.64)$$
$$(16)$$

The error of $g^{(8)}$ given in eq. (16) does not take into account effects of $SU3_{fl}$ symmetry breaking.

Ignoring any additional error beyond the one quoted in eq. (16) for $g^{(3)} = g_A$ and $g^{(8)}$ we find for the nonsinglet (NS) linear combination occuring in the spin moment $G_{\infty}^{p\, em}$ in eq. (15) :

$$G_{\infty}^{p\, NS} = \frac{1}{36} g^{(8)} + \frac{1}{12} g^{(3)} = 0.121 \pm 0.003 \tag{17}$$

It is hard to conceive that $SU3_{fl}$ breaking effects can change the error for $G_{\infty}^{p\, NS}$ in eq. (17) appreciably, beyond 0.005 say.

The result of the EMC collaboration for the spin moment at an inferred average value of $\langle Q^2 \rangle = 10\ GeV^2$ is [3] :

$$G_1^{p\, em} (\langle Q^2 \rangle / \Lambda^2) = 0.126 \pm 0.010 \pm 0.015$$

$$G_1^{p\, em} \approx \frac{1}{9} g^{(0)} + G_{\infty}^{p\, NS} = 0.121 \pm 0.003 + \frac{1}{9} g^{(0)} \tag{18}$$

$$\rightarrow \qquad g^{(0)} = 0.045 \pm 0.090 \pm 0.135$$

The remaining discussion is devoted to a theoretical estimate of the singlet axial current coupling constant $g^{(0)}$ as constrained by the angular momentum and quark spin structure of the proton [14] , [15] , [16] , [17] , [18] .

The link between quark spin and $g^{(0)}$ is provided by the relation in eq. (14).

Thus we decompose the total angular momentum in the following way :

$$\vec{J} = \vec{S}_q + \vec{R} \quad , \quad \vec{S}_q = \vec{S}_{123} + \vec{S}_{core} \quad ; \quad \left[\vec{S}_{123} \, , \, \vec{S}_{core} \right] = 0$$
$$\vec{S}_{q\,,\,123\,,\,core} = \vec{S}_{q\,,\,123\,,\,core}(t) \tag{19}$$

In eq. (19) $\vec{S}_{123}$ denotes the total spin of three valence quarks, $\vec{S}_{core}$ stands for the spin of quarks making up the core of hadrons, which we take to be $SU3_{fl}$ symmetric.

The proton wave function thus can be decomposed into basis states with respect to the operators (at time $t = 0$ say) :

$$\vec{J}^2 \rightarrow J(J+1) \quad , \quad J = 1/2 \quad ; \quad (J)_z \rightarrow m_J \quad , \quad m_J = 1/2$$

$$\vec{S}^{\,2}_{123} \rightarrow S_{123}(S_{123}+1) \quad ; \quad (S_{123})_z \rightarrow m_{123}$$

$$\left(\vec{R}'\right)^2 \rightarrow R'(R'+1) \quad ; \quad \left(\vec{R}'\right)_z \rightarrow m'$$

$$\vec{R}' = \vec{R} + \vec{S}_{core} \quad ; \quad m_J = m_{123} + m'$$

This decomposition is of the form :

$$\left| p \begin{array}{c} J \\ m_J \end{array} \right\rangle = \left| p \begin{array}{c} 1/2 \\ 1/2 \end{array} \right\rangle =$$

$$\sum C \left(\begin{array}{c|cc} 1/2 & S_{123} & R' \\ 1/2 & m_{123} & m' \end{array} \right) \left| \begin{array}{c} S_{123} \\ m_{123} \end{array} (\alpha) \right\rangle \left| \begin{array}{c} R' \\ m' \end{array} (\beta) \right\rangle \mathcal{A}^{S_{123}\,,\,R'}_{(\alpha)\,(\beta)}$$

$$\sum = \sum_{S_{123}\,,\,R'} \sum_{m_{123}\,,\,m'\,,\,(\alpha)\,,\,(\beta)} \Big|_{m_{123} + m' = 1/2} \tag{20}$$

In eq. (20) $C(\cdots)$ denotes the Clebsch Gordan coefficients. The labels (α) , (β) distinguish different states with given quantum numbers S_{123} , m_{123} and R' , m' respectively.

$\mathcal{A}^{S_{123}\,,\,R'}_{(\alpha)\,(\beta)}$ denotes the amplitude for the particular state with quantum numbers S_{123} , R' and labels (α) , (β) respectively. (α) runs over a finite set

$$(\alpha) = \begin{cases} m & \text{for } S_{123} = 3/2 \\ s, m, a & \text{for } S_{123} = 1/2 \end{cases} \tag{21}$$

The symbols s , m , a in eq. (21) refer to the symmetry class of the combined spin - flavor group of the three valence quarks : s = symmetric , m = mixed , a = antisymmetric.

The label (β) in eq. (21) represents orbital wave functions, which depend on the positions of valence and core quarks as well as gluons.

Angular momentum quantum numbers S_{123} , R' extend over the combinations :

$$(S_{123}, R') = \left\{ \left(1/2 \,\middle|\, \begin{matrix} 0 \\ 1 \end{matrix} \right) , \left(3/2 \,\middle|\, \begin{matrix} 2 \\ 1 \end{matrix} \right) \right. \tag{22}$$

We continue to assume exact $SU3_{fl}$ invariance for valence as well as core parts of the wave functions in eq. (20). Thus the following association holds :

$$\vec{S}_{123} = \left(\vec{S}_u + \vec{S}_d \right)_{val}$$

$$g_u - g_s = 2F = \left\langle p \begin{matrix} 1/2 \\ 1/2 \end{matrix} \middle| 2 (S_u)_{z\,val} \middle| p \begin{matrix} 1/2 \\ 1/2 \end{matrix} \right\rangle$$

$$g_d - g_s = F - D = \left\langle p \begin{matrix} 1/2 \\ 1/2 \end{matrix} \middle| 2 (S_d)_{z\,val} \middle| p \begin{matrix} 1/2 \\ 1/2 \end{matrix} \right\rangle \tag{23}$$

From eqs. (20 , 23) it follows :

$$g_q - g_s = \sum \left\{ \begin{array}{l} \left\langle \begin{array}{c} \tilde{S}_{123} \\ m_{123} \end{array} (\alpha) \middle| 2 (S_q)_{z\ val} \middle| \begin{array}{c} S_{123} \\ m_{123} \end{array} (\alpha) \right\rangle \\ C \left(\begin{array}{c} 1/2 \\ 1/2 \end{array} \middle| \begin{array}{cc} \tilde{S}_{123} & R' \\ m_{123} & m' \end{array} \right) C \left(\begin{array}{c} 1/2 \\ 1/2 \end{array} \middle| \begin{array}{cc} S_{123} & R' \\ m_{123} & m' \end{array} \right) \\ \varrho (\tilde{S}_{123} \mid S_{123} \ ; (\alpha) \ , \ R') \end{array} \right.$$

$$\varrho (\tilde{S}_{123} \mid S_{123} \ ; \ (\alpha) \ , \ R') = \sum_{(\beta)} \left(\mathcal{A}_{(\alpha)\ (\beta)}^{\tilde{S}_{123}\ ,\ R'} \right)^* \left(\mathcal{A}_{(\alpha)\ (\beta)}^{S_{123}\ ,\ R'} \right)$$

$$\sum = \sum_{\tilde{S}_{123}\ ,\ S_{123}\ ,\ m_{123}\ ,\ R',\ (\alpha)} \quad ; \quad m' = 1/2 - m_{123} \tag{24}$$

The symmetry class (α) , through Fermi statistics, is coupled with the symmetry class of the orbital valence quark variables. The sum over (β) in eq. (24) includes the orbital overlap of the associated wave functions. The latter vanishes for wave functions belonging to different orbital symmetry classes. Thus the density matrix ϱ in eq. (24) is diagonal with respect to (α) .

The matrix elements in eq. (24) for the case of mixed symmetry are :

$$\tau_q = \left\langle \begin{array}{c} \tilde{S}_{123} = 3/2 \\ m_{123} = \pm 1/2 \end{array} (\alpha) \middle| 2 (S_q)_{z\ val} \middle| \begin{array}{c} S_{123} = 1/2 \\ m_{123} = \pm 1/2 \end{array} (\alpha) \right\rangle \Bigg|_{(\alpha) = m}$$

$$\tau_q = (2/3) \sigma_q \quad ; \quad \sigma_q = \left\{ \begin{array}{ll} +1 & \text{for q = u} \\ -1 & \text{for q = d} \end{array} \right. \quad ; \quad R' = 1 \tag{25}$$

The nondiagonal contribution to the nonsinglet axial coupling constants $g_q - g_s$ thus are obtained collecting the appropriate factors in eqs. (24 , 25) :

$$g_q - g_s|_I = -\frac{8}{9}\sigma_q I \quad ; \quad \sigma_q = \begin{cases} +1 & \text{for q = u} \\ -1 & \text{for q = d} \end{cases}$$

$$I = c\,|\varrho_I| \quad ; \quad c = Re\left(\frac{\varrho_I}{|\varrho_I|}\right) \quad ; \quad -1 \leq c \leq 1$$

$$\varrho_I = \varrho\left(\tilde{S}_{123} = 3/2 \;\middle|\; S_{123} = 1/2 \;\; ; \;\; (\alpha) = m \;,\; R' = 1\right)$$

(26)

We introduce here the probalbilities corresponding to the diagonal elements of the density matrix ϱ :

$$p(S_{123}, (\alpha), R') = \varrho\left(S_{123} \;\middle|\; S_{123} \;\; ; \;\; (\alpha) \;,\; R'\right)$$

$$p[S_{123}, (\alpha), R'] = P[S_{123}] \; r_{(S_{123})}[(\alpha), R']$$

$$S_{123} = \begin{cases} 1/2 : R' = \begin{cases} \underline{0} & ; \quad (\alpha) = \underline{s}, m, a \\ \underline{1} & ; \quad (\alpha) = s, m, \underline{a} \end{cases} \\ 3/2 : R' = \begin{cases} \underline{2} \\ 1 \end{cases} & ; \quad (\alpha) = \underline{m} \end{cases} \tag{27}$$

In eq. (27) $P[S_{123}]$; $S_{123} = 1/2, 3/2$ denote the probability for the total valence quark spin to take on the value $1/2$ and $3/2$ respectively, whereas the quantities $r_{(S_{123})}[(\alpha), R']$ stand for the relative probabilities, given the value of S_{123}, of the quantum numbers (α), R' :

$$P[1/2] + P[3/2] = 1$$

$$\sum_{(\alpha), R'} \left\{ r_{(S_{123})}[(\alpha), R'] \right\} = 1 \quad ; \quad \text{for } S_{123} = 1/2, 3/2$$

The configurations , potentially dominating for a given value of S_{123} within the proton, are underlined in eq. (27) .
The probabilities of the configurations defined in eq. (27) are numbered in the following way :

$$\underline{w}_1 = P[1/2]\, r_{(1/2)}\, [\underline{s}, \underline{0}] \quad ; \quad w_2 = P[1/2]\, r_{(1/2)}\, [s, 1]$$

$$w_3 = P[1/2]\, r_{(1/2)}\, [m, 0] \quad ; \quad w_4 = P[1/2]\, r_{(1/2)}\, [m, 1]$$

$$w_5 = P[1/2]\, r_{(1/2)}\, [a, 0] \quad ; \quad \underline{w}_6 = P[1/2]\, r_{(1/2)}\, [\underline{a}, \underline{1}]$$

$$\underline{w}_7 = P[3/2]\, r_{(3/2)}\, [\underline{m}, \underline{2}] \quad ; \quad w_8 = P[3/2]\, r_{(3/2)}\, [m, 1] \qquad (28)$$

The diagonal elements of the valence quark spin operators in eq. (24) are of the form :

$$\left\langle \begin{matrix} S_{123} \\ m_{123} \end{matrix} (\alpha) \right| 2\,(S_q)_{z\,val} \left| \begin{matrix} S_{123} \\ m_{123} \end{matrix} (\alpha) \right\rangle$$

$$= D_q\,[S_{123}, (\alpha)] \left\langle \begin{matrix} S_{123} \\ m_{123} \end{matrix} \right| 2\,(S_{123})_z \left| \begin{matrix} S_{123} \\ m_{123} \end{matrix} \right\rangle \qquad (29)$$

$$q = u, d \quad ; \quad D_u\,[S_{123}, (\alpha)] + D_d\,[S_{123}, (\alpha)] = 1$$

The matrix elements of the total valence quark spin contain only diagonal terms. They are determined by appropriate Lande factors :

$$\left\langle p \begin{array}{c} 1/2 \\ 1/2 \end{array} \right| 2(S_{123})_z \left| p \begin{array}{c} 1/2 \\ 1/2 \end{array} \right\rangle_{S_{123},\,R'}$$

$$= g_L(J, S_{123}, R')|_{J=1/2} \left\langle p \begin{array}{c} 1/2 \\ 1/2 \end{array} \right| 2(J)_z \left| p \begin{array}{c} 1/2 \\ 1/2 \end{array} \right\rangle$$

$$g_L(J, J_1, J_2) = \frac{J(J+1) + J_1(J_1+1) - J_2(J_2+1)}{2J(J+1)} \tag{30}$$

The spin projection constants $D_q\,[S_{123}, (\alpha)]$ in eq. (29) and the Lande factors g_L are given in Table 2 below.

(α)	S_{123}	R'	w	g_L	D_u	D_d
$\underline{s}$	$1/2$	$\underline{0}$	$\underline{w}_1$	1	$4/3$	$-1/3$
		1	w_2	$-1/3$		
m	$1/2$	0	w_3	1	$2/3$	$1/3$
		1	w_4	$-1/3$		
$\underline{m}$	$3/2$	$\underline{2}$	$\underline{w}_7$	-1		
		1	w_8	$5/3$		
$\underline{a}$	$1/2$	0	w_5	1	0	1
		$\underline{1}$	$\underline{w}_6$	$-1/3$		

Table 2

From eqs. (26) - (30) and Table 2 we infer the following expressions for the nonsinglet axial coupling constants $g_u - g_s$, $g_d - g_s$ or equivalently $g_A = g_u - g_d$, $g_d - g_s$:

$$
\begin{aligned}
g_A &= \left[\begin{array}{l} \frac{5}{3}\,\underline{w}_1 - \frac{5}{9}\,w_2 + \frac{1}{3}\,w_3 - \frac{1}{9}\,w_4 \\[2ex] -\frac{1}{3}\,\underline{w}_7 + \frac{5}{9}\,w_8 - w_5 + \frac{1}{3}\,\underline{w}_6 \\[2ex] -\frac{16}{9}\,I \end{array} \right] \\
g_d - g_s &= \left[\begin{array}{l} -\frac{1}{3}\,\underline{w}_1 + \frac{1}{9}\,w_2 + \frac{1}{3}\,w_3 - \frac{1}{9}\,w_4 \\[2ex] -\frac{1}{3}\,\underline{w}_7 + \frac{5}{9}\,w_8 + w_5 - \frac{1}{3}\,\underline{w}_6 \\[2ex] +\frac{8}{9}\,I \end{array} \right]
\end{aligned}
\qquad (31)
$$

$$
I = \tilde{c}\,\sqrt{w_4\,w_8} \quad ; \quad -1 \le \tilde{c} \le 1
$$

$$
\sum_{i=1}^{8} w_i = 1
$$

From eq. (31) we derive an inequality for $g_d - g_s$ (among related inequalities like e. g. $-1 \le g_A \le 5/3$):

$$g_d - g_s \geq -\frac{1}{3}$$

$$\left(= -\frac{1}{3} \quad \leftrightarrow \quad w_2 = w_3 = w_4 = w_5 = w_8 = 0 \right) \tag{32}$$

Comparing $g_d - g_s = F - D$ with the corresponding quantity in Table 1 :

$$g_A / g_V \; (\Sigma^- \rightarrow n e^- \bar{\nu}_e) = F - D = -0.340 \pm 0.017 \tag{33}$$

it follows, that indeed (always assuming $SU3_{fl}$ symmetry) $g_d - g_s$ is very near to the lower limit $-1/3$.

The quark spin in the core of the proton

The following discussion shall be restricted to the extremal case in eq. (32) , whence eq. (31) implies :

$$\frac{4}{3} \underline{w}_6 + 2 \underline{w}_7 = \frac{5}{3} - g_A \approx 0.41$$

$$g^{(8)} = 3F - D = g_A - 2/3 \approx 0.59$$

$$\leftrightarrow \quad 0.58 \pm 0.10 \quad \text{from the } SU3_{fl} \text{ fit} \tag{34}$$

The remaining dominant contributions are given in Table 3 below, together with the Lande factors relative to the residual angular momentum R' , $g'_L = 1 - g_L$

and the spectroscopic $[\,SU6\,,\,R'\,]$ assignments [19] :

(α)	S_{123}	R'	w	g'_L	$[\,SU6\,,\,R'\,]$
$\underline{s}$	1 / 2	$\underline{0}$	$\underline{w}_1$	0	[56 ,0]
$\underline{m}$	3 / 2	$\underline{2}$	$\underline{w}_7$	2	[70 ,2]
$\underline{a}$	1 / 2	$\underline{1}$	$\underline{w}_6$	4 / 3	[20 ,1]

Table 3

We recall the angular momentum pieces defined in eqs. (19 - 20) :

$$\vec{J} = \vec{S}_q + \vec{R} \quad , \quad \vec{S}_q = \vec{S}_{123} + \vec{S}_{core} \quad , \quad \vec{R}\,' = \vec{R} + \vec{S}_{core}$$

Different values of R' belong to different symmetry classes as indicated in Table 3. Thus there are no nondiagonal contributions to the matrix elements of $\vec{S}_{core}$. The diagonal matrix elements of $\vec{S}_{core}$ are obtained from those of $\vec{R}\,'$:

$$\left| \begin{matrix} R' \\ m' \end{matrix} \; (\alpha)\,,\, S_c\,,\, R \right\rangle$$

$$= \sum \; C \left(\begin{matrix} R' \\ m' \end{matrix} \middle| \begin{matrix} S_c & R \\ m_c & m_R \end{matrix} \right) \left| \begin{matrix} S_c \\ m_c \end{matrix} \; (\alpha)\,,\,(\beta_1) \right\rangle \left| \begin{matrix} R \\ m_R \end{matrix} \; (\alpha)\,,\,(\beta_2) \right\rangle$$

$$\times \mathcal{A}^{\,S_{123}\,,\,R'\,,\,S_c\,,\,R}_{\,(\alpha)\,(\beta 1)\,(\beta 2)} \tag{35}$$

$$\left\langle \begin{matrix} R' \\ m' \end{matrix} (\alpha) \ , S_c \ , R \ \middle| \ 2\,(S_{core})_z \ \middle| \ \begin{matrix} R' \\ m' \end{matrix} (\alpha) \ , S_c \ , R \right\rangle$$

$$= g_L(R', S_c, R) \left\langle \begin{matrix} R' \\ m' \end{matrix} (\alpha) \ , S_c \ , R \ \middle| \ 2\,(R')_z \ \middle| \ \begin{matrix} R' \\ m' \end{matrix} (\alpha) \ , S_c \ , R \right\rangle \tag{36}$$

Thus the matrix element of the core spin in the proton generates the following contributions to the strange axial current coupling constant :

$$3\,g_s = \left\langle p \begin{matrix} 1/2 \\ 1/2 \end{matrix} \ \middle| \ 2\,(S_{core})_z \ \middle| \ p \begin{matrix} 1/2 \\ 1/2 \end{matrix} \right\rangle$$

$$3\,g_s = \left\{ \begin{matrix} \underline{w}_6\, g_L^{\prime(6)} \sum_{S_c,R} g_L(R' = 1, S_c, R)\, ww^{(6)}(S_c, R) \\ + \ \underline{w}_7\, g_L^{\prime(7)} \sum_{S_c,R} g_L(R' = 2, S_c, R)\, ww^{(7)}(S_c, R) \end{matrix} \right\}$$

$$g_L^{\prime(6)} = g_L(1/2, 1, 1/2) = 4/3 \quad ; \quad g_L^{\prime(7)} = g_L(1/2, 2, 3/2) = 2 \tag{37}$$

In eq. (37) the quantities $ww^{(j)}(S_c, R)$, $j = 6, 7$ denote the relative probabilities given the respective configurations denoted by the label $(j) = (6), (7)$ in Table 3, of all subconfigurations with quantum numbers S_c, R respectively :

$$\sum_{S_c,R} ww^{(j)}(S_c, R) = 1 \quad ; \quad \text{for } (j) = (6), (7)$$

From eq. (37) an inequality can be derived involving the mean value of the core spin S_c in the subspace of configurations characterized by $\underline{w}_6$ and $\underline{w}_7$:

$$\langle S_c \rangle_{(6,7)} = \sum_{S_c,R} S_c \frac{1}{\underline{w}_6 + \underline{w}_7} \left(\begin{matrix} \underline{w}_6\, ww^{(6)}(S_c, R) + \\ + \underline{w}_7\, ww^{(7)}(S_c, R) \end{matrix} \right) \tag{38}$$

The inequality reads :

$$- 3 g_s \leq \frac{2}{3} (\underline{w}_6 + \underline{w}_7) \langle S_c \rangle_{(6,7)} \tag{39}$$

Eqs. (37 , 39) allow , in all generality , arbitrary values for the core spin :

$$- \infty \leq 3 g_s \leq + \infty$$

Eq. (34) implies

$$\left(A = \frac{5}{3} - g_A \approx 0.41 \quad ; \quad A = \frac{4}{3} \underline{w}_6 + 2 \underline{w}_7 \right)$$

$$A / 3 = 0.136 \leq \frac{2}{3} (\underline{w}_6 + \underline{w}_7) \leq 0.205 = A / 2 \tag{40}$$

Combining eqs. (39) and (40) we obtain :

$$- 3 g_s \leq \frac{A}{2} \langle S_c \rangle_{(6,7)} = 0.205 \langle S_c \rangle_{(6,7)} \tag{41}$$

Finally eqs. (34 and 41) yield two relations, involving besides the core spin average $\langle S_c \rangle_{(6,7)}$ (eq. (38)) the corresponding mean value $\langle R \rangle_{(6,7)}$:

$$\left\{ \begin{matrix} \langle S_c \rangle_{(6,7)} \\ \\ \langle R \rangle_{(6,7)} \end{matrix} \right. \geq F (1 - g^{(0)}) \left. \begin{matrix} -2 \\ \\ -1 \end{matrix} \right\} \quad ; \quad F = \frac{2}{5 / 3 - g_A} = 4.92 \pm 0.05 \tag{42}$$

We set $F = 5$ in the following. This just simplifies the inequalities in eq. (42) without significantly changing any numerical conclusion relative to the exact range of values. Thus eq. (42) becomes :

$$\langle S_c \rangle_{(6,7)} \geq 3 - 5\, g^{(0)} \quad ; \quad g^{(0)} \geq \left(3 - \langle S_c \rangle_{(6,7)} \right) / 5$$

$$\langle R \rangle_{(6,7)} \geq 4 - 5\, g^{(0)} \quad ; \quad g^{(0)} \geq \left(4 - \langle R \rangle_{(6,7)} \right) / 5 \tag{43}$$

From eq. (43) we infer, that a vanishing value of $g^{(0)}$ implies large values for the angular momenta in the core of the proton [17] :

$$g^{(0)} = 0 \quad \rightarrow \quad \langle S_c \rangle_{(6,7)} \geq 3 \quad , \quad \langle R \rangle_{(6,7)} \geq 4 \tag{44}$$

I call the large and compensating values of the core angular momenta $\vec{S}_c$, $\vec{R}$ (eq. (44)) the double helix structure of the nucleon.
The clear mass separation of nucleon isobars with respect to total angular momentum does not show any evidence of the above double helix structure. A value of $\langle S_c \rangle_{(6,7)}$ as large as 3 cannot be excluded from first principles at present. Nevertheless I estimate

$$\langle S_c \rangle_{(6,7)} \lesssim 1 \quad \rightarrow \quad g^{(0)} \geq 0.4 \, (\pm 0.1) \tag{45}$$

The error in eq. (45) reflects the error from the $SU3_{fl}$ fit (eqs. (34 , 16) and Table 1) .

References

[1] J. D. Bjorken , *Phys. Rev.* **148** (1966) 1467 ,
Phys. Rev. **D1** (1971) 1376

[2] J. Ellis and R. L. Jaffe , *Phys. Rev* **D9** (1974) 1444 ,
Erratum *Phys. Rev* **D10** (1974) 1669

[3] J. Ashman et al. (EMC Collaboration) , *Phys. Lett.* **206B** (1988) 364
Nucl. Phys. **B328** (1989) 1

[4] (SLAC results) G. Baum et al. , *Phys. Rev. Lett.* **51** (1983) 1135
see also : V. W. Hughes et al. , *Phys. Lett.* **212B** (1988) 511

[5] E. Reya , in *Frontiers in Particle Physics*, Proceedings of the
XII Warsaw Symposium on Particle Physics ,
Z. Ajduk, S. Pokorski and A. K. Wróblewski edts. ,
World Scientific Publishing Company, Singapore, 1990
University of Dortmund preprint ,
DO-89/15 (1989)

[6] G. Altarelli , CERN preprint CERN/TH-5675/90 (1990)

[7] A. V. Manohar , University of California San Diego preprint ,
UCSD/PTH 90-28 (1990)

[8] J. Kodeira , *Nucl. Phys.* **B165** (1980) 129 ,
Phys. Rev. **D20** (1979) 627 ,
Nucl. Phys. **B159** (1979) 99

[9] G. Altarelli and G. G. Ross , *Phys. Lett.* **212B** (1988) 391

[10] A. V. Efremov and O. V. Teryayev , Dubna preprint , E2-88-287 (1988)

[11] R. D. Carlitz , J. C. Collins and A. H. Muller , *Phys. Lett.* **214B** (1988) 229

[12] R. L. Jaffe , *Phys. Lett.* **193B** (1987) 101

[13] M. Bourquin et al. , *Z. Phys.* **C21** (1983) 27

[14] J. Ellis and M. Karliner, *Phys. Lett.* **213B** (1988) 73

[15] H. Fritzsch , Max Planck Institute Munich report , MPI-PAE/PTh 18/89

[16] G. Veneziano , *Mod. Phys. Lett.* **A4** (1989) 1605

[17] H. J. Lipkin, *Phys. Lett.* **237B** (1990) 130

[18] G. M. Shore and G. Veneziano , CERN preprint , CERN-TH-5689/90 (1990)

[19] A. J. G. Hey and R. L. Kelly , *Phys. Lett.* **96C** (1983) 71

PION-NUCLEON UPDATE

M.E. SAINIO
Research Institute for Theoretical Physics, University of Helsinki,
Siltavuorenpenger 20 C, SF-00170 Helsinki, Finland

ABSTRACT

The low-energy pion-nucleon interaction is examined in view of recent developments. In particular, a new evaluation of the pion-nucleon sigma-term is discussed. The result $\sigma \simeq 45$ MeV implies for the strange quark content of the proton the value $y \simeq 0.2$ with a sizeable error.

1. Introduction

Pion-nucleon scattering can be described by four Lorentz invariant amplitudes ($A^{\pm}$,$B^{\pm}$) in the good isospin limit. If the isospin indices of the pions are denoted by (a, b), the amplitude $T^{ab}_{\pi N}$ has the form

$$\begin{aligned} T^{ab}_{\pi N} &= T^{+}\delta^{ab} + T^{-}\frac{1}{2}[\tau^a, \tau^b] \\ T^{\pm} &= \bar{u}(p', s')[A^{\pm}(\nu, t) + \frac{1}{2}(\not{q} + \not{q}')B^{\pm}(\nu, t)]u(p, s) \end{aligned} \tag{1}$$

where q (q') is the initial (final) pion momentum, $\nu = (s-u)/4m$ and m is the proton mass. The D-amplitude to be used below is defined as $D^{\pm} = A^{\pm} + \nu B^{\pm}$. The optical theorem relates ImD to the total cross section in the appropriate isospin channel. Therefore, the D-amplitude is directly connected to experimental information on pion-nucleon scattering.

In the low-energy region, the behaviour of the πN scattering amplitude is governed by chiral symmetry. The chiral effective Lagrangian[1] generates the leading contribution

$$\begin{aligned} A^{\pm}_{tree} &= A^{\pm}_{pv} = \frac{g^2}{m}\begin{pmatrix} 1 \\ 0 \end{pmatrix} \\ B^{\pm}_{tree} &= B^{\pm}_{pv} + \frac{1}{2F_\pi^2}\begin{pmatrix} 0 \\ 1 \end{pmatrix} \\ B^{\pm}_{pv} &= g^2\left(\frac{1}{m^2 - s} \mp \frac{1}{m^2 - u}\right) - \frac{g^2}{2m^2}\begin{pmatrix} 0 \\ 1 \end{pmatrix} \end{aligned} \tag{2}$$

where $F_\pi = 93.3$ MeV is the pion decay constant. For the pion-nucleon coupling the phenomenological value[2] $g^2/4\pi = 14.28$ is taken. The one-loop corrections to

Eq. 2 have been evaluated[3] in chiral SU(2)×SU(2). At order q^3 there are 8 low-energy constants occurring in the on-shell elastic pion-nucleon amplitude $T_{\pi N}$. The extensive work of the Karlsruhe group, summarized in ref.[2], gives a handle to the required low-energy constants[4]. However, the well known problem with the pion-nucleon sigma-term triggered a reanalysis of two of these constants. Reexamining the nucleon scalar form factor was also necessary for the understanding of that problem. The pion-nucleon sigma-term will be the topic of the rest of the discussion.

2. The sigma-term

The pion-nucleon sigma-term is a measure of the explicit chiral symmetry breaking due to the u and d quark masses, and it is defined as[5]

$$\sigma = \frac{\hat{m}}{2m} < p|\bar{u}u + \bar{d}d|p > \tag{3}$$

where $\hat{m} = \frac{1}{2}(m_u + m_d)$ and $|p>$ denotes a physical one-proton state. Algebraically σ can be written in the form

$$\sigma = \frac{\hat{m}}{2m} \frac{< p|\bar{u}u + \bar{d}d - 2\bar{s}s|p >}{1-y} \tag{4}$$

where the parameter

$$y = \frac{2 < p|\bar{s}s|p >}{< p|\bar{u}u + \bar{d}d|p >} \tag{5}$$

reflects the strange quark content of the proton. The numerator in Eq. 4 can be calculated using the quark mass expansion of the baryon octet and retaining terms up to $\mathcal{O}(m_q)$. The result for the sigma-term is

$$\sigma \simeq \frac{\hat{m}}{m_s - \hat{m}} \frac{m_\Xi + m_\Sigma - 2m_N}{1-y} \simeq \frac{26 \text{ MeV}}{1-y}, \tag{6}$$

where the quark mass ratio $m_s/\hat{m} \simeq 25$ derived from the kaon and pion masses is used. The Zweig rule leads one to expect[6] that $< p|\bar{s}s|p > \simeq 0$, i.e. $y = 0$, and therefore in this limit $\sigma \simeq 26$ MeV.

The strangeness content of the proton together with the related question of the valence quark contribution to the proton spin have been of considerable interest[7] recently. Experimental data on neutrino scattering indicate[8] that the proton momentum fraction carried by the strange quarks is small. On the other hand, some strange quark bilinear operators seem to have relatively large matrix elements in the proton[9], see also refs.[10] for other recent studies. One of the open questions[5,11,12] has been the size of the matrix element $< p|\bar{s}s|p >$ which dictates the contribution of the strange quarks to the proton mass.

Chiral symmetry relates[13,14] the sigma-term to the isospin even D-amplitude at the unphysical (on-shell) Cheng-Dashen point, denoted here by Σ,

$$\Sigma = F_\pi^2 \bar{D}^+(\nu = 0, t = 2\mu^2) \tag{7}$$

where μ is the pion mass. The pseudovector Born term has been subtracted from the amplitude, i.e. $\bar{D}^+ = D^+ - D^+_{pv}$. The extrapolation of the pion-nucleon amplitudes to the unphysical region has been extensively studied e.g. by the Karlsruhe group[2]. Hyperbolic dispersion relations yield the result[15] $\Sigma = 64 \pm 8$ MeV, where the uncertainty of 8 MeV reflects the internal consistency of the method used. The difference between σ and Σ is of the order $(m_q)^{3/2}$, and numerically the correction amounts[3,5,16] to $\Sigma - \sigma \simeq 5$ MeV in the one-loop level. The value of σ would then be 59±8 MeV, which is off by about a factor of two from the value extracted from the baryon spectrum (y=0). Of course, by choosing the value $y \simeq 0.5$ the problem would be solved. However, such a large value for the strange quark content would imply that the proton mass in the chiral limit is nowhere near the physical value.

It turns out, however, that there are significant corrections to the leading order formula of Eq. 6 from meson cloud effects corresponding to higher order terms in the quark mass expansion. These yield[5,17]

$$\sigma = \frac{35 \pm 5 \text{ MeV}}{1 - y} \tag{8}$$

in the one-loop level including all effects of the order $m_q^{3/2}$ and an estimate of the terms of the order m_q^2.

3. The scalar form factor

The relation between the σ-term and the Σ is crucial in the present analysis. The scalar form factor $\sigma(t)$ is defined as

$$\bar{u}'\sigma(t)u = < p'|\hat{m}(\bar{u}u + \bar{d}d)|p >; \qquad t = (p' - p)^2, \tag{9}$$

and the low-energy theorem of chiral symmetry can now be written in the form[14,16]

$$\Sigma = \sigma(2\mu^2) + \triangle_R \tag{10}$$

where $\triangle_R = \mathcal{O}(\mu^4 \log \mu^2)$. In the one-loop level of chiral perturbation theory one obtains[3] $\triangle_R = 0.35$ MeV. The key question is then the difference

$$\triangle_\sigma \equiv \sigma(2\mu^2) - \sigma(0) \tag{11}$$

where $\sigma(0) = \sigma$. The t-dependence of $\sigma(t)$ has to do with the $\pi\pi$ interaction for $t > 4\mu^2$, and it has to be constrained to be consistent with chiral symmetry[18,19]. A dispersion analysis yields then[19]

$$\triangle_\sigma = 15.2 \pm 0.4 \text{ MeV} \tag{12}$$

where the error bar reflects the estimated uncertainty in the $\pi\pi$ phase shifts. This value is by about 10 MeV larger than the result from chiral perturbation theory to one-loop quoted in the previous section. It turns out that the S-wave t-channel πN amplitude $f_+^0(t)$, which drives the magnitude of Δ_σ, is poorly described by the corresponding Born term away from $t = \mu^2$, and that is what the one-loop calculation of $\sigma(t)$ is, in fact, using for $t > 4\mu^2$.

The uncertainty of 8 MeV in Σ cited earlier shows only the scatter of values corresponding to different hyperbolae in the dispersion analysis of Koch[15]. Now it remains to check the importance of the experimental uncertainties in the error bar.

4. The analysis of the low-energy pion-nucleon data

A method, based on forward dispersion relations, was proposed recently for extracting the pion-nucleon Σ-term from low-energy data[20]. The aim was to simplify the analysis of the effect of the experimental uncertainties to the uncertainty of the Σ-term. At low energies πN scattering is dominated by six partial waves, denoted by S_{11}, S_{31}, P_{11}, P_{31}, P_{13} and P_{33}, which have to be extrapolated from the experimentally well known region down to the physical threshold. To determine the energy dependence of these partial waves six dispersion relations were written. These include the standard[2] equations for the invariant amplitudes $D^\pm$ and $B^\pm$ together with two further amplitudes $E^\pm$ defined by[20]

$$E^\pm = \frac{\partial}{\partial t}(A^\pm + \omega B^\pm)|_{t=0}, \tag{13}$$

where the partial derivative is taken at fixed s [$\nu = \omega$ for t=0]. Two subtraction constants remain, $D^+(\mu)$ and $E^+(\mu)$, which are proportional to the scattering length a_{0+}^+ and volume a_{1+}^+ respectively. They have to be fixed using experimental information at low energies.

Two kinds of input are needed in addition to the two scattering lengths a_{0+}^+ and a_{1+}^+. Firstly, it is assumed that $Im\, D^\pm$, $Im\, B^\pm$ and $Im\, E^\pm$ are known for $k_{LAB} \geq k_0$, where k_0, the matching momentum, represents the lower limit of the experimentally well explored momentum region. Secondly, the D-and F-wave amplitudes are assumed to be known for $k_{LAB} < k_0$. Here the input is taken from the Karlsruhe analysis[21] (KH.80). For the matching momentum the value $k_0 = 185$ MeV/c is taken (corresponding to $T_\pi \simeq 92$ MeV). The solution proceeds then by iteration. Of course, the results should be essentially independent of the value of k_0 for a given data set.

In the Taylor expansion of $\bar{D}^+$ in powers of ν^2 and t

$$\bar{D}^+ = d_{00}^+ + d_{10}^+\nu^2 + d_{01}^+ t + d_{20}^+\nu^4 + d_{11}^+\nu^2 t + d_{02}^+ t^2 + \ldots \tag{14}$$

the coefficients d_{00}^+ and d_{01}^+ have a simple connection to the forward amplitudes at $\omega = 0$, i.e. $d_{00}^+ = \bar{D}^+(0)$ and $d_{01}^+ = \bar{E}^+(0)$. Then, for given values of a_{0+}^+ and a_{1+}^+, the solutions of the dispersion relations fix d_{00}^+ and d_{01}^+ and the Σ-term is, according to Eq. 7,

$$\Sigma = F_\pi^2(d_{00}^+ + 2\mu^2 d_{01}^+) + \triangle_D \equiv \Sigma_d + \triangle_D \tag{15}$$

where $\triangle_D$ represents terms proportional to t^2 or higher in Eq. 14. Here the same $\pi\pi$ singularity as in the case of the form factor $\sigma(t)$ determines the t-dependence of the amplitude D^+ and, therefore, numerically the result[19]

$$\triangle_D = 11.9 \pm 0.6 \text{ MeV} \tag{16}$$

is not far from $\triangle_\sigma$. In chiral perturbation theory the curvature term from the one-loop contributions would be $\triangle_D = 6.1$ MeV, see however the discussion in refs.[20,22]

There remains then the task to fix a_{0+}^+ and a_{1+}^+ using directly the experimental pion-nucleon information in a χ^2-minimization[23]. The electromagnetic corrections for the observables are calculated according to the formalism of Tromborg et al.[24] In addition, the P_{33} π^-p phases are corrected for the isospin breaking due to the mass difference[21] $\Delta^{++} - \Delta^0$.

5. Results

In the analysis, published in ref.[25], all the elastic low-energy pion-nucleon data[26] since 1976 were considered. However, only the data sets of Bertin et al. and Frank et al. lead to stable solutions. The χ^2-minimum for the Bertin data (solution A) is found at $a_{0+}^+ = -0.0080\ \mu^{-1}$ and $a_{1+}^+ = 0.1317\ \mu^{-3}$ with Σ_d=48 MeV (χ^2/data point = 1.03). In Fig. 1, which displays the values of Σ_d, the curves $\chi^2_{MIN} + 1$ and $\chi^2_{MIN} + 4$ are shown in the (a_{0+}^+, a_{1+}^+)- plane with a solid dot denoting the minimum. To estimate the error for the resulting Σ_d is a difficult task. The statistical error is well defined as $\Delta\Sigma_d = \pm 4$ MeV, but the main problem is the remaining systematic error. To estimate that a number of minimizations were performed with different values for the matching momentum k_0. New data sets at higher energies were included in the analysis. The theoretical input of the present analysis, i.e. the imaginary parts $Im\, D^\pm$, $Im\, B^\pm$ and $Im\, E^\pm$ above k_0 and the D-and F-wave phase shifts below k_0, was allowed to vary and the effect on Σ_d was calculated. The result for the Bertin data (solution A) is then

$$\Sigma_d = 48 \pm 4 \pm 4 \pm 4 \text{ MeV} \tag{17}$$

where the errors refer to statistics, to data selection and to the input uncertainties respectively.

A similar analysis was performed for the Frank et al. data[26]. The quoted normalization uncertainties are generally quite large except for the π^+p data at 95 MeV/c. It turns out that this data set is inconsistent with the remainder and has to be removed. The minimum is found at $a_{0+}^+ = -0.0115\ \mu^{-1}$, $a_{1+}^+ = 0.1339\ \mu^{-3}$ with χ^2/data point = 1.20 (solution B). The corresponding Σ_d-value is

$$\Sigma_d = 50 \pm 3 \pm 7 \pm 4 \text{ MeV}$$

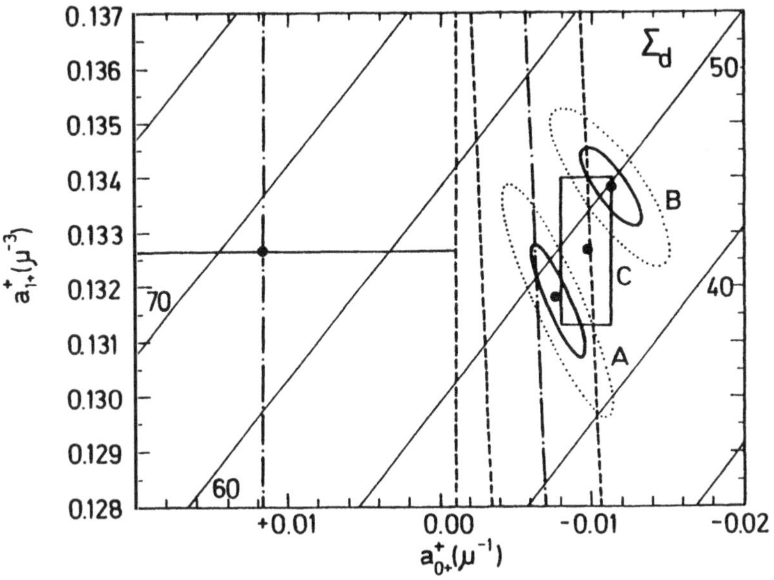

Figure 1: The value of the Σ_d-term[25] as a function of the parameters a^+_{0+} and a^+_{1+}. The ellipses on the right of the figure are for $\chi^2_{MIN}+1$ (solid line) and $\chi^2_{MIN}+4$ (dotted line) corresponding to the solutions A and B explained in the text. The KH.80 result is denoted by C. The error band on the right reflects the result of the π^-p atomic measurement[27], and the one on the left the $ReD^+(t=0)$ result[26].

where the errors were estimated as in the previous case, and the solution is denoted by B in Fig. 1. Both the solutions A and B are in good agreement with the Karlsruhe result[2] $\Sigma_d = 51 \pm 7$ MeV.

In Fig. 1. the rectangle C reflects the KH.80 solution with errors from ref.[2] The error band on the right displays the result of the π^-p atomic s-wave level shift measurement[27]

$$\Delta E = -7.12 \pm 0.32 \text{ eV}$$

corresponding to the scattering length $a^{\pi^-p}_{0+} = a^+_{0+} + a^-_{0+} = 0.086 \pm 0.004\ \mu^{-1}$ which now is consistent with the scattering data. Through measurements in the Coulomb interference region it is possible to determine the real part of the D-amplitude in the forward direction. Wiedner et al.[26] have measured $ReD^+(k_{LAB}, t=0)$ at 135 MeV/c and the error band on the left of Fig. 1 displays the result.

In the present analysis the low-energy data[28] on the charge exchange reaction $\pi^- p \to \pi^0 n$ is not used. The charge exchange reaction is proportional to the isospin odd channel and, therefore, constrains only weakly the isospin even channel which is of interest for the question of the Σ-term.

6. Conclusions

Combining the results from Eqs. 10, 12, 15 and 16 the low-energy theorem takes the form

$$\sigma = \Sigma_d + \triangle; \qquad \triangle = \triangle_D - \triangle_\sigma - \triangle_R,$$

where $\triangle \simeq -3$ MeV within an uncertainty of order 1 MeV. With $\Sigma_d \simeq 48$ MeV for the Bertin solution the sigma-term becomes $\sigma \simeq 45$ MeV which implies $y \simeq 0.2$, if Eq. 8 is used. The error estimate for y from the above errors yields about 0.2 which is certainly an underestimate. For Σ the result is 60 MeV in good agreement with the Karlsruhe result[15] $\Sigma = 64 \pm 8$ MeV.

The low-energy pion-nucleon information[26,29] is not, however, consistent at present. Therefore, more experimental effort is needed before the error estimate for the sigma-term and the strange quark content can be reduced.

Acknowledgements

I thank J. Gasser, G. Höhler and H. Leutwyler for numerous discussions on various aspects of the pion-nucleon interaction.

References

1. S. Weinberg, *Phys. Rev. Lett.* **18** (1967) 188.
2. G. Höhler, in *Landolt-Börnstein*, Vol. 9 b2, ed. H. Schopper (Springer, Berlin, 1983).
3. J. Gasser, M.E. Sainio and A. Švarc, *Nucl. Phys.* **B307** (1988) 779.
4. J. Gasser, *Nucl. Phys.* **B279** (1987) 65.
5. J. Gasser and H. Leutwyler, *Phys. Reports* **87C** (1982) 77.
6. T.P. Cheng, *Phys. Rev.* **D13** (1976) 2161.
7. M. Karliner, in *Proc. First Biennial Conference on Low Energy Antiproton Physics*, Stockholm 1990, eds. P. Carlson, A. Kerek and S. Szilagyi (World Scientific, Singapore, 1991) p. 330.
8. H. Abramowicz et al., *Z. Phys.* **C15** (1982) 19.
9. B.L. Ioffe and M. Karliner, *Phys. Lett.* **B247** (1990) 387.
10. D.B. Kaplan and A. Manohar, *Nucl. Phys.* **B310** (1988) 527;
    T.P. Cheng and L.-F. Li, *Phys. Rev. Lett.* **62** (1989) 1441;
    E.J. Beise and R.D. McKeown, *Comments Nucl. Part. Phys.* **20** (1991) 105.
11. J.F. Donoghue and C.R. Nappi, *Phys. Lett.* **168B** (1986) 105.
12. R.L. Jaffe and C.L. Korpa, *Comments Nucl. Part. Phys.* **17** (1987) 163.
13. T.P Cheng and R. Dashen, *Phys. Rev. Lett.* **26** (1971) 594.
14. L.S. Brown, W.J. Pardee and R.D. Peccei, *Phys. Rev.* **D4** (1971) 2801.
15. R. Koch, *Z. Phys.* **C15** (1982) 161.
16. H. Pagels and W.J. Pardee, *Phys. Rev.* **D4** (1971) 3335.

17. J. Gasser, *Ann. Phys.* (N.Y.) **136** (1981) 62.
18. J. Gasser and H. Leutwyler, *Ann. Phys.* (N.Y.) **158** (1984) 142.
19. J. Gasser, H. Leutwyler and M.E. Sainio, *Phys. Lett.* **B253** (1991) 260.
20. J. Gasser, H. Leutwyler, M.P. Locher and M.E. Sainio, *Phys. Lett.* **B213** (1988) 85.
21. R. Koch and E. Pietarinen, *Nucl. Phys.* **A336** (1980) 331.
22. J. Gasser, in *Proc. of the 2nd Int. Workshop on πN Physics*, Los Alamos 1987, eds. W.R. Gibbs and B.M.K. Nefkens, Los Alamos report LA-11184-C (1987), p. 266.
23. F. James and M. Roos, *Comput. Phys. Comm.* **10** (1975) 343.
24. B. Tromborg, S. Waldenstrøm and I. Øverbø, *Phys. Rev.* **D15** (1977) 725; *Helv. Phys. Acta* **51** (1978) 584.
25. J. Gasser, H. Leutwyler and M.E. Sainio, *Phys. Lett.* **B253** (1991) 252.
26. P.Y. Bertin et al., *Nucl. Phys.* **B106** (1976) 341;
    E.G. Auld et al., *Can. J. Phys.* **57** (1979) 73;
    J.S. Frank et al., *Phys. Rev.* **D28** (1983) 1569;
    B.G. Ritchie et al., *Phys. Lett.* **125B** (1983) 128;
    J.T. Brack et al., *Phys. Rev.* **C34** (1986) 1771; **C38** (1988) 2427; **C41** (1990) 2202;
    U. Wiedner et al., *Phys. Rev.* **D40** (1989) 3568.
27. W. Beer et al., *Phys. Lett.* **B261** (1991) 16.
28. J. Duclos et al., *Phys. Lett.* **43B** (1973) 245;
    D.H. Fitzgerald et al., *Phys. Rev.* **C34** (1986) 619;
    M. Salomon et al., *Nucl. Phys.* **A414** (1984) 493;
    A. Bagheri et al., *Phys. Rev.* **C38** (1988) 885.
29. E. Friedman et al., *Nucl. Phys.* **A514** (1990) 601.

FINITE TEMPERATURES AND VOLUMES

Restoration of Chiral Symmetry [1]

H. Leutwyler
Institute for Theoretical Physics
University of Bern
Sidlerstrasse 5, CH-3012 Bern, Switzerland

January 1992

Abstract

The lecture concerns the restoration of spontaneously broken symmetries at finite temperatures and volumes. In particular, the melting of the quark and gluon condensates at low T and the restoration phenomena related to the absence of spontaneous symmetry breaking at finite volume are discussed on the basis of chiral perturbation theory. Furthermore, it is shown that the effective Lagrangian analysis of QCD at finite volume allows one to predict some properties of the eigenvalue distribution of the Dirac operator.

[1]Lecture given at the Workshop on Effective Field Theories, Dobogokö, Hungary, 1991.
Work supported by Schweizerischer Nationalfonds

1 Introduction

Chiral symmetry flourished in the sixties and in the early seventies, but almost disappeared from sight when the standard model was discovered. In the last few years, however, chiral symmetry is being restored in the sense that the literature on the subject is again growing rapidly. One might be skeptical about the phenomenon, because history tells us that "restoration" often denotes rather deplorable developments. In my opinion, this skepticism is not warranted at all. Moreover, the present lecture does not concern this restoration of chiral symmetry. Instead, I will try to convince you that in the context of spontaneously broken symmetries, restoration is a fascinating phenomenon in its own right which has attracted considerable attention in solid state physics for many years and recently became popular also in particle physics, in particular in QCD.

In QCD, chiral symmetry is broken in two ways: (i) the quark masses *explicitly* break the symmetry of the Lagrangian and (ii) even if the quark masses are turned off, the symmetry is broken *spontaneously*, the ground state of the theory being asymmetric. Within the standard model, the quark masses represent phenomenological parameters which, a priori, could have any value. One can therefore study the effects due to explicit symmetry breaking by varying the quark masses which thus serve as symmetry breaking parameters. The Lagrangian does, however, not contain an analogous free parameter which would allow one to tune the strength of spontaneous symmetry breaking [2]. The magnitude of the quark condensate, e.g., which plays the role of an order parameter of the spontaneously broken symmetry, is determined by the scale of the theory. Nevertheless, the strength of spontaneous symmetry breaking can be influenced: it suffices to expose the system to a heat bath. Increasing the temperature of the bath, the condensate gradually melts; in the massless theory, it presumably disappears if the temperature reaches a critical

[2] Presumably, spontaneous symmetry breakdown only occurs if the number of light quark flavours is not too large. Hence, equipping the theory with sufficiently many flavours, the masses of the extra quarks could be used to interpolate between theories with and without spontaneous symmetry breaking.

value T_c and, for $T > T_c$, the symmetry is restored [1]. (The transition between the two phases may be of first order; if so, the condensate first gradually melts as the temperature is increased but then instantly drops to zero as T passes through T_c.)

In the following, I will concentrate on those aspects of symmetry restoration which can be analyzed by means of chiral perturbation theory. Although, in the context of a workshop on effective field theories, this selection criterion appears to be justified, I emphasize that, in recent years, significant information about the subject has also been gathered by other methods, in particular by means of numerical simulations on a lattice. For an overview, I refer to the proceedings of some recent topical conferences [2], [3], [4]. A few additional references will be served on the way, in the context of the specific phenomena encountered in the course of the present excursion.

In the first part of this talk, I will show that chiral perturbation theory provides significant information about the behaviour of the quark and gluon condensates below T_c [3]. In the second part, I will then consider a different category of symmetry restoration phenomena which originate in the fact that spontaneous symmetry breakdown does not occur if the volume of the system is finite. As we will see, the analysis of these phenomena is based on general symmetry considerations and does not invoke specific properties of the model. I will turn to some implications of this analysis for the structure of the underlying theory only in the last part of the talk, where I will show that in QCD, spontaneous symmetry breakdown reveals itself in the spectrum of the Dirac operator.

2 Melting of the quark condensate

The thermal expectation value of the operator $\bar{q}q$ is given by

[3] A more detailed discussion of this topic can be found in Peter Gerber's contribution these proceedings [5].

$$< \bar{q}q >_T = Z^{-1} \sum_n < n \mid \bar{q}q \mid n > \exp(-E_n/T) \tag{2.1}$$

where the sum extends over a complete set of eigenstates of the Hamiltonian, E_n is the corresponding eigenvalue and

$$Z = \sum_n \exp(-E_n/T) \tag{2.2}$$

is the partition function. At low temperatures, the Boltzmann factor restricts the sum to states of low energy, $E_n \lesssim T$. The behaviour of the condensate at low T is therefore determined by the matrix elements of the operator $\bar{q}q$ for states of low energy: vacuum, π, 2π, 3π, K, 4π, η, ... Let us for simplicity consider the chiral limit where N_f of the quark flavours are taken massless (the case $N_f = 2$, e.g., corresponds to $m_u = m_d = 0$, the remaining quark masses $m_s, m_c, \ldots$ being taken at their physical values). In this limit, states of low energy exclusively contain Goldstone bosons. The low energy behaviour of the matrix elements $< n \mid \bar{q}q \mid n >$ can therefore be worked out by means of chiral perturbation theory. The phase space integrals involved in the sum over states then convert powers of energy or momentum into powers of T. The vacuum generates a temperature independent contribution, while the one-pion state produces a term proportional to T^2 [6]. In the language of chiral perturbation theory, these contributions correspond to the tree and one-loop graphs, respectively. Including terms of order T^4 (two loops [7]) and T^6 (three loops [8], [9]), the low temperature expansion of the quark condensate takes the form

$$\begin{aligned} < \bar{q}q >_T = < 0 \mid \bar{q}q \mid 0 > \Bigg[1 &- \frac{(N_f^2-1)}{N_f} \frac{T^2}{12F^2} - \frac{(N_f^2-1)}{2N_f^2} \left(\frac{T^2}{12F^2} \right)^2 \\ &- N_f(N_f^2-1) \left(\frac{T^2}{12F^2} \right)^3 \ln \frac{\Lambda_q}{T} + O(T^8) + O\left\{ \exp\left(-\frac{M_1}{T} \right) \right\} \Bigg] \end{aligned} \tag{2.3}$$

where F is the value of the pion decay constant, $F_\pi \simeq 93$ MeV, in the chiral limit. The result (2.3) is exact. It shows that the condensate gradually melts with increasing temperature. At low T, the scale of the temperature dependence is set by the pion decay constant which determines the magnitude of the terms of order T^2 and T^4. At order T^6, the expansion involves a chiral logarithm. The scale Λ_q of this logarithm is determined by the coupling constants which occur in the effective Lagrangian at order p^4. Since the same combination of couplings also determines the $I = 0$, D-wave, $\pi\pi$ scattering length a_2^0, one can determine the value of the logarithmic scale from the phenomenology of $\pi\pi$-scattering, with the result $\Lambda_q \simeq 470$ MeV ($N_f = 2$) [9].

Numerically, using $F \simeq 88$ MeV [9], the representation (2.3) leads to $< \bar{q}q >_T = 0$ for $T \simeq 190$ MeV. It is clear, however, that this representation cannot be trusted there. Within chiral perturbation theory, the temperature dependence arises from the loop graphs - the tree graph contribution is given by the vacuum expectation value $< 0 \mid \bar{q}q \mid 0 >$ and is independent of T. The chiral expansion only makes sense at low T where the loop graphs represent corrections to the tree graph result. At the zero of the representation (2.3), the "corrections" cancel the leading term, i.e. this zero occurs at a temperature where the chiral expansion is not under control.

In fact, we are dealing with an asymptotic series here. The expansion in powers of T, even if carried to arbitrarily high order, does not account for terms which at low T are exponentially small. The fraction of the energy density stored in ρ-mesons, e.g., is of order $\exp(-M_\rho/T)$ and does not show up at any finite order of the series in powers of T. The occurrence of such additional contributions is characteristic of asymptotic expansions. In the expansion of QCD in powers of the coupling constant g, instantons generate analogous nonperturbative contributions,

$$A = a_0 + a_1 g^2 + a_2 g^4 + \ldots + O\left[\exp(-8\pi^2/g^2)\right] \qquad (2.4)$$

Similarly, in the low temperature expansion, the chiral perturbation series accounts for all powers of T, but misses contributions of order $\exp(-M_1/T)$ where M_1 is the mass of the lightest non-Goldstone state in the chiral limit (for $N_f = 2$, M_1 is the mass of the K-meson while for $N_f = 3$, $M_1 = M_\rho$). In the above comparison of the low temperature expansion with the standard perturbation series, the quantity T/M_1 plays a role analogous to $g^2/8\pi^2$.

In the low temperature expansion, the occurrence of exponentially small terms is related to the fact that a hadronic gas in thermal equilibrium does not exclusively consist of π, K, η, but also contains excited states: ρ, ω, K^*, N, ... [10]. The abundance of the various particle species is discussed in detail in ref. [9], [11], on the basis of the virial expansion. It turns out that, at a temperature of 100 MeV, the volume per particle is about 30 fm^3; at that temperature, 82% of all particles are pions, 12% are kaons or etas and only 6% are in an excited state. If the temperature, however, rises to 150 MeV, the volume par particle shrinks to about 4 fm^3 and 40% of the particles are in an excited state. Accordingly, the mean free path λ rapidly decreases, from $\lambda \simeq 15$ fm (T = 100 MeV) to $\lambda \simeq 2$ fm (T = 150 MeV) [12], [13], [11]. The exponential terms become increasingly important as one approches the critical temperature. Indeed, it would be strange if the phase transition were to take place from a configuration of well isolated hadrons, sitting in their ground state to a configuration where quarks and gluons are liberated. More likely, the transition only occurs at a temperature where the gas is dense and where a substantial fraction of the particles is in an excited state.

The exponential terms also affect the melting of the condensate. The magnitude of the contributions generated by these terms in $< \bar{q}q >_T$ is estimated in ref. [9]. It turns out that the effect is negligibly small below T = 140 MeV, but rapidly grows if the temperature is increased further. For details, in particular also for a discussion of the manner in which nonzero quark masses modify the low temperature expansion given in eq. (2.3), see refs. [5], [9].

A review of recent lattice results for the quark condensate can be found, e.g., in refs.[2], [14]. The literature contains many attempts at understanding the physics of QCD in the nonperturbative domain, based on instantons, monopoles, abelian fields, scalar degrees of freedom, analogies with superconductivity, ... [see, e.g., [15] - [22]]. The temperature dependence of the quark condensate can be studied within such models and the analysis of sum rules at nonzero temperature [4] also sheds some light on the issue [see, e.g., [24], [25], [26]]. In particular, the problem of whether or not the restoration of chiral symmetry and the liberation of colour occur at the same temperature and whether the chiral phase transition is of first or second order are extensively discussed in the literature which can be traced from the above references.

3 Trace anomaly and gluon condensate

Chiral perturbation theory also allows one to analyze the behaviour of the pressure P and of the energy density ε at low temperatures where the gas is dilute and almost exclusively consists of Goldstone bosons. As discussed above, the density of other particles is exponentially suppressed. In the chiral limit, the leading contribution occurring in the low temperature expansion of P or ε is the familiar free Bose gas term of order T^4. The interaction among the Goldstone bosons only shows up at order T^8 [8], [9],

$$P = \frac{\pi^2}{90}(N_f^2-1)T^4\left\{1+N_f^2\left(\frac{T^2}{12F^2}\right)^2 \ln\left(\frac{\Lambda_p}{T}\right)\right\}+$$
$$+O(T^{10})+O\left[\exp\left(-\frac{M_1}{T}\right)\right] \tag{3.1}$$

A term of order T^6 does not occur here, because - in the chiral limit - the $\pi\pi$-scattering amplitude vanishes in the forward direction, on account of an Adler zero. The logarithmic scale Λ_p is closely related to the scale Λ_q encountered at order T^6 in the expansion of the quark condensate:

$$\ln\frac{\Lambda_q}{\Lambda_p} = \frac{N_f^2+1}{6N_f^4}+0.491. \tag{3.2}$$

The numerical value, relevant for the limit $m_u = m_d = 0$ at fixed m_s is $\Lambda_p \simeq 275$ MeV [9].

[4]See Volodya Eletsky's contribution to these proceedings [23].

The formula (3.1) shows that, at low T, the interaction generates a very weak repulsion. The corresponding expressions for the entropy density and for the energy density are readily obtained with the thermodynamic relations

$$s = \frac{dP}{dT} \ , \ \varepsilon = Ts - P. \tag{3.3}$$

The thermal expectation value of the trace of the energy momentum tensor is therefore given by

$$< \Theta^\mu_\mu >_T = \varepsilon - 3P = T^5 \frac{d}{dT}\left(\frac{P}{T^4}\right). \tag{3.4}$$

The operator Θ^μ_μ is the divergence of the dilation current and vanishes in conformally invariant theories. For a free massless gas, the pressure is proportional to T^4 such that the formula (3.4) implies $< \Theta^\mu_\mu >_T = 0$, in agreement with the fact that a free gas of massless particles is conformally invariant. The interaction among the Goldstone bosons, however, does not respect conformal invariance [27]. The representation (3.1) for the pressure shows that in the thermal expectation value of Θ^μ_μ, the interaction starts showing up at order T^8:

$$< \Theta^\mu_u >_T = \frac{\pi^2}{3240} N_f^2 (N_f^2 - 1) \frac{T^8}{F^4} \left\{ \ln \frac{\Lambda_p}{T} - \frac{1}{4} \right\} + \ldots \tag{3.5}$$

Indeed, QCD is not conformally invariant. Even if the quark masses are set equal to zero, scale invariance is broken, because the coupling constant is scale dependent, i.e. because the β-function of the theory is different from zero,

$$\beta(g) = \mu \frac{dg}{d\mu} = -\frac{1}{48\pi^2}(11N_c - 2N_f)g^3 + O(g^5). \tag{3.6}$$

I write the explicit expression for the operator Θ^μ_μ in the form

$$\Theta^\mu_\mu = -G^2 + c\mathbb{1} \tag{3.7}$$

where G^2 stands for the familiar term proportional to the β-function and to the square of the gluon field strength,

$$G^2 \equiv \frac{(-\beta)}{2g^3} G^a_{\mu\nu} G^{\mu\nu a}. \tag{3.8}$$

The c-number term occurring in eq. (3.7) is fixed by the normalization condition $< 0 \mid \Theta_{\mu\nu} \mid 0 >= 0$,

$$c =< 0 \mid G^2 \mid 0 > \simeq \frac{9}{8} \times 0.012 \mathrm{GeV}^4 \simeq 2 \mathrm{GeV/fm}^3 \tag{3.9}$$

where I have inserted the numerical estimate of the gluon condensate obtained by Shifman, Vainshtein and Zakharov [28]. At finite temperatures, the gluon condensate is therefore given by

$$< G^2 >_T =< 0 \mid G^2 \mid 0 > - < \Theta^\mu_\mu >_T . \tag{3.10}$$

It is not a straightforward matter to reliably determine the magnitude of the gluon condensate at zero temperature and a considerable range of values can be found in the recent literature. The problems associated with the numerical value of $< 0 \mid G^2 \mid 0 >$ do, however, not affect the temperature dependence of the gluon condensate which is unambiguously determined by the thermal expectation value of Θ^μ_μ, i.e. by energy density and pressure.

The relations (3.10) and (3.5) show that the temperature dependence of the gluon and quark condensates is quite different. While the low temperature expansion of the quark condensate contains terms of order $T^2, T^4, \ldots$, the gluon condensate is nearly constant at low T, a temperature dependence only showing up at order T^8, through the term given in eq. (3.5). The sign of this term implies that the gluon condensate also melts with growing temperature, but the melting takes place much more slowly than in the case of the quark condensate.

It is not difficult to identify the origin of this qualitative difference. The operator $\bar{q}q$ transforms in a nontrivial manner under chiral transformations. Current algebra then implies that this operator readily emits or absorbs pions. In particular, the matrix element $< \pi \mid \bar{q}q \mid \pi >$ is different from zero: for pions of zero momentum, the magnitude of this matrix element is determined by the ratio

$< 0 \mid \bar{q}q \mid 0 > /F^2$. This is why the low temperature expansion of the quark condensate contains a term proportional to T^2/F^2. In contrast, the operators Θ^μ_μ and $G^a_{\mu\nu}G^{a\mu\nu}$ are chiral singlets and the corresponding one-pion matrix elements therefore vanish. Accordingly, the low temperature expansion of the gluon condensate does not contain a term of order T^2. In the framework of chiral perturbation theory, the quantity $< \Theta^\mu_\mu >_T$ vanishes up to and including two loops - the contribution of order T^8 given in eq. (3.5) originates in a three-loop graph [9].

Note that the suppression of the terms involving powers of T at the same time also implies that the exponentially small contributions - which chiral perturbation theory does not account for - become significant already at rather low temperatures. Equation (3.5) provides an adequate representation for the thermal expectation value of the trace only at very low T [9], [29].

The vacuum expectation value $< 0 \mid G^2 \mid 0 >$ may be viewed as a parameter associated with nonperturbative scale breaking effects. Asymptotic freedom indicates that these effects disappear at high temperatures where perturbation theory should be reliable, the effective coupling becoming small. Intuitively, one might therefore expect the gluon condensate to disappear above the critical temperature, together with the quark condensate. This is however not the case. The gluon condensate does not represent an order parameter. Conformal symmetry is broken also in the high temperature phase. Identifying the scale of the running coupling constant with T, the perturbative expansion in powers of g amounts to a series involving negative powers of $\ln(T/\Lambda_{\rm QCD})$. In the pressure, the leading term of this series is again proportional to T^4, now describing free gluons and quarks. The interaction lowers the pressure by a term proportional to $g^2T^4 \propto T^4/\ln T$. Inserting this in eq. (3.4), one concludes that, at high temperatures, the expectation value of Θ^μ_μ is of order $T^4/(\ln T)^2 \propto g^4T^4$ and is therefore small compared to the energy density of the plasma [30]. In absolute magnitude, the quantity $< \Theta^\mu_\mu >_T$ however grows without bounds as the temperature rises, such that - according to eq. (3.10) - the gluon condensate does not disappear but becomes negative and large. The two condensates thus behave in a qualitatively different manner also at high T.

The properties of the gluon condensate were intensively studied on the lattice [31], [32], [33]. Qualitatively, the data confirm the behaviour discussed above. More work is needed, however, to arrive at quantitative results, particularly in the presence of light quarks.

4 Symmetry restoration at finite volume

In the above discussion, I have tacitly assumed that the volume of the system is infinite. In this case, the magnitude of spontaneous symmetry breaking is controlled by the temperature, the symmetry being restored if T is sufficiently high. At finite volume, a qualitatively different category of symmetry restoration phenomena occurs which are related to the well-known fact that, in the mathematical sense of the term, spontaneous symmetry breakdown can only take place if the volume is infinite.

In particle physics, finite volumes may appear to be beyond reasonable interest. Note, however, that lattice simulations are necessarily performed at finite volume and an extrapolation to infinite volume is required to confront the results of these simulations with observation. This can be done in a reliable manner only if the finite size effects are understood. Their magnitude depends on the size L of the box and on the mass gap M occurring in the spectrum of the theory at infinite volume. If the box is large compared to the Compton wavelength of the lightest particle, $ML \gg 1$, then the finite size effects are small corrections of order $\exp(-ML)$. For a general analysis of these corrections, I refer you to Lüscher's work [34]. This analysis, however, does not apply if $ML \lesssim 1$. In particular, in theories which do not have a mass gap, $M = 0$, the box may strongly affect some of the observables - the finite size effects cannot be treated as small perturbations. Consider, e.g. the quark condensate $< \bar{q}q >_{T,L}$ where T is the temperature and L is the size of the three-dimensional box. In the chiral limit, this quantity tends to zero, because in this limit, QCD is invariant under chiral rotations and symmetries cannot break down spontaneously at finite volume, $< \bar{q}q >_{T,L} = 0$. On the other hand, if L

is sent to infinity before the chiral limit is taken, a nonzero condensate results because chiral symmetry does break down spontaneously at infinite volume, unless the temperature is too high. In the condensate, the presence of a box therefore manifests itself even if the box is very large: in the chiral limit, the finite size effect $<\bar{q}q>_{T,L} - <\bar{q}q>_{T,\infty}$ is of the same size as the quantity itself. A similar phenomenon also occurs in the two-point functions of the vector and axial currents. At infinite volume, these quantities are quite different - pion exchange, e.g., generates a pole in $<A_\mu A_\nu>$ which does not show up in $<V_\mu V_\nu>$. The difference $<A_\mu A_\nu> - <V_\mu V_\nu>$, however, disappears if the chiral limit is taken at finite volume, because the right- and left-handed combinations $V_\mu \pm A_\mu$ transform differently under the chiral group. Also, the Ward identity obeyed by the singlet axial current implies that the two-point function of the operator $G_{\mu\nu}\tilde{G}^{\mu\nu}$ exhibits strong finite size effects: if the chiral limit is taken at infinite volume, the topological susceptibility tends to zero with the first power of the quark mass while at finite volume it is proportional to the power m^{N_f}.

The main point of the following discussion is that for theories which undergo spontaneous symmetry breakdown at infinite volume, the origin of the leading finite size effects is known: they are generated by the lightest particles contained in the spectrum of the theory, by the Goldstone bosons. The propagation properties of these particles can be studied in detail by means of the effective Lagrangian and one can therefore calculate the leading finite size effects explicitly in terms of the coupling constants which occur in this Lagrangian [7].

Let us for definiteness again consider QCD. In this theory, the partition function can be represented as a euclidean path integral over the gluon field,

$$Z = \int [dG] \exp\{-\frac{1}{2g^2}\int_V dx \mathrm{tr}_c G_{\mu\nu}G_{\mu\nu}\} \det(-i\not{D} + \mathcal{M}) \tag{4.1}$$

where M is the quark mass matrix. The boundary conditions in the euclidean time direction are dictated by the thermal trace: in order for Z to represent the trace of the operator $\exp(-H/T)$, the quark (gluon) fields must be (anti)periodic in x^4 up to a gauge transformation, the period being set by the temperature. On the other hand, the boundary conditions imposed at the walls of the three-dimensional box are a matter of choice. In the logarithm of the partition function, the contributions which are proportional to the volume L^3 of the box are independent of this choice, but the boundary conditions do affect the surface terms which contribute at order L^2.

The behaviour of the partition function at large volume can be analyzed by means of an effective Lagrangian. The form of this Lagrangian, however, depends on the boundary conditions imposed in the underlying theory. In general, vertices occur which break euclidean invariance; in addition, the effective action may involve surface terms. In lattice simulations, the box is usually taken rectangular with sides L_1, L_2, L_3 and the boundary conditions imposed at the walls of the box are the same as those imposed in the time direction. In the continuum limit, the functional integral then extends over a four-dimensional torus, $V = L_1 \times L_2 \times L_3 \times L_4$ with $L_4 = 1/T$. In this case, the partition function is invariant under permutations of the four sides of the four-dimensional box. This property leads to a remarkable simplification of the effective Lagrangian: only euclidean invariant vertices occur and the effective action does not contain surface terms. Moreover, the effective coupling constants are independent of the size of the box - the effective Lagrangian coincides with the standard chiral Lagrangian describing the properties of the Goldstone bosons at infinite volume [35]. This allows one to analyze the leading finite size effects in a remarkably simple manner: for large volumes, low temperatures and small quark masses, the partition function of QCD coincides with the partition function of the effective pion field theory,

$$Z \int [dU] \exp\{-\int_V dx L_{\text{eff}}(U, \partial U, \partial^2 U, \ldots, \mathcal{M})\}. \tag{4.2}$$

The unitary matrix $\mathrm{U}(x) \in \mathrm{SU}(N_f)$ describes the $N_f^2 - 1$ Goldstone bosons and L_{eff} is the standard chiral Lagrangian. The temperature and the size of the box only enter through the boundary conditions obeyed by the meson field which lives on a four-dimensional torus, i.e. is periodic in all four directions of euclidean space. The representation (4.2) is exact in the sense of an asymptotic series; it determines the coefficients occurring in the expansion of, say, the energy density in powers of $T, 1/L$ and $\mathcal{M}$ to any finite order, but does not account for exponentially small terms of the type $\exp(-M_\rho/T)$ or $\exp(-M_\rho L)$.

The chiral perturbation series associated with eq. (4.2) is worked out in the standard manner. The expansion of the effective Lagrangian in powers of derivatives and powers of $\mathcal{M}$ is the same as at infinite volume. The leading term in this expansion,

$$L_{\text{eff}} = \frac{1}{4}F^2 \text{tr}\{\partial_\mu U^+ \partial_\mu U\} - \frac{1}{2}\Sigma \text{tr}\{\mathcal{M}(U + U^+)\} + \ldots \tag{4.3}$$

contains two effective coupling constants: F, Σ. The constant Σ is the vacuum expectation value of the operator $\bar{q}q$ in the chiral limit at infinite volume,

$$< 0 \mid \bar{u}u \mid 0 >=< 0 \mid \bar{d}d \mid 0 >= \ldots = -\Sigma \tag{4.4}$$

while the pion mass involves the ratio of the two constants,

$$M_{\pi^+}^2 = \frac{\Sigma}{F^2}(m_u + m_d) + O(m^2). \tag{4.5}$$

Next, one sets

$$U(x) = \exp i\frac{\pi(x)}{F} \tag{4.6}$$

and expands the effective Lagrangian in powers of the hermitean and traceless matrix $\pi(x)$. This gives rise to a kinetic term quadratic in $\pi(x)$ and a string of vertices involving four or more powers of $\pi(x)$. The functional integral over the pion field can then be evaluated in the standard perturbative fashion. The only modification brought about by the four-dimensional box is that the meson propagator is replaced by the Green's function associated with the torus. For quarks of equal mass, the propagator takes the form

$$G(x) = \frac{1}{V}\sum_p \frac{e^{ipx}}{M_\pi^2 + p^2} \tag{4.7}$$

where the components of the vector p_μ are integer multiples of $2\pi/L_\mu$. The counting of powers of momenta and masses is the same as at infinite volume and it is convenient to count the inverse box sides $1/L_\mu$ as quantities of order p.

In chiral perturbation theory, the logarithm of the partition function is described by connected graphs which do not contain external lines. Graphs which exclusively involve vertices from the leading effective Lagrangian specified in eq. (4.4) contribute to $\ln Z$ through a term of the form $(F\bar{L})^{2(1-l)}f$, where $\bar{L} = V^{1/4}$ is the mean size of the four-dimensional box and l is the number of loops. The coefficient f only depends on the four dimensionless variables $M_\pi L_1, \ldots, M_\pi L_4$. Graphs which contain vertices originating in the remainder of the effective Lagrangian are suppressed by powers of $1/\bar{L}$. The chiral perturbation series orders the vertices occurring in the effective Lagrangian according to the number of derivatives and quark mass factors and orders the graphs according to the number of loops. The above counting of powers therefore shows that the chiral series provides an expansion of the partition function in powers of $1/\bar{L}$. In this expansion, the pion mass is treated as a small parameter, the product $M_\pi \bar{L}$ counting as a quantity of order one.

Chiral perturbation theory allows one, in particular, to study the symmetry restoration phenomena occurring in the chiral limit. These phenomena are related to the fact that the loop expansion develops infrared singularities if the pion mass tends to zero. Indeed, the formula (4.7) shows that the contribution to the propagator from the mode with $p = 0$ is given by $1/M_\pi^2 V$ and therefore blows up for $M_\pi \to 0$. Accordingly, graphs where mesons with $p = 0$ propagate around a loop diverge in the chiral limit. The disease is of a technical nature and is easily cured. It originates in the assumption that the functional integral (4.2) can be evaluated by expanding the integrand around the Gaussian term which stems from the kinetic part of the effective Lagrangian. Since the mode with $p = 0$ does not carry kinetic energy if the pion mass is turned off, the fluctuations in this mode are not Gaussian. In fact, chiral symmetry implies that for $\mathcal{M} = 0$, the effective action vanishes to all orders if $\partial_\mu U(x) = 0$. For small quark masses, the action associated with a constant field is given by $-V\Sigma\text{Re tr}(\mathcal{M}U)$. It squeezes the field U to the vicinity of $U = 1$ only if the product $V\Sigma\mathcal{M}$ is large; if this is not the case, the different directions in group space acquire comparable probability. In particular, if $\mathcal{M}$ vanishes, they become equally likely - symmetry is restored. Note that the problem only concerns the zero mode. The amplitudes of the remaining modes do show up in the quadratic part of the action even if the pion mass vanishes. There is a standard technique which allows one to deal with problems of this type: one treats the zero mode amplitudes as collective variables, setting $U(x) = U_0 U_1(x)$, where U_0 is the space-independent field associated with the zero mode fluctuations while $U_1(x)$ describes the modes with $p \neq 0$. The latter do not acquire large amplitudes and can therefore be analyzed perturbatively. This technique leads to an explicit representation of the partition function as an ordinary integral over the collective variable U_0,

$$Z = \mathcal{N} \int_{\mathrm{SU}(N_f)} d\mu(U_0) \exp\{V\Sigma\text{Re tr}(\mathcal{M}U_0)\} \tag{4.8}$$

where $d\mu(U)$ is the Haar measure on the group. The representation is valid in the symmetry restoration region $V\Sigma\mathcal{M} \lesssim 1$ and holds up to corrections of order $1/\bar{L}^2$ [7], [36]. Since the quark condensate is related to the derivative of $\ln Z$ with respect to $\mathcal{M}$, this representation also predicts the manner in which $<\bar{q}q>_V$ disappears as $\mathcal{M} \to 0$. Note that if the volume is large, symmetry restoration only occurs if the quark masses are tiny, of order $1/V\Sigma$. For a box of mean size $\bar{L} = 10$ fm, e.g., the quark mass must be smaller than 0.01 MeV for the condensate to become small. The occurrence of spontaneous symmetry breakdown can therefore also be seen at large, but finite volumes, using a small quark mass to tell the condensate in which direction of group space to condense. If $\mathcal{M}$ is very small compared to the scale of the theory, the properties of the various observables are essentially the same as in the chiral limit at infinite volume, provided only that the volume is large enough, $V\Sigma\mathcal{M} \gg 1$. For a detailed analysis of the symmetry restoration phenomena, I refer to the literature [7], [36], [37], [38]. The beautiful data on the σ-model [39] confirm the predictions of chiral perturbation theory remarkably well.

Once lattice simulations of QCD will reach realistic pion masses on reasonably large boxes, chiral perturbation theory will likely become an efficient tool also for the analysis of these data.

5 Winding number and spectrum of the Dirac operator at finite volume

The spectrum of the Dirac operator is one of the central issues to be clarified on the way to a physical understanding of the vacuum structure in QCD. Indeed, the density of eigenvalues is known to be intimately related to the magnitude of the chiral order parameter $< 0 \mid \bar{q}q \mid 0 >$ [16]. At infinite volume, the relation is the following. Denote by $\rho(\lambda)d\lambda$ the number of eigenvalues λ contained in the interval $d\lambda$, per unit four-dimensional volume. The quark condensate is determined by the level density at $\lambda = 0$,

$$< 0 \mid \bar{q}q \mid 0 >= -\pi\rho(0). \tag{5.1}$$

The level density of free quarks tends to zero for $\lambda \to 0$, in proportion to $\mid \lambda \mid^3$. For chiral symmetry to be broken spontaneously with a nonzero quark condensate, the typical gluon field configurations must be such that small eigenvalues of the Dirac operator are more abundant than for free quarks. Indeed, several categories of field configurations for which $\rho(0)$ is different from zero have been found [17]. With the progress made in Monte Carlo techniques it has become possible to tackle the problem numerically, evaluating the lowest few hundred eigenvalues of the Dirac operator for a representative set of lattice gauge field configurations [see e.g. [40]].

In the preceding sections we have analyzed the properties of the l.h.s. of eq. (5.1) at finite temperatures and volumes. In this analysis, the specific structure of the underlying theory did not play an important role. We merely exploited the fact that massless QCD is symmetric under independent rotations of the right- and left-handed quark flavours and that this symmetry spontaneously breaks down at infinite volume. The relation (5.1) indicates that the properties of the quark condensate which follow from these general symmetry considerations must have immediate implications for the eigenvalue distribution of the Dirac operator. Let us now see what these implications are [41]. What becomes of the relation (5.1) at finite volume ?

To answer this question, it is convenient to slightly generalize the framework considered in the preceding sections, allowing for a nonzero vacuum angle Θ as well as for a complex mass matrix. The representation (4.1) of the partition function then becomes

$$Z = \int [dG] \exp\left\{-\frac{1}{2g^2}\int_V dx \mathrm{tr}_c G^2_{\mu\nu} + i\Theta \int_V dx\omega\right\} \det(-i\not{D} + \tilde{\mathcal{M}}) \tag{5.2}$$

where $\omega(x)$ denotes the winding number density

$$\omega = \frac{1}{16\pi^2}\mathrm{tr}_c G_{\mu\nu}\tilde{G}_{\mu\nu} \tag{5.3}$$

and $\tilde{\mathcal{M}}$ stands for

$$\tilde{\mathcal{M}} = \frac{1}{2}(1-\gamma_5)\mathcal{M} + \frac{1}{2}(1+\gamma_5)\mathcal{M}^+ \qquad (5.4).$$

The functional integral in eq. (5.2) extends over gluon field configurations on a torus. Only those configurations are admissible which can support antiperiodic fermion fields - the Dirac operator must be periodic up to a gauge transformation. As it is well-known, this requirement implies that the winding number

$$\nu = \int_V dx\omega \tag{5.5}$$

is an integer. Accordingly, the partition function consists of a sum of terms

$$Z = \sum_{\nu=-\infty}^{\infty} Z_\nu e^{i\nu\theta}. \tag{5.6}$$

Let us denote the eigenvalues of the Dirac operator associated with a given gluon field configuration by λ_n,

$$\not{D} u_n(x) = \lambda_n u_n(x). \tag{5.7}$$

If the winding number of the configuration is positive, then there are ν zero modes of positive chirality. The remaining eigenvalues occur in pairs $(\lambda_n, -\lambda_n)$. The Fourier coefficient Z_ν therefore takes the form

$$Z_\nu = \int [dG] \exp\left\{-\frac{1}{2g^2}\int_V dx \mathrm{tr}_c G_{\mu\nu}^2\right\} (\det_f \mathcal{M})^\nu \Pi_n{}' \det_f(\lambda_n^2 + \mathcal{M}\mathcal{M}^+). \tag{5.8}$$

Here, the integration only extends over field configurations with winding number ν and the product $\Pi_n{}'$ runs over pairs of nonzero eigenvalues [if ν is negative, the factor $(\det_f \mathcal{M})^\nu$ is to be replaced by $(\det_f \mathcal{M}^+)^{-\nu}$]. The product requires regularization. One may, e.g., replace the logarithm of the product by the sum of the logarithms of the individual terms, domesticated by a cutoff factor of the form $\exp(-\lambda_n^2/\Lambda^2)$. This regularization of Π' is invariant under chiral rotations, $\mathcal{M} \to V_R \mathcal{M} V_L^+$ with $V_R, V_L \in U(N_f)$.

The formulae (5.6) and (5.8) explicitly show that the partition function of QCD depends on the quark mass matrix and on the vacuum angle only through the combination $\mathcal{M}\exp(i\Theta/N_f)$. A change in the vacuum angle is equivalent to a change in the phase of the quark mass matrix - a well-known consequence of the anomalous Ward identity obeyed by the singlet axial current.

Let us now first consider QCD with a single quark flavour, $N_f = 1$. In this case the quark mass matrix reduces to a complex number m. The divergence of the axial current $\bar{q}\gamma_\mu\gamma_5 q$ contains an anomaly and hence fails to vanish even if m is set equal to zero. In other words, the theory does not have a chiral symmetry. There is an exact $U(1)$-symmetry related to baryon number, but this symmetry is not supposed to break down spontaneously. Accordingly, Goldstone bosons do not occur - the theory contains a mass gap M_0 which persists if the quark mass is set equal to zero. This immediately implies that the finite size effects are exponentially small and can be ignored if all sides of the four-dimensional box are large compared to $1/M_0$. The logarithm of the partition function is proportional to the volume

$V = L_1 L_2 L_3 L_4$,

$$Z = \exp\left\{-V\varepsilon(me^{i\Theta})\right\} \tag{5.9}$$

where ε is the energy density of the ground state. In fact, the function $\varepsilon(me^{i\Theta})$ is what remains of the effective Lagrangian - a Goldstone boson field $U(x)$ does of course not occur here. Since the theory does not contain massless physical states, the expansion in powers of the quark mass term does not give rise to infrared singularities but is an ordinary Taylor series,

$$\varepsilon(me^{i\Theta}) = \varepsilon_0 - \frac{1}{2}\Sigma(me^{i\Theta} + m^\star e^{-i\Theta}) + O(m^2). \tag{5.10}$$

I have used the same symbol for the low energy constant which specifies the term linear in m as in eq. (4.3): Σ determines the expectation values of the scalar and pseudoscalar quark currents for $m = 0$,

$$< 0 \mid \bar{q}q \mid 0 >= -\Sigma\cos\Theta \ ; \ < 0 \mid \bar{q}_0 i\gamma_5 q \mid 0 >= -\Sigma\sin\Theta. \tag{5.11}$$

Note that the finite size corrections in these quantities are exponentially small. For $N_f = 1$, the matrix element $< \bar{q}q >_V$ does not play the role of an order parameter associated with a spontaneously broken symmetry - the model does not have such a symmetry.

Taking the quark mass real and positive, dropping an overall normalization constant and disregarding terms of higher order in m, the partition function becomes

$$Z = \exp\left\{V\Sigma m\cos\Theta\right\}. \tag{5.12}$$

This explicit representation for the partition function holds provided the quark mass is small compared to the scale of the theory, while the box must be large, $m \ll M_0$, $L \gg 1/M_0$. The contribution generated by field configurations with winding number ν is obtained by inverting the Fourier series (5.6),

$$Z_\nu = I_\nu(V\Sigma m) \tag{5.13}$$

where $I_\nu(x)$ is the Bessel function of imaginary argument,

$$I_\nu(x) = \begin{cases} \frac{1}{\nu!}\left(\frac{x}{2}\right)^\nu & x \ll 1 \\ (2\pi x)^{-1/2} e^x & x \gg 1 \end{cases} . \tag{5.14}$$

The result shows that, in the limit $m \to 0$, the probability to encounter a gauge field configuration with nonzero winding number, Z_ν/Z, tends to zero in proportion to $m^{|\nu|}$. This agrees with the representation (5.8) where this factor originates in the zero modes of the Dirac operator. For massless quarks, only configurations with $\nu = 0$ occur.

In particular, if the integration over the gluon field is restricted to configurations with winding number zero, the partition function is given by $Z_0 = I_0(V\Sigma m)$. The corresponding expression for the quark condensate is determined by the logarithmic derivative with respect to the quark mass,

$$< \bar{q}q >_{\nu=0} = -\Sigma I_1(V\Sigma m)/I_0(V\Sigma m) \tag{5.15}$$

and therefore vanishes in proportion to m if m tends to zero.

These findings look queer. The winding number is a global concept and one should expect that, since the theory has a mass gap, the value of the condensate at a given point should only be sensitive to the behaviour of the gluon fields within a ball of size $1/M_0$ around this point. It should not be affected by the presence of a few instantons behind the moon and should therefore be independent of ν. In the above formulae, this expectation is borne out only if the product $V\Sigma m$ is large. The asymptotic representation of the Bessel function shows that Z_ν then indeed becomes independent of ν. In particular, the value of the condensate is then given by Σ, irrespective of whether we sum over all configurations or restrict ourselves to those with $\nu = 0$. We are thus dealing with a behaviour which strongly resembles the symmetry restoration phenomena discussed in the preceding section - despite the fact that there is no symmetry to be restored if $N_f = 1$.

The resolution of the puzzle is the following. Long range correlations are absent in the full theory, but they do occur if the integration over the gluon field is restricted to fixed winding number. In fact, the long range correlations are generated precisely by the constraint $\int_V dx\omega = \nu$. To see this, consider the two-point function of the operator $\bar{q}q$ in the sector with winding number ν,

$$\sigma_\nu(x-y) = < \bar{q}(x)q(x)\ \bar{q}(y)q(y) >_\nu . \tag{5.16}$$

The space integral of this quantity represents the second derivative of the partition function with respect to the quark mass,

$$\int_V dxdy\sigma_\nu(x-y) = \frac{1}{Z_\nu}\frac{\partial^2 Z_\nu}{\partial m^2}. \tag{5.17}$$

Consider, e.g., the sector $\nu = 0$. Using the explicit representation (5.13) for Z_0 and taking the limit $m \to 0$, this relation implies that the space integral of the correlation function is proportional to the volume,

$$\int_V dx\sigma_0(x) = \frac{1}{2}\Sigma^2 V. \tag{5.18}$$

Hence $\sigma_0(x)$ does not fall off within a distance of order $1/M_0$ - if it did, the space integral would become volume independent for large volumes. Note also that the result (5.18) is not due to a disconnected part; since we are restricting the gluon field configurations to winding number zero, the expectation value of $\bar{q}q$ vanishes if the quarks are massless [see eq. (5.15)].

The long range correlations originate in the fact that the Dirac operator admits small eigenvalues. For $N_f = 1$, the functional integral representation (5.8) of Z_0 involves the quark mass only through a factor of the form $\Pi_n(1+m^2/\lambda_n^2)$. If m is sufficiently small, the average of this factor over the gluon field yields the Bessel function $I_0(V\Sigma m)$. This indicates that eigenvalues λ_n of order $1/V\Sigma$ play the key role. In fact, the representation of the Bessel function in terms of its zeros $\xi_1, \xi_2, \ldots$

$$I_0(x) = \Pi_n(1 + x^2/\xi_n^2) \qquad ; \quad J_0(\xi_n) = 0 \tag{5.19}$$

shows that the result of the path integral over the gluon field is the same as if there was a single field configuration with eigenvalues distributed according to

$$\bar{\lambda}_n = \pm \frac{\xi_n}{V\Sigma}. \tag{5.20}$$

In reality, the eigenvalues of the Dirac operator depend on the gluon field under consideration. The explicit formula (5.20) which predicts essentially equidistant eigenvalues, $\bar{\lambda}_n \simeq n\pi/V\Sigma$, merely indicates that the spectrum fluctuates around these mean values.

For eigenvalues which are large compared to $1/V\Sigma$, the distribution (5.20) shows that the mean number of levels contained in the interval $\Delta\lambda$ is given by $\Delta\lambda V\Sigma/\pi$. Since Σ represents the expectation value of the operator $\bar{q}q$ at infinite volume (in the full theory where the winding number is not subject to restrictions), this result indeed agrees with the infinite volume formula (5.1). Note, however, that the distribution of the lowest few eigenvalues is sensitive to winding number. If $\nu \neq 0$, there are $\mid \nu \mid$ zero modes. As a compensation, the spectrum is diluted in the vicinity of $\lambda = 0$, the zeros of the Bessel function $J_\nu(x)$ being shifted towards larger values if $\mid \nu \mid$ is increased. At sufficiently large eigenvalues, however, the level density is independent of the winding number.

Actually, the above analysis only concerns those eigenvalues of the Dirac operator which are small compared to the scale of the theory. Our neglect of the higher order terms in eq. (5.10) is justified only if the quark mass is small, $m \ll \Lambda_{\text{QCD}}$. This restriction prevents us from analyzing the spectrum in the region $\lambda_n \gtrsim \Lambda_{\text{QCD}}$. On the other hand, it is not legitimate to simply ignore the occurrence of large eigenvalues. They are responsible for the fact that the product $\Pi_n(1 + m^2/\lambda_n^2)$ is divergent and thus formally even dominate the value of the product. Let me briefly explain why this problem does not affect the above analysis.

For $\mid \lambda \mid \rightarrow \infty$, the level density is given by phase space; for a given field configuration, the number of levels contained in the interval $(\lambda, \lambda + \Delta\lambda)$ tends to

$$\Delta n = \frac{N_c}{4\pi^2} V \mid \lambda \mid^3 \Delta\lambda \tag{5.21}$$

where N_c is the number of colours. Accordingly, the logarithm of $\Pi_n(1 + m^2/\lambda_n^2)$ contains a quadratic divergence proportional to m^2V as well as a logarithmic one of order m^4V. In the representation (5.9) of the partition function, these divergences manifest themselves as regularization dependent contributions to the vacuum energy. Denoting the cutoff by Λ, the quadratic divergence, e.g., shows up through a term of order $m^2\Lambda^2$ in eq. (5.10). In the corresponding result for Z_0, this term appears as a contribution of order $m^2\Lambda^2V$ and is to be compared with the term $(V\Sigma m)^2$ which occurs in the expansion of the Bessel function in powers of m. The point is that these two terms involve different powers of volume; at large volume, the term proportional to V^2 dominates. More precisely, our analysis concerns the limit $V \to \infty, m \to 0$, taken at a fixed value of the product Vm. In this limit, Z_0 tends to $I_0(\Sigma V m)$. At finite V, the correction is of order $1/V$ and could readily be evaluated by carrying the expansion (5.10) to order m^2. The low energy constants analogous to Σ which occur at this order of the expansion of course depend on the convention adopted to renormalize the quadratic divergence of the vacuum energy, but their value only matters if one wishes to analyze the subleading corrections to the large volume theorem (5.13).

In the correlation function $\sigma_\nu(x)$, large eigenvalues of the Dirac operator generate a short distance singularity $\propto\mid x \mid^{-6}$. The space integral (5.18) is therefore divergent. Short distances, however, only affect this integral through volume-independent terms which are of relative order $1/V$ compared to the leading long-range contribution. In the large volume limit the relation (5.18) represents an exact result.

Finally, let us see what happens if there are several massless flavours, $N_f \geq 2$. In this case, the low energy structure of the theory is more complicated, because the spectrum then contains Goldstone bosons. As discussed in the preceding sections, the partition function can be represented as a euclidean function integral over the Goldstone boson degrees of freedom which can be evaluated perturbatively in terms of an expansion in inverse powers of the box size. At leading order, the result boils down to an ordinary integral over the group SU(N_f), given in eq. (4.8). The only modification brought about by a nonzero vacuum angle and a complex quark mass matrix is that $\mathcal{M}$ is to be replaced by $\mathcal{M}^{+}\exp(-i\Theta/N_f)$. The vacuum angle and the collective meson field variable U_0 therefore only occur in the combination $U = U_0\exp(-i\Theta/N_f)$ such that the Fourier transform with respect to Θ merely extends the group integral from SU(N_f) to U(N_f),

$$Z_\nu = \int_{\mathrm{U}(N_f)} d\mu(U)(\det U)^\nu \exp\left\{V\Sigma \mathrm{Retr}(\mathcal{M}^{+}U)\right\}. \tag{5.22}$$

For $N_f = 1$, this integral can be done, yielding a Bessel function as discussed above. For a general value of N_f, the integral (5.22) cannot be performed explicitly. Note also that, unless the quark masses are set equal, the quantity Z_ν now depends on several variables. These complications are however inessential. The situation is basically the same as for a single quark flavour: we have an explicit representation for the contribution to the partition function generated by gluon field configurations of winding number ν in terms of an ordinary integral. The representation is valid provided the quark mass matrix is small compared to the scale of the theory and provided all four sides of the euclidean box are large compared to the inverse of this scale. The representation, in particular, shows that the mass matrix and the volume only enter through the product $V\Sigma\mathcal{M}$ where Σ is the value of the quark condensate at infinite volume and in the chiral limit. One readily checks that if ν is positive, the integral (5.22) is proportional to $(\det\mathcal{M})^\nu$, in accordance with the general representation (5.8). Hence nonzero winding numbers are suppressed in the chiral limit, even more strongly than in the case $N_f = 1$. However - and this is the main point of the present section - the suppression factor $(\det\mathcal{M})^\nu$ is accompanied by a corresponding power of $V\Sigma$ which enhances the result if the volume is large. The magnitude of Z_ν is determined by the product $V\Sigma\mathcal{M}$. In a basis where the quark mass matrix is diagonal with positive eigenvalues, nonzero winding numbers are suppressed only if these eigenvalues are small compared to $1/V\Sigma$. If the volume is very large, the integrand in eq. (5.22) is sharply peaked in the vicinity of $U = 1$, unless the quark masses are tiny. For $V\Sigma\mathcal{M} \gg 1$, the integral is proportional to $\exp\{V\Sigma\mathrm{tr}\mathcal{M}\}$ and the probability to encounter a gluon field configuration of winding number ν becomes approximately independent of ν. The factor $(\det U)^{\nu}$ occurring in eq. (5.22) significantly reduces the value of the integral only if ν is large, comparable to $(V\Sigma\mathcal{M})^{1/2}$. This shows that, outside the symmetry restoration region, the winding number is an irrelevant quantum number.

In the above discussion, the U(1)-anomaly plays a central role. It spoils the conservation of the singlet axial current, reducing the chiral symmetry of massless QCD from $\mathrm{U}(N_f)_L\times\mathrm{U}(N_f)_R$ to $\mathrm{SU}(N_f)_L\times\mathrm{SU}(N_f)_R\times\mathrm{U}(1)_V$. Accordingly, the spontaneous breakdown of chiral symmetry gives rise not to N_f^2 but only to N_f^2-1 Goldstone bosons. Indeed, QCD would be inconsistent with the observed mass spectrum if the limit $m_u = m_d = m_s = 0$ were to contain nine massless pseudoscalar mesons. The occurrence of an anomaly in the divergence of the singlet axial current is an essential feature of the theory - it explains why the mass of the η' remains different from zero in the chiral limit. This solution of the "U(1)-problem" has, however, repeatedly been questioned in the literature [42]. One argues that the volume integral of the U(1)-anomaly is the winding number. In the chiral limit, field configurations with nontrivial winding number do not occur. One concludes that the relevant matrix elements of the anomaly vanish in the chiral limit such that the mechanism which is supposed to be repsonsible for the mass of the η' actually fails. The above analysis shows why this argument is not tenable. The suppression of winding number is a finite size phenomenon. It only occurs if the quark masses are small compared to the inverse volume of the box, i.e. in the region where chiral symmetry is restored. The erroneous conclusion arises because finite volume concepts such as winding number are used in conjunction with infinite volume concepts such as spontaneous symmetry breakdown. A more detailed discussion of the issue is given in ref. [41].

6 Conclusion

(i) It is well-known that the low energy structure of QCD can be analyzed by means of an effective field theory involving pion fields rather than quarks and gluons. Chiral perturbation theory provides an explicit solution of this effective theory, based on an expansion in powers of energies and momenta.

(ii) The effective field theory can be used to investigate the thermal properties of QCD at low temperatures and the finite size effects occurring in lattice simulations can also be studied within this framework.

(iii) In the case of the partition function, the chiral perturbation series amounts to an expansion in powers of $\mathcal{M}$, T and $1/L$. All of these quantities must be small compared to the scale of the theory for the expansion to be useful.

(iv) The chiral expansion is an asymptotic series. Excited states generate corrections which at low T and large L are exponentially small.

(v) At infinite volume, both the convergence of the expansion in powers of T and the magnitude of the exponential corrections are under control only for $T < 150$MeV. In this temperature range, the melting of the quark and gluon condensates was studied in considerable detail. Note, however, that the behaviour in the vicinity of the chiral phase transition is beyond the reach of the chiral expansion.

(vi) In contrast, chiral perturbation theory fully covers the symmetry restoration phenomena associated with the fact that spontaneous symmetry breakdown does not occur at finite volume. The behaviour of various observables in the symmetry restoration region has been worked out in detail for various observables of interest.

(vii) The well-known suppression of gluon field configurations with nonzero winding number which occurs in the chiral limit at finite volume has much in common with the restoration of spontaneously broken symmetries. At sufficiently large volume, a tiny quark mass suffices to remove this suppression. Outside the symmetry restoration region, the winding number is an irrelevant quantum number.

(viii) When lattice simulations of QCD will reach reasonably small quark masses on reasonably large boxes, chiral perturbation theory will likely become a useful tool in the data analysis. In the meantime, it is instructive to carry out warm up exercises with scalar field theories where chiral perturbation theory has been shown to describe the finite size effects amazingly well.

Acknowledgements

It is a pleasure to thank the organizers for a stimulating workshop, demonstrating that recreation homes for party members can serve reasonable purposes.

References

[1] For a discussion of the physics underlying this picture, see e.g.
E.V. Shuryak, *The QCD Vacuum, Hadrons and the Superdense Matter*, Lecture Notes in Physics, Vol. 8 (World Scientific, Singapore, 1988).

[2] Proc. *Lattice '90* Nucl. Phys. B (Proc. Suppl.) **20** (1991).

[3] Proc. *QCD '90*, Nucl. Phys. B (Proc. Suppl.) **23** (1991).

[4] Proc. *Quark Matter '90*, Nucl. Phys. **A525** (1991).

[5] P. Gerber, in *Effective Field Theories of the Standard Model*, Dobogokö, Hungary, 1991, ed. U. Meissner (to be published by World Scientific).

[6] P. Binétruj and M.K. Gaillard, Phys. Rev. **D32** (1985) 931.

[7] J. Gasser and H. Leutwyler, Phys. Lett. **B184** (1987) 83; **B188** (1987) 477;
H. Neuberger, Phys. Rev. Lett. **60** (1988) 889.

[8] H. Leutwyler, Nucl. Phys. B (Proc. Suppl.) **4** (1988) 248.

[9] P. Gerber and H. Leutwyler, Nucl. Phys. **B321** (1989) 387.

[10] R. Hagedorn, Suppl. Nuovo Cim. **3** (1965) 147;
R. Hagedorn, in *Quark Matter '84*, Lecture Notes in Physics, Vol. **211** (Springer, Berlin, 1985);
R. Fiore et al., Nuovo Cim. **88A** (1985) 301.

[11] H. Bebie et al., preprint University of Bern, BUTP-91/14;
J. Goity, in *Effective Field Theories of the Standard Model*, Dobogokö, Hungary, 1991, ed. U. Meissner (to be published by World Scientific).

[12] J. Goity and H. Leutwyler, Phys. Lett. **B228** (1989) 517.

[13] A. Schenk, Nucl. Phys. **B363** (1991) 97.

[14] F. Karsch, in *Quark-Gluon-Plasma*, ed. R.C. Hwa (World Scientific, Singapore, 1991).
D. Toussaint, "Progress in lattice QCD-91", preprint Univ. of Arizona, AZPH-TH/91-33.

[15] E.V. Shuryak, Phys. Rep. **61** (1980) 72;
D. Gross, R. Pisarski and L. Yaffe, Rev. Mod. Phys. **52** (1981) 42.

[16] T. Banks and A. Casher, Nucl. Phys. **B169** (1980) 103;
E. Marinari, G. Parisi and C. Rebbi, Phys. Rev. Lett. **47** (1981) 1795.

[17] E. Floratos and J. Stern, Phys. Lett. **B119** (1982) 419;
D.I. Dyakonov and V. Yu. Petrov, Phys. Lett. **B147** (1984) 351; Nucl. Phys. **B272** (1986) 457;
V.V. Khoze and A. V. Yung, Z. Phys. **C50** (1990) 155;
Yu. A. Simonov, Phys. Rev. **D43** (1991) 3534; JETP Lett. **53** (1991) 9; Sov. J. Nucl. Phys. **54** (1991) 133.

[18] P. Minkowski, Czech. J. Phys. **40** (1990) 1003; Phys. Lett. **B102** (1981) 157.

[19] T. Hatsuda, in Proc. of *QCD '90*, Nucl. Phys. B (Proc. Suppl.) **23** (1991);
V. Bernard and U.-G. Meissner, Phys. Lett. **227** (1989) 465.

[20] A. Barducci et al., Phys. Lett. **B231** (1989) 463; **B240** (1990) 429; **B244** (1990) 311; Phys. Rev. **D41** (1990) 1610.

[21] Z. Frei and A. Patkós, Phys. Lett. **B222** (1989) 469; **B247** (1990) 381.

[22] J. Cleymans, A. Kocić and M.D. Scadron, Phys. Rev. **D39** (1989) 323.

[23] V.L. Eletsky, in *Effective Field Theories of the Standard Model*, Dobogokö, Hungary, 1991, ed. U. Meissner (to be published by World Scientific).

[24] A. I. Bochkarev and M.E. Shaposhnikov, Nucl. Phys. **B268** (1986) 220;

[25] V.L. Eletsky and B.L. Ioffe, Sov. J. Nucl. Phys. **48** (1988) 384;
J. Dey, M. Dey and P. Ghose, Phys. Lett. **B193** (1987) 98; **B221** (1989) 161;
M. Dey, V.L. Eletsky and B.L. Ioffe, Phys. Lett. **B252** (1990) 620.

[26] C.A. Dominguez and M. Loewe, Phys. Lett. **B233** (1989) 201; Z. Phys. **C51** (1991) 69.

[27] M.B. Voloshin and A.D. Dolgov, Sov. J. Nucl. Phys. **35** (1982) 120;
H. Leutwyler and M.A. Shifman, Phys. Lett. **B221** (1989) 384.

[28] M.A. Shifman, A.I. Vainshtein and V.I. Zakharov, Nucl. Phys. **B147** (1979) 385;
V. Novikov et al. Nucl. Phys. **B191** (1981) 301.

[29] C.A. Dominguez and M. Loewe, Z. Phys. **C51** (1991) 69.

[30] J. Kapusta, Nucl. Phys. **B148** (1979) 461.

[31] M. Campostrini and A. DiGiacomo, Phys. Lett. **B197** (1987) 403.

[32] A. Ukawa, Nucl. Phys. **A498** (1989) 227c.

[33] J. Engels et al., Phys. Lett. **B252** (1990) 625.

[34] M. Lüscher, in *Progress in Gauge Field Theory*, Cargèse, France 1983, eds. G. t'Hooft et al. (Plenum, New York 1984) and in *Fields, Strings and Critical Phenomena*, Les Houches, France, 1988, eds. E. Brézin et al. (North Holland, Amsterdam 1989); Commun. Math. Phys. **104** (1986) 177; **105** (1986) 133; Nucl. Phys. **B354** (1991) 531.

[35] J. Gasser and H. Leutwyler, Nucl. Phys. **B307** (1988) 763.

[36] P. Hasenfratz and H. Leutwyler, Nucl. Phys. **B343** (1990) 241.

[37] F.C. Hansen, Nucl. Phys. **B345** (1990) 685;
F.C. Hansen and H. Leutwyler, Nucl. Phys. **B350** (1991) 201.

[38] M. Göckeler and H. Leutwyler, Nucl. Phys. **B350** (1991) 228; Phys. Lett. **B253** (1991) 193; Nucl. Phys. **B361** (1991) 392.

[39] A. Hasenfratz et al., Z. Phys. **C46** (1990) 257; Nucl. Phys. **B356** (1991) 332; I. Dimitrović et al., Nucl. Phys. **B350** (1991) 893; Phys. Lett. **B268** (1991) 408;
M. Göckeler, K. Jansen and T. Neuhaus, preprint Univ. of Cal. San Diego, UCSD/PTH91-15.

[40] I. Barbour et al., Phys. Lett. **B136** (1984) 80;
M. Göckeler et al., Nucl. Phys. **B334** (1990) 527.

[41] H. Leutwyler and A. Smilga, in preparation.

[42] See, e.g.,
R.J. Crewther, Rivista Nuovo Cimento **2**, No. 8 (1979) 63;
H. Banerjee, in *Festschrift to Léon Van Hove*, eds. A. Giovannini and W. Kittel (World Scientific, Singapore, 1990);
H. Banerjee et al., "Electric Dipole Moment of the Neutron...", preprint Saha Institute, Calcutta SINP-TNP/89-7.

Quark condensate below the chiral phase transition [1]

P. Gerber
Institute for Theoretical Physics, University of Bern
Sidlerstrasse 5, CH-3012 Bern, Switzerland

Abstract

The leading coefficents of the low temperature expansion for the pressure and for the quark condensate are discussed. We estimate the critical temperature at which the chiral phase transition takes place, accounting for the exponentially small contributions generated by massive states.

1 Introduction

In the standard model colour is confined at low temperatures and chiral symmetry is spontaneously broken. At high temperatures asymptotic freedom predicts the interaction among quarks and gluons to be weak, such that colour is liberated and chiral symmetry is restored [1]. Presumably the restoration of chiral symmetry and the liberation of colour occur in a single phase transition from the hadronic to the quark gluon plasma phase. In the present article, we discuss the expansion of the pressure and energy density in powers of T and the temperature dependence of the quark condensate below the chiral phase transition. Extrapolating from lower temperatures, the melting of the quark condensate can be used as order parameter to estimate the critical temperature of the chiral phase transition.

At low temperatures the behaviour of the partition function is dominated by the contributions from the pions, the lightest particles occurring in the spectrum of the theory. In the chiral limit, the pions are massless. Particles which remain massive

[1]Work supported in part by Schweizerischer Nationalfonds

in this limit are exponentially suppressed by a factor $\exp(-M/T)$, where M is the mass of such a particle. The low energy properties of the pions are largely fixed by chiral symmetry, which implies that the interaction among low energy pions is proportional to the square of the energy. This property can be used to evaluate the coefficients of the low temperature expansion of the pressure, the energy density and the quark condensate in the hadronic phase [2].

2 Low temperature expansion

To analyze the low temperature expansion it is convenient to apply the effective lagrangian method [3] [4]. For definiteness, we consider QCD with two light quark masses (u, d) treated as perturbations. The partition function is determined by the euclidean functional integral

$$Tr\left[e^{-\frac{H}{T}}\right] = \int [dU] e^{-\int_\Gamma d^4x L_{eff}}, \tag{2.1}$$

where the integration covers all pion field configurations which are periodic in the euclidean time direction. In this picture multipion states are described in terms of a pion field $U(x)$ living on the group SU(2), i.e. the pion field is a map from the torus $\Gamma = R^3 \times S^1$ into SU(2):

$$U(\vec{x}, x_4 + \beta) = U(\vec{x}, x_4), \qquad \beta \equiv \frac{1}{T}. \tag{2.2}$$

The quantity L_{eff} in the path integral (2.1) denotes the euclidean form of the standard effective chiral lagrangian of Gasser and Leutwyler [4] [5].

$$L_{eff} = L^{(2)} + L^{(4)} + L^{(6)} + \ldots \tag{2.3}$$

The leading term $L^{(2)}$ of low energy dimension p^2 is the familiar lagrangian of the nonlinear σ - model:

$$L^{(2)} = \frac{1}{4} F^2 Tr\left[\partial_\mu U^\dagger \partial_\mu U - M^2(U + U^\dagger)\right] \tag{2.4}$$

where $F = (88.3 \pm 1.1)$ MeV stands for the pion decay constant in the chiral limit and M denotes the lowest order contribution to the pion mass: $M^2 = (m_u + m_d)B$.

The low temperature expansion of the partition function (2.1) can be evaluated using the conventional Feynman rules of chiral perturbation theory modified by the replacement of the propagator $\Delta(x)$ at zero temperature through the thermal propagator $G(x)$:

$$G(x) = \sum_{n=-\infty}^{\infty} \Delta(\vec{x}, x_4 + n\beta). \tag{2.5}$$

We use dimensional regularization and the general feature of finite temperature field theory that, if the coupling constants are renormalized such that the perturbative expansion is finite at $T = 0$, it is automatically also finite at nonzero temperature. With the renormalization procedure discussed in detail in ref. [6], we have calculated the free energy density z, defined by

$$z = -T \lim_{L \to \infty} L^{-3} \ln \left[Tr \exp(-H/T) \right] , \tag{2.6}$$

in the infinite volume limit to order $O(T^{10})$.

3 Massless quarks

In the chiral limit the formula for the free energy density is simplified to

$$z = \epsilon_0 - P, \qquad P = \frac{1}{30}\pi^2 T^4 \left(1 + \frac{T^4}{36F^4} \ln \frac{\Lambda_p}{T} + O(T^6) \right), \tag{3.1}$$

where the logarithmic scale $\Lambda_p = 275 \pm 65$ MeV can be phenomenologically determined [5] [7]. The quantity ϵ_0 denotes the vacuum energy density (T=0) and for an homogeneous system the pressure P is given by the temperature dependent part of the free energy density. The thermodynamic relations

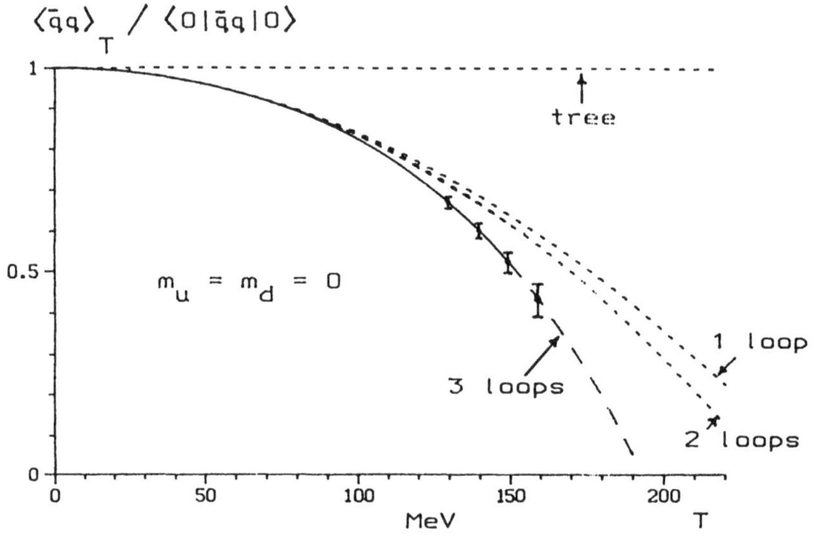

Figure 1: Quark condensate as a function of the temperature to three loops in the chiral limit.

$$s = \frac{\partial P}{\partial T}, \qquad u = Ts - P \tag{3.2}$$

deliver the corresponding expressions for the entropy density s and the energy density u

$$s = \frac{2}{15}\pi^2 T^3 \left[1 + \frac{T^4}{144F^4}\left(8 \ln \frac{\Lambda_p}{T} - 1\right) + O(T^6)\right]$$

$$u = \frac{1}{10}\pi^2 T^4 \left[1 + \frac{T^4}{108F^4}\left(7 \ln \frac{\Lambda_p}{T} - 1\right) + O(T^6)\right]. \tag{3.3}$$

The quark condensate is given by the logarithmic derivative of the partition function with respect to the quark mass,

$$\langle \bar{q}q \rangle = \langle \bar{u}u + \bar{d}d \rangle = \frac{\partial z}{\partial m}. \tag{3.4}$$

In the chiral limit the temperature dependence of the order parameter $\langle \bar{q}q \rangle$ has the form

$$\langle \bar{q}q \rangle = \langle 0|\bar{q}q|0 \rangle \left(1 - \frac{T^2}{8F^2} - \frac{T^4}{384F^4} - \frac{T^6}{288F^6} \ln \frac{\Lambda_q}{T} + O(T^8)\right) \tag{3.5}$$

where $\langle 0|\bar{q}q|0 \rangle$ stands for the vacuum expectation value and $\Lambda_q = 470 \pm 110$ MeV [2] [6]. Fig. 1 shows how the condensate gradually melts as the temperature grows. Being aware of the fact that the low temperature expansion cannot be trusted for $T \geq 150$ MeV, the vanishing of the order parameter at a temperature $T_0 \simeq 190$ MeV, nevertheless indicates the occurrence of a phase transition.

4 Massive states

The fact that the densities of massive particles are exponentially suppressed

$$n_i \sim e^{-\frac{M_i}{T}} \tag{4.1}$$

does not indicate much about the numerical size of their contribution at finite temperatures, because at low energies the density of levels is exponentially increasing [8]. Since the interaction among the massive particles manifests itself only at order $n_i n_k \sim \exp\left[-(M_i + M_k)/T\right]$ and since chiral symmetry suppresses the interaction between the massive particles and the pions by a factor T^2/F^2, at low temperatures the massive states can be treated as a free gas.

In order to estimate the effect generated by massive states, we perform the sum

$$\Delta z = -\frac{1}{2} T^4 \sum_i h_0\left(\frac{T}{M_i}\right)$$

$$h_0(\tau) = -\frac{1}{\pi^2 \tau^3} \int_0^\infty dx \, \sinh^2 x \cosh x \, \ln\left[1 - \exp(-\cosh x/\tau)\right] \tag{4.2}$$

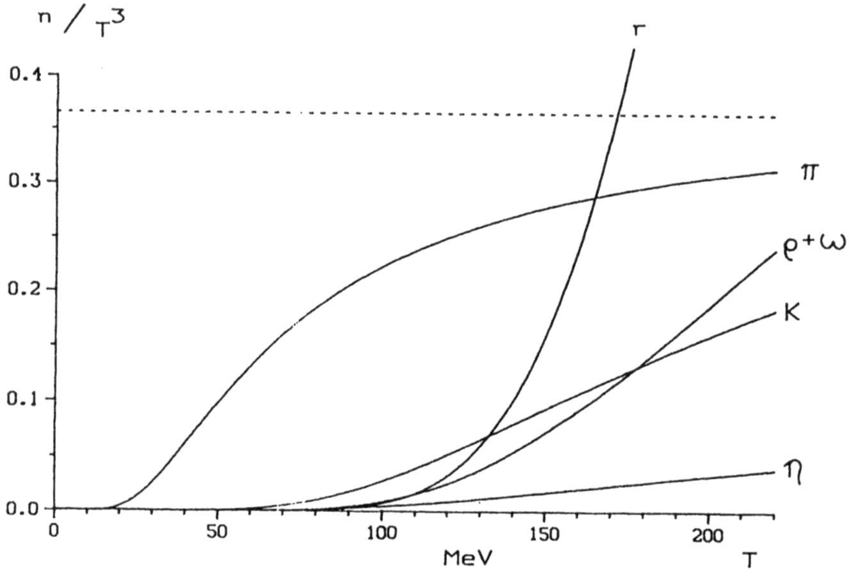

Figure 2: Particle densities of the lightest mesons and the remaining more massive states (r).

over the entries of the table provided by the particle data group [9], accounting for the multiplicity of the levels with the appropriate spin and isospin degeneracy factors.

Fig. 2 shows the density n_i/T^3 as a function of the temperature for the various components of the gas.

$$n_i = (2\pi)^{-3} \int d^3p \,[\exp(E_i/T) - 1]^{-1} \tag{4.3}$$

The dilute gas approximation for the massive particles is meaningful up to a temperature of the order of 150 MeV, where the mean distance between the massive particles is $d \approx 1.8$ fm. Beyond this point the density rapidly grows with temperature such that the interaction among the particles can no longer be neglected.

The temperature dependent contributions to the energy densities are illustrated in fig. 3. It is interesting to mention that already for temperatures $T \geq 130$ MeV the energy density contained in the massive states exceeds the corresponding value

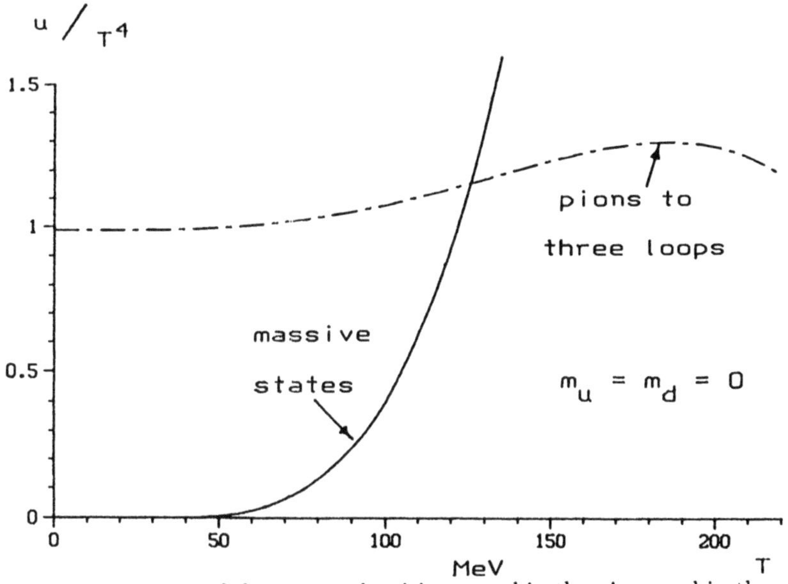

Figure 3: Comparison of the energy densities stored in the pions and in the massive states.

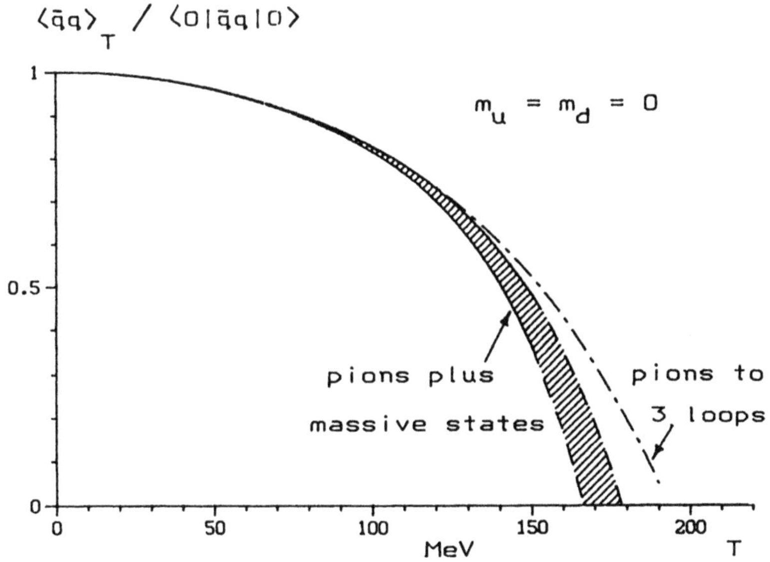

Figure 4: Superposition of the contribution of pions to the quark consensate in the chiral limit and of the effects generated by massive states.

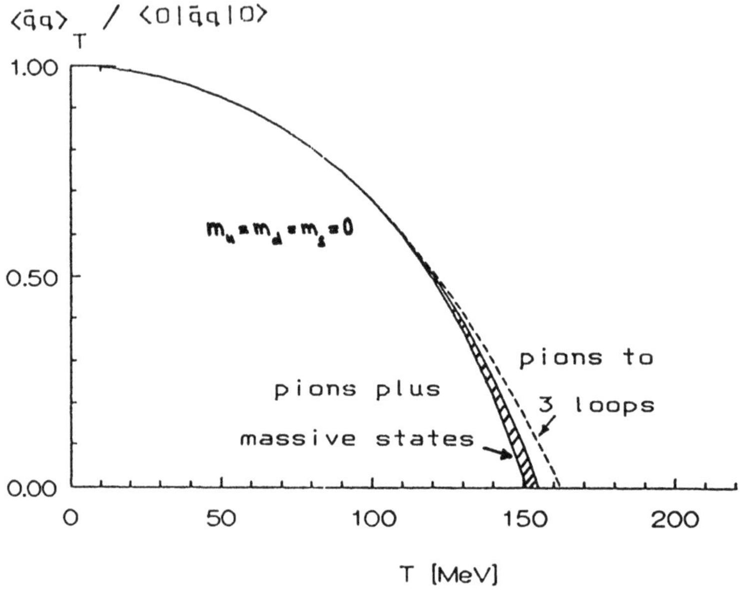

Figure 5: Quark condensate with the strange quark mass m_s set equal to zero too.

of the pion gas. This is due to the fact that the energy stored in the massive states predominantly consists of rest energy.

The effects generated by massive states to the quark condensate are estimated in fig. 4. The shaded region in the figure reflects the uncertainty in the quantity $\partial M_i/\partial m$ which we have naively estimated by the number of valence quarks of type u or d. We see that the order parameter is less affected by the massive states than the energy density.

The influence of setting the strange quark mass m_s equal to zero too, can be read off in fig. 5. Considering three massless quarks in a SU(3) model, shifts down the value of the critical temperature by about 20 MeV [10].

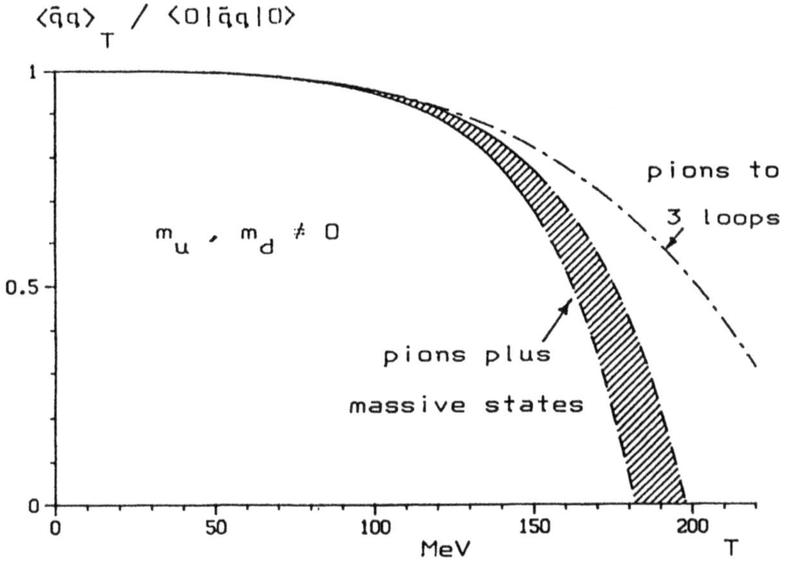

Figure 6: Temperature dependence of the condensate for nonzero quark mass.

5 Behaviour of the quark condensate for nonzero quark masses

We now turn to the evaluation of the low temperature representation at the physical value of the quark masses. In fig. 6 the shaded region is obtained by adding the contribution generated by the massive states to the result of the three loop calculation of ref. [6]. The figure indicates that the net effect of the quark masses is to put the critical temperature up by about 20 MeV. The critical temperature is beyond the range of validity of our formulae. Since the order parameter, however, rapidly falls at the upper end of this range, we nevertheless consider it meaningfull to make an estimate for the critical temperature:

$$T_c \simeq 190 MeV\,. \tag{4.4}$$

References

[1] E.V. Shuryak, Phys. Rep. **61** (1980) 72; **115** (1984) 151;
D. Gross, R. Pisarski and L. Yaffe, Rev. Mod. Phys. **53** (1981) 43;
O.K. Kalashnikov, Fortschr. Phys. **32** (1984) 525;
L. van Hove, *The quark-gluon plasma - a progress report*, CERN TH 5069/88

[2] J. Gasser and H. Leutwyler, Phys. Lett. **B184** (1987) 83

[3] S. Weinberg, Phys. Rev. **166** (1968) 1568;Physica **A96** (1979) 327;
S. Coleman, J. Wess and B. Zumino, Phys. Rev. **177** (1969) 2239;
C. Callan, S. Coleman, J. Wess and B. Zumino, Phys. Rev. **177** (1969) 2247;
R. Dashen and M. Weinstein, Phys. Rev. **183** (1979) 1291

[4] J. Gasser and H. Leutwyler, Nucl. Phys. **B307** (1988) 763

[5] J. Gasser and H.Leutwyler, Ann. Phys. **158** (1984) 142

[6] P. Gerber and H. Leutwyler, Nucl. Phys. **B321** (1989) 387

[7] J. Gasser and H. Leutwyler, Phys. Lett. **B125** (1983) 325;
J.L. Petersen, CERN yellow report 77-04 (1977);
M.M. Nagels et al., Nucl. Phys. **B147** (1979) 189;
J. Donoghue, C.Ramirez and G. Valencia, preprint University of Massachusetts, Amherst, UMHEP-296

[8] R. Hagedorn and J. Rafelski, Phys. Lett. **B97** (1980) 136;
R. Hagedorn, in *Quark matter '84, Lecture Notes in Physics*, Vol. 211 (Springer, Berlin, 1985)

[9] Review of Particle Properties, Phys. Lett. **B239** (1990)

[10] P. Gerber and H. Leutwyler, in preparation

PROPERTIES OF THE EXPANDING HADRONIC GAS

J. L. GOITY

Continuous Electron Beam Accelerator Facility

Newport News, VA 23606, U S A

ABSTRACT

The main features of the free-expansion of a hot hadronic gas are summarized. The phenomenon of loss of chemical equilibrium is emphasized, and the composition of the gas in the course of the expansion is analyzed in the adiabatic limit by means of the entropy per particle. Estimates of elastic and inelastic mean-free paths are discussed, as they are relevant in the determination of chemical and thermal freeze-out. The analysis of heavy-ion collision data in terms of thermalized hadronic matter is reviewed.

1 Introduction

High energy heavy-ion central collisions give rise to a high multiplicity final state consisting of a few hundred hadrons, mainly pions. For example, fixed target collisions of $O + Au$ ($S + S$) with projectile energy of $200\,GeV$/nucleon, lead to $\sim$ 400 (300) particles[1]. The presently available data on p_T and rapidity spectra[1], on relative abundances of stable hadrons[2] and on two-pion interferometry[3] (which measures the size of the zone from where the pions are emitted) call for a theoretical framework able to describe the main processes leading to such a final state. Different groups[4] have suggested to approach the problem by statistical mechanics methods as motivated by the relatively large number of particles in the system. In this report we review the main results in Refs.[5,6,7], where this general approach has been followed.

After the occurrence of a central collision, a high density hot central zone remains (initially resembling a pancake perpendicular to the collision axis, whose thickness rapidly grows) consisting of a quark-gluon plasma if the energy density is more than about 2 $GeV\,fm^{-3}$. As the hot zone expands and cools down, a phase transition to hadronic matter takes place. The characteristics of the phase transition have not been yet fully established. In our analysis, we assume that it happens without considerable supercooling, i.e., the phase transition is not strongly of first order and occurs in a short time as compared with the lifetime of the system. As the system further cools down , the stage is reached where the equation of state of the hadronic gas is known, and from there on it is, in principle, possible to follow its evolution. At this point, the state of the system would be characterized by the temperature T and the expansion velocity field. As the expansion proceeds, the inelastic processes which keep chemical equilibrium become ineffective, leading to an overpopulation of particles. Chemical potentials μ_π, μ_N and μ_K develop, which account for the departure from equilibrium. Finally, thermal freeze-out occurs when the mean free path for elastic collisions (defined as the mean distance a particle wave has to travel in order to be damped by a factor $1/e$) becomes similar to the radius of the system. Experimental data should then be described by a hadronic gas at the freeze-out temperature, with non-vanishing chemical potentials and an expansion velocity field built up in the course of the evolution.

There are uncertainties in the implementation of this picture, which should, therefore, be considered as a first approximation. The validity of statistical mechanics requires the temperature fluctuations to be small: $\Delta T/T \ll 1$. For a system in equilibrium, $\Delta T/T \sim 1/\sqrt{C_v} \sim 1/\pi\sqrt{N}$, where N is the total number of particles in the system. It is not clear yet that this same criterion can be used in our case, where the approach to equilibrium may be spoiled by the fast evolution. Likewise, the use of hydrodynamic equations may not be justified. Boundary effects, which become large as the system approaches freeze-out, may also lead to important departures from equilibrium. Certainly, there is still a large amount of work ahead until these aspects can be understood. One also expects these uncertainties to become less relevant with the increase of the collision energy leading to higher multiplicity systems.

The matters are presented herein as follows: section 2 is devoted to the equation of state of the hadronic gas, section 3 presents a discussion of thermal and chemical equilibrium[5,7], section 4 discusses the assignments of chemical potentials for the hadronic gas off chemical equilibrium[7], section 5 describes the main features of the expansion[7], section 6 confronts the approach with available data[6], and, section 7 presents some comments.

2 Equation of State

The equation of state of the hot hadronic matter is determined by the hadronic spectrum and the strong interactions. Throughout we will consider a situation with no net baryon number. For $T \leq 100\, MeV$ the gas consists mainly of pions and is nearly ideal because pions interact weakly at low energies. As T increases other stable hadrons ($K, N, \Sigma, ...$) and resonances ($\rho, \omega, K^*, N^*, ...$) become relevant. When chemical equilibrium prevails, at $T = 125\,MeV$ half of the energy density is stored in these states[8], at $T = 150\,MeV$ only half of the particles in the gas are pions, and at $T = 180\,MeV$ the particle number densities derived from Eq. (1) show that 1/4 of the particles are pions, 1/4 are excited pions ($\rho, \omega, ...$), 1/4 are strange mesons and 1/4 are baryons. This dramatic change in the composition of the gas is a consequence of the fast increase in the number of degrees of freedom with the mass of the states[9] which makes up for the exponential suppression of the population of individual degrees of freedom. It has been observed[8] that even at $T = 150\,MeV$ interactions only slightly affect the equation of state. For this reason we neglect interactions and consider the equation of state in the ideal gas approximation. The pressure is then given by:

$$P = T \sum_i \pm g_i \int \frac{d^3p}{(2\pi)^3} \ln\left[1 \pm \exp\left(\frac{\mu_i - E_i(p)}{T}\right)\right] \tag{1}$$

where i runs over all levels (we take those listed by the Particle Data Group[10] below $2\,GeV$), g_i are the statistical weights (e.g. $g_\pi = 3, g_K = 4, g_\rho = 9, ...$) and $+(-)$ refers to Bosons (Fermions). μ_i are chemical potentials, which we discuss in the next two sections. In our context, chemical equilibrium will mean $\mu_i = 0$.

According to (1), in chemical equilibrium and for $T \geq 150 MeV$ the ratio $b = \epsilon/P$ (the energy density is given by the usual thermodynamic relation $\epsilon = T\, dP/dT - P$) becomes almost T−independent. This implies that (1) can be approximated by a power law: $P = aT^{b+1}$. In the range $150\,MeV \leq T \leq 200\,MeV$ (1) is well approximated by[7] $b = 6.3$ and $a = T_1^{3-b}$ with $T_1 = 176\,MeV$. The ratio $\epsilon/T^4 \sim T^3$ rapidly approaches the quark-gluon plasma value of $19\pi^2/12$. Of course, as T approaches the critical temperature (1) becomes obsolete. However, (1) is likely to provide a decent approximation up to $T \sim 180\,MeV$. For further discussions on the equation of state see Ref.[8].

3 Thermal and Chemical Equilibrium

A system of finite volume which undergoes free expansion will remain in equilibrium (thermal and/or chemical) only if the appropriate processes occur at a high enough rate. Thermal equilibrium is maintained by elastic collisions. As the system expands and cools down the collision rate decreases, until the mean free path between collisions becomes comparable to the radius of the system, or the time between collisions becomes larger than the inverse expansion rate (for systems produced in a heavy-ion collision the expansion velocity is likely to remain smaller than the thermal velocity, in which case the first criterion applies). At this point the momentum distribution freezes-out into the distribution experimentally observed. For the hadron gas at $T \leq 120\, MeV$, thermalization mainly depends on the pion component of the gas since they are the majority. The relevant mean free-path is therefore the one for $\pi - \pi$ elastic collisions. In lowest order chiral perturbation theory, the resulting collision rate for a pion with momentum p_1 is given by[5]:

$$\begin{aligned} \gamma(p_1) &= \frac{\mathrm{sh}(\epsilon_1/2)}{2\omega_1} \int d\nu_2 d\nu_3 d\nu_4 \quad (2\pi)^4 \quad \delta^4(p_1 + p_2 - p_3 - p_4) \mid T \mid^2 \\ d\nu_i &\equiv \frac{d^3 p_i}{(2\pi)^3} \frac{1}{4\,\omega_i\, \mathrm{sh}(\epsilon_i/2)} \qquad\qquad \epsilon_i = (\omega_i - \mu_\pi)/T \\ \mid T \mid^2 &= [2(s^2 + t^2 + u^2) - 9M_\pi^2]/F_\pi^4 \qquad\qquad F_\pi = 93\, MeV \end{aligned} \tag{2}$$

where $|T|^2$ is the square of the scattering amplitude after summing over the isospin components of pions 2, 3 and 4. The mean free path is then given by $\lambda(p) = \frac{|p|}{\omega(p)} \gamma^{-1}(p)$. For $T \geq 100\, MeV$ the average relaxation time $\bar{\tau} = \bar{\gamma}^{-1}$ (where $\bar{\gamma}$ is defined by averaging $\gamma(p)$ over the thermal momentum distribution) is well approximated by :

$$\bar{\tau} \simeq 12 \, \frac{F_\pi^4}{T^5} \, \frac{n_\pi(\mu_\pi = 0)}{n_\pi(\mu_\pi)} \tag{3}$$

where the last factor involving the pion number density is the correction due to a non-vanishing chemical potential. In chemical equilibrium the T-dependence is easy to understand: the mean free path is inversely proportional to the density times the average over the thermal distribution of the elastic cross section. The first is proportional to T^3 and the second, because the cross section for pions is proportional to the square of the center of mass energy, leads to an extra factor T^2. Since the pion gas is relativistic for $T \geq 100\, MeV$, the average mean free path is: $\bar{\lambda} \simeq \bar{\tau}$. For $T > 150 MeV$, the gas contains a large fraction of heavy hadrons, and the formula mentioned above is not reliable. It is likely that it will overestimate the mean free path. This is however not very relevant because the systems of interest (see section 6) show a freeze-out temperature of 100 MeV or less, where (2) is a good approximation.

An improvement which considers the leading term in the virial expansion and a phenomenological input for the $\pi - \pi$ forward scattering amplitude[11] leads to corrections of the order of 20%. The elastic mean-free path for nucleons has also been analyzed[12].

Chemical equilibrium is more complex than thermal equilibrium because it is controlled by inelastic processes. As the system cools down, their rates rapidly decrease (faster than for elastic ones) and eventually chemical equilibrium is lost. When this occurs, the (properly defined) number of particles (e.g. pions) which is not conserved when chemical equilibrium prevails, becomes approximately conserved. This implies that the equation of state becomes a function of T and chemical potentials associated to such approximately conserved particle numbers (μ_π, μ_K, etc). Since for systems produced in a heavy ion collision chemical freeze-out is likely to take place at relatively high temperatures, where the population of the gas contains a large fraction of heavy hadrons, the estimate of the relevant inelastic mean-free paths is not very precise. Let us first consider the baryons; their number is controlled by annihilations $B\bar{B} \rightarrow \pi\pi, K\bar{K}, \pi\pi\pi$, etc. The mean-free path is given approximately by[7]: $\lambda_N \sim (2\sigma_{eff}\bar{n}_N)^{-1}$, where σ_{eff} is the annihilation cross section averaged over the thermal distribution and $\bar{n}_N$ is the number density of baryons plus antibaryons to be precisely defined in next section. We assume that the cross section is similar for all baryons. Data for $p\bar{p}$ annihilation at low energies give $\sigma_{eff} \sim 90\,mb$ for $140\,MeV \leq T \leq 180\,MeV$. In chemical equilibrium, one finds : $\lambda_N \sim 1.2; 3.5; 13 fm$ for $T = 180; 160; 140\,MeV$. For a hot zone with a radius of a few fm, the number of baryons plus antibaryons will therefore freeze-out around $T = 160\,MeV$. The number of strange plus anti-strange particles is mainly controlled by $K\bar{K}$-annihilations. Unfortunately, no data is available on these cross sections. We assume that they are similar to those for baryons (they are probably somewhat smaller). In this way, $\lambda_K \sim (2\sigma_{eff}.\bar{n}_K)^{-1}$, where $\bar{n}_K$ is the density of strange plus anti-strange hadrons. Since $\bar{n}_K/\bar{n}_N \sim 2; 3; 5.3$ for $T = 160; 140; 120\,MeV$[7], we expect that strangeness will freeze-out later than the baryons. Finally, for pions no good estimate exists. Their number can be changed by inelastic scattering like $\pi\pi \leftrightarrow \pi\pi\pi\pi$, $\pi\pi\rho$ and by the decays of resonances where the individual channels violate the conservation law for the number of pions (the ρ-meson, for instance, does not lead to such a violation, since the only relevant channel is $\rho \rightarrow \pi\pi$, and thus, ρ is equivalent to two pions; on the other hand, $\omega \rightarrow \pi\pi\pi$ and $\omega\pi \leftrightarrow \pi\pi$, as well as, $N^* \rightarrow N\pi$ and $N^* \rightarrow N\pi\pi$, are not compatible with the conservation law). The mean free-path for the first type of process behaves like T^{-9} and is, therefore, expected to become ineffective at a very early stage. The second type requires that the population of highly excited resonances be large in order to produce large enough fluctuations in the number of pions; they will also become ineffective as the population of such resonances rapidly decreases with the drop in T. We, therefore, expect that the number of pions will freeze-out at a rather early stage (earlier than the baryons). In our analysis, we have considered different freeze-out temperatures in order to have an idea of the uncertainties one may expect.

4 Chemical Potentials

The loss of chemical equilibrium is manifested in the equation of state by the build-up of chemical potentials for the different levels. These chemical potentials are not independent. The chemical potentials of excited states are given in terms of the chemical potentials of the stable states $\sigma : \{\pi, K, \eta, N, \Lambda, \Sigma, \Xi, \Omega\}$ according to[7]:

$$\mu_i = \sum_i d_i^\sigma \mu_\sigma \tag{4}$$

where, as illustration, d_i^π is the mean number of pions in the decay of level i (e.g.: $\mu_\rho = 2\mu_\pi, \mu_\Delta = \mu_\pi + \mu_N$, etc.). In the absence of net baryon number, $\mu_\sigma = \mu_{\bar{\sigma}}$, where the bar refers to antiparticle. It is possible to reduce the number of relevant stable particles, and thus reduce the number of independent chemical potentials. In the $SU(3)$ limit, the chemical potentials are the same for all the members of a multiplet, thus, we would have as independent potentials μ_π and μ_N ($\mu_{\eta'}$ does not play a rôle because η' participates only marginally in inelastic reactions). $SU(3)$ breaking produces intramultiplet splitting in the chemical potentials. In order to estimate them, we resort to the following assumption: the splitting is proportional to the number of strange quarks or antiquarks contained in the particle irrespective of the multiplet. This leads to :

$$\begin{aligned} \mu_\Lambda - \mu_N = \mu_\Sigma - \mu_N &= \mu_\Xi - \mu_\Sigma = \mu_K - \mu_\pi \\ \mu_\Omega - \mu_\Lambda &= 3(\mu_K - \mu_\pi) \end{aligned} \tag{5}$$

μ_η can be approximated by the ideal mixing relation $\mu_\eta = (4\mu_K - \mu_\pi)/3$. Deviations from this relation have negligible effects due to the low statistical weight of η. All chemical potentials are, therefore, given in terms of μ_π, μ_K and μ_N. The number densities conjugated to the chemical potentials are $\bar{n}_\sigma = \partial P/\partial\mu_\sigma$, $\sigma = \pi, K, N$. Accordingly, $\bar{n}_N$ is the density of baryons plus antibaryons, $\bar{n}_K$ is the density of s plus $\bar{s}$ valence quarks, and finally, $\bar{n}_\pi + \bar{n}_K$ counts the density of stable mesons, including those 'hidden' in resonances.

Naturally, there are reactions which tend to restore chemical equilibrium : for instance, $\omega \leftrightarrow \pi\pi\pi$ and $\omega\pi \leftrightarrow \rho\rho$ are consistent with the conservation of $N_\pi + 2N_\rho + 3N_\omega$, while the reaction $\pi\pi \leftrightarrow \pi\omega$ is not; decays like $N^* \rightarrow N\pi$ and $N^* \rightarrow N\pi\pi$ are also incompatible with the conservation law. The mere occurrence of these reactions does not however imply that chemical equilibrium will be sustained. As the density of resonances whose decays tend to restore equilibrium becomes much lower than the density of pions, they become ineffective in changing their number. On the other hand, their densities are easily kept in balance by the pions according to (4).

5 The Expansion

The expansion is determined by the conservation laws of energy-momentum and particle number. In the absence of dissipation (we constrain ourselves to this case), the energy momentum tensor is given by :

$$\Theta_{\mu\nu} = (\epsilon + P)u_\mu u_\nu - P g_{\mu\nu} \tag{6}$$

where $u_\mu(x)$ is the velocity of the comoving frame at x and $\epsilon(x)$ and $P(x)$ are defined by the usual thermodynamic expressions as functions of the temperature and chemical potentials in that frame. The equations of motion are the usual hydrodynamic equations:

$$\partial^\mu \Theta_{\mu\nu} = 0 \tag{7}$$

When chemical equilibrium prevails, all chemical potentials vanish and P and ϵ are only functions of T. Eq. (6) with appropriate boundary conditions (for a freely expanding gas it is sufficient to specify $T(\bar{x})$ and $u_\mu(\bar{x})$ at the initial time) completely determines the time evolution. As chemical equilibrium is lost, P and ϵ become also functions of μ_π, μ_K and μ_N. Eq. (6) together with the conservation equations for the number of particles:

$$\partial^\mu(\bar{n}_\sigma u_\mu) = 0 \tag{8}$$

fully determine the evolution ($\mu_\sigma(\bar{x})$ must be added as boundary condition).

It is easy to see that in both situations, i.e., when the system is in chemical equilibrium and when the number of particles is conserved, the flow is adiabatic : $\partial^\mu(s u_\mu) = 0$ (s is the entropy density given by $s = \partial P/\partial T$). During the transition from one situation to the other entropy will be produced. Since it is very hard to give even a rough estimate for this transient, we neglect it, assuming that it occurs rather suddenly, leading to minor entropy increase.

For a sufficiently homogeneous system, the conserved quantities $S = \int d^3x\,\gamma\, s$ and $N_\sigma = \int d^3x\,\gamma\,\bar{n}_\sigma$ ($\gamma = u_0(\bar{x})$ is the usual relativistic gamma-factor) imply that $s/\bar{n}_\sigma$ remains approximately constant during the evolution. This ratio will then determine the chemical composition of the gas in the course of the expansion[7]. For illustration, Fig. Ia shows the evolution of μ_π in a gas where baryons and strange particles have been omitted. Indicated are the values of the conserved ratio $s/\bar{n}_\pi$ for the different chemical freeze-out temperatures. As expected, there is a steady grow of μ_π as T drops. It is important to remark that the effect is enhanced with respect to an ideal case, where the gas consists only of pions, by the presence of the resonances $(\rho, \omega, \pi^*, ...)$. We will see that when baryons and strange particles are included a further enhancement of μ_π results.

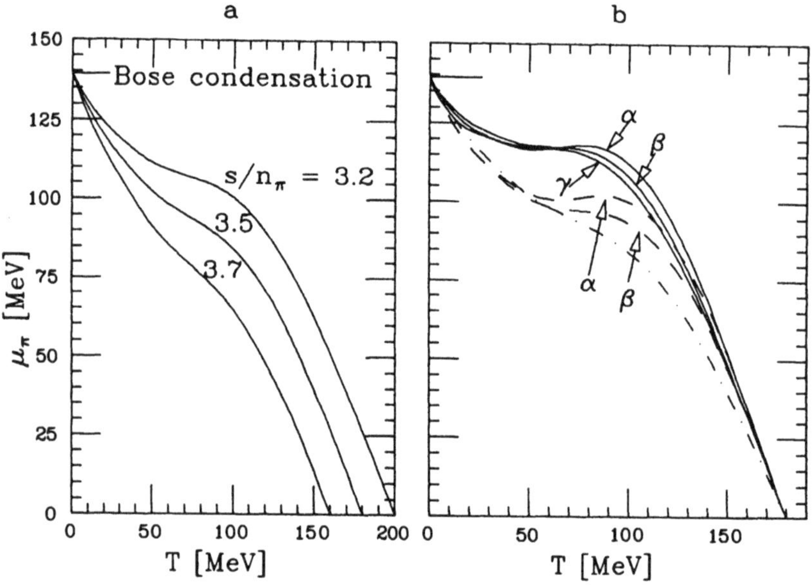

Fig. I

When baryons and strange particles are taken into account, chemical equilibrium is lost stepwise as mentioned in section 3. In a first partial equilibrium phase (α) only pions have lost chemical equilibrium and the chemical composition is characterized by μ_π. Since the reactions $N\bar{N} \leftrightarrow \pi\pi...\pi$, $K\bar{K} \leftrightarrow \pi\pi, \pi\pi\pi\pi$, etc. are still effective, μ_N and μ_K will not vanish as one would naively believe. They will instead be given by:

$\mu_N = \alpha_N \mu_\pi$ and $\mu_K = \alpha_K \mu_\pi$ where $2\alpha_{N(K)}$ is the average number of pions produced in $N\bar{N}$ $(K\bar{K})$ annihilations. In this phase, the conserved ratio is $s/(\bar{n}_\pi + \alpha_K \bar{n}_K + \alpha_N \bar{n}_N)$ rather than $s/\bar{n}_\pi$. In a second partial equilibrium phase (β) pions and baryons are out of equilibrium. There are now two independent chemical potentials, μ_π and μ_N, while $\mu_K = \alpha_K \mu_\pi + \alpha_K^N \mu_N$. Since this phase, in a realistic situation, would start when $T \sim 160\ MeV$ or lower, the reactions $K\bar{K} \leftrightarrow N\bar{N}$ are rare, and therefore α_K^N is very small (in the following we set it to zero). The conserved ratios are $s/(\bar{n}_\pi + \alpha_K \bar{n}_K)$ and $s/\bar{n}_N$. The final phase (γ) is the simplest and corresponds to conserved N_π, N_N and N_K. In this phase, the evolution will conserve the ratios $s/\bar{n}_\pi$, $s/\bar{n}_N$ and $s/\bar{n}_K$ separately. The coefficients α_N and α_K can, in general, depend on T. $p\bar{p}$-annihilation at low energies lead to $\alpha_N \sim 2.5$ (the average number of pions produced in the annihilation is around 5). The T dependence can be safely ignored. There is no data to extract α_K; there are however bounds: a kinematical bound (due to the fact that at least two pions are produced in the annihilation) is $\alpha_K \geq 1$, and an upper bound is given by the requirement that N_K does not grow as the system expands; this bound depends on T: e.g. $\alpha_K \leq 1.57(1.64)$ for $T = 180(140)\ MeV$. In the following, we will take into account this uncertainty. Fig. Ib shows the evolution of μ_π when chemical freeze-out occurs at $T = 180\ MeV$. The curves α, β and γ correspond to the situation where the system stays in the corresponding phase during the whole expansion. The real situation would be between these extreme cases. The solid curves correspond to $\alpha_K = 1.5$ and the dashed ones to $\alpha_K = 1$. For the sake of comparison, the curve with $s/\bar{n}_\pi = 3.5$ of Fig. Ia has been plotted with a dash-dotted line. We observe that μ_π developes rather rapidly, becoming of the order of $100\ MeV$ as T drops to $100\ MeV$. There is no indication that Bose-Einstein condensation can set in. This could eventually occur if μ_π would be large enough already at $T \sim 180\ MeV$. It seems, though, reasonable that chemical equilibrium will prevail in the early stages after hadronization if the collision energy is high enough. For such a situation, we do not expect condensation. We refer to[7] for details about the evolution of μ_N and μ_K. They play an important rôle in determining the relative abundances of nucleons and K mesons at low T. Fig. IIa shows these abundances defined by the ratios N/π and K/π, where π is the number density of pions after all excited states have decayed (likewise for K and N). The shaded region gives the uncertainties related to the chemical freeze-out temperature and α_K: (a) results when baryons and strangeness freeze-out simultaneously with the pions and (b) corresponds to $\alpha_K = 1$ and baryons and strangeness in chemical equilibrium down to $T = 140\ MeV$. The dash-dotted curves give the relative abundances in chemical equilibrium. At a typical freeze-out temperature of 100 MeV, K/π does not appreciably differ from the equilibrium case, while N/π does. It is clear that this enhancement could be used as a test of the general consistency of the approach.

The final step in the expansion is thermal freeze-out, which occurs when the elastic mean-free path $\bar{\lambda}$ becomes comparable to the radius of the system (for a spherical system, particles emitted at random within the sphere will travel in average a distance $\frac{3}{4}R$ in order to leave the sphere; in such a case freeze-out should occur when $\bar{\lambda} \sim \frac{3}{4}$). At this point the state of the system is specified by T, μ_π, μ_N, μ_K and the expansion velocity field $u_\mu(\bar{x})$. The last has not yet been investigated in full detail and involves explicit numerical solutions of the hydrodynamic equations. As next section suggests, the expansion velocity plays an important rôle in determining the shape of the transverse momentum and rapidity spectra.

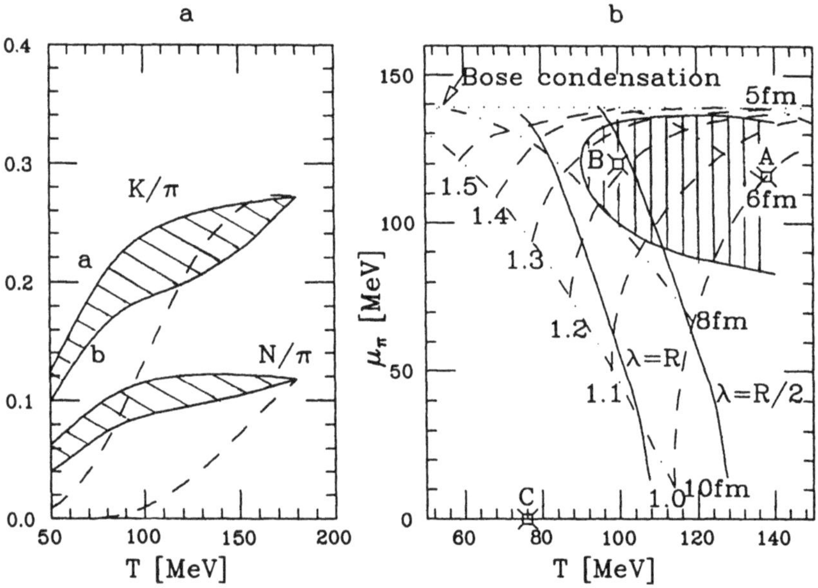

Fig. II

There are various effects that have been disregarded in our picture of the expansion. One of them is viscosity, which appears as extra terms proportional to the collective velocity gradients in the expression of $\Theta_{\mu\nu}$ [13]. The net effect is an increase of entropy in the course of the expansion, and an increase of $s/\bar{n}_{\pi,N,K}$, leading to smaller chemical potentials at freeze-out. We refer to[14,15] for a study of the viscosity coefficients and heat conductivity in a hadronic gas, and to[4] for estimates of entropy production in Bjorken's scaling solution[16]. Also the possibility of shock-wave onset can lead to corrections. To this respect, observe that the speed of sound in the hadronic gas for $T \geq 150\,MeV$ is rather small: $c_s^2 = \partial P/\partial\epsilon \sim 1/6$.

6 Confronting Data

This section is based on the analysis[6] of data from the NA35 collaboration[1,2,3] at CERN for $O + Au$ and $S + S$ fixed target collisions at $200\,GeV/nucleon$ projectile energy. We analyzed the normalized p_T-spectrum of negative particles (i.e. secondary particles) within a unit rapidity window around central rapidity. It is believed that this component of the spectrum is closest to thermalization. By fitting the spectrum we determined T, μ_π and the radius R of the system under the following assumptions: a) All negative particles are pions. Contamination with up to a total of 20% of antiprotons and K^- leads to small changes in the results. b) The system is spherically symmetric (and expands accordingly). This assumption is not compatible with the rapidity spectrum, which turns out to be too narrow as compared to the observed one. This fact suggests that the expansion is asymmetric, being faster in the axial direction. c) T and μ_π are constant over the volume of the system (they should be considered as average values). d) The expansion velocity profile is chosen, and checked that reasonable profiles lead to similar results. e) freeze-out is determined by the requirement that $\bar{\lambda} \sim \frac{3}{4}R$.

The main qualitative feature of the p_T-spectrum is the enhancement observed at low p_T ($\leq 250\,MeV$) as compared to the spectrum in $p + p$-collisions. This enhancement is largely responsible for the resulting non-vanishing μ_π. In fact, a positive chemical potential implies an excess of pions, which will tend to overpopulate the low energy component of the gas and thus lead to the mentioned enhancement. This explanation was suggested in Refs.[5,6,17].

The spectrum is fitted with the following expression :

$$dN^{(-)}(p_T) = 2\pi p_T \int_{y_0}^{y_1} dy \quad E(p_T, y)\, \eta^{(-)}(p_T, y) \; dp_T \tag{9}$$

$$y = \mathrm{arcth}(p_s/E) \qquad E(p_T, y) = \sqrt{\frac{M_\pi^2 + p_T^2}{1 - \mathrm{th}^2 y}}$$

where y is the rapidity and the interval $[y_0, y_1]$ is the rapidity window considered. The total number density of pions in momentum space is given by:

$$\eta^-(p) = \frac{1}{(2\pi)^3} \int_{Vol} d^3x \frac{1}{e^{\epsilon(\bar{x},p)} - 1} \qquad (10)$$
$$\epsilon(\bar{x}, p) = (p_\mu u^\mu(\bar{x}) - \mu_\pi)/T$$

The expansion mainly affects the result for T. The radial component of the expansion has the effect of shifting the population of pions to larger p_T; this is compensated by a decrease of the temperature, which shifts back the population. In this way, the expansion velocity can be adjusted until the freeze-out condition is met (since this is essentially determined by T due to the T-dependence of $\bar{\lambda}$). For $O + Au$ collisions, Fig. IIb shows in the μ_π vs T plane the 95% confidence level domain for the fit (shaded area). Its intersection with the band defined by $R/2 \leq \bar{\lambda} \leq R$ is the region where the freeze-out condition is met. Plotted are curves for R indicated in units of fm and $\bar{\gamma}$ (the average gamma-factor of the expansion) as they are obtained by performing a fit with given T and μ_π. (A) corresponds to no expansion and (C) to vanishing chemical potential. Both are inconsistent with the freeze-out condition [1]. The consistent point (B) is given by :

$$T = 100 \pm 10\,MeV \qquad \mu_\pi = 110 \pm 20\,MeV \qquad (11)$$
$$R = 7 \pm 1 fm \qquad \bar{\gamma} = 1.2 \pm 0.15 \qquad \chi^2 \sim 1.2 \text{ to } 1.6$$

This result does not depend appreciably on the chosen velocity profiles $v(r)$ (which must increase with r and satisfy $v(r = 0) = 0$). For different profiles, consistency with freeze-out is achieved when the resulting $\bar{\gamma}$ is about 1.2. The entropy per particle $s/\bar{n}_\pi$ resulting from (11) is ~ 3, which is considerably smaller than for a massless Bose gas, where it is 3.6. Comparison with the analysis of the previous section shows that this can only be understood if chemical freeze-out occurs in the very early stages of the expansion. It is interesting to notice that the resulting value of μ_π at $T \sim 100\,MeV$ implies that the particle number density is more than three times the density in chemical equilibrium and comparable to the density in nuclei. The radius can be compared with estimates stemming from two-pion interferometry measurements[3], which lead to $R \sim 7 \pm 2\,fm$ for the central rapidity region. Notice that $\bar{\lambda}$ is substantially shorter than R during the expansion. Therefore, every particle

[1] In Ref.[17] the fit was done considering that the system only expands axially. The resulting temperature of 160 MeV implies that $\bar{\lambda} \ll R$, which is not consistent with our freeze-out criterion

will suffer, in average, a few collisions before leaving the system. The results obtained from the analysis of $S + S$ collisions data are very similar.

Finally, the observed abundances of $\bar{p}$ and K-mesons are[2]: $\bar{p}/\pi \sim 0.10$ and $K/\pi \sim 0.16$. These numbers are consistent with the results shown in Fig. IIa for the case of an evolution off chemical equilibrium. In particular, at $T \sim 100\,MeV$ the equilibrium prediction for N/π is clearly inconsistent with data.

One should emphasize that the consistency found in this analysis is by no means a confirmation of the correctness of the picture. In fact, one would rather think that the approach to thermalization in the systems produced in present experiments is marginal. This can be clarified by a more precise theoretical analysis and the experimental study of larger systems.

7 Comments

The approach discussed in this report relies on the main premise that statistical mechanics provides a good first approximation to the description of a system consisting of the order of 10^3 particles undergoing free expansion. In particular, its evolution should then be given in terms of the usual hydrodynamic equations. A source of departure from equilibrium is the fact that during part of the evolution, a large fraction of the particles in the system are less than one mean-free path away from the boundary. Since particles moving in the outward direction can easily leave the system, equilibrium can not be maintained. This effect becomes more pronounced as the gas cools down and the mean-free path grows. On other hand, the hydrodynamic equations require the identification of the collective velocity $u_\mu(x)$, whose definition presumes that it can be assigned to volumes containing large numbers of particles. This is not the case for the systems we consider, where the collective velocity should vary substantially within the intraparticle distances, and thus loose its proper meaning. This also poses problems to the local definition of the thermodynamic variables. An analysis of these aspects requires the introduction of the more general framework of kinetic theory. Studies in this direction have been recently initiated[15,18].

8 Acknowledgements

The results contained in this report have been obtained in collaboration with H. Leutwyler, P. Gerber and H. Bebie. Thanks are due to them for useful and enlightening discussions, and, to the organizers of the workshop for providing an excellent atmosphere.

9 References

1. J. W. Harris, *Nucl. Phys.* **A498** (1989) 133.
2. A. Bamberger et al., *Z. Phys* **C43** (1989) 25.
   J. Bartke et al., *Z. Phys* **C48** (1990) 191.
   S. Wenig, Ph.D. Thesis, Univ. Frankfurt (1990).
3. A. Bamberger et al., *Phys. Lett.* **B203** (1988) 320.
   T. J. Humanic et al., *Z. Phys* **C38** (1988) 79.
4. G. Baym, in Proc. of *"Quark Matter '84"*, Ed. K. Kajantie, Springer Verlag, *Lecture Notes in Physics* 221 (1984) 39, and references therein.
5. J. L. Goity and H. Leutwyler, *Phys. Lett.* **B228** (1989) 517.
6. P. Gerber, H. Leutwyler and J. L. Goity, *Phys. Lett.* **B246** (1990) 513.
7. H. Bebie, P. Gerber, J. L. Goity and H. Leutwyler, "The role of entropy in an expanding hadronic gas", preprint BUTP-91/14, PSI-PR-09 (1991).
8. P. Gerber, and H. Leutwyler, *Nucl. Phys.* **B321** (1989) 387.
   P. Gerber, this Proceedings.
9. R. Hagedorn, in Proc. of *"Quark Matter '84"*, Ed. K. Kajantie, Springer Verlag, *Lecture Notes in Physics* 221 (1984) 57, and references therein.
10. Review of Particle Properties, *Phys. Lett.* **B239** (1990).
11. A. Schenk, *Nucl. Phys.* **B363** (1991) 97.
12. H. Leutwyler and A. Smilga, *Nucl. Phys.* **B342** (1990) 302.
13. L. D. Landau and E. M. Lifshitz, *Fluid Mechanics* (Pergamon, London, 1979).
14. S. Gavin, *Nucl. Phys.* **A435** (1985) 836 and Helsinki Preprint HU-TFT-90-50.
    S. Gavin and P. V. Ruuskanen, Helsinki Preprint HU-TFT-90-59.
15. M. Prakash, M. Prakash, R. Venugopalan and G. Welke, Stony Brook Preprint SUNY-NTG-91-2 (1991).
16. J. D. Bjorken, *Phys. Rev.* **D27** (1983) 140.
17. M. Kataja and P. V. Ruuskanen, *Phys. Lett.* **B243** (1990) 181.
18. S. Gavin, *Nucl. Phys.* **B351** (1991) 561.
    R. Venugopalan, M. Prakash, M. Kataja and V. Ruuskanen, Stony Brook Preprint SUNY-NTG-91-10 (1991).

ANOMALY MECHANISM AT FINITE TEMPERATURE

A.V.SMILGA

Institute for Theoretical Physics, Bern University, Sidlerstrasse 5, CH-3012 Bern, Switzerland

and

Institute for Theoretical and Experimental Physics B.Cheremushkinskaya 25, Moscow 117259, U.S.S.R.

ABSTRACT

The chirality breaking effects associated with the axial anomaly at finite temperature are studied. The exact results are obtained in the case when the interaction between fermions in heat bath is neglected. When the temperature is much greater than the characteristic frequencies and momenta of external fields, the dominant contribution to the anomaly is due not to the pair creation processes associated with level crossing mechanism but rather to the scattering of the existing thermal fermion excitations by the external fields.

A simple non-local field model bearing many assential features of the physical system in interest - the high T plasma with account of interaction effects - is also analyzed. The anomaly condition is saturated there due to singularities of vertices rather than of Green functions.

1. Introduction.

One of the most interesting still unresolved questions of the high energy physics on the "mundane" energy scale of several hundreds Gev is the problem of baryon number nonconservation at high temperature and/or high energies. The situation with temperature-induced baryon number nonconservation is particularly troubling.

The fact that at high temperature $T > T_c$ the probability of the processes not conserving baryon number is not suppressed exponentially $\propto \exp(-4\pi/\alpha_w)$, as is the case at low temperatures (and low energies), and large was discovered first by V.A.Kuzmin, V.A.Rubakov and M.E.Shaposhnikov in 1985 [1]. The exponential suppression disappears when the temperature is of order of the (temperature dependent) height of the barrier in free energy $F_{sph}(T) \propto m_W\sqrt{1-T^2/T_c^2}\,/\alpha_W$, separating vacua with different winding numbers. Numerous investigations performed in recent years [2] only confirmed this initial guess.

I want to note however that the all studies of fast thermal B + L violating transitions performed so far did not consider the detailed mechanism of quark $\longleftrightarrow$ lepton transitions induced by external gauge fields. All authors limited themselves to the remark that the divergence of the B + L - current is in fact anomalous:

$$\partial_\mu j_\mu^{B+L} = \frac{g^2 n_f}{16\pi^2} G^a_{\alpha\beta} \tilde{G}^a_{\alpha\beta} \qquad (1)$$

Integrating this exact relation,one may find that after a transition which changes the winding number of the gauge field

$$Q^{C.S.} = \frac{g^2}{32\pi^2} \int d\mathbf{x}\, \varepsilon_{ijk} (F^a_{ij} W^a_k - \frac{2g}{3} \varepsilon^{abc} W^a_i W^b_j W^c_k), \qquad (2)$$

the baryon charge should also be changed. This argument is solid enough but, as the question is indeed very important, it is desirable also to have a concrete dynamic picture of how this extra baryon charge is actually produced.

I still do not have answer to this question but a limited investigation performed in the present paper shows, at least, that at high temperature the standard level crossing picture does not describe adequately the dynamics of this process.

The standard level-crossing picture [3] amounts, in fact, to the creation of low-energy extra fermions with nonzero net baryon and lepton charge in a slowly varying external gauge field. A puzzling question which arises immediately is the following. The necessary condition for the description of this kind to be valid is the masslessness of the fermions. If the fermions are massive, *and* their mass is much larger than the characteristic frequency of the gauge field, it is known for a long time that no extra fermions are produced out of the vacuum [4]. The external field quanta just do not have enough energy to overcome the gap between negative energy and positive energy levels. So, in case $m\tau \gg 1$ where τ is the characteristic time scale of the transition, axial charge (the analog of baryon charge in nonchiral models) is conserved during the transition.

There is no contradiction with the anomaly arguments. If the fermions have mass, the axial current divergence involves also nonanomalous contribution $\propto m\, \bar{\psi}\gamma^5\psi$ which in the case $m\tau \gg 1$ neutralize completely the effects due to anomalous term [4].

However, at finite temperature fermions are not massless in true sense. The spectrum of collective fermion excitations involves a gap which for quarks is as large as $\propto g_s T$ where g_s is the strong coupling constant [5]. This gap has the same origin as the gap in gluon spectrum, alias plasma frequency (see sect.5 for more details). At $T \sim T_c$, this gap is parametrically much larger than the characteristic frequencies of sphaleron transition $\omega_{sph} \sim m_W \sim g_W T_c$. Thus, seemingly, the probability of the creation of new quark excitations during the sphaleron transition should be exponentially suppressed $\propto \exp(-g_s/g_W)$ and it is not immediately clear how the baryon excess is produced (the very end of the last reference in [2]).

By my opinion, clear understanding of this and related questions is necessary until one may assert with a 100% assurance that rapid baryon number violation takes place at high temperatures, indeed.

I am able now to provide the exact exhaustive analysis only in the most simple case of abelian theories without interaction. Some interesting specific features of $T \neq 0$ case are seen, however, already at this level. It is seen, in particular, that, though the total chirality of the system is duly changed as a result of the action of an external field according to the change of its topological charge, it is not due to the creation of extra fermions but rather to the scattering of the fermions being already present in the heat bath by the field.

What is actually studied is the structure of the thermal correlator $\langle j^5_\mu j_\nu \rangle_T$ in the two-dimensional massive Schwinger model (in case when the interaction between fermions are neglected - see discussion below) and also the structure of the thermal correlator $\langle j^5_\mu j_\nu \rangle_T$ *in the presence of external magnetic field* in the usual massive QED (it is equivalent to usual Adler triangle with one of the photon momenta set to zero). We preserve nonzero fermion mass as in this case the physical "infrared" interpretation of the anomaly may be given [6]. To acquire insight on the physics of real plasma where interaction effects are essential, we study a toy two-dimensional nonlocal model bearing many essential features of the physical system in interest. We discuss the implications of our analysis for physics in the end of the paper.

Note that anomalous correlators at finite temperature were studied earlier by Itoyama and Mueller [7]. However, they were interested mainly not at the Adler triangle itself but rather at its *divergence*. They showed that the divergence is the same as at zero temperature (and, as usual, higher order corrections do not contribute). This is very natural, of course, as the anomalous divergence is an essentially ultraviolet effect, and ultraviolet structure of the theory is not at all affected by the temperature. We shall be concentrated, however, on the study of the correlator $\langle j^5_\mu j_\nu \rangle$, as such, to unravel its nontrivial infrared dynamics.

2. Two-dimensional QED

To begin with, let us establish two-dimensional notations. γ-matrices are chosen as

$$\gamma_0 \equiv \sigma_2 , \qquad \gamma_1 \equiv i\sigma_1, \qquad \gamma_5 \equiv \sigma_3 \tag{3}$$

The following useful relations are valid:

$$\mathrm{Sp}\left\{\gamma_5\gamma_\mu\gamma_\nu\right\} = 2\ \varepsilon_{\mu\nu} \tag{4a}$$

$$\mathrm{Sp}\left\{\gamma_5\gamma_\mu\gamma_\nu\gamma_\alpha\gamma_\beta\right\} = 2\ \left(\varepsilon_{\mu\nu}g_{\alpha\beta} + \varepsilon_{\alpha\beta}g_{\mu\nu} \right) \tag{4b}$$

$$g_{\alpha\beta}\varepsilon_{\mu\nu} + g_{\mu\beta}\varepsilon_{\nu\alpha} + g_{\nu\beta}\varepsilon_{\alpha\mu} = 0 \qquad (4c)$$

$$\gamma_\mu\gamma_5 = -\varepsilon_{\mu\nu}\gamma_\nu \qquad (4d)$$

It follows from eq.(4d) that in two dimensions vector current $j_\mu = \bar\psi\gamma_\mu\psi$ and the axial current $j^5_\mu = \bar\psi\gamma_\mu\gamma_5\psi$ are kinematically related.

A) Zero temperature.

Axial anomaly in the Schwinger model has been studied by many authors. In the Feynman graph language, it comes from the simple "diangle" depicted in fig.1.

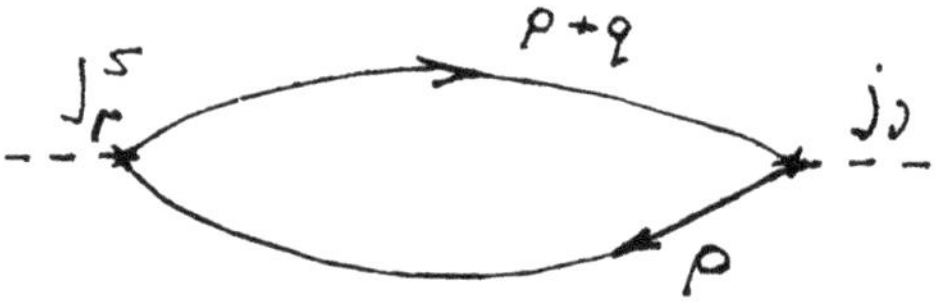

Fig.1. Anomalous diangle

After the necessary ultraviolet regularization, the result of the calculation for the correlator can be written in the following form

$$T_{\mu\nu} = \frac{1}{2\pi}\left[\varepsilon_{\mu\nu} - \frac{(\varepsilon_{\mu\alpha}q_\nu + \varepsilon_{\nu\alpha}q_\mu)q_\alpha}{q^2}\right]\left[1 - m^2 I(m^2,q^2)\right] \qquad (5)$$

where

$$I(m^2,q^2) = \frac{4}{\sqrt{Q^2(Q^2+4m^2)}} \ln\left[\sqrt{1+Q^2/4m^2} + \sqrt{Q^2/4m^2}\right], \qquad (6)$$

$Q^2 = -q^2$. As $m^2 \to 0$ and Q^2 is fixed, the term $m^2 I(m^2,Q^2)$ vanishes, and the amplitude is just

$$T^{m=0}_{\mu\nu} = -\frac{q_\mu\varepsilon_{\nu\alpha}q_\alpha}{\pi q^2} \qquad (7)$$

(the relation (4c) has been used). In the massless limit this result is exact as all higher order corrections vanish (massless Schwinger model is, in fact, a free theory)[1]. The

---

1 Only one-particle irreducible graphs are considered here. In the

amplitude $T_{\mu\nu}$ vanishes when multiplied by q_ν (the vector current is conserved) whereas when multiplied by q_μ it gives

$$q_\mu T^{m=0}_{\mu\nu} = -\frac{1}{\pi}\varepsilon_{\nu\alpha}q_\alpha \tag{8}$$

exactly as required by the anomalous axial current divergence

$$\partial_\mu j^{\mu 5} = \frac{1}{2\pi}\varepsilon^{\alpha\beta}F_{\alpha\beta} \tag{9}$$

What we are going to do now is to use the *dispersive* approach to reproduce the result (6) for $T_{\mu\nu}$. The subsequent treatment is exactly parallel to the Dolgov-Zakharov treatment of triangle anomaly in 4-dimensional QED [6].

Let us find the imaginary part of the correlator depicted in Fig.1.

It is not very difficult to obtain the answer

$$\text{Im } T_{\mu\nu} = \frac{2m^2}{\sqrt{q^2(q^2-4m^2)}}\frac{q_\mu\varepsilon_{\nu\alpha}q_\alpha}{q^2} \tag{10}$$

If setting here $m^2 = 0$, while keeping q^2 fixed, the imaginary part vanishes. This is quite natural as the process of creation of e^+e^- - pair moving in the opposite directions involves chirality breaking (In two dimensions, chirality of massless fermions is rigidly connected with the direction of their motion, with the right-going electrons and left-going positrons having the chirality +1 , and the left-going electrons and right-going positrons - the chirality -1). And it is absent in the massless case.

However, one may be easily convinced that, as long as $m^2 \neq 0$, the imaginary part, though negligibly small at $q^2 \gg m^2$, is very large and significant in the low q^2 region $q^2 \gtrsim 4m^2$. In fact,

$$\int_{4m^2}^{\infty} \frac{m^2 ds}{s^{3/2}\sqrt{s-4m^2}} = \frac{1}{2} \tag{11}$$

independently of m so that

$$\lim_{m^2\to 0} \text{Im } T_{\mu\nu} = \delta(q^2)\, q_\mu\varepsilon_{\nu\alpha}q_\alpha \tag{12}$$

Restoring the full amplitude by use of dispersion relation, we arrive obviously at the result (7) (and, needless to say, the exact $m \neq 0$ amplitude (5) can be restored by its imaginary part (10) with an equal ease). The physical interpretation is clear - the correlator is saturated by the process of creation

---

massless Schwinger model with *several* flavours, the result (7) describes the whole correlator of *non-singlet* external currents.

of e^+e^- - pairs with very low momenta of order of fermion mass.

B). Finite temperature.

Consider now the same correlator but at finite temperature. As we are interested especially in its imaginary part in Minkowsky space, we are in a position to use the real time diagram technique [8]. There are several ways to define Green functions at finite temperature. The most important physically are the *retarded* Green functions

$$\int d^4x\, e^{ipx}\, \theta(x^0)\, Z^{-1}\, \mathrm{Sp}\left\{ e^{-\beta H}\, [A(0),\, B(x)]_{\mp} \right\} \qquad (13)$$

(the commutator for Bose fields and anticommutator for Fermi fields) which have at finite temperature a clear physical interpretation. Such a function describes the response of the system in terms of an observable B(x) for the action of the external current A(0) being exerted *at earlier time,* and isrelated to a generalized susceptibility. And this is exactly which we are interested in - the response of the system, displayed in the change of its chirality, for the action of external electromagnetic field. In the momentum representation, retarded Green functions have simple analytic properties (no sigularities in the upper p_0 half-plane). Also, they are the functions whose poles determine the spectrum of collective excitations in the system.

Skipping all technical details, we present here only the result for the imaginary part of the correlator $<j^5_\mu j_\nu>^{Ret}$ at finite temperature :

$$\mathrm{Im}\, T^R_{\mu\nu} = \frac{1}{2}\int d^2p\, \delta(p^2 - m^2)\, \delta\,[(p+q)^2 - m^2]\left[\mathrm{th}\,\frac{p_0+q_0}{2T} - \mathrm{th}\,\frac{p_0}{2T}\right]$$

$$\varepsilon(p_0)\varepsilon(p_0+q_0)\,\left[m^2\varepsilon_{\mu\nu} + (\varepsilon_{\mu\alpha}g_{\nu\beta} + \varepsilon_{\nu\beta}g_{\mu\alpha})\, p_\alpha (p+q)_\beta\right] \qquad (14)$$

The factor

$$\frac{1}{2}\left[\mathrm{th}\,\frac{p_0+q_0}{2T} - \mathrm{th}\,\frac{p_0}{2T}\right] = n_F(p_0) - n_F(p_0+q_0) =$$

$$= \quad n_F(p_0+q_0)\,(1 - n_F(p_0)) - n_F(p_0)\,[1 - n_F(p_0+q_0)] \qquad (15a)$$

(and the whole structure of the expression (14)) has also a rather clear physical meaning [9]. In the case $p_0 > 0$, $p_0 + q_0 > 0$ (this is a kinematically forbidden region at T = 0 !) the corresponding contribution to Im $T^R_{\mu\nu}$ is due to the scattering of the thermal excitations in the medium by the external field so that the probabilities of the direct and the inverse processes are subtracted. Similar interpretation may be given for the "decay channel" where p_0 and p_0+q_0 have the opposite signs. Let $p_0 = -\varepsilon_1 < 0$, $p_0+q_0 = \varepsilon_2$. Then

$$n_F(p_0)-n_F(p_0+q_0) = 1-n_F(\varepsilon_1)-n_F(\varepsilon_2) = \left[1 - n_F(\varepsilon_1)\right]\left[1 - n_F(\varepsilon_2)\right] - n_F(\varepsilon_1)\, n_F(\varepsilon_2) \qquad (15b)$$

which is again the difference of probabilities of direct and inverse processes. Naturally, as $T \longrightarrow 0$, $\mathrm{th}(p_0/2T) \longrightarrow \varepsilon(p_0)$, and the zero temperature Cutcovsky rules are reproduced.

The interesting case for us is when $T \gg q_{0,1}$. The temperature effects are essential only in this region. Consider first the "decay channel". Let for simplicity $q_1 = 0$. Then kinematical conditions imply $-p_0 = q_0/2 \ll T$, and the factor in eq. (14) is just $q_0/4T \ll 1$. Thus, we see that the decay channel contribution to Im $T^R_{\mu\nu}$ has exactly the same form as Im $T_{\mu\nu}$ at $T = 0$, but suppressed with a small factor $q_0/4T$:

$$\mathrm{Im}\ T^R_{\mu\nu}(q^2 \geq 4m^2) = \left[\frac{q_0}{4T} + O\left(\frac{q_0^2}{4T^2}\right)\right] \frac{2m^2}{\sqrt{q^2(q^2-4m^2)}} \frac{q_\mu \varepsilon_{\nu\alpha} q_\alpha}{q^2} \qquad (16)$$

The origin of this suppression is rather clear from eq.(15b) : the probability of the direct process of the creation of e^+e^- pair by the external field nearly cancels the probability of the inverse process where e^+e^- pairs being present in the heat bath are annihilated (see also a discussion in the last section). If restoring the corresponding contribution to the whole amplitude by use of dispersion relations, and setting afterwards $m = 0$, we would reproduce the result (7) but with a suppression $q_0/4T$. Obviously, such an amplitude would not satisfy the anomalous Ward identity (8).

The point is that, at $T \gg q$, the main contribution to Im $T^R_{\mu\nu}$ comes from forbidden at $T = 0$ kinematic region $q^2 < 0$ which corresponds to *scattering* of the medium particles by external field.

It may be shown that

$$\mathrm{Im}\ T^R_{\mu\nu}\ (q^2 < 0) = \frac{1}{(q_1^2 - q_0^2)^{3/2}}\ \mathrm{ch}^{-2} \frac{mq_0}{2T\sqrt{q_1^2 - q_0^2}} \left[1 - \frac{q_0}{2T}\ \mathrm{th}\ \frac{mq_0}{2T\sqrt{q_1^2 - q_0^2}} + O\left(\frac{q_0^2}{T^2}\right) \right] \frac{mq_0 q_\mu \varepsilon_{\nu\alpha} q_\alpha}{2T} \qquad (17)$$

If setting here $m = 0$, while keeping q_μ fixed, the imaginary part Im $T^R_{\mu\nu}$ vanishes. However, when $m \neq 0$, the imaginary part is very large in the small $|q^2|$ region. To be more exact, the characteristic values of $|q^2|$ giving the main contribution to the dispersive integral for the whole amplitude are $m^2q_0^2/T^2$ [2]. One can be readily convinced that, as $m^2 \longrightarrow 0$, the whole expression (17) tends to

$$\lim_{m^2\to 0} \text{Im}\, T^R_{\mu\nu}(q^2 < 0) = \left(1 - \frac{q_0}{4T} + O\left(\frac{q_0^2}{T^2}\right) \right) \delta(q^2)\, q_\mu \varepsilon_{\nu\alpha} q_\alpha \qquad (18)$$

Combined with the decay channel contribution (which, as mentioned earlier, has the same structure, but with the factor $q_0/4T$) it leads to the same result as in $T = 0$ case, and the anomalous Ward identity is satisfied.

The same analysis has recently been independently done by R.Baier and E.Pilon [10]. They also checked the fulfillment of anomalous Ward identity in any order in q_0/T .

3. Four-dimensional QED.

In four-dimensional case, the anomaly exhibits itself in a triangle graph. The kinematics of the triangle graph is considerably more complicated than that of the Schwinger model diangle. It is especially so in $T \neq 0$ case where many new tensor structures do appear [7]. The additional complication is that the properly defined 3-point vertices are expressed via corresponding spectral densities by the double dispersive relations [9]. For our purposes, however, it is sufficient to consider a simplified problem when *one of the photon momenta is set to zero.* Actually, what we are going to study is the two-point correlator $\langle j^5_\mu j_\nu \rangle_T$ in the *external magnetic field.* We shall see that the structure of this correlator is very much similar to what we had in d=2 QED.

A) Zero temperature

We start with the expression for the fermion propagator in external magnetic field:

$$G(p) = \frac{i}{\hat{p} - m} - e\, \varepsilon_{ijk} H_k \frac{\gamma_i (\hat{p} - m) \gamma_j}{2(p^2 - m^2)^2} \qquad (19)$$

For us, only the linear in H_k terms are relevant. Not dwelling here on details, we present here the result :

---

[2] The region $q_1^2 - q_0^2 \ll m^2q_0^2/T^2$ is cut off by the factor ch^{-2} in the r.h.s. of eq.(25). The reason is clear : if $|q^2|$ is very small, only the scattering of the particles with momenta $p_0 \gg T$ is kinematically allowed, which are just absent in the heat bath.

$$\mathrm{Im}\, T^{H}_{\mu\nu} = - \frac{eHm^2}{\pi\sqrt{q^2(q^2-4m^2)}} \frac{q_\mu \tilde{\varepsilon}_{\nu\alpha} q_\alpha}{q^2} \tag{20}$$

($\tilde{\varepsilon}_{\nu\alpha}$ acts in two-dimensional (03) - space so that $\tilde{\varepsilon}_{03} = -\tilde{\varepsilon}_{30} = 1$, and $\tilde{\varepsilon}_{\perp\alpha} = 0$) It coincides exactly with the result (12) for two-dimensional QED up to the factor $-eH/2\pi$!

This remarkable fact has a very clear physical explanation. In the kinematics considered, the anomaly is associated with the n=0 branch of Landau electron levels in the external magnetic field - the only branch which has no degeneracy in chirality. And this branch has exactly the same structure and properties as the fermion energy levels pattern in two-dimensional QED [3-4]. Hence, the amplitude $T^{H}_{\mu\nu}$ involves the pole, and, multiplying it by q_μ , we restore the anomaly equation

$$\partial_\mu j^5_\mu = \frac{e^2}{2\pi^2} (\mathbf{EH}) = \frac{e^2}{8\pi^2} F_{\alpha\beta} \tilde{F}_{\alpha\beta} \tag{21}$$

We see that all the results are exactly parallel to what we had in two-dimensional QED, and also to the 4-dimensional QED in the usual kinematics with two on-mass-shell photons [6]. But, as was already mentioned above, the kinematics considered here brings about serious simplifications for finite temperature case.

B) Finite temperature.

The finite temperature electron propagator in external magnetic field can also be easily found. The finite temperature calculation for the retarded correlator in the presence of external magnetic field gives the result

$$\mathrm{Im}\, T^{H,T\neq 0}_{\mu\nu} = - \frac{emH\, q_0 q_\mu \tilde{\varepsilon}_{\nu\alpha} q_\alpha}{4\pi T} \left\{ \frac{m\, \theta(q_0^2 - \mathbf{q}^2 - 4m^2)}{\sqrt{(q_0^2-\mathbf{q}^2)^3 (q_0^2-\mathbf{q}^2-4m^2)}} + \frac{\theta(\mathbf{q}^2-q_0^2)}{(\mathbf{q}^2-q_0^2)^{3/2}}\, \mathrm{ch}^{-2} \frac{mq_0}{2T\sqrt{\mathbf{q}^2-q_0^2}} \left[1 - \frac{q_0}{2T}\, \mathrm{th} \frac{mq_0}{2T\sqrt{\mathbf{q}^2-q_0^2}} \right] \right\} + \text{some contributions not related to anomaly} \tag{22}$$

(only the two first terms in the expansion on q_0/T are written.) It is just the sum of eqs.(16) and (17), with $\varepsilon_{\nu\alpha}$ being substituted by $\tilde{\varepsilon}_{\nu\alpha}$, and multiplied by the overall factor $-eH/2\pi$!

For sure, all the properties and physical interpretation of this contribution are identically the same as in two-dimensional case. The main contribution to the correlator stems from the scattering of external field by the thermal fermions populating the *lowest Landau band.*

4. Interaction effects: a nonlocal toy model.

Thus far, we only considered free fermions in external field. Can one make any judgment on the real physical situation where the interaction between the fermions is important ? What occurs, for example, in the quark-gluon plasma placed in external electromagnetic field ?

The adequate study of these questions is difficult technically and has not been done yet. Still, we think we may provide some preliminary guesses right now. These guesses are based mainly on the observations following from the analysis of a toy model presented in this section. But first - few more words about physical features of quark-gluon plasma system.

As was mentioned earlier, the spectrum of collective excitations in the high-temperature quark-gluon plasma involves a gap ω_0 even if the lagrangian quark mass is zero. At high temperature $T \gg \Lambda_{conf}$, $\omega_0 \sim g_s T$. There are, in fact, two branches of collective excitations - the "transverse" and the "longitudinal". At $k = 0$, $\omega_\parallel(0) = \omega_\perp(0) = \omega_0$ but the dependence $\omega_\perp(k)$ and $\omega_\parallel(k)$ is rather intricate.

The question of how to reconcile the presence of the gap *and* the absence of chirality breaking with the presence of anomaly arises, in fact, only if trying to treat the dispersion curves with a maximal seriousness - as the energy levels of some effective hamiltonian. Indeed, only in this case the question of how the energy levels of this hamiltonian behave under the action of external field may be posed. This hamiltonian is not, of course, especially simple. If taking into acoount the imaginary parts of the dispersion curves (i.e. the damping), the effective hamiltonian is even non-hermitian. But even in the naive one-loop approximation of ref.[5], the effective hamiltonian is highly nonlocal. A good guess for $\mathcal{L}^{eff}$ is

$$\mathcal{L}^{eff} \propto \bar{\psi}\, G^{-1\ (\text{one loop})}\, \psi \qquad (23)$$

where (in the momentum space)

$$G^{-1\ (\text{one loop})}(\omega,\mathbf{k}) = \omega\gamma^0 - \mathbf{k}\boldsymbol{\gamma} - \frac{\omega_0^{\,2}}{\omega} F(\omega/k)\,\gamma^0 + \frac{\omega_0^{\,2}\,(\mathbf{k}\boldsymbol{\gamma})}{k^2}\left[F(\omega/k) - 1 \right], \qquad (24)$$

with $k = |\mathbf{k}|$, $\omega_0^{\,2} = g_s^2 T^2 c_F/8$ and

$$F(x) = \frac{x}{2} \ln \frac{x+1}{x-1} \qquad (25)$$

(see [5]).

One can see immediately that $\mathcal{L}^{eff}$ of eq.(23) is nonlocal and singular at the point $\omega = k = 0$. The zeroes of (24) correspond to the modes $\omega_{\perp}(k)$ and $\omega_{\parallel}(k)$. As G^{-1} (and G) anticommute with γ^5 , the chirality is conserved.

A. The model.

Consider the two-dimensional zero-temperature model with the fermion Green function of an unusual form

$$-iG(p) = \frac{\hat{p}}{p^2 - m^2} \tag{26}$$

It differs from the standard massive fermion Green function by the absence of the term m in the numerator so that chirality is now conserved. From the other hand, the Green function (26) has the poles on the curve $p^2 = m^2$ corresponding to the massive particles. The Green function (26) corresponds to the following nonlinear lagrangian

$$\mathcal{L} = i\,\bar{\psi}\,\hat{\partial}\,(1 + m^2/\partial^2)\,\psi \tag{27}$$

In order to calculate the anomalous correlator $\langle T\{A_\mu V_\nu\}\rangle_0$ (in two dimensions it is a diangle rather than triangle), one should couple the free fermions of eq.(27) to the external vector and axial current. That can be done either by Nöther method or by covariantizing derivatives. There is some ambiguity in the procedure related to the presence of nonzero commutator $[D_\mu, D_\nu]$ but we do not want to discuss it here as the final result does not depend on the particular way of covariantizing. The most simple choice is

$$\mathcal{L}^{\psi, A_\mu} = i\,\bar{\psi}\left(\hat{D} + \frac{m^2}{\hat{D}}\right)\psi \tag{28}$$

The vector vertex has then the form

$$\Gamma_\mu = -\,i\left[\gamma_\mu + m^2\,\frac{1}{\hat{p}'}\,\gamma_\mu\,\frac{1}{\hat{p}}\right] \tag{29}$$

It satisfies, of course, the Ward identities following from the gauge invariance of the lagrangian (28) :

$$q_\mu \Gamma_\mu(p', p;\, q) = G^{-1}(p') - G^{-1}(p) \tag{30}$$

with G(p) of eq.(26). The lagrangian (28) implies also the existence of the generalized seagulls $\Gamma_{\mu\nu}$,$\Gamma_{\mu\nu\rho}$ etc. involving two and more gauge field legs.

The axial 3-point vertex Γ^5_μ and the axial-vector seagull $\Gamma^5_{\mu\nu}$ (which is important for the subsequent calculation) are chosen in the form: $\Gamma^5_\mu = \Gamma_\mu \gamma^5$, $\Gamma^5_{\mu\nu} = \Gamma_{\mu\nu}\gamma^5$.

This choice provides the conservation of the axial current on the lagrangian level , and this is after all the only thing we need.

Consider now the anomalous correlator $\langle T\{A_\mu V_\nu\}\rangle_0$. The two graphs depicted in Fig.2 contribute.

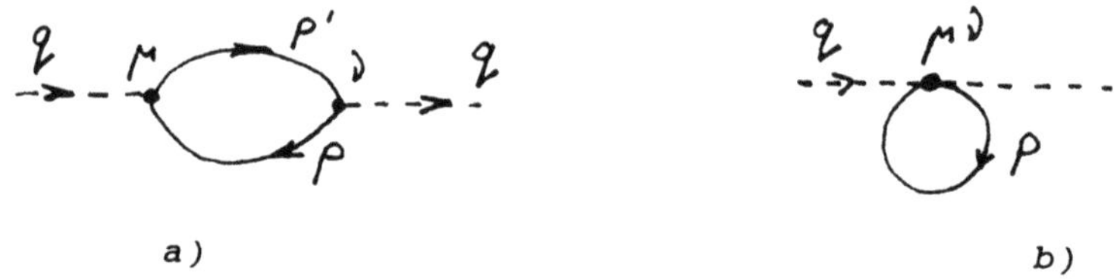

Fig.2. Contributions to anomalous correlator in nonlocal model

Performing the calculations, we arrive at the wonderful result: the expression for $T_{\mu\nu}$ is exactly the same as that in the massless Schwinger model (see eq.(7)). This, of course, can be anticipated in advance as the deep reason for the anomalous behaviour of correlators is the ultraviolet structure of the theory which is exactly the same here as in Schwinger model.

Technically, the reason for the pole in $T_{\mu\nu}$ to appear is the infrared singularities in the vector and axial vertices at $p^2 = 0$ and/or at $p'^2 = 0$. These singularities bring about the additional contribution to the imaginary part of $T_{\mu\nu}$ *besides* the normal contributions due to the massive poles of the propagators G(p) and G(p+q) in the graph in Fig.2a. The singularities in the vertices do the same job as the presence of the massless fermions did in the standard Schwinger model and provide for the pole in the axial-vector correlator $T_{\mu\nu}$ at $q^2 = 0$. Note that the singular behaviour of vertices is the immediate corollary of the singular behaviour of $G^{-1}(p)$ and the Ward identities. These singularities can also be traced back to the presence of special ghost states inherent for any nonlocal model [11].

6.Discussion.

What are the implications to be drawn from the analysis of different models performed in this paper for quark-gluon plasma - the physical system of interest ?

A general lesson is that the anomaly conditions of eq.(1) type are very "clever". In any particular physical system they find a particular nontrivial way to be satisfied. At zero temperature the necessary singularity of the anomalous correlator is provided by physical massless particles - massless fermions in the usual perturbation theory or massless goldstones in theories of QCD type involving spontaneous chiral symmetry breaking

In the thermal bath of free fermions, the anomaly is saturated mainly by the scattering processes. And, for nonlocal theories imitating the physical features of quark-gluon plasma, the anomaly is saturated due to singularities of vertices.

In guessing about the particular mechanism for saturating the anomaly in quark-gluon plasma, the analysis of the nonlocal model performed in the preceding section is, perhaps, of most immediate interest.

The real problem may be formulated as follows: In what cases is it possible to *change effectively the ground state of perturbation theory* and to consider instead of free particles the plasmon collective excitations as the zero-order approximation. We know that in some problems this trick works well - for example it allows one to calculate the damping of the physical transverse excitations at large momenta [12]. But in other cases (the anomaly problem is one of them) this procedure *is* dangerous and may lead to senseless results.

A more accurate analysis displays that whenever this kind of approach is successful (like in [12]) it is due essentially to the fact that the *corrections to vertices* are not essential in that particular problem in the kinematic region considered. Whereas in the region of small external momenta relevant for anomaly it is not true.

We have seen that in our nonlocal model. We cannot provide the quantitative results for the dressed vertices in quark-gluon plasma, but the fact that they should also be singular follows immediately from Ward identities and the singularities of the inverse thermal fermion propagator (24).

Unfortunately, the singularities of $G^{-1}(\omega,k)$ at real ω,k disappear if taking into account also damping effects which appear at higher orders. That means that, though corrections to vertices are large, they are not singular anymore. And that means that the anomalous correlator is also not singular at small real momenta.[3]

To conclude with, our guess is that, in the approximation where damping effects are ignored, anomaly is saturated due to singularities of vertices (perhaps, these singularities are physically related to the so called quasigoldstino mode [13]).

To give a nice physical interpretation for the mechanism of chirality change in the real physical situation (and, which is even more important - to describe in details the mechanism of baryon number change in sphaleron transitions with accounting for plasma effects) is still a challenge.

Anyway, the paradox does not look at present very sharp and dangerous. It is not quite clear by now how, precisely, the anomaly is saturated in hot plasma, but it is clear that the arguments based solely on collective excitation pattern are too naive, simplistic and unreliable.

I acknowledge the very useful discussions with R.Baier, B.L.Ioffe, V.Kiselev, V.V.Lebedev, L.McLerran, H.Leutwyler, and A.H.Mueller .

---

[3] One may note, however, that at $T \neq 0$ the anomaly condition does *not* imply the singularity of the correlator $T_{\mu,\alpha\beta}$ at $q^2 = 0$. The matter is that at, $T \neq 0$, the Lorentz invariance is lost and many new possible tensor structures may contribute in $T_{\mu,\alpha\beta}$ [7].

References

1. V.A.Kuzmin, V.A.Rubakov and M.E.Shaposhnikov, *Phys. Lett.*, **155B** (1985) 36.
2. V.A.Matveev et al., *Sov. Phys. Usp.* **31** (1988) 916 ; P.Arnold and L.McLerran, *Phys. Rev.* **D28** (1987) 581 ; J.Ambjorn et al., *Phys. Lett.* **244B** (1990) 479 ; A.V.Smilga, *Nucl. Phys.* **B335** (1990) 569.
3. A.Blair, N.Christ and J.Tang, *Phys. Rev. Lett.* **49** (1981) 1364 ; H.B.Nielsen and M.Ninomia, *Phys. Lett.* **130B** (1983) 389.
4. J. Ambjorn, J.Greensite and C.Peterson, *Nucl. Phys.* **B221** (1983) 381.
5. O.K.Kalashnikov, *Fort. Phys.* **32** (1984) 525, and references therein.
6. A.D.Dolgov and V.I.Zakharov, *Nucl. Phys.* **B27** (1971) 525.
7. H.Itoyama and A.H.Mueller, *Nucl. Phys.* **B218** (1983) 349.
8. N.P.Landsman and Ch.G. van Weert, *Phys. Repts.* **145** (1987) 141, and references therein.
9. R.Kobes, *Phys. Rev.* **D42** (1990) 562.
10. R.Baier and E.Pilon, *Zeit. Phys.* **C52** (1991) 339.
11. A.Pais and G.E.Uhlenbeck, *Phys. Rev.* **79** (1950) 145.
12. V.V.Lebedev and A.V.Smilga, *Phys. Lett.* **253B** (1991) 231 ; Bern Univ. Prepr. BUTP-90/38.
13. V.V.Lebedev and A.V.Smilga, *Ann. Phys.* **202** (1990) 229.

On the pion induced finite volume effects

V.L.Eletsky, M.I.Polikarpov
ITEP, B. Cheremushkinskaya 25, 117 259 Moscow, USSR

Abstract

Correlator of currents with quantum numbers of the a_1 meson is considered at finite temperature and volume. We argue that $\rho\pi$ states which couple to the currents modify behaviour of the correlator at large euclidean time and may be important in extracting a_1 mass from lattice calculations.

A correlator of currents with quantum numbers of hadrons is a useful object to provide information on masses and couplings of hadrons. The two standard approaches are QCD sum rules and lattice simulations of QCD. In the former case the correlators are estimated "theoretically" using the operator product expansion with condensates accounting for non-perturbative effects, while in the latter case the correlators are calculated numerically. In both approaches a fitting procedure is then applied based on the fact that the dominant contribution to the correlator at large imaginary time should come from the lowest intermediate state with quantum numbers of the currents in question. Strictly speaking this applies to the case of correlators defined as vacuum matrix elements of products of currents. However, the lattices used in simulations are of course finite in all four dimensions. This means that lattice simulations are in fact always done at finite temperature, which for a typical value of lattice spacing $a^{-1} = 2GeV$ and size in the temporal direction $N_t = 40$ is equal to $T = 1/aN_t = 50MeV$. At finite temperature the spectral density of the correlator differs from the zero temperature case so that it may not be dominated by the lowest intermediate state with quantum numbers of the currents at large t. This was demonstrated in ref.[1] where finite temperature correlators of vector and axial currents V^a_μ, A^a_μ

$$\begin{aligned} C^V_{\mu\nu}(x,T) &= \sum_n < n|TV^a_\mu(0)V^a_\nu(x)e^{(\Omega-H)/T}|n > \\ C^A_{\mu\nu}(x,T) &= \sum_n < n|TA^a_\mu(0)A^a_\nu(x)e^{(\Omega-H)/T}|n > \end{aligned} \quad (1)$$

($e^{-\Omega} = \sum_n < n|e^{-H/T}|n >$) were considered in an approach similar to QCD sum rules.

Using PCAC and current algebra it was shown that at $xT \ll 1$ finite temperature correlators are expressed through zero temperature ones:

$$\begin{aligned} C^V_{\mu\nu}(x,T) &= (1-\epsilon)C^V_{\mu\nu}(x,0) + \epsilon C^A_{\mu\nu}(x,0) \\ C^A_{\mu\nu}(x,T) &= (1-\epsilon)C^A_{\mu\nu}(x,0) + \epsilon C^V_{\mu\nu}(x,0) \end{aligned} \quad (2)$$

The mixing parameter ϵ is given by phase space integral for thermal pions

$$\epsilon = \frac{4}{F_\pi^2}\int \frac{d^3k}{(2\pi)^3 2E}\frac{1}{e^{E/T}-1} \quad (3)$$

where $E = (m_\pi^2 + k^2)^{1/2}$ and $F_\pi = 93MeV$ is the pion decay constant. ϵ is an expansion parameter in powers of the pion density and it is assumed that $\epsilon \ll 1$. Low temperature expansions of this type were introduced by Gasser and Leutwyler[2]. In the chiral limit $\epsilon = T^2/6F_\pi^2$. For massive pions at $T \ll m_\pi$, $\epsilon = \frac{1}{F_\pi^2}\left(\frac{m_\pi T^3}{2\pi^3}\right)^{1/2} exp(-m_\pi/T)$ (corrections to PCAC of order $m_\pi T/M_h^2$, where $M_h \sim m_\rho$ is a typical hadronic mass scale, are neglected in the preexponential factor).

It was demonstrated in[1] that if one naively tries to saturate each of these correlators by a single low lying resonance with temperature dependent mass and coupling supplemented by a continuum model of higher states, then m_{a_1} rapidly decreases with the rise of T, while m_ρ remains almost constant. However, the physical masses should not change in the leading order in ϵ by Adler theorem. What happens is that the small admixture of ρ-dominated vector channel in $C^A(T)$ becomes important at intermediate momenta Q^2, reducing the effective mass of a_1 obtained from behaviour of $C^A(T)$ in this region. There is obviously no such effect in the vector channel, since a_1 is considerably heavier than ρ. If ρ and a_1 mesons were degenerate, the effect would be absent. Note, that the mixing of correlators at $T \neq 0$ is due to a possibility for the axial (vector) current to absorb a thermal pion and to emit a vector (axial) intermediate state.

Formally this mixing just comes from one-loop diagram with ρ and π coupling to axial currents at finite temperature. It is well known[3,4] that at $T > 0$ the spectral density of this diagram receives a contribution corresponding to a new cut which appears in the complex energy plane and runs from $-(m_\rho - m_\pi)$ to $+(m_\rho - m_\pi)$ (in the c.m. reference frame). Spectral density accosiated with this cut is called scattering spectral density and vanishes at $T = 0$ as $exp(-m_\pi/T)$. It corresponds to the process in which a pion from the heat bath is absorbed and a ρ is produced (at $E > 0$) by the axial current. The usual spectral density corresponding to production of the two particles also gets additional contribution of order $exp(-m_\pi/T)$. In euclidean formulation these finite T contributions correspond to pions in the loop going around the box of temporal size $1/T$ forward (production) and backwards (scattering) in euclidean time. It is clear that due to lightness of pion these two-particle contributions may be important at large imaginary time and provide an additional finite size effect in lattice calculations of correlators with *dynamical* quarks. This effect is purely kinematical if the masses of ρ and π are considered as fixed. However, they also may obtain dynamical corrections due to interactions in the box.

Dynamical finite size corrections to particle masses were first discussed by Lüscher in [5]. He showed that the mass shift of a *stable* particle is related to the forward elastic scattering amplitudes involving light particles in the theory and is of order $exp(-mL)$, where m is the mass of the lightest particle, i.e. pion in case of QCD. Finite size effects induced by Goldstone bosons were considered in chiral perturbation theory in refs.[6,7]. For *unstable* particles the situation is more complicated. E.g. it was shown in two recent papers[8,9] that energy spectrum in the ρ sector in a finite box below two-particle threshold is non-trivial due to avoided level crossing and depends strongly on the *spatial* size of the box. Thus, determination of mass in lattice calculations with unstable particles may be a complicated problem and we do not discuss it here.

Let us consider a correlator of two axial currents in a $L^3 \times 2L$ box at points $(0, \vec{0})$ and $(t, \vec{0})$ and take into account contributions of a_1 and

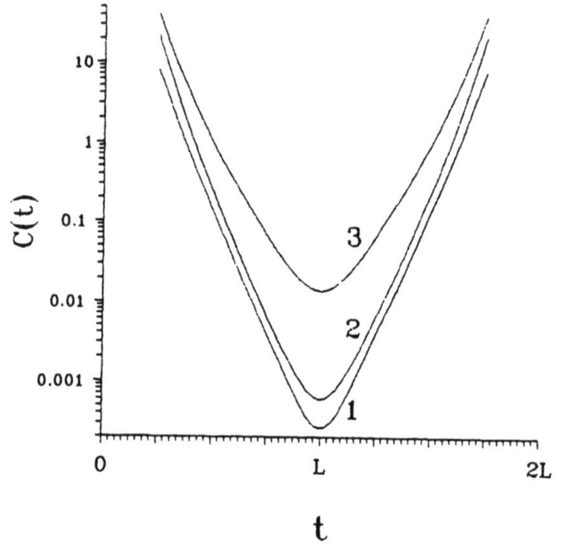

Figure 1: Correlator of two axial currents in a $L^3 \times 2L$ box, $L = 10GeV^{-1}$, $m_{a_1} = 1.26GeV$, $m_\rho = 0.77GeV$. 1: pure a_1 contribution; 2: $\rho\pi$ states taken into account, $m_\pi = 0.44GeV$ (modern lattices); 3: same as 2 with $m_\pi = 0.14GeV$ (physical mass)

two-particle $\rho\pi$ state to the structure $\sim \delta_{\mu\nu}$. Then we have

$$C(t) \sim D_{m_{a_1}}(t) + \frac{1}{8\pi^2(F_\pi L)^2} D_{m_\rho}(t) D_{m_\pi}(t) \tag{4}$$

PCAC was used here to estimate the coupling of $\rho\pi$ state to axial current and the couplings of a_1 and ρ to axial and vector currents are taken to be equal which is a good approximation[1]. Finite volume propagator $D_m(t)$ is

$$D_m(x_i) = \sum_{n_i=-\infty}^{\infty} D_m^{(0)}(x_i + n_i L_i) \tag{5}$$

where $D_m^{(0)}(x_i)$ is the usual infinite volume propagator of a scalar particle with mass m. Periodic boundary conditions are implied here and the integers n_i count windings around the box. Using the integral representation

$$D_m^{(0)}(x_i) = \frac{1}{16\pi^2}\int_0^{\infty}\frac{d\alpha}{\alpha^2}exp\left(-\alpha m_\pi^2 - \frac{x^2}{4\alpha}\right) \tag{6}$$

we computed $C(t)$ for a typical lattice size $L = 10GeV^{-1}$, at physical values $m_{a_1} = 1.26GeV$, $m_\rho = 0.77GeV$ and two values of pion mass, $m_\pi = 0.44GeV$ (typical for modern lattices) and $m_\pi = 0.14GeV$ (physical mass). The result given in Figure shows that on modern lattices the effect is noticable only close to the point $t = L$ and may affect a $cosh(L-t)m_{a_1}$ fit. However, in case of physical pion the effect is large for all t and should be taken into account in exponential fits in future calculations. with light *dynamical* quarks on large lattices.

Finally, we note that two-particle states containing a pion, such as πN, $\pi\Delta$, $\pi\Lambda$ may also be important in case of baryonic correlators in a finite box.

We are grateful to S. Gottlieb and G. Martinelli for useful discussions.

References

[1] M. Dey, V.L. Eletsky and B.L. Ioffe, Phys.Lett. 252B (1990) 620.

[2] J. Gasser and H. Leutwyler, Phys.Lett. 184B (1987) 83; Nucl.Phys. B307 (1988) 763.

[3] A. Weldon, Phys.Rev. D28 (1983) 2007.

[4] R.L. Kobes and G. Semenoff, Nucl.Phys. B260 (1985) 714.

[5] M. Lüscher, in *Progress in Gauge Field Theory*, G. 't Hooft et al, eds., (Plenum, New York, 1984); Comm. Math. Phys. 104 (1986) 227.

[6] P. Hasenfratz and H. Leutwyler, Nucl.Phys. B343 (1990) 241.

[7] F.C. Hansen and H. Leutwyler, Nucl.Phys. B350 (1991) 201.

[8] T.A. DeGrand, Phys. Rev. D43 (1990) 2296.

[9] M. Lüscher, Signatures of Unstable Particles in Finite Volume, Preprint DESY 91-052, May 1991.

SKYRMIONS ON THE HYPERSPHERE

ANDREAS WIRZBA
NORDITA, Blegdamsvej 17
DK-2100 Copenhagen Ø, Denmark

ABSTRACT

When skyrmion-configurations realized in infinite periodic arrays or on a compact 3-(hyper)sphere are compressed to high densities, one encounters transitions from well-localized baryon solutions to solutions which can be characterized by their delocalization with respect to energy and baryon density, by increased symmetries and by their tendency to restore chiral invariance. The solutions on the hypersphere show the restoration of a continuous symmetry at high densities, with all the signatures of chiral symmetry restoration, such as parity doubling of the fluctuation spectrum and the vanishing of the three massless Goldstone modes from the spectrum.

We show by shrinking the volume for one skyrmion localized on a 3-sphere that such phase transitions are common to all lagrangians which involve soliton-solutions of the skyrmion-type and which are manifestly chiral invariant: The phase transitions persist when the stabilizing fourth order term of the usual Skyrme model is replaced by any polynomial of second-order, fourth-order and/or sixth-order terms which leads to stabilization and by vector-meson terms which include ρ and ω mesons, etc. The high-density behavior of skyrmions stabilized by of the vector-meson terms is the same as the one resulting from higher-order contact terms. The vector-meson profile functions become identical to their corresponding pionic currents.

1. Introduction

The Skyrme model [1] and its variants [2] are effective models of strong interaction hadron physics which are formulated as field theories with meson degrees of freedom and which allow for non-trivial topological excitations which are identified as baryons. Built on input from low-energy $\pi\pi$ scattering, they give a qualitatively successful description of the structure of baryons [3, 4] (nucleon and Δ static properties with about 30% accuracy, baryon resonances and meson-baryon scattering [5, 4]) *and* the interactions between the baryons (asymptotic one-pion exchange, short-range repulsion, nucleon-nucleon channels) [6, 4]. There is no difference in principle between the stabilization and the interaction physics, the models are just applied in different topological sectors: in the baryon (=winding) number $B = 1$ sector for the structure physics, and in the $B > 1$ sector for the interaction physics. In its simplest form, the Skyrme model generates the energy density of a given configuration just by its field gradients. The task for finding energetically favorable configurations is therefore to reduce the field gradients as far as possible and as allowed by the topology. Thus the minimization of the energy of the skyrmion configurations is in fact a minimization of the "frustrations" in the corresponding field gradients. Since the baryon structure and interaction are treated on the same footing, such models have the built-in feature to lead to radical changes when baryonic matter is compressed to high densities, *e.g.* the baryon matter which at low densities consists of well-separated clearly defined baryons might become more or less uniform at high densities making it impossible to isolate single baryons any longer. There are at least two complementary ways to study these phenomena.

The first approach involves numerical studies of crystalline arrays of skyrmions in the usual (3+1)-dimensional space-time [7–11]. There one tries to find classical (static) field configurations which minimize the energy per cell volume of the array for a given fixed baryon charge per cell. The non-reduced equations of motion (which are nonlinear partial differential equations) are solved numerically on a lattice grid. The only constraints are the form of the crystal itself (*e.g.* simple cubic, face-centered cubic, hexagonal etc.) and the various 'twisted' periodic boundary conditions [7, 9] between the array cells which are there to guarantee the periodic baryonic structure and to reduce the frustrations in the field gradients between the skyrmions in neighboring cells as far as possible. At low densities each skyrmion is well localized in its cell and interacts only via its 'tail' under the influence of the twisted boundary conditions with the neighboring skyrmions. The periodic array is in a phase of rather weakly interacting baryons. With increased density, however, the skyrmions grow in size as measured by their baryonic *rms*-radius until at a critical density a transition to a distinct high-density phase is encountered, where it is impossible to identify single baryons. The signals of this phase are: (a) the approximate uniformity of the baryonic as well as the energy density, (b) an additional symmetry[#1], the "half-skyrmion symmetry" [12], and (c) – as a consequence of this new symmetry – the vanishing of the unit cell averages of the σ and $\vec{\pi}$ fields.

These calculations are of numerical nature and furthermore involve ad hoc assumptions about the choice of the crystalline arrangement and about the form of the twisted boundary conditions. For instance, the order of the phase transition can depend on these choices [8, 9]. (The phase transition itself, however, has been found in all studied situations, furthermore the critical densities turned out to be approx-

[#1]The fields reveal families of planes on which the scalar field, σ, is zero. The pionic fields are reflection symmetric about these planes while the scalar field is reflection antisymmetric.

imately the same.) Fortunately, Nick Manton [13, 14] found a complementary way to study the same phenomena by replacing the periodic arrays in flat space $\mathbf{R}^3$ by few-skyrmion systems or even a single skyrmion on the compact manifold $\mathbf{S}^3(L)$. The idea is – by shrinking the hypersphere radius and thus making use of the curvature and finite size effects of the hypersphere – to increase the interaction mechanism between skyrmions relative to the stabilization mechanism. A skyrmion on a large hypersphere has basically the same properties as an isolated skyrmion in flat space or in a large cell of an infinite skyrmion array. By decreasing the hypersphere radius one forces the skyrmions to readjust itself because of the curvature and the finite size effects. The single skyrmion acts now similarly to a skyrmion in a decreasing cell of an infinite array in flat space which "feels" the presence of the neighboring skyrmions. Finally, the hypersphere radius can become small enough, such that the interaction mechanism becomes comparable in size with the stabilization mechanism smearing (=delocalizing) the skyrmion completely over the whole hypersphere.

Both approaches can be linked via the identification of the averaged baryon density in flat space with $B/(2\pi^2 L_{\rm ph}^3)$ on the hypersphere [15, 16], where B is the baryon charge and $L_{\rm ph}$ the (dimensioned) radius of the hypersphere. The $\mathbf{S}^3(L)$ approach has the advantage that it is technically far simpler and that it allows for mathematical rigorous results for the "high-density phase" of a single baryon on the hypersphere. The disadvantage is obvious: The physical space is not a 3-sphere. Both approaches are therefore complementary: The first is realized in normal space-time and for infinite systems (a precondition for a phase transition), but is technically complicated and plagued by the ambiguities in choosing the array and the boundary conditions; the other is simple and even analytic for the the "high-density regime", but it is set up in an unphysical space and limited to finite systems which correspond to just one cell of the infinite array. The results of the $\mathbf{S}^3(L)$ calculations should therefore not be identified with the total array in $\mathbf{R}^3$, but just with *one* array cell.

In this paper, we will argue that the phenomena of phase transitions (described above) are not limited to the Skyrme model itself (which is the sum of the non-linear σ model and the stabilizing fourth-order Skyrme term) but are a property of any Skyrme-type model as long as the lagrangian is explicitly chirally invariant and guarantees the stabilization of a skyrmion. We will show in complete generality that the same phenomena occur when the fourth-order Skyrme term is replaced by a sixth-order term or even any polynomial of second, fourth- and sixth-order terms which stabilizes the soliton. The position where these phase transitions will occur (*i.e.* the critical density) will of course dependent on the stabilization mechanism. Furthermore, we will show – by specializing to the so-called hedgehog ansatz – that phase transitions still occur when the contact-like fourth- or sixth-order stabilizing terms are replaced by stabilizing terms involving vector-mesons of finite range, *e.g.*

ω and/or ρ meson terms.

The paper is organized as follows. In sect. 2, we review some of the results for the $B = 1$ sector of the usual Skyrme model on $\mathbf{S^3}(L)$. In sect. 3, we argue that the phase transitions should be interpreted as chiral symmetry restoration. Generalizations to extended Skyrme models where the fourth order term is replaced by a sixth order term or any polynomial of positive integer powers of second, fourth and/or sixth order terms are listed in sect. 4. In this part we also investigate ω and ρ meson stabilized lagrangians which are based on the hidden gauge construction. In sect. 5 we discuss the order of the phase transition under the incorporation of the trace anomaly. Sect. 6 contains the discussion and final remarks.

2. The Skyrme Model on a 3-sphere

In this chapter we briefly report on the results of ref. [14] and ref. [15] about the normal Skyrme model

$$\mathcal{L}_{2,4} = \frac{f_\pi^{\,2}}{4}\mathrm{Tr}\left(\partial_\mu U^\dagger \partial^\mu U\right) + \frac{\epsilon_4^{\,2}}{4}\mathrm{Tr}\left[U^\dagger\partial_\mu U, U^\dagger\partial_\nu U\right]^2 , \tag{1}$$

on the 3-sphere $\mathbf{S^3}(L)$ with hypersphere radius L.

Hedgehog configurations are conveniently described in terms of the conventional 'polar' coordinates μ, θ, ϕ on $\mathbf{S^3}(L)$ with $0 \le \mu, \theta \le \pi$ and $0 \le \phi \le 2\pi$, so that a typical point on $\mathbf{S^3}(L)$ has Cartesian coordinates

$$(L\cos\mu,\ L\sin\mu\cos\theta,\ L\sin\mu\sin\theta\cos\phi,\ L\sin\mu\sin\theta\sin\phi). \tag{2}$$

In these coordinates the metric is $ds^2 = L^2(d\mu^2 + \sin^2\mu\, d\theta^2 + \sin^2\mu\ \sin^2\theta\, d\phi^2)$ and the volume element is

$$dV = L^3\sin^2\mu\, d\mu\ \sin\theta\, d\theta\, d\phi\ . \tag{3}$$

The field components of a hedgehog of baryon number B on $\mathbf{S^3}(L)$ have then the form

$$\begin{aligned}\Phi^0 &= \cos f(\mu)\\ \Phi^z &= \sin f(\mu)\cos\theta\\ \Phi^x &= \sin f(\mu)\sin\theta\cos\phi\\ \Phi^y &= \sin f(\mu)\sin\theta\sin\phi\quad ,\end{aligned} \tag{4}$$

where the "radial" profile function $f(\mu)$ is subject to the boundary condition $f(0) = 0$ and $f(\pi) = B\pi$, B integer. The topological winding-number B gives the baryon charge of the corresponding field configuration. In Eq.(4) we used a cartesian representation $\{\Phi^\alpha\} = (\Phi^0,\ \Phi^z,\ \Phi^x,\ \Phi^y)$ with the imposed constraint $\Phi^\alpha\Phi^\alpha = 1$. The field components, $f_\pi \Phi^\alpha$ (f_π is the pion-decay constant) should be identified with the familiar σ, π^z, π^x, π^y fields. The relation to the usual quaternion representation of the Skyrme model is given by

$$\begin{aligned} U &= \Phi^0 + i\vec{\tau}\cdot\vec{\Phi} \\ &= \exp\left(i\tau\cdot\hat{r}(\theta,\phi)f(\mu)\right) \qquad \text{for the hedgehog} \end{aligned} \tag{5}$$

where $\hat{r}(\theta,\phi)$ is the usual radial $\mathbf{S}^2$ unit vector.

The energy of the hedgehog field configuration (5) on $\mathbf{S}^3(L)$ is #2

$$\begin{aligned} E &= 4\pi L \int_0^\pi \sin^2\mu d\mu \left(f'^2 + 2\frac{\sin^2 f}{\sin^2\mu}\right) \\ &+ 4\pi\frac{1}{L}\int_0^\pi \sin^2\mu d\mu \left(\frac{\sin^2 f}{\sin^2\mu}\left[2f'^2 + \frac{\sin^2 f}{\sin^2\mu}\right]\right), \end{aligned} \tag{6}$$

where f' stands for $\frac{df}{d\mu}$. Variation of (6) with respect to the radial profile function $f(\mu)$ leads us to the Euler equation

$$\begin{aligned} &\frac{d^2 f}{d\mu^2} + \frac{\sin 2\mu}{\sin^2\mu}\frac{df}{d\mu} - \frac{\sin 2f}{\sin^2\mu} \\ &+\frac{2}{L^2}\frac{\sin^2 f}{\sin^2\mu}\frac{d^2 f}{d\mu^2} + \frac{\sin 2f}{L^2\sin^2\mu}\left\{\left(\frac{df}{d\mu}\right)^2 - \frac{\sin^2 f}{\sin^2\mu}\right\} = 0\,. \end{aligned} \tag{7}$$

Because of the general principle concerning reduced variational equations for symmetric fields [17], solutions of (7) yield solutions of the full Euler-Lagrange equations of the Skyrme model on $\mathbf{S}^3(L)$ when substituted into (4) or (5). It is very simple to check that one solution of Eq.(7) (in the case $B = 1$) is the identity map $f(\mu) = \mu$ which corresponds to the uniform mapping of $\mathbf{S}^3(L)$ onto the isospin manifold $\mathbf{S}^3(1) \cong SU(2)$. This is a solution for all L, but has no finite $L \to \infty$ limit. The energy associated with the identity map is

$$\frac{E}{12\pi^2} = \frac{1}{2}\left(L + \frac{1}{L}\right). \tag{8}$$

The topological lower bound, $E \geq 12\pi^2 B$, is saturated when $L = 1$. For $L = 1$, the uniform solution is therefore the absolute minimum of the $B = 1$ sector. The same is true for $L < 1$ where there exist even the more stringent lower bound (8) [14, 18]. The identity map is not stable for all L, however. Whereas for $1 < L < \sqrt{2}$, the uniform map is still at least a local minimum, for $L > \sqrt{2}$ there exist a lower energy configuration which represents a skyrmion localized about one point, *i.e.* $f(\mu) \neq \mu$ in the hedgehog parameterization (4). In fact in this parameterization the identity

#2We have adopted a dimensionless form of the Skyrme lagrangian. To obtain the dimensioned quantities, one multiplies L in (6) by $2\sqrt{2}\epsilon_4/f_\pi$ and E by $\sqrt{2}\epsilon_4 f_\pi$.

map bifurcates into two solutions of the same energy concentrated around the north or south pole of the hypersphere which are related by [15]

$$f_N(\mu) = \pi - f_S(\pi - \mu). \tag{9}$$

In the limit $L \to \infty$, one recovers the usual flat space skyrmion [15]. The identity map has O(4) symmetry (*see* Eq.(4)) and is therefore completely uniform in energy and baryon density. For $L > \sqrt{2}$ the symmetry of the energetically most favorable solution is broken to O(3) indicating that the solution is localized in energy and baryon distribution. The bifurcation at $L = \sqrt{2}$ is of the standard "pitchfork" type corresponding to a second order phase transition. In Landau's scheme the bifurcation is characterized by a symmetric quartic polynomial whose quadratic term changes sign.

In refs. [14, 18] the stability of the identity map was analyzed in complete generality without referring to a particular parameterization of the skyrmion. For this purpose it is useful to introduce the strain tensor

$$\begin{aligned} M_{ij} &\equiv \partial_i\Phi^a\partial_j\Phi^a \qquad (\text{with} \quad \{\Phi^\alpha\} = (\sigma, \pi^z, \pi^x, \pi^y)) \\ &= -\tfrac{1}{4}\mathrm{Tr}\{L_i, L_j\}_+, \end{aligned} \tag{10}$$

where i and j label the space coordinates. This is a symmetric 3×3 matrix with three positive eigenvalues which we will denote as λ_a^2, λ_b^2, λ_c^2. Note that for the hedgehog configuration (in spherical coordinates) the eigenvalues of the strain tensor are simply $\lambda_a^2 = f'^2$, $\lambda_b^2 = \lambda_c^2 = \sin^2 f/r^2$ in $\mathbf{R^3}$ and $\lambda_a^2 = f'^2/L^2$, $\lambda_b^2 = \lambda_c^2 = \sin^2 f/(L^2\sin^2\mu)$ on $\mathbf{S^3}(L)$. For the special case of the uniform map ($f(\mu) = \mu$) on $\mathbf{S^3}(L)$ all three eigenvalues are even equal: $\lambda_a^2 = \lambda_b^2 = \lambda_c^2 = 1/L^2$. Manton [14] has shown that the λ_i have a simple geometrical interpretation. They correspond to the length changes of the images of any orthonormal system in the space manifold (*e.g.*, $\mathbf{R^3}$ or $\mathbf{S^3}(L)$) under the conformal map, U, onto the group manifold $SU(2) \cong S_3$. Among the invariants of M_{ij}, three are fundamental[#3] and have a simple geometric meaning:

$$\begin{aligned} \mathrm{Tr}(M) &= \lambda_a^2 + \lambda_b^2 + \lambda_c^2 &&= \textstyle\sum \text{length}^2 \\ \tfrac{1}{2}\left\{(\mathrm{Tr}(M))^2 - \mathrm{Tr}(M^2)\right\} &= \lambda_a^2\lambda_b^2 + \lambda_b^2\lambda_c^2 + \lambda_c^2\lambda_a^2 &&= \textstyle\sum \text{area}^2 \\ \det(M) &= \lambda_a^2\lambda_b^2\lambda_c^2 &&= \text{volume}^2\ . \end{aligned} \tag{11}$$

With the help of Eq.(10) it is easy to see that these invariants are (modulo normalization factors) the static energy densities of the second-order non-linear σ model, the fourth-order Skyrme term (*see* Eq.(1) for both), and the sixth-order term to be introduced in Eq.(35) in section 4. The static energy of a general skyrmion configuration

[#3]All other invariants can be constructed from them.

on the hypersphere has in this language the form

$$\begin{aligned} E &= \int_{S^3(L)} dV\,(\lambda_a^2+\lambda_b^2+\lambda_c^2+\lambda_b^2\lambda_c^2+\lambda_c^2\lambda_a^2+\lambda_a^2\lambda_b^2) \\ &= \int_{S^3(L)} dV\,(\lambda_a-\lambda_b\lambda_c)^2+(\text{cycl. perm.'s})+6\int_{S^3(L)} dV\,\lambda_a\lambda_b\lambda_c \qquad (12) \end{aligned}$$

where the last last term is just $12\pi^2 B$ since $\lambda_a\lambda_b\lambda_c/(12\pi^2)$ is the Jacobian of the map $\mathbf{S^3}(L)\rightarrow\mathbf{S^3}(1)\cong SU(2)$. From the form (12) its is obvious that the identity map $\lambda_a = \lambda_b = \lambda_c = 1/L$ satisfies the topological lower bound for a $B=1$ skyrmion at $L=1$ independent of the parameterization of the skyrmion. Secondly, it is now simple to show in full generality [14, 18] that for $L<\sqrt{2}$ the identity map is stable for all allowed ($\delta B = 0$) small-amplitude perturbations $\lambda_i \rightarrow (1/L)+\delta\lambda_i$, whereas it becomes a saddle for $L>\sqrt{2}$.

In order to discriminate the delocalized (high-density) phase from the localized (low-density) phase in a way which can be generalized to few-skyrmion systems on the hypersphere or to flat space array-calculations, different order parameters were introduced in ref. [15]: One suggestion is the integrated squared deviation of the local energy/baryon density from the averaged density normalized by the squared averaged density times the hypersphere volume. It converges to the value 1 for the localized solution in the limit $L\rightarrow\infty$, for smaller L it decreases and shows a square-root bifurcation at $L=\sqrt{2}$ to the value zero belonging to the identity map. The interpretation would be that this order parameter signalled deconfinement in the energy and baryon distribution. This is, however, a special property of the $B=1$ system on $\mathbf{S^3}(L)$. For few-skyrmion systems [15, 19] and flat space arrays [9, 11] this parameter never becomes zero in the high-density phase, although there is still a sizable decrease in its value from the low-density to the high-density phase. The second suggestion of ref. [15] on the other hand works also in these cases. It is the squared chiral expectation value

$$\langle\sigma/f_\pi\rangle^2+\langle\vec{\pi}/f_\pi\rangle^2. \qquad (13)$$

where the σ and pion fields are averaged over the hypersphere volume (or the cell volume for periodic arrays). Because of the residual $O(3)$ symmetry for the hedgehog (or discrete symmetries in the array calculations) which cause $\langle\vec{\pi}\rangle$ to be zero it has the same content as the parameter $\langle\sigma\rangle$ alone. For the identity map it is obvious (*see* Eq.(4)) that the parameter (13) vanishes. In the case of the localized solution it approaches the value 1 with increasing L, since in most of the space the U field of the localized skyrmion is approximately close to the vacuum value $U_0 = 1$ and only markedly deviates from this in the small region where the skyrmion is located. The interpretation is that a zero in this order parameter signals chiral symmetry restoration or at least "chiral democracy", since there is no escape from the "chiral-circle" constraint $\sigma^2+\vec{\pi}^2=f_\pi^2$. But nevertheless as shown for few-skyrmion systems on the hypersphere [15, 19] and for flat space arrays [9, 11] the σ and pion profiles of the high-density phase are so equally distributed about the chiral circle that their averages vanish.

Let me summarize the $B = 1$ results on $\mathbf{S^3}(L)$: There is a second-order transition at a baryon density $1/(2\pi^2 L_c{}^3)$ for $L_c = \sqrt{2}$ from a localized and chirally broken low-density solution to a "chirally restored" high-density solution. Restoring dimensionful parameters into the Skyrme lagrangian (by using the empirical value of the pion decay constant $f_\pi = 93$ MeV and an $\epsilon_4 = 0.0743$ guaranteeing a reasonable value for g_A) one finds

$$\rho_c = \left(\frac{f_\pi}{2\sqrt{2}\epsilon_4}\right)^3 \frac{1}{\sqrt{2}^3} \frac{1}{2\pi^2} \approx 0.20\text{fm}^{-3}, \tag{14}$$

which is far too low[#4] (remember that the nuclear matter density is $\rho_{\text{nm}} = 0.16\text{fm}^{-3}$). The important result, however, is that the $\mathbf{S^3}(L)$ value for ρ_c is more or less the same than the one found in the periodic array calculations [8, 9, 11] under the same input parameters, approximately $\rho_c = 0.17\text{fm}^{-3}$. Furthermore, the corresponding transition densities ρ_c in the few-skyrmion calculation [15, 19] interpolate between these two numbers. Thus the identification of the hypersphere formalism with the periodic array calculation seems to be justified even quantitatively.

3. Indications for a Chiral Symmetry Restoration

In this section I will list the arguments (*see* refs. [21, 22] for more details) why the transition reviewed in the preceeding section should be interpreted as chiral symmetry restoration. As mentioned the complete uniformity of the high-density $B = 1$ solution on $\mathbf{S^3}(L)$ is linked to the strong condition $f(\mu) = \mu$ and does not generalize to few-skyrmion systems [15, 19] on $\mathbf{S^3}(L)$ or periodic arrays [8, 9, 11] in flat space. However, the "chiral democracy" as discussed at the end of the preceeding section can even be achieved by the weaker condition

$$f(\mu) = \pi - f(\pi - \mu) \tag{15}$$

which indicates that the high-density phase has a symmetry about the hypersphere equator between the "northern" and the "southern" hemisphere. In fact this is the half-skyrmion symmetry [12] which is a common feature of the high-density phase for few-skyrmion systems on the hypersphere as well as for periodic arrays in flat space. So the chiral symmetry restoration seems to be prior to the delocalization. In

---

[#4] This might improve if the stabilization term is replaced by more realistic ones. Furthermore all contributions from kinetic terms are neglected. Finally, there is the problem of the missing central attraction [6] which is probably linked to loop corrections [20]. Any additional attractive term tends to localize the skyrmion and will therefore move the transition point up to higher densities.

fact this can be tested by adding a (pion mass) term which explicitly breaks chiral symmetry [21]:

$$\mathcal{L}_{\mathrm{SB}} = \frac{m_\pi{}^2 f_\pi^2}{4} \mathrm{Tr}(U + U^\dagger - 2). \tag{16}$$

When this term is added, the phase transition is not sharp any longer, but smeared out. In fact the chiral order parameter (13) still approaches zero for higher and higher densities, but never actually touches it as it is obvious from the explicit breaking of the chiral symmetry.

In ref. [21] it is furthermore argued that the vanishing of the order parameter $\langle\sigma\rangle$ (for the Skyrme model without explicit symmetry breaking of course) implies the vanishing of the matrix element for pion decay $\langle 0|A_\mu^a(x)|\pi^b\rangle$ in case the axial current A_μ^a is calculated in the mean-field approximation in the frame-work of a σ-model, and therefore the vanishing of the "effective" pion decay constant and the quark condensate.

The main arguments why the in section 2 discussed transition for the $B = 1$ hedgehog should be identified with a chiral symmetry restoration are the following two:

(a) The transition from the low-density phase to the high-density phase is accompanied by a parity doubling in the fluctuation spectrum

(b) and by the vanishing of the three pionic Goldstone modes from the spectrum.

These two points are treated at length in ref. [22] where the local stability of the uniform and the localized skyrmion solution on $\mathbf{S}^3(L)$ and thus their small-amplitude normal modes are investigated. If all normal modes have positive energy, the solution is locally stable. If some normal modes have negative energy, the solution is a saddle point. The study of the small-amplitude fluctuations about the $B = 1$ solutions on the hypersphere revealed the following scenario:

1. In general, the modes about a hedgehog solution have just the usual $O(3)$ degeneracies, *i.e.* a degeneracy $2N + 1$ in terms of the principal quantum number $N \geq 1$ of the modes.

2. The modes about the identity map on the other hand can be classified according to the following representations of the group $O(4)$: the symmetric tensor representation $(N, 0)$ which has $(N + 1)^2$ degenerate states and as static eigenvalues

$$\lambda_{N,0} = L\{N(N + 2) - 4\} + 2\{N(N + 2) - 2\}/L, \tag{17}$$

and the $(N, 1)$ representation with a $2((N+1)^2-1)$ degeneracy and the static eigenvalues

$$\lambda_{N,1} = \{N(N+2)-3\}(L+1/L). \tag{18}$$

Thus their degeneracy exceeds by far the one of the modes about localized hedgehog solutions. Furthermore for each mode about the identity map with a given parity, there exists at least one degenerate mode of opposite parity. In other words there is *parity doubling* in the fluctuation spectrum. There are six zero modes, the $(N=1,1)$ modes of $O(4)$.

3. At high densities ($L < \sqrt{2}$) all normal modes about the identity map have positive energies, the identity map is stable. There are four degenerate low-lying modes in the spectrum, one with σ quantum numbers and three with pion ones indicating that the σ and all three π-fields are treated on the same footing. They have the static eigenvalue

$$\lambda_{N=1,0} = -L + 2/L. \tag{19}$$

4. The four $(N = 1, 0)$ modes cross zero at $L = \sqrt{2}$ and become negative for $L > \sqrt{2}$ signalling there the instability of the identity map.

5. At $L = \sqrt{2}$ there is the bifurcation into the normal localized hedgehog solution with the reduced $O(3)$ symmetry. All modes about the localized solution are stable for $L \geq \sqrt{2}$. The modes with $N \geq 1$ have (with the exception of some accidental degeneracies for special values of L) only the usual $O(3)$ degeneracy $2N+1$. However, there exist 9 zero modes[#5]. The three additional zero modes (compared with the number of zero modes for the identity map) correspond to the three unstable pionic $(N = 1, 0)$ modes of the identity map. Their fourth partner, the σ-like mode, corresponds to a positive stable excitation about the localized hedgehog. The breakdown of the continuous symmetry $O(4)$ of the identity map at $L = \sqrt{2}$ to the $O(3)$ symmetry of the localized for $L > \sqrt{2}$ energetically favorable hedgehog solution is therefore closely linked to the appearance of three new zero modes.

6. These three zero modes should be in fact interpreted as the three pionic Goldstone modes linked to the spontaneous breakdown of chiral symmetry. The arguments for this identification are the following: Whereas six of the nine zero modes stay normalizable even in the limit $L \to \infty$, three have no finite $L \to \infty$ limit. This is consistent with the fact that the flat space hedgehog has only six zero modes. Furthermore when the chiral symmetry breaking term (16) is added to the Skyrme lagrangian, the above mentioned normalizable zero modes

[#5] Note that a hedgehog in flat space has only 6 zero modes: 3 rotational (which cannot be separated from iso-rotational ones because of the hedgehog symmetry) and 3 translation ones.

stay zero modes, whereas the three additional zero modes are shifted in energy by the pion mass. These are the three pionic Goldstone modes which become massive when an explicitly chiral symmetry breaking term is added and which show up as non-normalizable plain wave excitations of the vacuum in the flat space limit.

Since the three Goldstone modes disappear for the high-density delocalized phase, the chiral symmetry restoration is established.

4. Generalization

4.1. Contact Terms

Is the form of the Skyrme model (1), especially the stabilization by the fourth-order Skyrme term

$$\mathcal{L}_4 = \frac{\epsilon_4^2}{4}\mathrm{Tr}[U^\dagger\partial_\mu U, U^\dagger\partial_\nu U]^2 \tag{20}$$

a necessary precondition for the very existence of the chiral phase transition? The answer to this question is no. Let us for instance replace the fourth-order stabilizing term (20) by a stabilizing term of sixth-order (*see e.g.* ref. [23])

$$\mathcal{L}_{2,6} = \frac{{f_\pi}^2}{4}\mathrm{Tr}\left(\partial_\mu U^\dagger\partial^\mu U\right) - c_6 B_\mu B^\mu \tag{21}$$

(note $c_6 > 0$) where B_μ is the topological baryon current

$$B^\mu = \frac{\varepsilon^{\mu\nu\alpha\beta}}{24\pi^2}\mathrm{Tr}\left(U^\dagger\partial_\nu U\, U^\dagger\partial_\alpha U\, U^\dagger\partial_\beta U\right). \tag{22}$$

As shown in section 2 one can assign to the fourth-order Skyrme term (20) as well as to the sixth-order term and the second-order non-linear sigma model term in (21) a geometrical interpretation with the help of the strain tensor language. Therefore one can apply Manton's general machinery [14] to show that the sixth-order term allows for the identity map as a solution and to construct the critical value of L where the identity map of the model (21) becomes unstable and bifurcates into a localized solution. In fact one encounters exactly the same phenomena as discussed before, only the critical hypersphere radius L_c where the chiral restoration occurs is for the stabilization by the sixth-order term (21) at

$$L_c = (3)^{\frac{1}{4}} L_{\min} \tag{23}$$

instead of at $L_c = \sqrt{2}L_{\min}$ as it was the case for the usual Skyrme model with the Skyrme-term (20) stabilization [15].

In fact any lagrangian based on a polynomial of the fundamental invariants (11) of the strain tensor which allows for (locally) stable $B = 1$ Skyrmion solutions, leads to a chirally restored phase at high densities [24]. Furthermore the position of the phase transition can be uniquely derived from the lagrangian itself without solving any equation of motion. Consider for this purpose a lagrangian of the form

$$\mathcal{L}_{\mathcal{G}} \equiv -\mathcal{G}(-\mathcal{L}_2, \mathcal{L}_4, \mathcal{L}_6) = -\sum_{j_1,j_2,j_3=0}^{\infty} A_{j_1j_2j_3}\,(-\mathcal{L}_2)^{j_1}(-\mathcal{L}_4)^{j_2}(-\mathcal{L}_6)^{j_3} \tag{24}$$

where j_1, j_2, j_3 are integer indices and $A_{j_1j_2j_3}$ Taylor coefficients. The arguments of the analytic function $\mathcal{G}$ and the signs are chosen in such a way, that the corresponding static energy density has the simple form

$$\mathcal{E} \equiv \mathcal{G}(\mathcal{E}_2, \mathcal{E}_4, \mathcal{E}_6) = \sum_{j_1,j_2,j_3=0}^{\infty} A_{j_1j_2j_3}\,\mathcal{E}_2{}^{j_1}\mathcal{E}_4{}^{j_2}\mathcal{E}_6{}^{j_3}. \tag{25}$$

The function $\mathcal{G}$ should be of course positive definite to ensure a positive energy density and $\mathcal{G}(0,0,0) \equiv 0$ (*e.g.*, $A_{000} = 0$) to ensure triviality in the $B = 0$ sector. Note that this form of the energy density is the most general symmetric function of the squared eigenvalues, λ_i^2, of the strain tensor M_{ij} (10) and includes all forms which can come form lagrangians which involve arbitrary combinations of powers of first derivatives of the fields to even order.

In the chirally restored (uniform) regime on the hypersphere the energy densities (modulo a prefactor which can be incorporated into the Taylor coefficients $A_{j_1j_2j_3}$) have the simple form

$$\mathcal{E}_{2i} = \frac{3}{L^{2i}}, \qquad i = 1,2,3. \tag{26}$$

In ref. [24] the static eigenvalues of all possible small amplitude fluctuations around the identity map were obtained in a closed analytical form:

$$\lambda^{\mathcal{G}}_{N,a} = \left\{ \sum_{i=1,2,3} \lambda_{i,N,a}\frac{\partial}{\partial\mathcal{E}_{2i}} + \lambda_{3,N,a}\frac{2L^2}{3}\left(\frac{\partial}{\partial\mathcal{E}_2} + \frac{2}{L^2}\frac{\partial}{\partial\mathcal{E}_4} + \frac{3}{L^4}\frac{\partial}{\partial\mathcal{E}_6}\right)^2 \right\} \mathcal{G}(\mathcal{E}_2, \mathcal{E}_4, \mathcal{E}_6) \tag{27}$$

with

$$\lambda_{i,N,a} = i\left\{N(N+2) + a^2 - 4 - \frac{i-1}{2}\left(\{a(N+1)\}^2 - 4\right)\right\} L^{3-2i}. \tag{28}$$

where the indices $N \geq 1$ and $a = 0, 1$ characterize the allowed $O(4)$ representations[#6]: $(N, a = 0)$, the symmetric tensor representations which have $(N+1)^2$ degenerate states and the $(N, a = 1)$ representations which have a $2\{(N+1)^2 - 1\}$ degeneracy

[#6] *See* section 3 for the corresponding results for the usual Skyrme model (1).

(*see* refs. [22, 24] for further details). Note that in the case $a = 1$ the expression for $\lambda^{\mathcal{G}}_{N,a=1}$ simplifies to

$$\lambda^{\mathcal{G}}_{N,a=1} = L\{N(N+2)-3\}\left(\frac{\partial}{\partial \mathcal{E}_2} + \frac{1}{L^2}\frac{\partial}{\partial \mathcal{E}_4}\right)\mathcal{G}(\mathcal{E}_2, \mathcal{E}_4, \mathcal{E}_6). \tag{29}$$

In the regime where the identity map is stable (*i.e.* all $\lambda^{\mathcal{G}}_{N,a} \geq 0$), the monotonic increase of the eigenvalues $\lambda^{\mathcal{G}}_{N,a}$ with increasing N is guaranteed. For a given explicit form of $\mathcal{G}$ the critical L_c where the identity map becomes unstable and therefore the position of the phase transition can be found from determining the lowest value of L where the first $\lambda^{\mathcal{G}}_{N,a}$ in Eq.(27) becomes negative.

Finally, let

$$\mathcal{G}(\mathcal{E}_2, \mathcal{E}_4, \mathcal{E}_6) \to \mathcal{E}_2 \qquad \text{for } \mathcal{E}_{2i} \to 0 \tag{30}$$

and

$$\mathcal{G}(\mathcal{E}_2, \mathcal{E}_4, \mathcal{E}_6) \to C_1 {\mathcal{E}_2}^2 + C_2 \mathcal{E}_4 \qquad \text{for } \mathcal{E}_{2i} \to \infty \tag{31}$$

where i runs from 1 to 3 and $C_1 \geq 0$ and $C_2 > 0$ are fixed positive coefficients. The condition (30) enforces that asymptotically for low densities (large L) the lagrangian (24) is dominated by the non-linear sigma model lagrangian, whereas for high densities (small L) it scales as a free Fermi gas (*i.e.* $\mathcal{E}_{\mathcal{G}} \propto 1/L^4$). The low density behavior is of course indisputable, the high density constraint on the other hand requires the additional input that such effective models can be applied even for densities where the underlying theory, QCD, becomes to leading order a free Fermi gas.

Back to the stability analysis: Note that the signs of the static eigenvalues $\lambda^{\mathcal{G}}_{N,a}$ and therefore the stability of the uniform regime follow from the signs of the derivatives acting on $\mathcal{G}$ and from the signs of the coefficients $\lambda_{i,N,a}$ (28). From all possible $\lambda_{i,N,a}$ with $i = 1, 2, 3$, $N \geq 1$ and $a = 0, 1$ only the term $\lambda_{i=1,N=1,a=1}$ can become negative for sufficiently large L, all the other coefficients are guaranteed to be larger than zero or at least equal zero. (To the latter category belong the six $O(4)$ zero modes with $N = 1$ and $a = 1$. Furthermore all the terms $\lambda_{i=3,N=1,a=1}$ are zero indicating that the sixth-order term alone has infinitely many zero modes.) Now taking the high-density (31) and the low-density behavior (30) into account, we can conclude that (a) at high densities the uniform $B = 1$ skyrmion solution on the hypersphere is bound to be stable whereas at sufficiently low densities where the non-linear σ model term becomes the dominant one, the uniform solution is unstable since $\lambda_{i=1,N=1,a=1}$ becomes negative eventually and dominates above all other terms in (27). That means in other words that the existence of a critical value of L is guaranteed where the uniform solution becomes unstable and where there is a bifurcation from the uniform solution which is chirally restored to the usual localized and chirally broken $B = 1$ hedgehog solution. So even for the most general *geometric* lagrangian (24) built form first order field derivatives the existence of the chiral phase transition (as discussed at length in the preceeding chapters for the normal Skyrme model) is guaranteed under the assumption of a few reasonable asymptotic constraints, Eqs. (30) and (31), on the form of the lagrangian.

4.2. Stabilization by Vector Mesons

So far we have discussed lagrangians which ensured the stabilization of skyrmions by higher-order (contact) terms in the field derivatives. Is the existence of the chiral phase transition linked to the stabilization by *contact* terms and is it destroyed if the the skyrmions are stabilized by the action of *finite range* vector mesons, as *e.g.* ω (and ρ) mesons? The answer is again no. Let us take as an example the skyrmion stabilized by an ω meson: The corresponding lagrangian has the structure [25]

$$\mathcal{L}_{2,\omega} = \frac{f_\pi^2}{4}\mathrm{Tr}(\partial_\mu U^\dagger \partial^\mu U) - \frac{1}{4}\omega_{\mu\nu}\omega^{\mu\nu} + \frac{m_\omega^2}{2}\omega_\mu\omega^\mu + g_\omega \omega_\mu B^\mu \tag{32}$$

where the first term is the usual non-linear σ model lagrangian, the second term the ω kinetic term involving the ω field-strength tensor $\omega_{\mu\nu} = \partial_\mu\omega_\nu - \partial_\nu\omega_\mu$, the third term is the ω mass term and the last term is the coupling of the ω field to the topological baryon current (22). Note that the $B = 1$ hedgehog solution is stable against (radially) rescaling the soliton profile: The hedgehog form necessarily induces a non-zero B_0 term which in turn induces a non-zero ω_0 component. The latter has no corresponding conjugate momentum. Thus the ω_0-profile is "nailed" to the solution of the ω_0 equation of motion $\frac{\delta\mathcal{L}}{\delta\omega_0} = 0$. In other words, there is a constraint to the full quantum hamiltonian, $\frac{\delta\mathcal{H}}{\delta\omega_0} = 0$ and the ω_0-component cannot be scaled to zero [26]. Thus the ω coupling to the baryon current provides for stabilization against rescaling of the hedgehog to zero size. (The scaling to infinite size is prevented by the non-linear σ term.)

Now it can easily be shown that the equations of motion for the $B = 1$ hedgehog on the $\mathbf{S}^3(L)$-sphere can always be satisfied by the uniform solution

$$f(\mu) = \mu \tag{33}$$

for the hedgehog profile and by

$$\omega_0 = -\frac{g_\omega}{m_\omega^2}B_0 = -\frac{g_\omega}{m_\omega^2}\frac{1}{2\pi^2 L^3} \tag{34}$$

for the ω_0-component, *i.e.* the ω_0-component is in this case spatially constant and only depends on the hypersphere radius L. (The spatial components ω_i are identically zero because of the static hedgehog form of the soliton profile.) When the uniform solutions

(33–34) are reinserted into the corresponding energy density of the lagrangian (32), one recovers the same structure as for a model stabilized by a sixth-order term

$$\mathcal{L}_{2,6} = \frac{f_\pi^2}{4}\mathrm{Tr}(\partial_\mu U^\dagger \partial^\mu U) - \frac{1}{2}\left(\frac{g_\omega}{m_\omega}\right)^2 B_\mu B^\mu. \tag{35}$$

This does not come as a complete surprise, since for $m_\omega, g_\omega \to \infty$ with fixed ratio g_ω/m_ω the lagrangian (32) converges to the lagrangian (35). The high-density behavior of the static sector of both lagrangians is the same. However, the value L_c of the bifurcation-point and the low-density behavior of both lagrangians are different. For increasing parameters with fixed ratio g_ω/m_ω the critical density and the low density phase of the ω stabilized model will approach the corresponding quantities of the model (35). The physical reason for the identical high-density behavior of the static solutions of both lagrangians is that at high densities it is energetically favorable that the ω_0-component becomes spatially constant, otherwise the "price" in energy from a finite spatial gradient term in ω_0 becomes higher and higher with decreasing hypersphere-radius L. This together with the equation of motion for the ω_0-component necessarily lead to Eq.(34) which in turn guarantees the identical high-density behavior of both lagrangians. The important point is that also for the ω-stabilized Skyrme model there exist a critical hypersphere radius L_c where the uniform (chirally restored) phase on the hypersphere becomes unstable and bifurcates in a localized (chirally broken) phase which has the usual $B = 1$ hedgehog as flat space limit.

The behavior is similar in case the $\mathcal{L}_{2,\omega}$ model is extended to the coupling of ρ mesons. This can be done in different ways. For our purpose the only essential precondition is that the total lagrangian is still explicitly chiral symmetric. For simplicity let us consider in the following the hidden ρ meson coupling à la Bando et al. [27] in a minimal way [28], *i.e.*

$$\begin{aligned}\mathcal{L}_{2,\omega,\rho} &= \frac{f_\pi^2}{4}\mathrm{Tr}(\partial_\mu U^\dagger \partial^\mu U) - a\frac{f_\pi^2}{4}\mathrm{Tr}\left(\sqrt{U^\dagger}\partial_\mu\sqrt{U} - \sqrt{U}\partial_\mu\sqrt{U^\dagger} - 2ig\rho_\mu\right)^2 \\ &\quad - \frac{1}{4}\rho_{\mu\nu}\rho^{\mu\nu} - \frac{1}{4}\omega_{\mu\nu}\omega^{\mu\nu} + \frac{m_\omega^2}{2}\omega_\mu\omega^\mu + g_\omega\omega_\mu B^\mu \\ &\quad + (\omega\rho\pi \text{ coupling terms}).\end{aligned} \tag{36}$$

where the second term is responsible for the generation of the ρ mass and the $\rho\pi\pi$ coupling with a coupling constant g. (The standard choice for the parameter a is $a = 2$, *see* ref. [27].) Note that the hedgehog ansatz for the chiral profile $U = \exp(i\tau\cdot\hat{r}f(\mu))$ induces that only the spatial ρ_i components are excited, whereas the ρ_0 component is non-excited and therefore left at zero in the static situation. Again it is easy to show that the equations of motion of this model are satisfied by the uniform profiles

$$f(\mu) = \mu \tag{37}$$

$$\omega_0 = -\frac{g_\omega}{m_\omega^2} B_0 \tag{38}$$

$$\rho_i = \frac{1}{2ig}\left(\sqrt{U^\dagger}\partial_i U - \sqrt{U}\partial_i U^\dagger\right) \qquad \text{with } U = \exp(i\tau\cdot\hat{r}\mu). \tag{39}$$

The last equation guarantees that the second term in (36) vanishes and that the ρ kinetic term involving the ρ field-strength tensor $\rho_{\mu\nu} = \partial_\mu\rho_\nu - \partial_\nu\rho_\mu - ig[\rho_\mu, \rho_\nu]$ behaves as fourth-order Skyrme term (20) with a coefficient $\epsilon_4^2 = 1/(8g^2)$. The lagrangian (36) (without extra $\omega\rho\pi$ couplings) has therefore the same high density behavior as the simple contact lagrangian

$$\mathcal{L}_{2,4,6} = \frac{f_\pi^2}{4}\mathrm{Tr}(\partial_\mu U^\dagger \partial^\mu U) + \frac{1}{32g^2}\mathrm{Tr}[U^\dagger\partial_\mu U, U^\dagger\partial_\nu U]^2 - \frac{1}{2}\left(\frac{g_\omega}{m_\omega}\right)^2 B_\mu B^\mu. \tag{40}$$

This behavior exemplified for the simplest $\omega\rho\pi$-model is generic – only the coefficients in (40) may change. For all models with vector-meson stabilization which have been studied so far [2, 29, 30] the same is true: It can be numerically shown that

(a) at high densities the only solution is the uniform (delocalized) hedgehog one, the static solutions are identical to those of contact-term lagrangians with suitable coefficients and the static vector meson fields are equal to their driving pionic currents (*see e.g.* (34) or (39)),

(b) there exist a critical model-dependent hypersphere-radius L_c where the uniform solution becomes unstable and bifurcates into the usual localized configuration and

(c) at low densities the solutions on the hypersphere $\mathbf{S}^3(L)$ converge asymptotically to the corresponding solutions in the flat space.

The essential conditions on these models are two-fold:

(i) they should be explicitly chiral symmetric,

(ii) they should allow for locally stable $B = 1$ hedgehog solitons.

If these two conditions are met, there will be necessarily a chiral phase transition at high densities with the same features as discussed before. However, under realistic values for the mesonic input parameters (*e.g.* f_π, ϵ_4, c_6, *see* ref. [23], or g_ω and g, *see* refs. [28, 30, 2])) the transition densities ρ_c come out systematically too low, $\rho_c \approx \rho_{\rm nm}$ (0.16 fm^{-3}). What apparently is missing in these models are terms which generate additional attraction. Note that the hedgehog mass under these realistic parameters is systematically too high (~ 1.6 GeV) whereas there is not sufficient central attraction in the $B = 2$ system [6]. In case one would tune the stabilizing parameters – keeping $f_\pi = 93$MeV fixed – such that the classical hedgehog mass would be 0.87 GeV (leading to a nucleon mass of 0.94 GeV) then the transition density ρ_c would be in turn in the order of 3 times the nuclear matter density.

5. Second- Versus First-Order Transitions

So far we have only encountered second order transitions between the delocalized and the localized $B = 1$ phase. This does not necessarily have to be the case: Our Skyrme-type lagrangians which all explicitly break scale invariance do not "know" yet about the trace anomaly in QCD. Let us therefore try to incorporate the same scaling behavior as in QCD into these lagrangians (*see* refs. [31, 32, 33]), *e.g.* consider the lagrangian

$$\begin{aligned}\mathcal{L}_{2,4,\chi} &= \frac{f_\pi^2}{4}\frac{\chi^2}{\chi_0{}^2}\mathrm{Tr}(\partial_\mu U^\dagger \partial^\mu U) + \frac{\epsilon_4^2}{4}\mathrm{Tr}[U^\dagger\partial_\mu U, U^\dagger \partial_\nu U]^2 \\ &+ \frac{1}{2}(\partial_\mu\chi)^2 - B_B\left(1 + \left(\frac{\chi}{\chi_0}\right)^4 \log\left(\frac{\chi^4}{e\chi_0{}^4}\right)\right) \end{aligned} \tag{41}$$

as simplest extension of the usual Skyrme lagrangian. The scalar field χ with the vacuum expectation value $\chi_0 = \langle 0|\chi|0\rangle$ is introduced with the purpose of making the first term in (41) scale invariant (the Skyrme term and the χ-kinetic term are already scale invariant), whereas the last term in (41), the χ-potential term, is adjusted to fit the trace anomaly of QCD. B_B is the "bag constant" which can be expressed in terms of the gluon condensate as $B_B = (9/32)\langle 0|(\alpha_s/\pi)G^2|0\rangle$. For values of the parameters χ_0 and B_B *see* ref. [32]. (Note that the fluctuations of the χ-field correspond to glueball-excitations.) In ref. [34] it was numerically shown that the lagrangian (41) leads again to a phase transition to a uniform delocalized solution at high densities with $f(\mu) = \mu$ and a constant χ profile which – depending on the choice of parameters – can be even $\chi = 0$. What the authors of ref. [34] missed because of insufficient accuracy in their numerical calculations is the fact that the phase transition is of first order. This is more or less obvious from the form of the χ-potential term. The χ-profile in the tail regime on the hypersphere does not smoothly change from its low density asymptotic value $\chi = \chi_0$ to (in the extreme case) $\chi = 0$ all over the hypersphere, but does this in a jump. Thus by incorporating a scalar field à la Schechter [31] in the Skyrme model (or its extensions) the second order phase transition can be easily changed to a first order one. *See* also ref. [21] for a different mechanism to achieve the same thing. Note that it is now possible to tolerate a chiral symmetry breaking pion mass term at low densities without endangering the phase transition to the high density phase, if the pion mass term is modified to [31, 32]

$$\mathcal{L}_{\chi SB} = \frac{f_\pi^2 m_\pi^2}{4} \left(\frac{\chi}{\chi_0}\right)^3 \mathrm{Tr}(U + U^\dagger - 2). \tag{42}$$

It is rather easy to extend the lagrangian (24) to the most general with the QCD trace anomaly consistent geometrical form involving only first oder derivatives to even order:

$$\begin{aligned}\mathcal{L}_{\mathcal{G},\chi} &= -\left(\frac{\chi}{\chi_0}\right)^4 \mathcal{G}\left(-(\frac{\chi_0}{\chi})^2\mathcal{L}_2, -(\frac{\chi_0}{\chi})^4\mathcal{L}_4, -(\frac{\chi_0}{\chi})^6\mathcal{L}_6\right) \\ &\quad + \frac{1}{2}(\partial_\mu\chi)^2 - B_B\left(1 + \left(\frac{\chi}{\chi_0}\right)^4 \log\left(\frac{\chi^4}{e{\chi_0}^4}\right)\right).\end{aligned} \tag{43}$$

Note that the restoration of scale invariance, $\chi = 0$, is only consistent with the lagrangian (43) if $\mathcal{G}$ has the asymptotics (31), *i.e.*

$$\left(\frac{\chi}{\chi_0}\right)^4 \mathcal{G}\left(-(\frac{\chi_0}{\chi})^2\mathcal{L}_2, -(\frac{\chi_0}{\chi})^4\mathcal{L}_4, -(\frac{\chi_0}{\chi})^6\mathcal{L}_6\right) \to C_1(-\mathcal{L}_2)^2 + C_2(-\mathcal{L}_4) \quad \text{if } \chi \to 0. \tag{44}$$

The restoration of scale invariance, the Fermi-gas limit, at high densities is of course natural. We know from what was said before that one can expect in this case a first-order phase transition to a chirally restored phase. The restoration of scale invariance, the high-density limit and the restoration of chiral symmetry are therefore in this class of models intrinsically tied together. If a lagrangian is considered which includes *any* term higher than fourth-order in the derivatives of the pion fields, terms of *all orders* have to be included as well both to satisfy the high-density asymptotic behavior (44) and to ensure the stability of the chiral symmetric high-density phase [24]. The necessity for infinitely many terms is consistent with the large N_c philosophy [35, 36]. Because of asymptotic freedom, the effective model which follows from QCD to leading order in the $1/N_c$ expansion must involve infinitely many mesons. The hope is – as shown for the ω and ρ meson case – that the high density phase of this infinite tower of meson resonances can be approximated by infinitely many suitable contact terms. This might lead to "dreams" of Regge trajectories.

Finally a note on flavor extensions: So far we have discussed only $SU(2)_f$ skyrmions. The hypersphere scheme with the strain tensor machinery seems to be intrinsically linked to $SU(2)_f$, because it employs the map from the spatial 3-sphere $\mathbf{S}^3(L)$ onto the isospin manifold $\mathbf{S}^3(1) \cong SU(2)$, and therefore at first sight not directly applicable to $SU(3)_f$, for instance. But note that $SU(3)_f$ is definitely broken in nature. Furthermore approaches where this breakdown is incorporated only perturbatively into the Skyrme model seem to run into severe problems with phenomenology [4], whereas the bound-state approach of Callan and Klebanov [37] where the strange and heavy flavor baryons are described as bound states of kaons (or heavy-flavored mesons) in an $SU(2)$-background work rather successfully [38, 39]. In the latter case, however, the hypersphere-machinery is again directly applicable, *see e.g.* refs. [40, 21].

6. Discussion

Even effective models (without any quark degrees of freedom) show as a rule phase transitions to a high-density phase with restored chiral symmetry. There are essentially just two conditions: The effective model must have the relevant symmetries (especially the lagrangian must be explicitly chirally invariant) *and* it must treat the baryon structure and interaction on the same footing. The phase transitions are normally of second order, but can easily be changed to first-order ones by incorporating the QCD trace anomaly for instance. With the input from the meson sector the transition densities are predicted systematically too low, $\rho_c \approx \rho_{\rm nm}$. This problem may be connected with two other deficiencies which plague the Skyrme-like models: the missing central attraction in the $B = 2$ channel and the predicted high value for the hedgehog mass. Quantum (higher loop) corrections may have a chance to cure all three problems at the same time by providing extra non-local terms which generate attraction.

The fact that the *de facto* order parameter $\langle\sigma\rangle$ decreases and finally vanishes with increasing densities supports the claim [33] that meson masses should be taken as medium dependent and should "run" with the density. Thus the possibility of having phase transitions in effective theories could have important consequences for the design of heavy–ion experiments to detect a transition to chirally restored or deconfined matter. New degeneracies and running masses on the way to chiral symmetry restoration should be taken as a signal that the entropy of the hadronic phase increases relative to its vacuum value and may become comparable to the quark-gluon one making it unlikely to find a large value for the latent heat.

The author would like to acknowledge discussions with A. D. Jackson, L. Castillejo, N. S. Manton, H. Bang, N. Kaiser and H. B. Nielsen and thank Ulf Meißner for "twisting his arm" to write up this presentation.

References

1. T.H.R. Skyrme, *Nucl. Phys.* **31** (1962) 556.
2. U.-G. Meißner, *Phys. Rep.* **161** (1988) 213, and references therein.

3. G.S. Adkins, C.R. Nappi and E. Witten, *Nucl. Phys.* **B228** (1983) 552.
4. *for a review and further references see:*
I. Zahed and G.E. Brown, *Phys. Rep.* **142** (1986) 1;
U.-G. Meißner and I. Zahed, *Adv. Nucl. Phys.* **17** (1986) 143;
G. Holzwarth and B. Schwesinger, *Rep. Prog. Phys.* **49** (1986) 825;
G. S. Adkins *in:* Chiral Solitons, ed. K.F. Liu (World Scientific, Singapore, 1987).
5. A. Hayashi, G. Eckart, G. Holzwarth and H. Walliser, *Phys. Lett.* **147 B** (1984) 5;
M.P. Mattis and M. Karliner, *Phys. Rev.* **D31** (1985) 2833.
6. A. Jackson, A.D. Jackson and V. Pasquier, *Nucl. Phys.* **A432** (1985) 567.
7. I. Klebanov, *Nucl. Phys.* **B262** (1985) 133.
8. E. Wüst, G.E. Brown and A.D. Jackson, *Nucl. Phys.* **A468** (1987) 450.
9. A.D. Jackson and J.J.M. Verbaarschot, *Nucl. Phys.* **A484** (1988) 419.
10. T.S. Walhout, *Nucl. Phys.* **A484** (1988) 397;
M. Kugler and S. Shtrikman, *Phys. Lett.* **208 B** (1988) 491.
11. L. Castillejo, P.S. Jones, A.D. Jackson, J.J.M. Verbaarschot and A. Jackson, *Nucl. Phys.* **A501** (1989) 801.
12. A.S. Goldhaber and N.S. Manton, *Phys. Lett.* **198 B** (1987) 231.
13. N.S. Manton and P.J. Ruback, *Phys. Lett.* **181 B** (1986) 137.
14. N.S. Manton, *Commun. Math. Phys.* **111** (1987) 469.
15. A.D. Jackson, A. Wirzba and L. Castillejo, *Nucl. Phys.* **A486** (1988) 634 and *Phys. Lett.* **198 B** (1987) 315.
16. A.D. Jackson, C. Weiss, A. Wirzba and A. Lande, *Nucl. Phys.* **A494** (1989) 523.
17. R.S. Palais, *Commun. Math. Phys.* **69** (1979) 19.
18. M. Loss, *Lett. Math. Phys.* **14** (1987) 149.
19. A.D. Jackson, N.S. Manton and A. Wirzba, *Nucl. Phys.* **A495** (1989) 499.
20. I. Zahed, A. Wirzba and U.-G. Meißner, *Phys. Rev.* **D33** (1986) 830.
21. H. Forkel, A.D. Jackson, M. Rho, C. Weiss, A. Wirzba and H. Bang, *Nucl. Phys.* **A504** (1989) 818.
22. A. Wirzba and H. Bang, *Nucl. Phys.* **A515** (1990) 571.
23. G.E. Brown, M. Rho and W. Weise, *Nucl. Phys.* **A454** (1986) 669;
A. Wirzba and W. Weise, *Phys. Lett.* **188 B** (1987) 6.
24. A.D. Jackson, C. Weiss and A. Wirzba, *Nucl. Phys.* **A529** (1991) 741.
25. Adkins and Nappi, *Phys. Lett.* **137 B** (1984) 251.
26. C. Adami, Diplomarbeit, Bonn University (1988), unpublished.

27. M. Bando, T. Kugo, S. Uehara, K. Yamawaki and T. Yanagida, *Phys. Rev. Lett.* **54** (1985) 1215;
M. Bando, T. Kugo and K. Yamawaki, *Nucl. Phys.* **B259** (1985) 493;
Y. Igarashi, M. Johmura, A. Kobayashi, H. Otsu, T. Sato and S. Saweda, *Nucl. Phys.* **B259** (1985) 721;
T. Fujiwara, T. Kugo, H. Terao, S. Uehara and K. Yamawaki, *Prog. Theor. Phys.* **73** (1985) 926.

28. U.-G. Meißner, N. Kaiser, A. Wirzba and W. Weise, *Phys. Rev. Lett.* **57** (1986) 1676.

29. N.N. Scoccola, M. Rho and D.P. Min, *Nucl. Phys.* **A489** (1988) 612.

30. P. Jain, R. Johnson, U.-G. Meißner, N.W. Park and J. Schechter, *Phys. Rev.* **D37** (1988) 3252;
U.-G. Meißner, N. Kaiser, H. Weigel and J. Schechter, *Phys. Rev.* **D39** (1989) 1956.

31. H. Gomm, P. Jain, R. Johnson and J. Schechter, *Phys. Rev.* **D33** (1986) 3476;
P. Jain, R. Johnson and J. Schechter, *Phys. Rev.* **D35** (1987) 2230.

32. B.A. Campbell, J. Ellis and K.A. Olive, *Nucl. Phys.* **B345** (1990) 57 and *Phys. Lett.* **235 B** (1990) 325.

33. G.E. Brown and M. Rho, *Phys. Rev. Lett.* **66** (1991) 2720.

34. H. Reinhardt and B.V. Dang, *Phys. Rev.* **D38** (1988) 2881.

35. G. 't Hooft, *Nucl. Phys.* **B72** (1974) 461, *Nucl. Phys.* **B75** (1974) 461.

36. E. Witten, *Nucl. Phys.* **B160** (1974) 57.

37. C.G. Callan and I. Klebanov, *Nucl. Phys.* **B262** (1985) 365.

38. N.N. Scoccola, H. Nadeau, M.A. Novak and M. Rho, *Phys. Lett.* **201 B** (1988) 425;
C.G. Callan, H. Hornbostel and I. Klebanov, *Phys. Lett.* **202 B** (1988) 269;
U. Blom, K. Dannbom and D.O. Riska, *Nucl. Phys.* **A493** (1989) 384.

39. M. Rho, D.O. Riska and N.N. Scoccola, *Phys. Lett.* **251 B** (1990) 597.

40. H. Forkel, A.D. Jackson, M. Rho and N.N. Scoccola, *Nucl. Phys.* **A509** (1990) 673.

STRONGLY COUPLED HIGGS SECTOR

APPLYING EFFECTIVE LAGRANGIANS TO TEST THE STANDARD MODEL

DOMÈNEC ESPRIU

CERN, Geneva

and

Departament d'Estructura i Constituents de la Matèria
Universitat de Barcelona
Diagonal, 647
E-08028 Barcelona

ABSTRACT

We parametrize different possibilities for the symmetry breaking sector of the electroweak theory in terms of the coefficients of dimension four operators of an effective non-linear sigma model that retains only the light degrees of freedom. Assuming custodial symmetry to be exact, all relevant information for physics below the multi-TeV region is encoded in just four numbers. We study how to compute radiative corrections to physical processes in an unambiguous manner even if the symmetry breaking sector is described by an effective theory. We apply our techniques to LEP physics.

1. Introduction

Few would dispute the statement that the most interesting piece of the electroweak theory yet to be tested —and the one where more surprises may be in store for us— is the symmetry breaking sector, giving the W and Z bosons a mass. Indeed, while many features of the electroweak theory are well established by now such as the gauge symmetry, the coupling to fermions, or even the family structure, the Standard Model (SM) has succeeded so far in hiding any relevant information on its symmetry breaking sector (SBS).

The SM (with the $U(1)_Y$ interaction switched off) has an exact global $SU(2)_L \times SU(2)_R$ symmetry. In the minimal SM the above symmetry is broken when some scalar field, belonging to a $SU(2)$ doublet, acquires a vacuum expectation value. Three real fields are swallowed by the W and Z. The one remaining excitation is a physical particle, the Higgs, assumed to be elementary. At tree level its mass is an unconstrained free parameter (the current experimental limit for direct observation from the four LEP experiments lies around 50 GeV). Veltman's screening theorem[1] ensures that the Higgs mass appears only logarithmically (at the one loop level). This, combined with the usual loop suppression factors, makes the Higgs contribution to radiative corrections very small. If the Higgs is a light particle of less than a few hundred GeV it will be found either at LEP II or in the next generation of pp colliders. If the Higgs is heavy, somewhere in the TeV scale, the scalar sector of the SM is strongly interacting and there should be some signals of a strongly interacting sector once we reach the TeV region. Of course, this minimal SM can be arbitrarily enriched by adding more doublets of scalar fields or even higher representations, provided that the ρ-parameter is kept close enough to one.

A completely different possibility is that there is no such thing as an elementary Higgs. If this is the alternative chosen by Nature, the $SU(2)_L \times SU(2)_R$ global symmetry must be broken by some dynamical mechanism. Composite models or some variant of technicolor theories would be required to provide the dynamics leading to the symmetry breaking. The typical scale involved must again lie somewhere in the TeV region, as we will emphasize later. In any case, either the Higgs is found somewhat below the 1 TeV scale or else something fundamentally new must happen once we begin exploring the multi TeV region. This discussion is often referred to as the "no-lose" corollary[2].

Meanwhile, it seems pertinent to investigate at what level the differences between the various symmetry breaking alternatives would set in in ongoing high precision experiments, singularly at LEP. The level of precision being attained is quite amazing[3]. For instance, for the parameter Δr, the present accuracy is set to the level of just 1.7×10^{-2} and is essentially due to the uncertainty in the measurement of M_W, $M_W/M_Z = 0.8807 \pm 0.0031$. The future measurement of M_W at LEP II and CDF will lead to an accuracy in $\delta(\Delta r)$ at the 10^{-3} level. The effective Weinberg angle, currently $\overline{\sin}^2\theta_W = 0.2317 \pm 0.0030$ U9 , will reduce its error to $\delta\overline{\sin}^2\theta_W = 4 \times 10^{-4}$or less with polarized beams[4].

The information that can be obtained from present and planned e^+e^- machines complements what will be learned from the scattering of longitudinal W and Z bosons. In fact, assuming an exact custodial symmetry in whatever underlying physics triggers the breaking $SU(2)_L \times SU(2)_R \rightarrow SU(2)_V$, the whole SBS can be parametrized by just four numbers, the L_i's of Gasser and Leutwyler[5]. Two of them can be determined from e^+e^- experiments, the other two from pp machines. (It is worth pointing out that another interesting source of constraints on the parameters of the effective chiral lagrangian is provided by atomic parity violation experiments[6].)

We want thus to embark in the task of providing a unified description of "low energy" physics in the SBS; that is, physics in the ≤ 1 TeV region. This description must contain only "light" fields with respect to the natural scale in the problem and it must be general enough so that theories so different as a heavy Higgs and a technicolor model both fit in. And, above all, we must know how to make *precise* predictions with it, even if we are dealing with an effective theory; order-of-magnitude estimates being no longer acceptable given the level of precision that will be achieved in e^+e^- machines. Chiral lagrangians[7] are ideally suited for this task[8] since the only physical input they use is the symmetry breaking pattern, which must at least contain $SU(2)_L \times SU(2)_R \rightarrow SU(2)_{L+R}$.

Recently, some works[9] have appeared using "model independent" parametrizations of LEP physics, closely related to the present effective lagrangian approach. Let us make here contact with the notation employed in the work of these authors

$$S = -16\pi L_{10} \qquad \epsilon_3 = \frac{\alpha}{4s^2}S = -g^2 L_{10} \tag{1}$$

Also, $\alpha T = \Delta\rho$ and $\epsilon_1 = \alpha T = \Delta\rho$. On the contrary, there is no equivalent term (at the order we are working in the custodially symmetric lagrangian we use) to ϵ_2 (or U).

2. Chiral Lagrangians and the Symmetry Breaking Sector

After the symmetry breaking the global group $SU(2)_L \times SU(2)_R$ is broken to its diagonal subgroup $SU(2)_V$, the so-called custodial symmetry, responsible for keeping the ρ parameter exactly equal to one if one does not switch g' on. The Goldstone theorem implies the existence in the broken phase of 3 massless bosons, which will eventually become the longitudinal degrees of freedom of the $W^\pm$ and Z. The interactions amongst the goldstones can be described, entirely on symmetry grounds, by a lagrangian of the form

$$\mathcal{L} = \mathcal{L}^{(2)} + \mathcal{L}^{(4)} + \ldots \tag{2}$$

The first term of this expansion is

$$\mathcal{L}^{(2)} = \frac{v^2}{4} tr D_\mu U D^\mu U^\dagger \tag{3}$$

where the covariant derivative $D_\mu U$, that incorporates the local $SU(2)_L \times U(1)_Y$ gauge symmetry of the electroweak theory is defined as $D_\mu U = \partial_\mu U + U L_\mu - R_\mu U$. Here, L_μ and R_μ are the left and right-handed gauge fields $L_\mu = ig\sigma^a W_\mu^a/2$ and $R_\mu = ig'\sigma^3 B_\mu/2$. v is a dimensionful constant to be determined, and the U field is a unitary matrix that can be parametrized as $U = \exp(i\sigma^a w^a/v)$ where the three w^a fields would be longitudinal components of the intermediate bosons.

The next term in the expansion is

$$\mathcal{L}^{(4)} = L_1 tr D_\mu U D^\mu U^\dagger tr D_\nu U D^\nu U^\dagger + L_2 tr D_\mu U D_\nu U^\dagger tr D^\mu U D^\nu U^\dagger$$

$$-iL_9 tr(F_R^{\mu\nu} D_\mu U D_\nu U^\dagger + F_L^{\mu\nu} D_\mu U D_\nu U^\dagger)$$

$$+L_{10} tr U^\dagger F_R^{\mu\nu} U F_{L\mu\nu} + H_1(F_R^{\mu\nu} F_{R\mu\nu} + F_L^{\mu\nu} F_{L\mu\nu}) \tag{4}$$

Terms which vanish by the equations of motion can be neglected at this order. Notice that the gauge fields always appear through covariant derivatives or field strengths. This reflects the requirement that the unknown SBS, whose dynamics determines the $O(p^4)$ coefficients, is $SU(2)_L \times U(1)_Y$ gauge invariant.

Since only symmetry arguments have been used in choosing the different terms, the above effective lagrangian is completely general. The information about the underlying physics is contained in a number of parameters (such as the coefficients of the $\mathcal{O}(p^4)$ terms L_i). If we keep terms only up to $\mathcal{O}(p^4)$ we have a surprisingly economical description of the SBS. L_1 and L_2 are, by virtue of the Equivalence Theorem[10], relevant in scattering of longitudinally polarized W's and Z's and have been amply discussed in the literature before in connection to pp machines[11]. L_9 and L_{10} are relevant for e^+e^- physics. As we will review in section 3, L_{10} enters in the oblique corrections to the process $e^+e^- \to \bar{f}f$ and along with L_9 also contributes to the γWW and ZWW effective vertices. The term proportional to H_1 does not appear in any observable quantity.

Eqs. (3) and (4) are formally quite similar to the type of effective lagrangians that have been widely used in pion physics, with v replaced by $f_\pi \simeq 93$MeV. To determine v in the electroweak theory we take advantage of the gauge invariance of the model and choose the unitary gauge, $U = 1$. Then, after the usual rotation to the physical basis,

$$\mathcal{L}^{(2)} = \frac{v^2}{4} g^2 W_\mu^+ W_\mu^- + \frac{v^2 g^2}{8\cos^2\theta_W} Z_\mu Z_\mu \tag{5}$$

Here $\sin^2\theta_W$ is defined as $\sin^2\theta_W \equiv 1 - M_W^2/M_Z^2$ and g and g' are defined at any order as $g = e/s$ and $g' = e/c$, with $s = \sin\theta_W$ and $c = \cos\theta_W$. Eq. (3) reduces to the mass term of the SM in the unitary gauge. From here one reads $v = 246$ GeV. This argument also shows that $\mathcal{L}^{(2)}$ is, in fact, universal. Any underlying theory whose long distance behaviour is that of the SM admits a derivative expansion whose first term is $\mathcal{L}^{(2)}$.

Higher dimensional operators, of $\mathcal{O}(p^6)$ and beyond, are suppressed by powers of the energy over some scale (M_H or $4\pi v$). In the minimal SM, with a heavy Higgs, the chiral lagrangian (2.1) provides us with a simple book-keeping device to expand observables in powers of p^2/M_H^2. For large M_H the scalar sector is strongly interacting (recall that the scalar quartic coupling $\lambda \sim M_H^2/M_W^2$) and not tractable by ordinary perturbation theory. However, in that limit all higher orders beyond $\mathcal{O}(p^4)$ which are, by power counting, suppressed by powers of p^2/M_H^2 can be ignored. The effective lagrangian consisting of $\mathcal{L}^{(2)} + \mathcal{L}^{(4)}$ thus provides a non-perturbative treatment of the strongly interacting sector. On the other hand, in composite models we expect, by analogy with pion physics, that the scale that normalizes the contribution of higher dimensional operators is[7,12], roughly speaking, $\Lambda_C \sim 4\pi v$. Since $4\pi v = 3.1\text{TeV}$, neglection of the terms beyond $\mathcal{L}^{(4)}$ is again fully justified for application to "low energy" physics.

Let us now consider what choice of $\mathcal{O}(p^4)$ constants corresponds to the minimal SM with a heavy Higgs. We introduce an effective action obtained after integrating out the Higgs field σ

$$\exp i\Gamma(A, w) = \int d\sigma \exp iS_{SM}(A, w, \sigma) \tag{6}$$

where A generically denotes the gauge fields and w^a are the Goldstone bosons (we omit fermions and ghosts for simplicity.) After performing the functional integrals over the remaining fields A_μ and w, Γ will reproduce the Green functions of the SM.

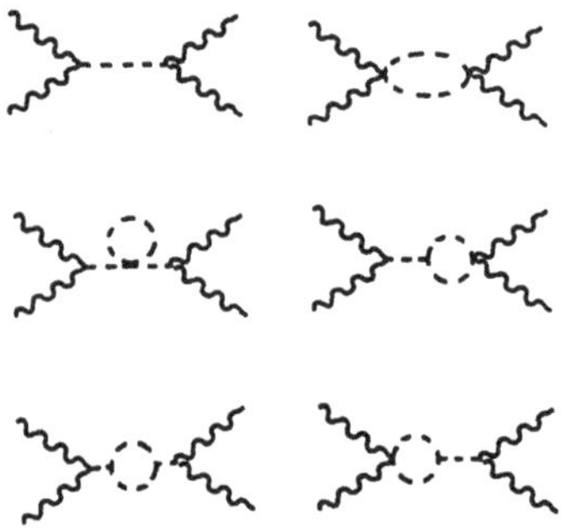

Fig. 1

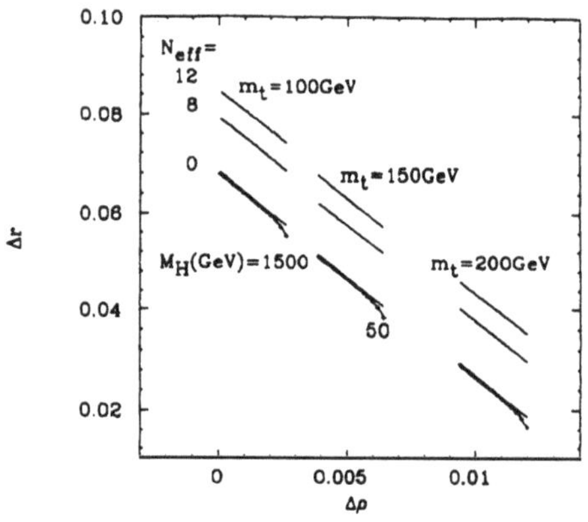

Fig. 2

Diagrammatically, Γ consist of the sum of all the graphs of the theory that contain σ (and only σ) as internal particle, as is, for instance, represented in Fig. 1 for the 4-point function. The resulting effective action Γ can be expanded in inverse powers of M_H. On symmetry grounds $\Gamma(A, w)$ must be of the form given by (2). The value of the coefficients

L_i that one obtains after this calculation at one loop (in the field σ) are

$$L_1^{MSM} = \frac{v^2}{8M_H^2} + \frac{1}{16\pi^2}\left(\frac{9\pi}{16\sqrt{3}} - \frac{47}{48}\right) \qquad L_2^{MSM} = L_9^{MSM} = L_{10}^{MSM} = 0 \tag{7}$$

This result is not exact for L_1, which gets further contributions at higher orders in the Higgs self-coupling (large when $M_H \to \infty$). $L_9^{MSM} = L_{10}^{MSM} = 0$ remains true at any order of perturbation theory. This is a unique[13] feature of the minimal SM.

In order to compute the value of an observable in the minimal SM we have now to use the lagrangian $\mathcal{L}$ with the values for the L_i just quoted. We must also include, of course, the rest of the action, namely the pure Yang-Mills piece (W^2 and B^2), the ghost part, and the gauge fixing term, which we take to be the 't Hooft-Feynman gauge. None of these pieces have been modified by removing the Higgs particle. (The fermion part of the lagrangian is modified, but this plays no important role, at leading order, neither in pp nor on e^+e^- machines, at least for light quarks.) The Higgs field is, of course, not explicitly present since its effects have been *exactly* replaced by the effective lagrangian $\mathcal{L}$. This poses at once the problem of ultraviolet divergences, which are much more severe in the non-linear effective theory than in the (linear) minimal SM.

Before discussing how we can get rid of these divergences we would like to turn to alternative theories, such as technicolor. The basic idea is very much the same. One integrates out the heavy fields that trigger the dynamical breakdown of the $SU(2)_L \times SU(2)_R$ symmetry and, in this way, obtains finite values for the $\mathcal{O}(p^4)$ coefficients. For theories with a $SU(N)_{TC}$ QCD-like SBS, the value of these constants can be approximately derived by a combination of analogies with strong interaction and large N_{TC} calculations. If we simply take the leading (in α_s) calculation of L_1, L_2, L_9 and L_{10} given e.g. in[12], we have

$$L_1 = \frac{N_{TC}}{384\pi^2} \quad L_2 = \frac{N_{TC}}{192\pi^2} \quad L_9 = \frac{N_{TC}}{48\pi^2} \quad L_{10} = -\frac{N_{TC}}{96\pi^2} \tag{8}$$

which are close enough to the scaled values for QCD. Simple inspection of Eqs. (7) and (8) reveals that the general pattern of the coefficients is quite different in theories with interchange of scalars or in QCD-type models. We assume that there is only one doublet under $SU(2)_L$. If not the above formulae should be modified accordingly. L_{10} is related to the spectral function $\Pi_{LR}(0) = \Pi_{VV}(0) - \Pi_{AA}(0)$ that is positive definite in theories with matter in vector-like representations (and, in fact, proportional to the number of $SU(2)_L$ representations) In this type of theories L_{10} basically counts the number of heavy particles. Contributions to L_{10} from additional pseudo-Goldstone bosons have been considered in[14].

The order of magnitude of all these coefficients dictates at what order departures from the minimal SM can be expected. On the whole, the contributions from $\mathcal{O}(p^4)$ at moderate energies are comparable to "normal" electroweak radiative corrections in the SM. Therefore it is necessary to include consistently one loop corrections, as expected from the chiral counting rules.

As we pointed out, due to the non-linear nature of the effective lagrangian $\mathcal{L}$ there are additional divergences which are not absorbed by the usual renormalization of α, M_Z and M_W To make sense of the model one is thus forced to introduce additional counterterms. Some of the divergences have the same structure that the original $\mathcal{O}(p^4)$ lagrangian and can be elimininated by way of redefining the L_i's that appear in $\mathcal{L}^{(4)}$, i.e. $L_i \to L_i + \delta L_i$.

These counterterms have been extensively studied in chiral perturbation theory and are well known[5] (for instance, $\delta L_{10} = (1/16\pi^2)(1/6\epsilon)$). This is not enough, however. The $U(1)_Y$ interactions explicitly break the global $SU(2)_L \times SU(2)_R$ symmetry of the chiral lagrangian. Accordingly, there are also other $SU(2)_L \times U(1)_Y$ gauge invariant - but not $SU(2)_L \times SU(2)_R$ chiral invariant - counterterms that have to be included when $U(1)_Y$ gauge loops are taken into account. The origin of these counterterms can be traced back to mixed loops, in which gauge particles and Goldstone bosons are interchanged in the same diagram. There is only one counterterm of this type[15]

$$\delta\mathcal{L}^{(2)} = \frac{v^2}{4}\delta\beta(trU\sigma^3U^\dagger D_\mu UU^\dagger)^2 \qquad \text{with} \qquad \delta\beta = \frac{g^2}{16\pi^2}\frac{s^2}{c^2}\frac{3}{4\epsilon} \tag{9}$$

When this counterterm is added, all one loop diagrams can be made finite. We note that from a custodially-preserving SBS such a term is never generated in the process of integrating out the heavy fields. Although all observables are now finite, they are dependent on some unknown renormalization scale.

Since $\mathcal{L}^{(2)}$ is universal, i.e. independent of the underlying SBS, we know for sure that all the counterterms that are required to make contributions from $\mathcal{L}^{(2)}$ finite are also universal (and, in a way, not very interesting for us). All the relevant information is hidden in the values of the L_i's of the effective action (6), which only enter at tree level. We want to identify observables which are sensitive to the *finite* parts of the L_i's and *not* to the chiral infinities themselves. In short, we have to look for quantities which are renormalization group invariants with respect to divergences in the non-linear sigma model. In pion physics one is used to the fact that unambiguous predictions from chiral perturbation theory are those where higher dimensional operators do not contribute at all. While this is generally the case, there are exceptions. Observables where higher dimensional operators contribute in some specific combinations, such as $L_1 - L_2/2$ or $L_9 + L_{10}$, are known to be free from chiral divergences. We will see examples of such observables in the present context in the coming sections. Notice that mixed loops induce yet another counterterm $\delta\beta$ that will be instrumental in finding finite combinations. Since it is well known[15] that there is a one-to-one correspondence between divergences in the non-linear theory and $\log M_H$ terms in the (linear) minimal SM, to identify observables which are free from chiral infinities we have simply to look for observables that are finite in the large M_H limit. For those observables that cancel the divergent logs in the large M_H limit, we can take the limit of a heavy Higgs mass safely. Now, in the minimal SM we can compute such observables in two different ways[13]:

1) Work consistently at the one loop level with a propagating physical Higgs and then expand in inverse powers of M_H^2 and retain only the constant piece. This is what is usually done.
2) Work with the non-linear sigma model, including the appropriate values of the L_i^{MSM}. It is not correct to work with the lowest dimensional operator of the non-linear sigma model $\mathcal{L}^{(2)}$ and ignore the terms of $\mathcal{O}(p^4)$.

For this type of observables it is quite easy now to determine its value in a different theory; we just have to replace L_i^{MSM} by the appropriate L_i's. The recipe is

$$A^X = A^{MSM} - A^{(4)}(L_i^{MSM}) + A^{(4)}(L_i^X) \tag{10}$$

where A is some observable that in the MSM is *finite* when we take $M_H \to \infty$ and $A^{(4)}$ is the contribution from $\mathcal{L}^{(4)}$. Since the difference between the value of the observable A in the model X and in the minimal SM with a large Higgs mass involves only $\mathcal{O}(p^4)$ terms, it can be worked out by a tree level calculation only. This is quite remarkable.

The above expression is quite useful because many observables have been computed in the minimal SM already. It would be equally correct (always for observables which are finite in the non-linear theory) to compute A^X directly in the non-linear theory, according to the recipe

$$A^X = A^{(2)} + A^{(4)}(L_i^X) \tag{11}$$

where $A^{(2)}$ is the contribution from the effective lagrangian $\mathcal{L}^{(2)}$ (plus gauge, ghost and matter part eventually) and $A^{(4)}$ is the contribution from $\mathcal{L}^{(4)}$.

3. Application to $e^+e^- \to \bar{f}f$

We will work throughout in the on-shell scheme and use the 't Hooft-Feynman gauge. Any effects in $e^+e^- \to \bar{f}f$ that are sensitive to the SBS are contained, to a very good approximation, in three basic process-independent quantities[16].

1) Δr, appearing in the relation between the Fermi constant G_F (relevant in muon decay) and the W mass.

$$G_F = \frac{\pi\alpha}{\sqrt{2}M_W^2 \sin^2\theta_W} \frac{1}{1-\Delta r} \tag{12}$$

$\Delta r = \Delta r(0) +$ *vertex corrections* $+$ *box corrections*, where $\Delta r(q^2)$ is defined in terms of unrenormalized Green functions as

$$\Delta r(q^2) = -\frac{\Sigma_{WW}(q^2) - \Sigma_{WW}(M_W^2)}{q^2 - M_W^2} + \Sigma'_{\gamma\gamma}(0) + \frac{c^2}{s^2}\left(\frac{2s}{c}\frac{\Sigma_{\gamma Z}(0)}{M_Z^2} - \frac{\Sigma_{ZZ}(M_Z^2)}{M_Z^2} + \frac{\Sigma_{WW}(M_W^2)}{M_W^2}\right) \tag{13}$$

2) ρ, the ratio between neutral and charged currents at $q^2 = 0$. If we write $\rho = 1 + \Delta\rho$

$$\Delta\rho = \frac{\Sigma_{ZZ}(0)}{M_Z^2} - \frac{\Sigma_{WW}(0)}{M_W^2} - 2\frac{s}{c}\frac{\Sigma_{\gamma Z}(0)}{M_Z^2} \tag{14}$$

3) $\Delta\kappa$, entering in the ratio between J_{em} and J^3 in the one-loop corrected Born approximation, i.e. in the effective Weinberg angle $\bar{s}^2(q^2) = \kappa(q^2)s^2 = s^2(1 + \Delta\kappa(q^2))$.

$$\Delta\kappa(q^2) = -\frac{c}{sq^2}(\Sigma_{\gamma Z}(q^2) - \Sigma_{\gamma Z}(0)) - \frac{c^2}{s^2}\left(\frac{2s}{c}\frac{\Sigma_{\gamma Z}(0)}{M_Z^2} - \frac{\Sigma_{ZZ}(M_Z^2)}{M_Z^2} + \frac{\Sigma_{WW}(M_W^2)}{M_W^2}\right) \tag{15}$$

Like for $\Delta r(q^2)$, this equation includes only the self-energy ("oblique") corrections. The (subtracted) vertex corrections are largely independent of the SBS.

The "oblique" corrections to Δr, $\Delta\rho$ and $\Delta\kappa$ can be separated into a fermionic and a bosonic part. For instance we write $\Delta\rho = \Delta\rho^{(f)} + \Delta\rho^{(b)}$. In turn in $\Delta\rho^{(b)}$, $\Delta\kappa^{(b)}$ and $\Delta r^{(b)}$ we distinguish between the part coming from $\mathcal{L}^{(2)}$ (plus the gauge piece) at one loop, and the part coming from $\mathcal{L}^{(4)}$ that we have to use only at tree level. For instance,

$$\Delta\kappa^{(b)} = \Delta\kappa^{(b)}_{(2)} + \Delta\kappa^{(b)}_{(4)} \tag{16}$$

To illustrate the discussion in the previous section, we quote from [13] the universal part for $\Delta r^{(b)}$, namely $\Delta r^{(b)}_{(2)}$

$$\Delta r^{(b)}_{(2)} = vertex \; + \; box \; + \Delta\alpha - 2\frac{c^2}{s^2}\delta\beta$$

$$+\frac{g^2}{16\pi^2}(\frac{11}{12}N_\epsilon - \frac{11}{12}\log\frac{M_W^2}{\mu^2} + \log c^2(\frac{5}{c^2} - 1 + 3\frac{c^2}{s^2} - \frac{17}{4s^2c^2}) - s^2(3+4c^2)F(M_Z^2, M_W, M_W)$$

$$+I_2(c^2)(1-\frac{c^2}{s^2}) + \frac{c^2}{s^2}I_1(c^2) + \frac{1}{8c^2}(43s^2 - 38) + \frac{1}{18}(154s^2 - 166c^2) + \frac{1}{4c^2} + \frac{1}{6})$$

$$N_\epsilon = \frac{2}{\epsilon} - \gamma_E + \log 4\pi \qquad \Delta\alpha = 0.0602 + \frac{40}{9}\frac{\alpha}{\pi}\log\frac{M_Z}{92(GeV)}$$

$$vertex \; + \; box \; = (\frac{g^2}{16\pi^2})(6 + \frac{7-4s^2}{2s^2}\log c^2) \tag{17}$$

and explicit expressions for the I_1,I_2 and F functions can be found in [16,17]. We refer the interested reader to [13] for the analogous expressions for $\Delta\rho$ and $\Delta\kappa$.

On the other hand, $\mathcal{L}^{(4)}$ gives the following contributions to Δr

$$\Delta r^{(b)}_{(4)} = -2g^2(L_{10} + \delta L_{10}) \tag{18}$$

When we put together (17) and (18) we notice that the divergences due to the non-linear nature of the theory cancel. Unfortunately, for each individual observable we have to trade this divergence for the knowledge of the renormalized $L_{10}(\mu)$ and $\beta(\mu)$ at some scale μ, extracting them from the experiment. In other words, beyond the minimal SM each of the quantities $\Delta\rho$, $\Delta\kappa$ and Δr are not really predictable by itself. The renormalized coefficients $L_i(\mu)$, $\beta(\mu)$ are the ones that should be compared with the coefficients of Gasser and Leutwyler, in exact parallelism with what is done in pion scattering. It should be clear to the reader that if we simply ajust the renormalized coefficients $L_i(\mu)$ and $\beta(\mu)$ so as to reproduce the values for Δr, etc. (which can certainly be done) and do nothing else we are throwing away precious information about the SBS. The dependence on the underlying theory is *only* in the finite parts and not in the chiral logs. For this reason our definition of the effective action is the only natural one in this context, even if, physically, one is more confortable with a renormalized action in phenomenological applications.

The way out is to look for combinations of observables that are renormalization group invariant. These are the combinations that in the minimal SM do not depend on $\log M_H$. For instance, in the combinations

$$O_1 = \Delta\rho - \frac{9s^2}{10c^2}\Delta\kappa \qquad O_2 = \Delta r + \frac{11}{10}\Delta\kappa \qquad O_3 = \Delta\rho + \frac{9s^2}{11c^2}\Delta r \tag{19}$$

the counterterms δL_{10} and $\delta\beta$ cancel, but L_{10} remains.

Remarkably, in the observables O_i there is an almost exact cancellation of the leading quadratic dependence on m_t. A simple exercise shows that an experimental uncertainty of 10 GeV in the top mass propagates to an uncertainty in O_i which is well below the expected errors in measurements of O_i. Fig. 2 shows the relation between Δr and $\Delta\rho$ for a variety of underlying theories in the SBS. The curved solid line corresponds to the SM (calculated in the linear theory for a variety of M_H values and three different m_t values). We see that for those observables that cancel the $\log M_H$ piece the result is in perfect agreement with the one loop calculation in the non-linear theory ($\mathcal{L}^{(2)}$) plus $L_{10}^{MSM} = 0$. The latter corresponds to the straight line with $N_{eff} = 0$ We refer the reader to [13] for more details.

4. $e^+e^- \to W^+W^-$ at High Energies

The study of the W and Z self interactions in general, and the precise measurement of the γWW and ZWW couplings in particular are some of the most important experimental tasks ahead. Due to the subtle cancellations taking place in the SM among the s-channel γ,Z exchange and t-channel ν exchange, one expects this process to be particularly sensitive to deviations from the SM. "Big" deviations of these couplings from their values in the $SU(2) \times U(1)$ gauge theory would very quickly produce observable effects like for instance an anomalous high energy behaviour of the $e^+e^- \to W^+W^-$ amplitude. However, these couplings enter in a number of radiative corrections, well tested by now, and "big" deviations can be already excluded. In other words, there is no reason to expect below 1 TeV, sizeable deviations from the tree-level amplitudes, rather one will have to deal with deviations that are typically of the same order of magnitude as the electroweak radiative corrections.

The chiral lagrangian formalism provides a natural framework to make qualitative and quantitative predictions in a systematic way for the deviations to the γWW and ZWW effective couplings from both types of contributions, i.e. loops with known particles of the SM and effective operators incorporating the effect of unknown ones.

The most general CP preserving VWW coupling ($V = \gamma$, Z), consistent with the symmetries and kinematical relations of the problem, is[18]

$$\begin{aligned}\mathcal{L}_{VWW}/g_{VWW} =& ig_1^V(W^+_{\mu\nu}W^{\mu}V^{\nu} - W^+_{\mu}V_{\nu}W^{\mu\nu}) + i\kappa^V W^+_{\mu}W_{\nu}V^{\mu\nu}+\\ & i\frac{\lambda^V}{M_W^2}W^+_{\lambda\mu}W^{\mu}_{\nu}V^{\nu\lambda}+\\ & \frac{g_5^V}{q^2}\partial^{\mu}\tilde{V}^{\rho\nu}\Big(W^+_{\mu}(\partial_{\rho}W_{\nu}) - (\partial_{\rho}W^+_{\mu})W_{\nu} + W_{\mu}(\partial_{\rho}W^+_{\nu}) - (\partial_{\rho}W_{\mu})W^+_{\nu}\Big)\end{aligned} \quad (20)$$

where W_{μ} denotes the W^-_{μ} field, $W_{\mu\nu}$, $V_{\mu\nu}$ are the corresponding field strengths and $\tilde{V}_{\mu\nu} = \frac{1}{2}\epsilon_{\mu\nu\rho\sigma}V^{\rho\sigma}$. The overall constants g_{VWW} can be chosen as in the SM $g_{\gamma WW} = -e$, $g_{ZWW} = e\cot\theta_W$ In eq.(4.1) the effective couplings $g_1^V(q^2)$, $\kappa^V(q^2)$ and $\lambda^V(q^2)$ respect the two discrete symmetries C and P separately whereas $g_5^V(q^2)$ is C and P violating but CP preserving.

The phenomenological lagrangian in eq.(20) is a parametrization in *a* particular gauge; therefore, it need not be *manifestly* gauge invariant. In fact in eq.(20) only the term in λ_V is gauge invariant in an obvious manner. What should be done instead is to demand this invariance at the level of the "parent" effective lagrangian; before fixing the gauge. This is the case of the chiral lagrangian in eq.(4) that parametrizes different SBS's. Thus one can start with the invariant lagrangian in eq.(4) and end up with the lagrangian in eq.(20) in the unitary gauge $U = 1$. There will be some relations among the L_9 and L_{10}

of eq.(4) and the g_1^V and κ^V of eq.(20). Note that in the language of chiral lagrangians the terms in λ^V and in g_5^V will not appear up to $O(p^6)$, and therefore they will be suppressed respect to g_1^V and κ_V by one power of $p^2/4\pi v^2$.

The experimental sensitivities to $\delta\kappa$ and $\delta\lambda$ that can be reached are[19]: 0.2 at LEPII (200 GeV, 1.2 fb^{-1}), 0.01 at NLC (500 GeV, 10 fb^{-1}) and 0.0025 at NLC (1 TeV, 40 fb^{-1}). In addition the study of ref.[20] shows that a dedicated $\gamma\gamma$ collider with 350 GeV and 10 fb^{-1} will be sensitive through the $\gamma\gamma \to W^+W^-$ process to a deviation in the photon couplings of 0.03-0.05. From the above quoted values one can conclude that although LEPII will, most likely, not be sensitive enough to the small deviations we are concerned with, the next generation of linear colliders will indeed start being sensitive to them.

Within the SM the values of the effective couplings at tree level are $g_1^\gamma = g_1^Z = 1$, $\kappa^\gamma = \kappa^Z = 1$ and zero the rest. Once radiative corrections are switched on all these effective couplings will pick contributions both from standard and non-standard physics. The former will be given by $\mathcal{L}^{(2)}$, whilst the latter will be parametrized in terms of $\mathcal{L}^{(4)}$. It is by no means obvious that all radiative corrections can be included in such effective couplings, but this turns out to be the case. This is discussed in great detail in [21].

The differential cross-section of the process $e^+e^- \to W^+W^-$ with polarized electrons up to $O(\alpha^3)$ is[22]

$$\left(\frac{d\sigma}{d\Omega}\right)^\sigma = \left(\frac{d\sigma}{d\Omega}\right)_0^\sigma (1+\delta_B) + 2Re(T_0^\sigma \delta T^\sigma)\frac{\alpha}{4\pi}\frac{1}{4q^2}\sqrt{1-\frac{4M_W^2}{q^2}} \tag{21}$$

where, δ_B denotes the contribution from soft photon bremsstrahlung, $\left(\frac{d\sigma}{d\Omega}\right)_0^\sigma$ is the Born cross-section, T_0^σ is the lowest order amplitude and δT^σ parametrizes the total contribution from virtual corrections. The explicit expressions for the lowest order amplitudes corresponding to left-handed electrons, T_0^-, and right-handed electrons, T_0^+, are the following

$$T_0^- = T_0^{-\nu} + T_0^{-\gamma} + T_0^{-Z} \qquad T_0^+ = T_0^{+\gamma} + T_0^{+Z}$$

$$T_0^{-\nu} = \mathcal{M}_1^- e^2 (g_1^{e\nu})^2 \frac{1}{p_\nu^2}$$

$$T_0^{-\gamma} = -(\mathcal{M}_2^- - \mathcal{M}_3^-)e^2\frac{2}{q^2} \qquad T_0^{-Z} = (\mathcal{M}_2^- - \mathcal{M}_3^-)(e^2\frac{c}{s}g_e^-)\frac{2}{q^2-M_Z^2}$$

$$T_0^{+\gamma} = -(\mathcal{M}_2^+ - \mathcal{M}_3^+)e^2\frac{2}{q^2} \qquad T_0^{+Z} = (\mathcal{M}_2^+ - \mathcal{M}_3^+)(e^2\frac{c}{s}g_e^+)\frac{2}{q^2-M_Z^2} \tag{22}$$

where the Z boson couplings to left and right handed electrons and the $e\nu W$ coupling are given by

$$g_e^- = \frac{1}{sc}(-\frac{1}{2}+s^2) \ , \ g_e^+ = \frac{s}{c} \ , \ g_1^{e\nu} = \frac{1}{\sqrt{2}s} \tag{23}$$

Explicit expressions for the Dirac matrix elements $\mathcal{M}_i^\pm(\lambda_+,\lambda_-)$ can be found in ref. [22]. The total virtual corrections include self-energy, vertex and box corrections that we shall denote as follows

$$\delta T^\sigma = \delta T_S^\sigma + \delta T_V^\sigma + \delta T_B^\sigma \tag{24}$$

The $\log M_H$ dependence of the differential cross-section in the large M_H limit comes entirely from the virtual corrections and more precisely from the self-energy, δT_S^σ, and vertex corrections, δT_V^σ. We separate the box and bremsstrahlung pieces because they contain neither a significant contribution from the SBS nor information about radiative corrections to the triple gauge vertex. As a result of not including the soft photon bremsstrahlung, the effective constants (κ^V and g_1^V in particular) will depend on an infrared cut-off that is traded by the experimental resolution ΔE. We take $\Delta E = 0.1E$.

In the large M_H limit the different amplitudes go as[23]

$$\delta T^{-\nu} \sim T_0^{-\nu}\left(\frac{g^2}{16\pi^2}\left(\frac{11}{12}\right)\log M_H^2\right)$$

$$\delta T^{-Z} \sim T_0^{-Z}\left(\frac{g^2}{16\pi^2}\left(\frac{1}{24c^2}+\frac{5}{6}\frac{1}{1-2s^2}\right)\log M_H^2\right)$$

$$\delta T^{+Z} \sim T_0^{+Z}\left(\frac{g^2}{16\pi^2}\left(\frac{1}{24c^2}\right)\log M_H^2\right) \tag{25}$$

The ones independent of $\log M_H$ are not listed. This behaviour is reflected in the following dependence of the effective couplings

$$\frac{\delta g_1^{e\nu}}{g_1^{e\nu}} \sim \frac{g^2}{16\pi^2}\frac{11}{24}\log M_H^2 \qquad \frac{\delta g_1^Z}{g_1^Z} \sim \frac{g^2}{16\pi^2}\frac{1}{24c^2}\log M_H^2 \qquad \frac{\delta\kappa^Z}{\kappa^Z} \sim \frac{g^2}{16\pi^2}\frac{1}{24c^2}\log M_H^2 \tag{26}$$

Note that there is no $\log M_H$ dependence in the effective γWW couplings. It is also remarkable that these logarithmical pieces are the same for the two effective ZWW couplings g_1^Z and κ^Z. Besides, there are no $\log M_H$ contributions to the λ^V or g_5^V couplings.

Considering now *all* the radiative corrections one arrives at the following effective couplings

$$g_1^\gamma = 1 - \frac{\Sigma_{\gamma\gamma}(q^2)}{q^2} - \frac{\partial\Sigma_{WW}(q^2)}{\partial q^2}\Big|_{q^2=M_W^2} + \Sigma'_{\gamma\gamma}(0) + V_1^{\gamma WW}(q^2) - \frac{q^2}{2M_W^2}V_2^{\gamma WW}(q^2)$$

$$\kappa^\gamma = 1 - \frac{\Sigma_{\gamma\gamma}(q^2)}{q^2} - \frac{\partial\Sigma_{WW}(q^2)}{\partial q^2}\Big|_{q^2=M_W^2} + \Sigma'_{\gamma\gamma}(0) + V_3^{\gamma WW}(q^2) - V_1^{\gamma WW}(q^2) - \left(1-\frac{q^2}{2M_W^2}\right)V_2^{\gamma WW}(q^2)$$

$$g_1^Z = 1 + \frac{s^2-c^2}{sc}\frac{\Sigma_{\gamma Z}(q^2)}{q^2} - \frac{\Sigma_{ZZ}(q^2)-\Sigma_{ZZ}(M_Z^2)}{q^2-M_Z^2} - \frac{\partial\Sigma_{WW}(q^2)}{\partial q^2}\Big|_{q^2=M_W^2} + \Sigma'_{\gamma\gamma}(0) + V_1^{ZWW}(q^2) - \frac{q^2}{2M_W^2}V_2^{ZWW}(q^2)$$

$$\kappa^Z = 1 + \frac{s^2-c^2}{sc}\frac{\Sigma_{\gamma Z}(q^2)}{q^2} - \frac{\Sigma_{ZZ}(q^2)-\Sigma_{ZZ}(M_Z^2)}{q^2-M_Z^2} - \frac{\partial\Sigma_{WW}(q^2)}{\partial q^2}\Big|_{q^2=M_W^2} + \Sigma'_{\gamma\gamma}(0) + V_3^{ZWW}(q^2) - V_1^{ZWW}(q^2) - \left(1-\frac{q^2}{2M_W^2}\right)V_2^{ZWW}(q^2) \tag{27}$$

The unrenormalized vertex functions V in these formulae are defined by

$$
\begin{aligned}
\Gamma^{\alpha\beta\mu}_{VWW} &= V_1^{VWW}(k_- - k_+)^\mu g^{\alpha\beta} - \frac{V_2^{VWW}}{M_W^2}(k_- - k_+)^\mu q^\alpha q^\beta \\
&+ V_3^{VWW}(q^\alpha g^{\mu\beta} - q^\beta g^{\alpha\mu}) + iV_5^{VWW}\epsilon^{\mu\alpha\beta\nu}(k_- - k_+)_\nu
\end{aligned}
\tag{28}
$$

Terms proportional to $\Sigma_{\gamma Z}(0)$ (a pure divergence in dimensional regularization) have been omitted. $\Sigma_{\gamma Z}(0)$ is needed to render these couplings finite, but it is otherwise irrelevant. In addition to these effective couplings one also needs to introduce effective $ee\gamma$ and eeZ couplings to sum up all radiative corrections. These contain no information on the SBS.

The effective couplings λ^V and g_5^V do not get contributions from self energies or renormalization counterterms, although they are certainly non-zero due to radiative vertex corrections. g_5^V is a interesting object. The operator it accompanies should be expressable in terms of gauge and U fields in a gauge invariant manner. Being C and P violating it should be contained in the abnormal parity effective lagrangian derived recently by Bijnens (and presented in this Workshop). However, it is not there. The reason is that to construct this term one must use necessarily a mass matrix M. This implies that g_5^V must go like $m_t - m_b$ in the SM.

For the t-channel contributions we proceed in a similar fashion. Due to the fact that the neutrino is not on-shell a new spinorial structure $\mathcal{M}_4^-$ and a new form factor $g_2^{e\nu}$ appear. For the structure that is already present at tree level, we introduce the effective coupling $g_1^{e\nu}$, defined as $g_1^{e\nu} \to g_1^{e\nu} + \delta g_1^{e\nu}$

$$
\frac{\delta g_1^{e\nu}}{g_1^{e\nu}} = -\frac{1}{2}\hat{\Sigma}_\nu(p_\nu^2) + \frac{1}{2}\Delta r(M_W^2) + \hat{V}_1^{e\nu}(k_-, p_\nu) + \hat{V}_1^{e\nu}(k_+, p_\nu)
\tag{29}
$$

The hats denote renormalized Green functions.

The way to estimate the values of κ^V, g_1^V, $g_1^{e\nu}$ and the rest of parameters within the framework of an effective theory has been explained in detail in sections 2 and 3. We take advantage of the fact that different symmetry breaking alternatives differ from the result in the minimal SM by terms of $\mathcal{O}(p^4)$ that, to the chiral order to which we are working, enter only at tree level. Knowing the predictions of the minimal SM in the limit of large M_H and the appropriate values of the L_i is enough to have a *precise* prediction in any other theory. In particular, the contributions from $\mathcal{L}(p^4)$ to the VWW effective couplings are

$$
\begin{aligned}
g_1^\gamma &= 1 \qquad \kappa^\gamma = 1 + g^2 L_{10} + g^2 L_9 \\
g_1^Z &= 1 + g^2\frac{1}{2c^2}(L_9 + \delta L_9) - g^2\frac{1}{c^2}(L_{10} + \delta L_{10}) \\
\kappa^Z &= 1 - g^2\frac{s^2 - c^2}{2c^2}(L_9 + \delta L_9) - g^2\frac{1 + s^2}{c^2}(L_{10} + \delta L_{10}) \\
g_1^{e\nu} &= \frac{1}{\sqrt{2}s}\left(1 - g^2(L_{10} + \delta L_{10}) - \frac{c^2}{s^2}\delta\beta\right)
\end{aligned}
\tag{30}
$$

The rest of the effective couplings are independent of the $\mathcal{O}(p^4)$ coefficients and of $\delta\beta$. Accordingly, they must be finite in the large M_H limit (and they are). By the same token they must also be universal in the limit of large M_H. Notice that in g_1^γ and κ^γ the chiral divergences cancel and, as a consequence, the effective couplings for the photon vertex are predictable on their own in the effective theory. This is not quite the case for the Z couplings, but it is easy to see that $\kappa^Z - g_1^Z$ is predictable since its $\mathcal{O}(p^4)$ contribution is

$$\kappa^Z - g_1^Z = -g^2 \frac{s^2}{c^2}(L_9 + L_{10}) \tag{31}$$

Recall that $L_9 + L_{10}$ is independent of chiral logs. $g_1^{e\nu}$ is expressable in terms of Δr.

Using these expressions it is easy to keep track of the $\log M_H$ terms. We know already that a $\log M_H$ corresponds to a $1/\epsilon$ in the non-linear theory. These $1/\epsilon$ are cancelled by δL_{10}, δL_9 and $\delta\beta$. Therefore it is enough to keep track of the counterterms δL_{10}, δL_9 and $\delta\beta$ to identify $\log M_H$'s. We reproduce in this way the $\log M_H$ behaviour quoted in (25), but with a much smaller calculational effort than the one required to trace the $\log M_H$'s in the Green functions of the minimal SM.

The radiative corrections are very much dominated by the unknown top mass. In our numerical analysis we have used the value $m_t = 130$ GeV, not very different from the central value of the range allowed from an analysis of radiative corrections to $e^+e^- \to \bar{f}f$.

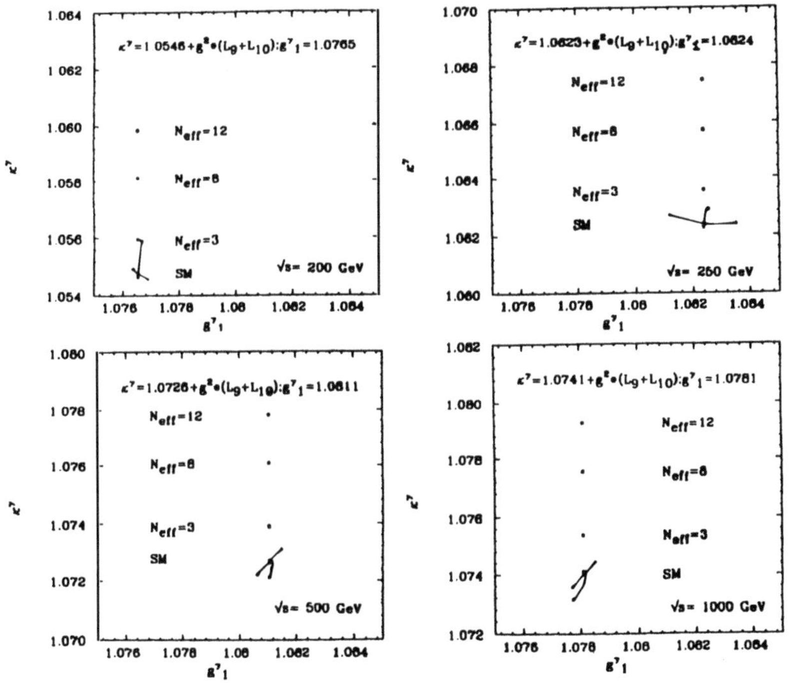

Fig. 3

We include in our results an error bar of $m_t = \pm 10$ GeV. We do not believe that it is really feasible to analyze rigorously the contribution of the SBS before the uncertainties due to the top quark mass are removed. Since we take as input the M_W mass in our analysis and this is not very well measured, we have to decide on a value for $\sin^2\theta_W$. We have taken $\sin^2\theta_W = 0.230$, which, with the value of the Z mass $M_Z = 91.185 \pm 0.020$ GeV, gives $M_W = 80.015$ GeV. The numerical results for the photon effective couplings are shown in Fig.3 for various choices of the center-of-mass energy. In this case the prediction of the effective theory is a point in the $\kappa^\gamma - g_1^\gamma$ plane. The numerical values for κ^γ and g_1^γ written on the top correspond to the universal contribution coming from $\mathcal{L}^{(2)}$ to one-loop level (plus gauge, ghost and matter part). We show four different points. One is the prediction of the minimal SM for a very large Higgs mass, which, as pointed out is a finite limit for the photon couplings. We also show, for this point, a line showing the variations due to effects $\mathcal{O}(1/M_H^2)$ for the range $M_H = 200 - 3000$ GeV (the line in the plot with crossed points). As expected the variations are tiny. An error in the top quark mass of ± 10 GeV around the assumed value of $m_t = 130$ GeV is also represented for the particular choice of $M_H = 1000$ GeV. The modifications induced by technicolor theories can be included by adding the appropriate L_i's. We use the values given in (8). Fig.3 shows the results for $N_{TC} = 3$, $N_{TC} = 8$ and $N_{TC} = 12$. We take N_d=1, but the analysis can be easily extended.

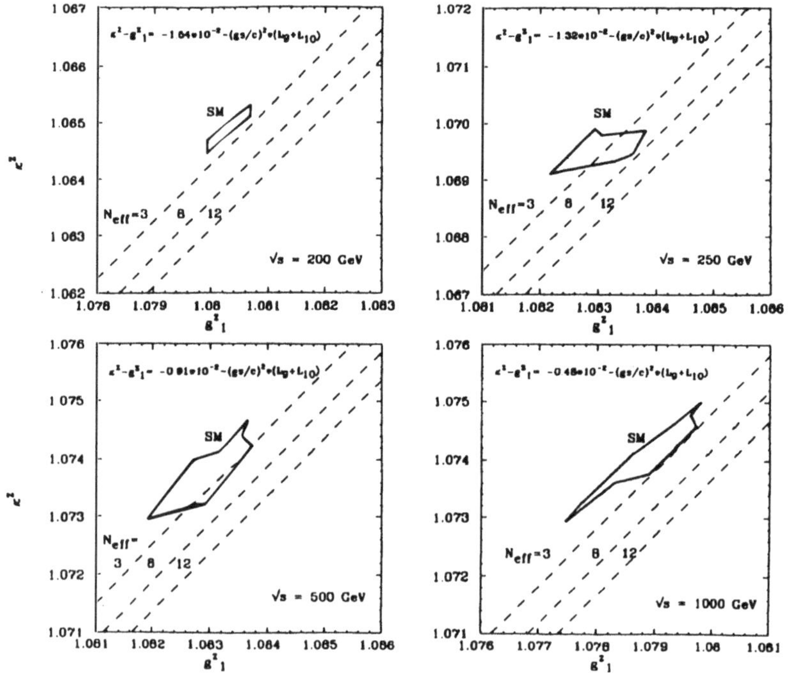

Fig. 4

For the Z couplings the situation is a little less clear. If we set $L_9 = L_{10} = 0$ we have the minimal SM result. A variation of the Higgs mass in the range $M_H = 200 - 3000$ GeV combined with a top mass in the range $m_t = 130 \pm 10$ GeV leads to the allowed regions in Fig.4 In this case, the same type of technicolor theories as considered above lead only to relations between κ^Z and g_1^Z, which correspond to the straight lines of Fig.4.

Recent related work can be found in[24].

7. Conclusions

Let us summarize the salient points of our analysis.

1) The most efficient and elegant way to discriminate amongst competing possibilities for the SBS of the electroweak theory (technicolor, minimal SM, supersymmetry,...) is provided by a Chiral Lagrangian.

2) The choice of which are the "right" observables to distinguish amongst various possibilities is dictated by the Chiral Lagrangian itself. The privileged observables are those where chiral logs drop out and are, accordingly, free from cut-off ambiguities. From these observables, the value of the higher-order coefficients in the Chiral Lagrangian can be extracted from experiment and compared with theoretical predictions. In the linear theory, these are observables where the logarithmic dependence on the Higgs mass cancels.

3) The relevant coefficients are those of $\mathcal{O}(p^4)$. These coefficients accompany operators of dimension four and, therefore, they are *not* suppressed by any power of some large scale. This is why they are so important. At "low energies" the non-universal information about the SBS is contained in these coefficients (four if the SBS is custodially symmetric). It is nice that everything is so economically described by a few numbers. On the other hand, it may be difficult to open a window on the new physics beyond the SM with only the knowledge of these four quantities.

4) For LEP I the results are very promising. We do not believe that it is possible to conclude much yet, but it is certainly true that by the end of the run of LEP (assuming, of course, that the Higgs or something else is not found before) it will be possible to extract very interesting conclusions on the SBS, specially if the polarization or high luminosity options are implemented. LEP II, on the other hand, will not be able to shed much light on possible realizations of the symmetry breaking except, indirectly, through a precise measure of the W mass.

Acknowledgements

It is a pleasure to thank A.Dobado and M.J.Herrero for an enjoyable collaboration. Discussions with F. del Aguila, J.Bernabeu, J.Bijnens, B.Gavela, M. Martinez, E. de Rafael and A. de Rújula are gratefully acknowledged. W.Beenakker and T.Sack provided us with their code computing the electroweak radiative corrections to the amplitude for the process $e^+e^- \to W^+W^-$. Finally, I would like to thank the organizers and participants of this Workshop. The financial support of the EEC Science Twinning Grant SCI-000337 and CICYT Grant AEN89-0347 is gratefully acknowledged.

References

1. M.Veltman, *Acta Phys. Polon.* **B8** (1977) 475
2. See e.g. M.Chanowitz, *Ann. Rev. Nucl. Part. Sci.* **38** (1988) 323

3. See for instance: H.Burkhardt and J.Steinberger, CERN preprint CERN-PPE /91 - 50, (1991); G.Altarelli, R.Barbieri and S.Jadach, CERN preprint CERN-TH-6124/91, June 1991
4. R.Kleiss and D.Treille, *Preliminary Report to the LEPC from the Working Group on the High luminosities at LEP* (May 1990) p.180
5. J.Gasser and H.Leutwyler, *Ann. Phys.* (NY) **158** (1984) 142; *Nucl. Phys.* **B250** (1985) 465
6. W.Marciano and J.Rosner, *Phys. Rev. Lett.* **65** (1990) 2963
7. S. Weinberg,*Physica* **96 A**(1979)327; H.Georgi, *Weak interactions and Modern Particle Theory.* Benjamin/Cummings Publishing Company, Inc., Menlo Park, California, 1984
8. A.Dobado, D.Espriu and M.J.Herrero, *Phys. Lett.* **B255** (1991) 405; H.Georgi, *Nucl. Phys.* **B 363** (1991) 301
9. M.Peskin and T.Takeuchi,*Phys. Rev. Lett.* **65** (1990) 964; B.Holdom and J.Terning, *Phys. Lett.* **B247** (1990) 88; G.Altarelli and R.Barbieri, *Phys. Lett.* **B253** (1991) 161
10. See e.g. M.S.Chanowitz, M.K.Gaillard, *Nucl. Phys.* **B261** (1985) 379
11. A.Dobado and M.J. Herrero, *Phys. Lett.* **B228** (1989) 495; *ibid* **B233** (1989) 505; J.F. Donoghue and C. Ramirez, Phys. Lett. **B234** (1990) 361; S.Dawson and G.Valencia, *Nucl. Phys.* **B352** (1991) 27
12. D.Espriu, E. de Rafael and J.Taron, *Nucl. Phys.* **B345** (1990) 22
13. A.Dobado, D.Espriu and M.J.Herrero, *Phys. Lett.* **B255** (1991) 405
14. M.Golden and L.Randall, *Nucl. Phys.* **B361** (1991) 1; M.Dugan and L.Randall, Harvard Univ. preprint HUTP/91-A10 (1991)
15. A.Longhitano, *Nucl. Phys.* **B188** (1981) 118
16. See e.g. M. Consoli and W. Hollik, in Z Physics at LEP 1, CERN Yellow Report (1989), Ed. by G. Altarelli et al.
17. W.J.Marciano and A.Sirlin, *Phys. Rev.* **D22** (1980) 2697
18. K.Hagiwara, R.D.Peccei and D.Zeppenfeld, *Nucl.Phys.* **B282** (1987) 253
19. See for instance: *Proceedings of the Workshop Physics with e^+e^- linear colliders*, Annecy, June 10-11,1991
20. E.Yehudai, SLAC-preprint, SLAC-PUB-5495, April 1991
21. D.Espriu and M.J.Herrero, CERN preprint 6186 (1991)
22. M.Bohm et al., *Nucl. Phys.* **B304** (1988) 463
23. M.Lemoine and M.Veltman, *Nucl. Phys.* **B164** (1980)
24. B.Holdom, Fermilab preprint Fermilab-pub-90/263-T, 1990; A.Falk, M.Luke and E.Simmons, Harvard preprint HUTP-91/A021; A. de Rujula et al. CERN preprint 6272 (1991)

TOP QUARK CONDENSATE*

Koichi Yamawaki
Department of Physics, Nagoya University
Nagoya 464-01, Japan

ABSTRACT

We outline basic idea of the top quark condensate model (Top Mode Standard Model) proposed by Miransky, Tanabashi and Yamawaki and by Nambu, in which the dynamical electroweak symmetry breaking is due to a top quark condensate. Based on the ladder Schwinger-Dyson equation, we explicitly obtain spontaneous chiral-symmetry-breaking solutions of the gauged Nambu-Jona-Lasinio models (gauge theories plus four-fermion interactions) and argue the phase structure associated with them. We find a very large anomalous dimension in these theories, particularly $\gamma_m \simeq 2$ in QCD plus four-fermion interactions, which is a crucial ingredient to the Top Mode Standard Model. The simplest version of the model predicts a very large mass for the top quark, typically of $m_t \simeq 250\text{GeV}$, and also existence of a Higgs boson as a $\bar{t}t$ composite with the mass $M_H \simeq \sqrt{2}m_t$. Possible modifications of these values, particularly made by Bardeen, Hill and Lindner, are also discussed.

1. Introduction

As it stands now, the Standard Model (SM) is a very successful framework for describing elementary particles in the low energy region, say, less than 100GeV. However one of the most mysterious parts of the theory, *the Origin of Mass*, has long been left unexplained. Actually, mass of *all* particles in the SM is attributed to a *single* order parameter, the vacuum expectation value (VEV) of the Higgs doublet ($\simeq$ 250GeV). Thus the problem of the origin of mass is simply reduced to understanding the dynamics of the Higgs sector.

Recently, the experiment has shown that the number of light neutrino species is three, which strongly suggests that the number of families is also three. This implies that top quark is the last quark to be found. Also the lower bound of the top quark mass m_t is nearly a hundred GeV. Some indirect measurements suggest even larger value than a hundred GeV. This implies only the top quark has a mass close to the weak scale 250GeV: top quark is a special quark. Thus we may infer a distinguished

*Work supported in part by the Kurata Foundation and the Takeda Science Foundation.

role of the top quark in the mass generation of *all* particles. In the SM language a large top quark mass in fact corresponds to a large Yukawa coupling, which is nothing but a reminiscent of the Goldberger-Treiman relation in hadron physics. This would be a good sign of the dynamical symmetry breaking (DSB), i.e., symmetry breaking due to a composite order parameter like fermion pair condensate.

As a simple picture to understand such a large top quark mass, Miransky, Tanabashi and Yamawaki (MTY)[1,2] proposed an explicit DSB model in which the top quark mass is regarded as a *dynamical mass* due to a *top quark condensate* $\langle\bar{t}t\rangle \neq 0$. The top quark condensate actually produces composite Nambu-Goldstone (NG) bosons which give rise to the mass of W and Z bosons, and at the same time communicates to the mass of other quarks and leptons. Thus the top quark condensate entirely takes the place of elementary Higgs VEV in the SM.* We may call it "Top Mode" Standard Model (TMSM) in contrast to the SM (may be called "Higgs Mode" Standard Model).

At first glance the obstacle to such a scenario would be obvious. We need a large isospin violation in the condensate, $\langle\bar{t}t\rangle \gg \langle\bar{b}b\rangle$, to give a realistic mass difference $m_t \gg m_b$. The problem thus would be two fold:

1. Is there any *explicit* dynamical model to give a large isospin violation $m_t \gg m_b$?

2. How can one guarantee the observed approximate weak isospin symmetry $\rho \equiv m_W^2/m_Z^2 \cos^2\theta_W \simeq 1$ ($\delta\rho \equiv 1 - 1/\rho \lesssim O(10^{-2})$) for such a large isospin symmetry violation $m_t \gg m_b$?

These were the problems that MTY[1,2] simultaneously solved in the concrete dynamical model, "gauged Nambu-Jona-Lasinio (NJL) model" (gauge theory plus four-fermion interactions), through the ladder Schwinger-Dyson (SD) equation. The first question can be answered by the existence of the *critical line*[4,5], while the answer to the second one is given by the appearance of *large anomalous dimension*[6] $\gamma_m \simeq 2$, corresponding to the very slowly damping solution[4,5] of the SD equation.

TMSM is actually based on the QCD plus four-fermion interactions, in which the ladder SD equation yields a very slowly (only logarithmically) damping dynamical mass function. Plugging *explicit* form of such a solution into the Pagels-Stokar type formula for $F_\pi(\simeq 250\text{GeV})$, MTY predicted [1,2] a *very large mass* of the top quark (as a decreasing function of the cutoff Λ), typically

$$m_t \simeq 250\text{GeV} \tag{1.1}$$

at $\Lambda \simeq 10^{19}$GeV (the largest physical cutoff). MTY also predicted a scalar $\bar{t}t$ bound state (Higgs boson) with a mass (based on a naive guess from NJL model[7])

$$M_H \simeq 2m_t. \tag{1.2}$$

*Nambu suggested earlier a close relation among the top quark, Higgs boson and W/Z bosons in the "bootstrap symmetry breaking".[3]

The idea that the Higgs boson is a $\bar{t}t$ composite was also proposed independently by Nambu[8], with the above NJL-type mass relation $M_H = 2m_t$ being introduced as a new input in the "bootstrap symmetry breaking" proposed earlier[3].

Numbers of groups[9,10,11,12,13], most notably Bardeen, Hill and Lindner (BHL)[10], have further studied consequences of the TMSM. BHL made an elegant formulation of the TMSM so as to incorporate additional loop effects of the composite Higgs boson (and $SU(2)_L \times U(1)_Y$ interactions) and confirmed large m_t (as a decreasing function of cutoff), e.g., $m_t \simeq 220$GeV at $\Lambda \simeq 10^{19}$GeV, which is similar to the original MTY value[1,2]. In fact, it is explicitly shown[14] that *BHL becomes identical to MTY*, when such additional loop effects are simply switched off.

In this talk I would like to outline basic idea of the TMSM in the original MTY[1,2] language of the SD equation.[15] I shall also present results of our recent works on the phase structure of the gauged NJL model (renormalization, effective theory, mass of composite Higgs boson, etc.) done in collaboration with Kondo, Shuto and Tanabashi[16,17,14,18]. In the ladder approximation we explicitly obtain a renormalized continuum theory on the critical line at $\Lambda \to \infty$, which has a *large anomalous dimension* $1 < \gamma_m < 2$ near the critical line in the renormalized coupling space *in both the symmetric and the spontaneously broken phases* of the chiral symmetry.[18] This sheds some lights on the possible renormalizability of the four-fermion interactions and also on the renormalization property of the higher dimensional interactions (eight-fermion interactions, etc.) *in the presence of gauge interaction.* The running effect of the QCD coupling is also incorporated in the phase diagram of the two-coupling space, leading to two-fixed-points structure [17].

We also calculate[16] M_σ (mass of the composite scalar (Higgs) boson) in terms of the fermion dynamical mass m in the ladder gauged NJL model. For very small (but *non-zero*) gauge coupling we have $M_\sigma \simeq \sqrt{2}m$ near the critical line, in contrast to the famous result of the pure NJL model[7] $M_\sigma = 2m$. Similarly, in the case of running QCD coupling in the TMSM we have[16]

$$M_H \simeq 1.35 m_t (\simeq \sqrt{2} m_t) \tag{1.3}$$

at $\Lambda \simeq 10^{19}$GeV, in contrast to the original prediction (1.2). The difference is due to the effect of small (but *non-zero*) QCD gauge coupling. The result again coincides with the RG calculation of BHL[10], when the loop effects of composite Higgs boson (and $SU(2)_L \times U(1)_Y$ gauge interactions) are switched off.

The concrete prediction (1.1) and (1.3) might be subject to some modifications in more refined calculations. Nevertheless, the basic idea that the top quark condensate has a significant role in the mass generation of *all* the particles may survive experimental tests in 1990's and/or SSC, which I hope will open the window to deeper understanding of *the Origin of Mass.*

2. Top Mode Standard Model

The Model

Let me first explain the original version of the top quark condensate model (TMSM) proposed by MTY[1,2]. The model consists of the standard three families of quarks and leptons with the standard $SU(3)_C \times SU(2)_L \times U(1)_Y$ gauge interactions but *without Higgs doublet*. Instead of the standard Higgs sector we introduce $SU(3)_C \times SU(2)_L \times U(1)_Y$-invariant *four-fermion interactions among quarks and leptons*, the origin of which is expected to be a new physics not specified at this moment. The new physics specifies the UV scale (cutoff Λ) of the model, in contrast to the infrared (IR) scale (weak scale $F_\pi \simeq 250$GeV) determined by the mass of W/Z bosons.

MTY [1,2] introduced a general form of such four-fermion interactions among the top and bottom quarks $\psi \equiv (t, b)$:

$$\begin{aligned}\mathcal{L}_{4f} \; = \; & \Big[G^{(1)}(\bar{\psi}_L^i \psi_R^j)(\bar{\psi}_R^j \psi_L^i) \\ & + G^{(2)}(\bar{\psi}_L^i \psi_R^j)(i\tau_2)^{ik}(i\tau_2)^{jl}(\bar{\psi}_L^k \psi_R^l) \\ & + G^{(3)}(\bar{\psi}_L^i \psi_R^j)(\tau_3)^{jk}(\bar{\psi}_R^k \psi_L^i)\Big] + \text{h.c.}, \qquad (2.1)\end{aligned}$$

where i, j, k, l are the weak isospin indices and $G^{(1)}$, $G^{(2)}$ and $G^{(3)}$ are the four-fermion coupling constants.

The symmetry structure (besides $SU(3)_C$) of those four-fermion interactions in (2.1) is $SU(2)_L \times SU(2)_R \times U(1)_V \times U(1)_A$, $SU(2)_L \times SU(2)_R \times U(1)_V$ and $SU(2)_L \times U(1)_Y \times U(1)_V \times U(1)_A$, respectively. As will be discussed later, $G^{(2)}$ term is vital to the mass of the bottom quark in the TMSM.[1,2] In the absence of the $G^{(2)}$-term, (2.1) possesses a $U(1)_A$ symmetry which is explicitly broken only by the color anomaly and plays the role of the Peccei-Quinn (PQ) symmetry.[2,12] It is straightforward[1,2] to include other families and leptons into the form of (2.1).

Let us disregard the $G^{(2)}$ term for the moment, in which case (2.1) simply reads

$$\mathcal{L}_{4f} = G_t(\bar{\psi}_L t_R)^2 + G_b(\bar{\psi}_L b_R)^2 + h.c., \qquad (2.2)$$

with $G_t \equiv G^{(1)} + G^{(3)}$ and $G_b \equiv G^{(1)} - G^{(3)}$.

Dynamics for $m_t \gg m_b$

We now explain one of the key points of the TMSM, i.e., explicit dynamics which gives rise to a large isospin violation in the condensate $\langle \bar{t}t \rangle \gg \langle \bar{b}b \rangle$ $(m_t \gg m_b)$. MTY found[1,2] that such a dynamics does exists, based on the spontaneous-chiral-symmetry-breaking (SχSB) solution of the ladder SD equation for the gauged NJL model.

For simplicity we first consider the ladder SD equation with the *non-running* QCD coupling and the four-fermion coupling (2.2). It is possible to show that the ladder SD equation possesses the solution without $U(1)_{em}$ breaking ($\langle \bar{t}b \rangle = \langle \bar{b}t \rangle = 0$).* In such a

*It has recently been shown[20] that such a solution is really the ground state solution.

case, top and bottom quarks satisfy decoupled SD equations. In Euclidean space, the ladder SD equation for each quark propagator $S_i^{-1}(p) = A_i(p^2)(\not p - \Sigma_i(p^2))(i = t, b)$ in Landau gauge takes the form (after angular integration)

$$\Sigma_i(x) = \frac{g_i}{\Lambda^2}\int_0^{\Lambda^2} dy \frac{y\Sigma_i(y)}{y+\Sigma_i(y)^2} + \int_0^{\Lambda^2} dy \frac{y\Sigma_i(y)}{y+\Sigma_i(y)^2} K(x,y), \tag{2.3}$$

where $x \equiv p^2$, $K(x,y) = \lambda/\max(x,y)$ and Λ is a UV cutoff (a scale of new physics). We have defined *dimensionless* four-fermion couplings $g_i \equiv (N_c\Lambda^2/4\pi^2)G_i$ and $\lambda \equiv (3C_2(\boldsymbol{F})/4\pi)\alpha_{QCD}$, with $N_c(= 3)$ and $C_2(\boldsymbol{F}) = (N_c^2 - 1)/2N_c(= 4/3)$ being the number of color and the quadratic Casimir of the fermion color representation $\boldsymbol{F}(= \mathbf{3})$, respectively. (Note that $A_i(p^2) = 1$ in Landau gauge in the ladder approximation.) The dynamical mass function is normalized as $\Sigma_i(m_i^2) = m_i$.

When the QCD coupling is switched off ($K(x,y) = 0$), this SD equation is simply reduced to the gap equation of the NJL model[7] in the large N_c limit. Although the NJL model is not a sensible model (actually a trivial or free theory) in the continuum limit ($\Lambda \to \infty$), essence of the MTY mechanism[1,2] for having maximal isospin violation is most transparent in this model. In fact, in the NJL model there exists a nontrivial solution $\Sigma(p^2) \equiv$ const. $= m \neq 0$ for $g > g^* = 1$. In view of the existence of such a critical coupling $g = g^* = 1$, it is easy to see that our four-fermion interactions (2.2) can give a *maximal isospin violation* in dynamical mass, $m_t \neq 0$ and $m_b = 0$, if $g_t > g^* = 1 > g_b$.

Now, the argument can easily be extended to the gauged NJL model, (2.3). Eq.(2.3) was first studied by Bardeen, Leung and Love[19] for the strong gauge coupling region $\lambda > \lambda_c = 1/4$. A full set of S$\chi$SB solutions of (2.3) in the whole (λ, g) plane and the *critical line* were found by Kondo, Mino and Yamawaki[4] and independently by Appelquist, Soldate, Takeuchi and Wijewardhana[5]. The critical line in the (λ, g) plane is a generalization of the critical coupling in NJL model. It is the line of the second-order phase transition separating spontaneously broken ($m/\Lambda \neq 0$) and unbroken ($m/\Lambda = 0$) phases of the chiral symmetry (Fig.1)[4,5]:

$$\begin{aligned} g &= \frac{1}{4}(1+\omega)^2 \equiv g^*, \qquad \omega \equiv \sqrt{1-\lambda/\lambda_c} \qquad (0 < \lambda < \lambda_c), \\ \lambda &= \lambda_c \qquad (g < \frac{1}{4}). \end{aligned} \tag{2.4}$$

It was further shown by Miransky and Yamawaki [6] that the model has a large anomalous dimension

$$\gamma_m = 1 + \omega \tag{2.5}$$

at the critical line for $0 < \lambda < \lambda_c$. Through the operator product expansion (OPE) and the renormalization-group (RG) equation, such a large anomalous dimension actually corresponds to a slowly damping SχSB solution[4,5] to the ladder SD equation:

$$\Sigma(p^2) \underset{p \gg m}{\simeq} \frac{m^3}{p^2}\left(\frac{p}{m}\right)^{\gamma_m} \simeq m\left(\frac{p}{m}\right)^{-1+\omega} \underset{\lambda \ll 1}{\simeq} m\left(\frac{p}{m}\right)^{-2\lambda}. \tag{2.6}$$

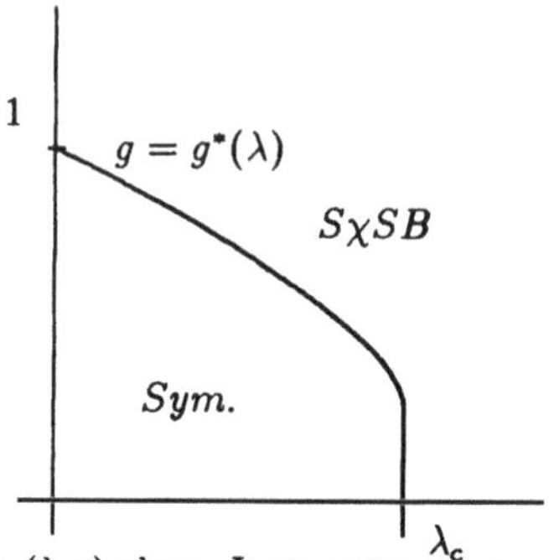

Fig.1 Critical line in (λ, g) plane. It separates spontaneously broken ($S\chi SB$) phase and unbroken phase (*Sym.*) of the chiral symmetry.

Now, the critical coupling of g on the critical line does depend on the value of gauge coupling λ and vice versa. This means that *even a tiny difference (symmetry violation) of λ (g) for the same g (λ) can cause amplified effects on the dynamical mass; $m = 0$ (below the critical line) or $m \neq 0$ (above the critical line).* In view of the critical line (2.4), MTY[1,2] indeed found a solution with *amplified isospin symmetry violation* for a small (*however small*) violation in the coupling constants. Actually, our SD equation (2.3) includes isospin-violating four-fermion couplings $g_t \neq g_b$ ($g^{(3)} \neq 0$). Thus we have an SχSB solution with *maximal isospin violation*, $m_t \neq 0$ and $m_b = 0$, when

$$g_t > g^* = \frac{1}{4}(1+\omega)^2 > g_b \tag{2.7}$$

(g_t is above the critical line and g_b is below it)*.†

One can easily take account of the *running effects* of the QCD coupling in the ladder SD equation ("improved ladder SD equation")[25] by replacing λ in (2.3) by the one-loop running one $\lambda(p^2)$ parametrized as follows:

$$\lambda(p^2) = \begin{cases} \lambda_\mu & (p^2 < \mu_{IR}^2) \\ \frac{A/2}{\ln(p^2/\Lambda_{QCD}^2)} & (p^2 > \mu_{IR}^2) \end{cases}, \tag{2.8}$$

where $A = 18C_2(\boldsymbol{F})/(11N_c - 2N_f)$ $(= 24/(33-2N_f))$ and $\lambda_\mu (= (A/2)/\ln(\mu_{IR}^2/\Lambda_{QCD}^2))$ are constants and $\mu_{IR} (= O(\Lambda_{QCD})$ an artificial "IR cutoff" of otherwise divergent running coupling constant (We choose $\lambda_\mu > 1/4$ so as to trigger the SχSB already in the

*We need *not* to set $G_b = 0$[10] in the four-fermion interactions (2.2)!

†This amplification phenomenon of the symmetry violation called "catalyzed symmetry violation" was first found for $\gamma_m \simeq 1$[21,22] and was shown to be a general feature of the DSB with large anomalous dimension.[1,2] Recently, it has also been re-emphasized[23,24] under the name of "critical enhancement" or "critical instability" in the TMSM with horizontal gauge boson exchanges instead of the four-fermion interactions.

pure QCD). Then the SD equation takes the form

$$\Sigma_i(x) = \frac{g_i}{\Lambda^2}\int_0^{\Lambda^2} dy \frac{y\Sigma_i(y)}{y+\Sigma_i^2(y)} + \int_0^{\Lambda^2} dy \frac{y\Sigma_i(y)}{y+\Sigma_i^2(y)} \boldsymbol{K}(x,y), \tag{2.9}$$

where $\boldsymbol{K}(x,y) \equiv \lambda(\max(x,y,\mu_{IR}^2))/\max(x,y)$. Note that the non-running case is regarded as the "standing" limit $A \to \infty$ (with $\lambda_\Lambda \equiv \lambda(\Lambda^2)$ fixed) of the "walking" coupling ($A \gg 1$).[27,17]

The SχSB solution of (2.9) is logarithmically damping[6,28], essentially the same as (2.6) with the small power $\lambda(\sim \lambda_\Lambda) \ll 1$:

$$\Sigma(p^2) \simeq m\left[\frac{\lambda(p^2)}{\lambda(m^2)}\right]^{\frac{A}{2}}, \quad A = \frac{8}{7} \quad (N_f = 6). \tag{2.10}$$

In the case of the pure QCD ($g = 0$), such a very slowly damping solution ("irregular asymptotics") is the explicit chiral-symmetry-breaking solution due to the quark bare mass.[26,25] However, Miransky and Yamawaki[6] pointed out that it can be the SχSB solution *in the presence of the additional four-fermion interaction.* The solution corresponds to a very large anomalous dimension $\gamma_m \simeq 2 - 2\lambda_\Lambda$ (compare with (2.5)) near the "critical line"[29,17,24]

$$g = g^* \simeq 1 - 2\lambda_\Lambda \tag{2.11}$$

at $\lambda_\Lambda \ll 1$. (There is no critical line in the exact sense in this case, since SχSB takes place in the whole coupling region due to pure QCD dynamics, yielding dynamical mass $m = m_{QCD} = O(\Lambda_{QCD})$.) Note that (2.11) coincides with the critical line in the non-running case (2.4), $g = \frac{1}{4}(1+\omega)^2 \simeq 1 - 2\lambda$, at $\lambda \ll 1$. Actually, we can obtain exact expression for "critical line", which becomes identical with the entire critical line (2.4) in the limit $A \to \infty$.[17]

In view of the "critical line", we again have an SχSB solution with *maximal isospin violation*, $m_t \neq 0, m_b = 0$ (apart from m_{QCD}) , under a condition similar to (2.7); $g_t > g^*(\simeq 1 - 2\lambda_\Lambda) > g_b$.

Isospin Relation and The Top Quark Mass Prediction

Now we come to the central part of the TMSM, relating the dynamical mass of the condensed fermion (top quark) to the mass of W/Z bosons. MTY[1,2] gave a dynamical mechanism by which the large anomalous dimension $\gamma_m \simeq 2$ can save the weak isospin relation $\rho \equiv m_W^2/m_Z^2 cos^2\theta_W \simeq 1$ ($\delta\rho \equiv 1 - \frac{1}{\rho} \ll 1$) for $m_t \gg m_b$, and at the same time determined the top quark mass m_t and ρ *as the (decreasing) function of the cutoff* Λ.

The top quark condensate $\langle \bar{t}t \rangle$ indeed yields a standard gauge symmetry breaking pattern $SU(2)_L \times U(1)_Y \to U(1)_{em}$ to feed the mass of W and Z bosons. Actually, the mass of W and Z bosons in the TMSM is generated via dynamical Higgs mechanism as in technicolor[30];

$$m_W^2 = (\frac{g_2}{2}F_{\pi^\pm})^2, \qquad m_Z^2 \cos^2\theta_W = (\frac{g_2}{2}F_{\pi^0})^2, \tag{2.12}$$

where g_2 is the $SU(2)_L$ gauge coupling, and $F_{\pi^\pm}$ and F_{π^0} are the decay constants of the composite NG bosons $\pi^\pm, \pi^0$ to be absorbed into W and Z bosons, respectively. $F_\pi(\simeq 250\text{GeV})$ determines the IR scale of the model and plays a central role in the TMSM. Then ρ is simply given by $\rho = F_{\pi^\pm}^2/F_{\pi^0}^2$, which would measure the isospin violation $F_{\pi^\pm} \neq F_{\pi^0}$.

Those composite NG bosons are tightly bound states of the doublet fermions (top quark and bottom quark) and their antifermions due mainly to the short range four-fermion interactions. In the case of maximal isospin violation, $m_t \neq 0$ and $m_b = 0$, the NG boson contents become very simple $\pi^+ \sim i\bar{b}(1+\gamma_5)t/\sqrt{2}$, $\pi^- \sim i\bar{t}(-1+\gamma_5)b/\sqrt{2}$ and $\pi^0 \sim i\bar{t}\gamma_5 t$ (or, in the language of standard Higgs doublet ϕ, $\phi \sim \bar{\psi}_L t_R$, where $\psi \equiv (t,b)$.) *

Now, the decay constants of those composite NG bosons may be calculated in terms of the dynamical mass $\Sigma(p^2)$ of the condensed fermions through the famous Pagels-Stokar formula[31], this time a form generalized to the $SU(2)$-asymmetric case $m_t \neq m_b$[1,2,32]. In the most general case $m_{t,b} \neq 0$, MTY[1,2] gave formulas for F_π's

$$\begin{aligned} F_{\pi^\pm}^2 &= \int_0^{\Lambda^2} dx I_\pm(\Sigma_t, \Sigma_b) = \frac{N_c}{8\pi^2}\int_0^{\Lambda^2} dx x \cdot \\ &\cdot \frac{(\Sigma_t^2+\Sigma_b^2) - \frac{x}{4}(\Sigma_t^2+\Sigma_b^2)' + \frac{x}{2}(\Sigma_t^2-\Sigma_b^2)\left[\frac{1+(\Sigma_t^2)'}{x+\Sigma_t^2} - \frac{1+(\Sigma_b^2)'}{x+\Sigma_b^2}\right]}{(x+\Sigma_t^2)(x+\Sigma_b^2)} \end{aligned} \tag{2.13}$$

$$\begin{aligned} F_{\pi^0}^2 &= \int_0^{\Lambda^2} dx I_0(\Sigma_t, \Sigma_b) \\ &= \frac{N_c}{8\pi^2}\int_0^{\Lambda^2} dx x \cdot \left[\frac{\Sigma_t^2 - \frac{x}{4}(\Sigma_t^2)'}{(x+\Sigma_t^2)^2} + \frac{\Sigma_b^2 - \frac{x}{4}(\Sigma_b^2)'}{(x+\Sigma_b^2)^2}\right]. \end{aligned} \tag{2.14}$$

Let us consider the extreme case, the maximal isospin violation mentioned above, $\Sigma_t(p^2) \neq 0$ and $\Sigma_b(p^2) = 0$. The isospin violation $F_{\pi^\pm} \neq F_{\pi^0}$ solely comes from the different propagators having different $\Sigma_i(p^2)$, essentially the IR quantity. One would naively expect a large isospin violation for the maximal isospin-violating dynamical mass. However, this is not the case for a very slowly damping mass function $\Sigma_t(p^2) \simeq$ const.[32,1,2]; in fact integrals (2.13) and (2.14) are dominated by the UV region $p^2 \gg m_t^2$, where the difference in the denominators in (2.13) and (2.14) can be ignored, and the contribution from the terms with derivative Σ_t' is also suppressed. Actually, the formulas (2.13) and (2.14) imply the following two important facts;[1,2]

1. $F_{\pi^\pm}^2 \simeq F_{\pi^0}^2$ for a very slowly damping mass function $\Sigma_t(p^2)$.
2. Top quark mass is determined as a *decreasing function* of cutoff Λ, once $\Sigma_t(p^2)$ is explicitly given.

*The bottom quark condensate in addition to the top quark condensate does not yield additional NG bosons (as far as $g^{(2)} \neq 0$) but does change the contents of the above NG bosons.[2]

This is most easily seen in the extreme case $\Sigma_t(p^2) \equiv$ const. (pure NJL limit). In this case (2.13) and (2.14) are both *logarithmically divergent* at $\Lambda/m_t \to \infty$ *with the same coefficient*:

$$F_{\pi^\pm}^2 = \frac{N_c}{8\pi^2} m_t^2 \left[\ln \frac{\Lambda^2}{m_t^2} + \frac{1}{2} \right], \tag{2.15}$$

$$F_{\pi^0}^2 = \frac{N_c}{8\pi^2} m_t^2 \ln \frac{\Lambda^2}{m_t^2}. \tag{2.16}$$

Thus we have $F_{\pi^\pm} \simeq F_{\pi^0}$ and

$$\delta\rho \equiv \frac{F_{\pi^\pm}^2 - F_{\pi^0}^2}{F_{\pi^\pm}^2} = \frac{N_c m_t^2}{16\pi^2 F_{\pi^\pm}^2} \simeq \frac{1}{2\ln \frac{\Lambda^2}{m_t^2}} \ll 1 \tag{2.17}$$

for $\Lambda \gg m$. Then the problem of weak isospin relation can in principle be solved *without custodial symmetry*. This is the essence of the "dynamical mechanism" of MTY to save the isospin relation $\rho \simeq 1$.* Furthermore, if one fixes $F_\pi \simeq 250$GeV in order to have a correct m_W through (2.12), then (2.15) can "predict" m_t as a *decreasing function of cutoff* Λ: $m_t \simeq 145$GeV at $\Lambda = 10^{19}$GeV, $m_t \to 0$ at $\Lambda \to \infty$. Incidentally, this also implies triviality (no interaction) of the NJL model:$g_Y \equiv m_t/F_\pi \to 0$ at $\Lambda \to \infty$.

Now in the gauged NJL model, QCD plus four-fermion interaction (2.2), we have a very slowly damping $\Sigma(p^2)$, (2.10), as an SχSB solution to the ladder SD equation (2.9). Actually, MTY[1,2] predicted m_t and $\delta\rho$ as the *decreasing function of cutoff* Λ, by plugging (2.10) into (2.13) and (2.14):

$$F_{\pi^\pm}^2 = \int_0^{\Lambda^2} dx I_\pm(\Sigma_t(x), 0) = m_t^2 \int_0^{\frac{\Lambda^2}{m_t^2}} dx I_\pm(\sigma_t(x), 0) \equiv \frac{m_t^2}{g_Y^{(\pm)2}}, \tag{2.18}$$

$$F_{\pi^0}^2 = \int_0^{\Lambda^2} dx I_0(\Sigma_t(x), 0) = m_t^2 \int_0^{\frac{\Lambda^2}{m_t^2}} dx I_0(\sigma_t(x), 0) \equiv \frac{m_t^2}{g_Y^{(0)2}}, \tag{2.19}$$

with $\sigma_t(x) \equiv \Sigma_t(x)/m_t$. It is easy to see that the effective Yukawa couplings $g_Y^{(\pm)}$ and $g_Y^{(0)}$ are given as slowly *decreasing functions of* Λ/m_t;

$$g_Y^{(\pm)}, \quad g_Y^{(0)} \searrow \quad \text{as} \quad \Lambda \nearrow. \tag{2.20}$$

Since we fix $F_\pi \simeq 250$ GeV, this implies that m_t *is also a decreasing function of cutoff* Λ;

$$m_t \searrow \quad \text{as} \quad \Lambda \nearrow. \tag{2.21}$$

*In the alternative formulation made by BHL[10] this dynamical consequence is tacitly incorporated into their assumption to take the *renormalizable form* for the effective theory of the composite Higgs (pure Higgs sector). Actually, it is impossible to write down a renormalizable pure Higgs Lagrangian having isospin violation $F_{\pi^\pm} \neq F_{\pi^0}$ (it is possible in the nonlinear sigma model).

The largest physically sensible Λ (new physics scale) would be the Planck scale $\Lambda = 10^{19}$GeV, for which we predict[1,2†]

$$m_t \simeq 250\text{GeV}, \tag{2.22}$$

$$\delta\rho \simeq 0.02, \tag{2.23}$$

$$g_Y \simeq 1. \tag{2.24}$$

Contrary to the logarithmic divergence in the pure NJL limit, (2.15), this time we have a *convergent* integral for F_π and hence a *finite* m_t at $\Lambda \to \infty$ *in the case of* $A > 1$ ($A = 8/7$ for $N_f = 6$). Then we have a nontrivial (interacting) theory $g_Y = m_t/F_\pi \neq 0$ in the continuum limit *iff* $A > 1$; presence of the gauge interaction can change the trivial theory (NJL model) into a nontrivial theory (gauged NJL model)![17*]

It will be more convenient to write an analytical expression for F_π. Neglecting the derivative terms with $\Sigma_t(x)'$ and using (2.10), we may approximate (2.13) as[9]

$$\begin{aligned} F_\pi^2 &\simeq \frac{N_c}{8\pi^2}\int_{m_t^2}^{\Lambda^2} dx \frac{\Sigma_t^2}{x} \\ &\simeq \frac{N_c m_t^2}{16\pi^2}\frac{A}{A-1}\frac{(\lambda(m_t^2))^{A-1} - (\lambda(\Lambda^2))^{A-1}}{(\lambda(m_t^2))^A}, \end{aligned} \tag{2.25}$$

which yields the same result as (2.22) at $\Lambda \simeq 10^{19}$GeV ($A = 8/7$). As was already noted, the integral for F_π^2 is convergent if $A > 1$. For $A = 8/7$ the above expression in fact yields $m_t \simeq 115$GeV at $\Lambda \to \infty$ ($\lambda(\Lambda^2) \to 0$).[9]

Also, one may directly look at $\delta\rho$ in which the UV contributions are canceled:

$$\delta\rho \equiv \frac{F_{\pi\pm}^2 - F_{\pi^0}^2}{F_{\pi\pm}^2} = \frac{N_c}{8\pi^2 F_{\pi\pm}^2}\int_0^{\Lambda^2} dx \frac{\Sigma_t^4 + \frac{x}{4}(\Sigma_t^4)'}{2(x+\Sigma_t^2)^2}. \tag{2.26}$$

The integral is now an IR dominant integral quite insensitive to the UV behavior and yields $\simeq m_t^2/2$ ($= m_t^2/2$ for $\Sigma(p^2) \equiv$ const.) for very slowly damping $\Sigma(p^2)$. Thus we have

$$\delta\rho \simeq \frac{N_c m_t^2}{16\pi^2(250\text{GeV})^2}. \tag{2.27}$$

Note that (2.27) is the same as the top quark contribution to $\delta\rho$ in the SM, implying that the same constraint as in the SM should be imposed on m_t in the TMSM.

---

†One may substitute into (2.10) the *numerical* solution (instead of the analytical one (2.10)) of the ladder SD equation (2.9), the result being the same as (2.22).[33]

*In the non-running case ($A \to \infty$) the integral for F_π^2 is more rapidly convergent, since $\Sigma(p^2)$ is power damping instead of logarithmic damping.

3. Gauged NJL Model— Phase Structure and Effective Theories

Having explained the essential features of the TMSM, we here discuss field theoretical aspects of its underlying dynamics, i.e., gauged NJL model, with particular emphasis on the phase structure (fixed points, large anomalous dimension, etc.) and the effective theories.[18,17,14] This is more than simply an academic problem, since some people[34,35,36,37] have recently discussed predictability of the TMSM on the basis of certain versions of the four-fermion theories having higher dimensional operators but *no gauge interactions.*

TMSM is actually based on the *gauged* NJL model, QCD plus four-fermion interactions. Here I will emphasize a possibility that the *presence of the gauge interactions* may change drastically the four-fermion theories into *renormalizable theories* without introducing higher dimensional operators. In fact, in the ladder approximation (with non-running gauge coupling, or $A \to \infty$ limit) the gauged NJL model has a large anomalous dimension $2 > \gamma_m = 1 + \omega > 1$, which suggests the four-fermion interactions may be relevant operators $2 < d_{(\bar{\psi}\psi)^2} = 2(3 - \gamma_m) = 4 - 2\omega < 4$.[6] Accordingly, possible *higher dimensional interactions*, $(\bar{\psi}\psi)^4$, $\partial_\mu(\bar{\psi}\psi)\partial^\mu(\bar{\psi}\psi)$, etc., may be *irrelevant operators* ($d > 4$ due to $d_{\bar{\psi}\psi} > 1$), in contrast to the case without gauge interactions where these operators may be marginal ones ($d = 4$ due to $d_{\bar{\psi}\psi} = 1$)). Moreover, as already seen in the previous section, $\gamma_m < 2$ ($\lambda \neq 0$) guarantees a convergent integral for F_π^2 and hence a finite effective Yukawa coupling, $g_Y \equiv m/F_\pi \neq 0$, in the continuum limit $\Lambda \to \infty$. This is in sharp contrast to the pure NJL model with $\gamma_m = 2$ ($g_Y = 0$, trivial or Gaussian). Then the theory has a sensible (interacting) continuum limit at the critical line which is identified with a nontrivial UV fixed line[4,38,39].

To demonstrate such a possibility more explicitly, we shall construct a renormalized effective potential/action for the gauged NJL model ($A \to \infty$) without higher dimensional operators *in the continuum limit.*[18] The renormalized continuum theory (effective theory) also has a nontrivial UV fixed line and a large anomalous dimension at the fixed line.[18]

Gauged NJL Model—Non-Running Gauge Coupling ($A \to \infty$)

Let us consider renormalization of QED plus chiral-invariant four-fermion interaction $(G_0/2)[(\bar{\psi}\psi)^2 + (\bar{\psi}i\gamma_5\psi)^2]$ in the ladder approximation as the simplest gauged-NJL model with $A \to \infty$.[18] This model illustrates essence of the dynamics of TMSM and is also interesting in its own right, since it provides us with a dynamical model for the walking/standing technicolor with anomalous dimension $\gamma_m \simeq 1$[40,41,42,43] and the strong-ETC technicolor[40,6,44,45,1] having larger anomalous dimension $1 \lesssim \gamma_m \lesssim 2$.

By using the auxiliary fields σ and π, we can rewrite the original Lagrangian into an equivalent one:

$$\mathcal{L} = \bar{\psi}i\not{D}\psi - \bar{\psi}(\sigma + i\gamma_5\pi)\psi - V_{\text{classical}}(\sigma, \pi) - \frac{1}{4}F_{\mu\nu}F^{\mu\nu}, \tag{3.1}$$

where the classical contribution $V_{\text{classical}}(\sigma,\pi)$ is defined by $G_0 V_{\text{classical}} \equiv (\sigma^2+\pi^2)/2 - m_0\sigma$ with m_0 being the bare mass of the fermion. It is rather straightforward to derive the CJT effective potential[46] in two-loop approximation (in Landau gauge):

$$
\begin{aligned}
-4\pi^2 V[\Sigma,\sigma) &= \int_0^{\Lambda^2} dx x \left[\frac{1}{2}\ln(1+\frac{\Sigma^2}{x}) - \frac{\Sigma^2}{x+\Sigma^2} + \frac{\sigma\Sigma}{x+\Sigma^2}\right] \\
&+ \frac{1}{2}\int_0^{\Lambda^2} dx dy \frac{x\Sigma(x)}{x+\Sigma^2}\frac{y\Sigma(y)}{y+\Sigma^2}K(x,y) - \Lambda^2 g^{-1}(\frac{\sigma^2}{2} - m_0\sigma), \quad (3.2)
\end{aligned}
$$

with $g \equiv G_0\Lambda^2/4\pi^2$ and $\lambda \equiv 3e^2/(4\pi)^2$ (e is the QED coupling), where we have set $\pi = 0$, $\sigma > 0$.

Stationary condition for Σ (gap equation), $\delta V/\delta\Sigma = 0$, yields the ladder SD equation

$$
\Sigma(x) = \sigma + \int_0^{\Lambda^2} dy K(x,y)\frac{y\Sigma}{y+\Sigma^2}, \quad (3.3)
$$

which is the same form as the ladder SD equation of pure QED, with the auxiliary field σ now acting as if the "bare mass". If we further used the solution of the stationary condition for σ, $\partial_\sigma V[\Sigma,\sigma) = 0$, we could have solved away σ from (3.3), arriving at the same ladder SD equation as (2.3) (except for the bare mass m_0). Hereafter we shall keep the explicit σ field.

Substituting the solution Σ^{sol} of the stationary condition (3.3) back into (3.2), we obtain $V(\sigma) \equiv V[\Sigma^{\text{sol}},\sigma)$ as[18]

$$
\begin{aligned}
-4\pi^2 V(\sigma) &= \Lambda^2 g^{-1} m_0\sigma + \Lambda^2\left(g^{*-1} - g^{-1}\right)\frac{\sigma^2}{2} \\
&+ \Lambda^4 C\frac{2-\omega}{4\lambda}\left(\frac{\sigma}{\Lambda}\right)^{\frac{4}{2-\omega}} + \Lambda^4 \mathcal{O}((\frac{\sigma}{\Lambda})^{\frac{4+2\omega}{2-\omega}}, (\frac{\sigma}{\Lambda})^4), \quad (3.4)
\end{aligned}
$$

where $g^* \equiv (1+\omega)^2/4$ and C is a certain constant depending on ω. Eq.(3.4) was first derived by Bardeen and Love through a different method.[47] It is evident that the phase is determined by the sign of the coefficient of σ^2 term, i.e., $g = g^*$ is the critical line (2.4). The critical line is actually a nontrivial UV fixed line [4,38,39], with the line of constant gauge coupling being identified with the RG flow in the SχSB phase.* It should be emphasized that as is clear from the derivation, this potential holds *both in the symmetric and the SχSB phases.*

Now, we renormalize the above effective potential as follows:[18]

$$
\begin{aligned}
\Lambda^{1-\omega}\sigma &= \mu^{1-\omega}\sigma_R, && (3.5) \\
\Lambda^{2\omega}(g^{*-1} - g^{-1}) &= \mu^{2\omega}(g_R^{*-1} - g_R{}^{-1}), && (3.6) \\
\Lambda^{1+\omega} g^{-1} m_0 &= \mu^{1+\omega} g_R{}^{-1} m_R, && (3.7)
\end{aligned}
$$

*This is in accord with the usual expectation that the gauge coupling may not be renormalized in the absence of vacuum polarization in the SχSB phase as well as the symmetric one in the ladder approximation. Here we consistently consider the renormalization of only the four-fermion coupling in both phases.

where σ_R, m_R (current mass) and g_R are renormalized at the renormalization point μ. (Note that g_R^* ($0 < g_R^* < \infty$) is left arbitrary and the simplest choice would be $g_R^* = g^*$.) Through this renormalization we in fact have a renormalized effective potential ($\Lambda \to \infty$) and a renormalized scalar vertex:[18]

$$-4\pi^2 V(\sigma_R) = \mu^2 g_R{}^{-1} m_R \sigma_R + \mu^2 \left(g_R^{*-1} - g_R{}^{-1}\right) \frac{\sigma_R^2}{2} + \mu^4 C \frac{2-\omega}{4\lambda} \left(\frac{\sigma_R}{\mu}\right)^{\frac{4}{2-\omega}} \tag{3.8}$$

$$\Gamma_S^{(R)}(p) \equiv \left.\frac{\partial \Sigma^{\rm sol}(p^2)}{\partial \sigma_R}\right|_{\sigma_R=0} = \frac{2}{1+\omega} \left(\frac{p^2}{\mu^2}\right)^{-(1-\omega)/2} . \tag{3.9}$$

Thus we have obtained a sensible (interacting) continuum theory defined on the critical line, $g \to g^*$ as $\Lambda \to \infty$. We need *no higher dimensional operators* like $(\bar\psi\psi)^4$ ("counter term") to renormalize the theory. It is very important that *this argument breaks down in the absence of gauge interactions* $\lambda = 0$ ($\omega = 1$), in which case logarithmic divergence appears in the σ^4 term (signaled by $1/\lambda$ singularity in (3.8)) to be renormalized by such a higher dimensional operator.

From (3.6) and (3.7) we immediately obtain $\beta(g_R)$ and $\gamma_m(g_R)$ by using the μ independence of the bare parameters:[18]

$$\beta(g_R) = 2\omega g_R \left(1 - \frac{g_R}{g_R^*}\right), \tag{3.10}$$

$$\gamma_m(g_R) = 1 - \omega + 2\omega \frac{g_R}{g_R^*}. \tag{3.11}$$

It is also evident from (3.6) and (3.7) that $\beta(g)$ and $\gamma_m(g)$ in the bare coupling take the same form as those in the renormalized coupling *in both phases.*

Thus the theory has been renormalized at zero momentum ($q_\mu = 0$) of the auxiliary field. The continuum theory does have a nontrivial UV fixed line $g_R = g_R^*$. It should be emphasized again that this renormalization *holds both in the symmetric ($g_R < g_R^*$)and the SχSB ($g_R > g_R^*$) phases.* Accordingly, the anomalous dimension becomes very large, $1 < \gamma_m \simeq 1 + \omega < 2$, near the fixed line without discontinuity *in both phases.* It is explicitly checked that the OPE is consistent with this renormalization and the large anomalous dimension.[18]

As to $q_\mu \neq 0$, we need to know the propagator of σ. Recently, Appelquist, Terning and Wijewardhana[48] have made an interesting approximation to sum up an infinite series of higher derivative terms in the symmetric phase. Remarkably enough, their σ propagator is also renormalized via our renormalization condition mentioned above:[18]

$$i(4\pi^2)\Delta_R^{-1}(-q^2) = -\mu^2 (q^2/\mu^2)^\omega / (4g^*\omega\lambda) + \mu^2 (g_R^{*-1} - g_R{}^{-1}) \tag{3.12}$$

at $\Lambda \to \infty$. Then the (induced) kinetic term of the σ has also been renormalized without higher dimensional operators like $\partial_\mu(\bar\psi\psi)\partial^\mu(\bar\psi\psi)$ ("counter term"). Again this renormalization breaks down at $\lambda \to 0$ (NJL limit) due to the appearance of logarithmic divergence (signaled by $1/\lambda$ singularity in (3.12)) corresponding to the logarithmic divergence of F_π^2 ("kinetic term" of π and σ).

As was suggested by Kikukawa and Yamawaki[49], all these features are in complete analogy with those in the four-fermion theory (without gauge interactions) in D dimensions ($2 < D < 4$) which is renormalizable in $1/N$ expansion.

We are also interested in the mass of σ a composite Higgs boson. In the symmetric phase the σ pole is read off from the above propagator: $M_\sigma^2 = [4g^*\omega\lambda]^{1/\omega}[(g_R^{*-1} - g_R^{-1})^{1/2\omega}\mu]^2$, which is indeed μ-independent. In the SχSB phase ($g_R > g_R^*$), on the other hand, it is rather difficult to compute directly the σ propagator. Here we use[16] partially conserved dilatation current (PCDC) hypothesis to compute M_σ ("massive dilaton") in the SχSB phase:

$$M_\sigma^2 = -16\frac{d_\theta E}{d_\sigma^2 F_\pi^2}, \tag{3.13}$$

where E is the vacuum energy, d_σ and d_θ are the scale dimensions of σ ($\sim \bar{\psi}\psi$) and the trace of the energy-momentum tensor $\theta_{\mu\nu}$ ($\langle\theta_\mu^\mu\rangle = 4E$), respectively. E is given[38] in terms of $\Sigma(p^2)$ through the CJT effective potential $E = V[\Sigma^{sol}, \sigma^{sol}) - V[0,0)$, at the stationary point of Σ, and we have $d_\sigma = d_{(\bar\psi\psi)} = 3 - \gamma_m = 2 - \omega$ and $d_\theta = d_{(\bar\psi\psi)^2} = 2d_{(\bar\psi\psi)} = 2(2-\omega)$. Note that the scale dimension of θ_μ^μ is determined *not* by the gauge boson operator $(\beta(\lambda)/\lambda)F_{\mu\nu}^2$ *but* by the four-fermion operator $g(2\pi^2/\Lambda^2)[(\bar\psi\psi)^2 - (\bar\psi\gamma_5\psi)^2]$, since the RG-flow is the fixed-λ line as mentioned earlier. PCDC calculation gives a correct answer[16] in the four-fermion theory in D dimensions ($2 < D < 4$) where the σ propagator is explicitly calculated both in the symmetric and the SχSB phases[49]. Putting everything together in (3.13), we obtain[16]

$$M_\sigma/m \simeq \sqrt{2} \tag{3.14}$$

for $\Lambda \gg m(= \Sigma(m^2))$ and very small λ (but non-zero), in contrast to the pure NJL result, $M_\sigma/m = 2$. Quite recently, the same result as (3.14) has also been obtained by the RG method.[47] This result has direct relevance to the mass of the composite Higgs boson (" technidilaton"[40,50]) in the strong-ETC technicolor.

Gauged NJL Model — Running QCD Coupling

The essential feature of the above renormalization remains true also in the case of running QCD coupling with $A > 1$, although convergence is weaker in this case. We show a "phase diagram" in (λ_Λ,g) plane (Fig.2(a))[17] based on the numerical solution of (2.9). The "RG flows" (lines of m=const.)* flow out of two UV fixed points, (0,1) and (0,0), corresponding to the pure NJL and pure QCD points, respectively. We also depicted in Fig.2b the case with larger A (= 10), "walking coupling", in comparison with Fig.4a, the 6-flavored QCD ($A = 8/7$). In either case there exist two "critical

*The "RG flow" of F_π =const. does not exactly coincide with that of m = const. Actually, as shown in section 2, m is a slowly decreasing function of cutoff Λ along the F_π = const line. Nevertheless, general structure of the "phase diagram" is essentially the same in either case.

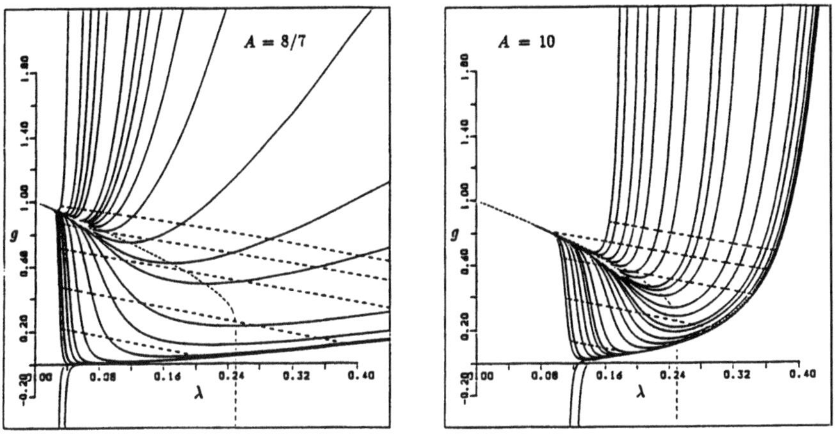

Fig.2 (a) "RG flows" with $m = \text{const.}$ in (λ_Λ, g) plane are depicted by the solid lines for $A = 8/7$. The lines get downward as we decrease m, with all the lines of $m > m_{QCD}$ being degenerate at "critical line", (2.11) at $\Lambda \gg m$ ($\lambda_\Lambda \simeq 0$). Dashed line curve is the critical line of the non-running case ($A \to \infty$), (2.4). Dashed parallel straight lines are equi-γ_m lines ($\gamma_m = 0.5, 1.0, 1.5, 1.8$ and 2.0 from below). (b) The same as (a) for $A = 10$.

lines", one being (2.11) and the other $g \simeq \lambda_\Lambda^2$, for $\lambda_\Lambda \ll 1$. In the $A \to \infty$ limit we can show[17] numerically and analytically that they actually coincide with the critical line (2.4) (dashed line in Fig.2) and $g = (1-\omega)^2/4$ (no longer meaningful in this limit), respectively.

In Fig.2 the lines of $\gamma_m = \text{const.}$ calculated numerically are also depicted (dashed parallel lines), which lie on the lines[17*]

$$g = -\lambda_\Lambda + \frac{\gamma_m}{2}. \tag{3.15}$$

It clearly shows that $\gamma_m \simeq 2 - 2\lambda_\Lambda (\simeq 2)$ near (0,1) and $\gamma_m \simeq 2\lambda_\Lambda (\simeq 0)$ near (0,0) along the respective "critical lines".

Thus we have observed two-UV-fixed-points structure, (0,1) and (0,0), in the gauged NJL model with QCD running coupling. The above property near (0,1) is the essential ingredient of the TMSM. In the same way as the non-running gauge coupling case the theory may be renormalizable without higher dimensional operators which

*This is an A-independent result and actually can be analytically derived in the $A \to \infty$ limit.[18] This is slightly different from (3.11) off the critical line.

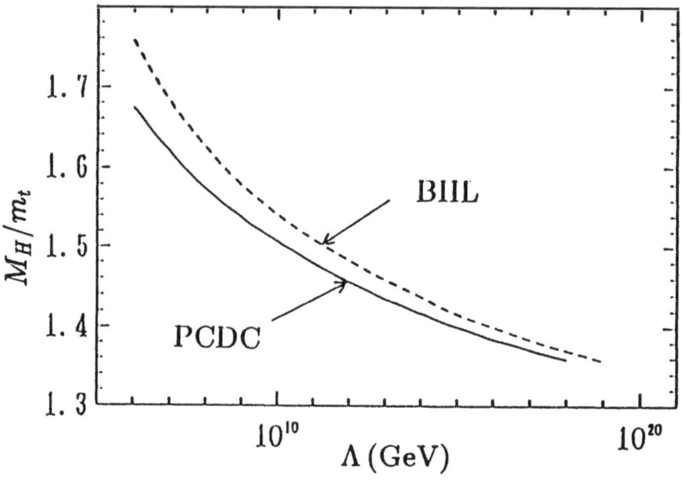

Fig.3 Dependence of M_H/m_t on cutoff Λ obtained through PCDC. Also depicted is the result of BHL method within the same approximation as ours (switching off the loop effects of the composite Higgs and the $SU(2)_L \times U(1)_Y$ gauge interactions).

will be irrelevant due to the logarithmic corrections of gauge interactions ($\gamma_m \simeq 2 - 2\lambda < 2$) in the $\Lambda \to \infty$ limit. As already noted the theory seems to be nontrivial $g_Y \neq 0$ at $\Lambda \to \infty$ for $A > 1$ at least in this approximation.[17,14]

We can also calculate Higgs boson ("massive dilaton") mass M_H through the PCDC in TMSM in the same way as in the non-running gauge coupling case (Fig.3).[16] In particular, for $\Lambda \simeq 10^{19}$GeV we have a similar result to (3.14):

$$m_H \simeq 1.35 m_t (\simeq \sqrt{2} m_t), \tag{3.16}$$

which is contrasted to the original prediction $m_H \simeq 2m_t$[1,2,8].

4. Conclusion and Discussion

In the light of new experimental situation that the top quark is very heavy, we have discussed the top quark condensation (Top Mode Standard Model) proposed by Miransky, Tanabashi and Yamawaki[1] and independently by Nambu[8]. This is actually the simplest idea to understand the large top quark mass and the *Origin of Mass* of *all* the existing particles, identifying the two missing ingredients of the Standard Model, the top quark and the Higgs boson, as strongly correlated objects.

The *explicit* SχSB solution to the ladder SD equation for QCD plus four-fermion interaction was used to yield m_t and m_H as *decreasing function of cutoff* Λ. Since $\delta\rho$

is proportional to m_t^2, we can minimize it at (physically sensible) largest Λ, 10^{19}GeV, where we have $m_t \simeq 250$GeV and $m_H \simeq \sqrt{2}m_t$. Possible modifications of this result including other effects than QCD and (2.1) might change it to somewhat smaller values but the gross structure would remain intact.

BHL versus MTY

Here we comment on the BHL formulation[10] of the TMSM. BHL can actually be shown[14] to be equivalent to MTY[1,2], besides the point that BHL newly introduced loop effects of the composite Higgs boson and of the $SU(2)_L \times U(1)_Y$ gauge interactions in addition to the MTY analysis.

BHL start with the SM Lagrangian which includes the explicit Higgs field at the Lagrangian level:

$$\mathcal{L}_{Higgs} = -y_t(\bar{\psi}_L^i t_R \phi_i + \text{h.c.}) + \left(D_\mu \phi^\dagger\right)\left(D^\mu \phi\right) - m_H^2 \phi^\dagger \phi - \lambda_4 \left(\phi^\dagger \phi\right)^2, \qquad (4.1)$$

where y_t and λ_4 are Yukawa coupling of the top quark and quartic interaction of the Higgs, respectively. BHL imposed "compositeness condition" on y_t and λ_4 in such a way that (4.1) becomes the MTY Lagrangian (2.2) (with $G_b = 0$):

$$\frac{1}{y_t^2} \to 0, \quad \frac{\lambda_4}{y_t^4} \to 0 \qquad \text{as } \mu \to \Lambda, \qquad (4.2)$$

where μ is the renormalization point above which the composite dynamics are integrated out to yield an effective theory (4.1). Thus the compositeness condition implies the divergence of Yukawa coupling of the top quark at $\mu = \Lambda$.

Now, in the one-loop RG equation the beta function of y_t is given by

$$\beta(y_t) = \frac{y_t^3}{(4\pi)^2}\left(N_c + \frac{3}{2}\right) - \frac{y_t}{(4\pi)^2}\left(8g_3^2 + \frac{9}{4}g_2^2 - \frac{17}{12}g_1^2\right), \qquad (4.3)$$

where g_1, g_2 and g_3 are the gauge couplings of $U(1)_Y$, $SU(2)_L$ and $SU(3)_C$, respectively. Since it is rather complicated to deal with the loop effects of the composite Higgs boson and the $SU(2)_L \times U(1)_Y$ interactions within the SD equation approach of MTY, we here compare BHL with MTY by switching off these effects.[9,33,18] Then (4.3) becomes simply

$$\frac{dy_t}{d\mu} = \beta(y_t) = N_c \frac{y_t^3}{(4\pi)^2} - \frac{8y_t g_3^2}{(4\pi)^2} \qquad (4.4)$$

Within the same approximation the beta function of QCD gauge coupling reads

$$\frac{dg_3}{d\mu} = \beta(g_3) = -\frac{1}{A}\frac{8g_3^3}{(4\pi)^2}. \qquad (4.5)$$

Solving (4.4) and (4.5) by imposing the compositeness condition at $\mu = \Lambda$, we arrive at

$$y_t^2(\mu) = \frac{2(4\pi)^2}{N_c}\frac{A-1}{A}\frac{\left(\lambda(\mu^2)\right)^A}{\left(\lambda(\mu^2)\right)^{A-1} - \left(\lambda(\Lambda^2)\right)^{A-1}}. \tag{4.6}$$

Noting the usual relation $m_t^2 = y_t^2(\mu = m_t)v^2/2$ ($v = F_\pi$), we obtain

$$\frac{m_t^2}{F_\pi^2}(\equiv g_Y^2) = \frac{y_t^2(m_t)}{2} = \frac{(4\pi)^2}{N_c}\frac{A-1}{A}\frac{\left(\lambda(m_t^2)\right)^A}{\left(\lambda(m_t^2)\right)^{A-1} - \left(\lambda(\Lambda^2)\right)^{A-1}}, \tag{4.7}$$

which is precisely the same as (2.25) obtained in the MTY approach based on the SD equation.[14*]

Also, the Higgs boson mass can be calculated within the same approximation switching off the loop effects of the Higgs boson and the $SU(2)_L \times U(1)_Y$ gauge interactions (Fig.3). At $\Lambda = 10^{19}$ GeV we have $M_H/m_t \simeq 1.36$, which is compared with (3.16). The agreement with the PCDC calculation[16] based on the SD equation is remarkable.

Having convinced ourselves of the validity of compositeness condition of BHL approach, we now look into a full analysis of RG equation including the loop effects of the composite Higgs and the $SU(2)_L \times U(1)_Y$ gauge couplings. BHL obtained[10]

$$m_t = 218 \pm 3, \qquad \text{at } \Lambda = 10^{19}\text{GeV}, \tag{4.8}$$

which is very large similarly to the MTY value[1,2]. The mass of Higgs boson can also be calculated through the full RG equation of λ_4, the result being[10]

$$M_H = 239 \pm 3 (M_H/m_t \simeq 1.1) \qquad \text{at } \Lambda = 10^{19}\text{GeV}. \tag{4.9}$$

Remarks

1. If m_t were found to be less than 200 GeV as is currently believed by many authors through the ρ parameter argument, then the simplest version described above would get in trouble. In such a case, MTY[2] already suggested some modification including the condensate of the almost degenerate fourth family quarks with comparable to the top quark condensate, which implies m_t would be roughly reduced by factor $1/\sqrt{2}$, or $m_t \simeq 180$GeV. Further analysis[10,9] confirmed the above result. MTY also suggested[1] to include $K-1$ almost degenerate technifermion doublets, in which case m_t is also reduced by the factor $1/\sqrt{K}$.

2. Also, there may exist finite cutoff effects ($\Lambda < 10^{19}$GeV) of higher dimensional interactions to reduce m_t substantially.[34,35] Although they are irrelevant operators

*Alternatively, we may define $F_\pi^2(\mu^2) \equiv 2m_t^2/y_t^2(\mu)$ which coincides with the integral (2.25) with the IR end m_t^2 simply replaced by μ^2. Then the compositeness condition (4.2) reads $F_\pi^2(\mu^2 = \Lambda^2) = 0$ (no kinetic term of the Higgs).

and suppressed by the gauge interactions in the continuum limit $\Lambda \to \infty$ as discussed in Section 3, the QCD running coupling ($A = 8/7$) may not be powerful enough to suppress (only logarithmically) those effects at $\Lambda < 10^{19}$GeV. Then a certain class of underlying theories at Λ may give rise to higher dimensional operators large enough to survive the suppression effects of QCD interaction.

3. The bottom quark can acquire mass from the $\bar{t}t$ condensate

$$\langle \bar{t}t \rangle_\Lambda = -\frac{N_c}{4\pi^2} \int^{\Lambda^2} dx \frac{x\Sigma_t(x)}{x + \Sigma_t(x)^2} \simeq -\frac{N_c}{4\pi^2} \left(\frac{m_t}{\Lambda}\right)^{2-\gamma_m} \cdot \Lambda^2 m_t, \tag{4.10}$$

through the perturbative $G^{(2)}(= g^{(2)} 4\pi^2/(N_c\Lambda^2))$ interaction in (2.1):[1,2]

$$m_b = -G^{(2)} \langle \bar{t}t \rangle_\Lambda \simeq g^{(2)} \left(\frac{m_t}{\Lambda}\right)^{2-\gamma_m} \cdot m_t \simeq g^{(2)} m_t, \tag{4.11}$$

for $\gamma_m \simeq 2$. This also coincides with the result of the nonperturbative $G^{(2)}$ interaction.[2]

4. Similarly, quarks of the first and the second families also acquire mass from the $\bar{t}t$ condensate through the perturbative $G^{(a)}$ interactions[1,2]. In fact, mass matrices of the "up" and the "down" quarks are given by $m_{(u)} \sim (g^{(1)} + g^{(3)}) m_t$ and $m_{(d)} \sim g^{(2)} m_t$, respectively. It is now evident that our dimensionless four-fermion couplings $g^{(a)}$'s play the role of Yukawa couplings of the SM. Obviously the above mass spectrum leads to a realistic Kobayashi-Maskawa matrix. Also the lepton masses are acquired in the same way as those of the "down" quarks through the $g^{(2)}$ type couplings $m_{(\ell)} \sim g^{(2)} m_t$. Due to the large Λ (could be 10^{15}GeV-10^{19}GeV) and the small anomalous dimension of $\bar{\psi}\psi$ other than $\bar{t}t$, our four-fermion interactions of order O($1/\Lambda^2$) are perfectly free from the FCNC's problem.

5. An alternative possibility for the bottom quark mass would be that the bottom quark condensate is formed on the same footing as the top quark condensate.[2] Solving the coupled SD equations for Σ_t and Σ_b including $G^{(2)}$ term, we have a new critical line[2], with g in (2.4) being simply replaced by $\bar{g} \equiv g^{(1)} + \sqrt{{g^{(2)}}^2 + {g^{(3)}}^2}$. Similar situation is also possible in the running λ case. If $g_t > g_b \simeq g^*$, the number of light spinless bound states changes from 4 into the full components 8 out of $(\bar{t}, \bar{b}) \times (1, \gamma_5) \times (t, b)$, namely, the low energy effective theory becomes the two Higgs doublet model ("non-minimal model")[11,12]. This opens an interesting possibility for the experimental signature of the model. In such a case, when $G^{(2)} = 0$, there arises an axion due to the spontaneously broken PQ symmetry in the original Lagrangian (2.1).[2] Such an axion is already ruled out, unfortunately. Thus the $G^{(2)}$ term is really a necessity to obtain m_b anyway.

6. It is an interesting question to ask what is the origin of the four-fermion interaction. It is also connected with the higher dimensional operators in the underlying theories. One might immediately think of heavy vector (gauge) boson exchange similar to ETC. However, as we emphasized in Ref.13, such a model does possess PQ

symmetry mentioned above and hence never gives rise to the $G^{(2)}$ term which explicitly breaks the PQ symmetry. This implies that m_b cannot be acquired through the top quark condensate, or we have a trouble of axion, if we assume the bottom quark condensate in addition to the top quark condensate. A possible way-out within the vector boson exchange model would be an invisible axion scenario.[51] An alternative possibility would be the scalar boson exchange, i.e., the strong coupling Yukawa interaction[13], which easily breaks the PQ symmetry and gives rise to $G^{(2)}$ term. At any rate, in the framework of the TMSM the strong CP problem and the generation of m_b are the different sides of the same coin.

7. We have argued that the TMSM gives the same results as the SM all the way up to the Planck scale, except that we can *predict* m_t and M_H in the former model. This may be an oversimplification, however. Since the top quark condensate is the composite theory, there might be a possibility for light vector bound states or more. Such a possibility might be realized if one includes ($1/N_c$ dominant) vector-exchange type four-fermion interaction in addition to (2.1). If this is indeed the case, the low energy physics will be affected substantially, which can be tested through precise electroweak experiments in the near future. The low energy effective theory of such vector bound states will well be described by the notion of the Hidden Local Symmetry[52].

I would like to thank K.-I. Kondo, S. Shuto and M. Tanabashi for collaborations and discussions on the recent results presented in this talk.

References

1. V.A. Miransky, M. Tanabashi and K. Yamawaki, Phys. Lett. **B221** (1989) 177.
2. V.A. Miransky, M. Tanabashi and K. Yamawaki, Mod. Phys. Lett. **A4** (1989) 1043.
3. Y. Nambu, *in Proc. XI Warsaw Symposium on Elementary Particle Physics, Kazimierz, Poland, 1988*, ed. Z. Ajduk, S. Pikorski and A. Trautman (World Scientific, Singapore, 1989); *Proc. International Workshop on New Trends in Strong Coupling Gauge Theories, Nagoya, 1988*, eds. M. Bando, T. Muta and K. Yamawaki (World Scientific, Singapore, 1989).
4. K.-I. Kondo, H. Mino and K. Yamawaki, Phys. Rev. **D39** (1989) 2430;
5. T. Appelquist, M. Soldate, T. Takeuchi and L.C.R. Wijewardhana, *in Proc. Johns Hopkins Workshop on Current Problems in Particle Theory 12, Baltimore, 1988*, eds. G. Domokos and S. Kovesi-Domokos (World Scientific, Singapore, 1988).
6. V.A. Miransky and K. Yamawaki, Mod. Phys. Lett. **A4** (1989) 129.
7. Y. Nambu and G. Jona-Lasinio, Phys. Rev. **122** (1961) 345
8. Y. Nambu, Chicago preprint EFI 89-08 (1989).
9. W.J. Marciano, Phys. Rev. Lett. **62** (1989) 2793; Phys. Rev. **D41** (1990) 219.
10. W.A. Bardeen, C.T. Hill and M. Lindner, Phys. Rev. **D41** (1990) 1647.

11. M. Suzuki, Phys. Rev. **D41** (1990) 3457.
12. M. Luty, Phys. Rev. **D41** (1990) 2893.
13. K.-I. Kondo, M. Tanabashi and K. Yamawaki, *in Proc. 1989 Workshop on Dynamical Symmetry Breaking, Nagoya, 1989,* eds. T. Muta and K. Yamawaki (Nagoya University, 1990); D.E. Clague and G.G. Ross, Nucl Phys. **B364** (1991) 43.
14. K.-I. Kondo, S. Shuto, M. Tanabashi and K. Yamawaki, in preparation.
15. K. Yamawaki, *in Proc. 1990 International Workshop on Strong Coupling Gauge Theories and Beyond, Nagoya, 1990,* eds. T. Muta and K. Yamawaki (World Scientific Singapore, 1991).
16. S. Shuto, M. Tanabashi and K. Yamawaki, *in Proc. 1989 Workshop on Dynamical Symmetry Breaking, Nagoya, 1989,* eds. T. Muta and K. Yamawaki (Nagoya University, 1990); also in preparation.
17. K.-I. Kondo, S. Shuto and K. Yamawaki, Mod. Phys. Lett. **A6** (1991) 3385.
18. K.-I. Kondo, M. Tanabashi and K. Yamawaki, Nagoya/Chiba preprint DPNU-91-20/Chiba-EP-53; also in preparation.
19. W.A. Bardeen, C.N. Leung and S.T. Love, Phys. Rev. Lett. **56** (1986) 1230; C.N. Leung, S.T. Love and W.A. Bardeen, Nucl. Phys. **B273** (1986) 649.
20. M. Harada and N. Kitazawa, Phys. Lett. **B257** (1991) 383; N. Kitazawa, to appear in Mod. Phys. Lett. **A**; E.V. Gorbar, Journ. Nucl. Phys. **53** (1991) 1034 (in Russian).
21. J. Kogut, E. Dagotto and A. Kocić, Phys. Rev. Lett. **62** (1989) 1001.
22. R.S. Chivukula, Phys. Rev. Lett. **61** (1988) 2657.
23. Y. Nagoshi, K. Nakano and S. Tanaka, Prog. Theor. Phys. **85** (1991) 131.
24. M. Bando, T. Kugo and K. Suehiro, Prog. Theor. Phys. **85** (1991) 1299.
25. V.A. Miransky, Sov. J. Nucl. Phys. **38** (1983) 280; K. Higashijima, Phys. Rev. **D29** (1984) 1228.
26. K. Lane, Phys. Rev. D **10** (1974) 2605; H.D. Politzer, Nucl. Phys. **B117** (1976) 397.
27. M. Bando, T. Morozumi, H. So and K. Yamawaki, Phys. Rev. Lett. **59** (1987) 389.
28. V.A. Miransky, T. Nonoyama and K. Yamawaki, Mod. Phys. Lett. **A4** (1989) 1409.
29. T. Takeuchi, Phys. Rev. **D40** (1989) 2697.
30. For a review, E. Farhi and L. Susskind, Phys. Rep. **74** (1981) 277.
31. H. Pagels and S. Stokar, Phys. Rev. **D20** (1979) 2947.
32. S. Englert and R. Brout, Phys. Lett. **49B** (1973) 77; A. Carter and H. Pagels, Phys. Rev. Lett. **43** (1979) 1845.
33. S.F. King and S.H. Mannan, Phys.Lett. **B241** (1990) 249; F.A. Barrios and U. Mahanta, Phys. Rev. **D43** (1991) 284.
34. M. Suzuki, Mod. Phys. Lett. **A5** (1990) 1205.

35. A. Hasenfratz, P. Hasenfratz, K. Jansen, J. Kuti and Y. Shen, Nucl. Phys. **B365** (1991) 79.
36. E.A. Paschos and V.I. Zakharov, Phys. Lett. **B272** (1991) 105.
37. J. Zinn-Justin, Nucl. Phys. **B367** (1991) 105.
38. T. Nonoyama, T.B. Suzuki and K. Yamawaki, Prog. Theor. Phys. **81** (1989) 1238.
39. A. Kocić, S. Hands, J.B. Kogut and E. Dagotto, Nucl. Phys. **B347** (1990) 217.
40. K. Yamawaki, M. Bando and K. Matumoto, Phys. Rev. Lett. **56** (1986) 1335.
41. T. Akiba and T. Yanagida, Phys. Lett. **B169** (1986) 432.
42. T. Appelquist and L.C.R. Wijewardhana, Phys. Rev. **D36** (1987) 568.
43. B. Holdom, Phys. Lett. **B150** (1985) 301; **B198** (1987) 535.
44. K. Matumoto, Prog. Theor. Phys. **81** (1989) 277.
45. T. Appelquist, T. Takeuchi, M. Einhorn and L.C.R. Wijewardhana, Phys. Lett. **B220** (1989) 223.
46. J. Cornwall, R. Jackiw and E. Tombolis, Phys. Rev. **D10** (1974) 2428.
47. W.A. Bardeen and S.T. Love, Purdue/Fermilab preprint PURD-TH-91-08-/FERMILAB-PUB-91/233-T.
48. T. Appelquist, J. Terning and L.C.R. Wijewardhana, Phys. Rev. **D44** (1991) 871.
49. Y. Kikukawa and K. Yamawaki, Phys. Lett. **B234** (1989) 497.
50. M. Bando, K. Matumoto and K. Yamawaki, Phys. Lett. **B178** (1986) 308.
51. Y. Achiman and A. Davidson, Phys. Lett. **B261** (1991) 431.
52. M. Bando, T. Kugo, S. Uehara, K. Yamawaki and T. Yanagida, Phys. Rev. Lett. **54** (1985) 1215; M. Bando, T. Kugo and K. Yamawaki, Nucl. Phys. **B259** (1985) 493; Phys. Rep. **164** (1988) 217.

THE GENERALIZED NAMBU–JONA-LASINIO MECHANISM AND THE STANDARD MODEL

Julius Kuti

University of California, San Diego, Dept. of Physics B-0319, 9500 Gilman Dr., La Jolla, CA, 92093-0319

The generalized Nambu–Jona-Lasinio (NJL) model which has three independent terms in the Lagrangean is shown to be renormalizable within the framework of the 1/N expansion. Physics below the cut-off generated by the NJL mechanism is equivalent to the renormalizable Higgs-Yukawa sector of the Standard Model: the three independent parameters of the generalized NJL model are mapped into the vacuum expectation value (*vev*) of the Higgs field, the Higgs self-coupling, and the Yukawa coupling. Although the original NJL model with a single coupling constant is also known to be renormalizable in the 1/N expansion, the model is more short distance sensitive than simple renormalizable theories and new parameters are hidden in the cut-off procedure. The compositeness of the Higgs particle only appears on the cut-off scale and no constraint exists between the Higgs mass and top quark mass in the generalized NJL model without further assumptions on the three independent parameters of the Lagrangean. The "naturalness" of the assumptions will be discussed in Hill's toy gauge model which was introduced as a minimal extension of the Standard Model to illustrate the role of the NJL condensate in the top quark mode as advocated by Nambu, and Bardeen, Hill, Lindner (BHL). Universality arguments suggest that the analysis of the generalized NJL model presented here should remain valid beyond the 1/N expansion which only serves as a useful technical tool in the discussions.

1. Introduction

In several interesting papers it was suggested that the symmetry breaking mechanism of the Standard Model (SM) occurs dynamically by the formation of a top quark condensate [1–4] in analogy with the BCS, or Nambu-Jona-Lasinio mechanism [5]. No fundamental Higgs scalar is introduced. At low energies the Standard Model emerges with some constraints leading to non-trivial predictions. Most interestingly, the mass of the Higgs boson (a "composite" particle in this scheme) is related to the top quark mass and both mass values were predicted [2] to be close to the electroweak symmetry breaking scale set by $vev \sim 250\ GeV$.

In a recent paper [6], in collaboration with A. Hasenfratz, P. Hasenfratz, K. Jansen, and Y. Shen, we investigated the generalized NJL model in the large-N approximation. The generalization is necessary, because higher dimensional operators have a non-trivial effect on the renormalization procedure: all independent terms have to be included in a general NJL Lagrangean when used as an effective theory at some high-energy scale Λ. Three independent parameters were found without the predictive power of the original NJL mechanism which only included one interaction term. The renormalization properties of the NJL model were later discussed in an independent paper [7] and similar conclusions to ours were reached. I will also make some comments on Hill's interesting top color model [8] which was proposed in defense of the predictive power of the top quark condensate idea. In the conclusions I will summarize the controversial issues.

2. The Nambu–Jona-Lasinio Model

I will present the most important results on the original NJL model in a simplified formulation where no gauge fields are included and the top quark is considered only. The basic conclusions are not expected to change when gauge fields are included perturbatively. A more realistic version within the context of Hill's top color gauge model [8] will be presented later.

The Euclidean action of the NJL model with U(1) chiral symmetry is given by

$$S^{\Lambda}_{NJL} = \int d^4x\{\overline{\psi}(x)\gamma^{\mu}\partial_{\mu}\psi(x) - \frac{1}{2}G[(\overline{\psi}(x)\psi(x))^2 - (\overline{\psi}(x)\gamma_5\psi(x))^2]\} \quad (1)$$

where the superscript Λ designates some momentum cut-off procedure when the partition function and correlation functions are calculated. The dimension of the coupling constant G is carried by the appropriate power of Λ so that $g = \Lambda^2 \cdot G$ is dimensionless. The top quark field $\psi^a(x)$ carries an index $a = 1, ..., N$ for the convenience of the 1/N expansion.

The model has a simple phase diagram which is depicted in Fig.1. The critical point g_c separates the chiral symmetric phase with massless fermions from the phase with broken chiral symmetry. In the broken phase the fermion is massive and there is a massless Nambu-Goldstone particle. The ground state of the broken phase is understood in terms of BCS fermion pair condensation which drives the NJL mechanism of dynamical symmetry breaking. The mass m of the fermion is determined from the gap equation, and one also finds in the spectrum a scalar (Higgs) particle with mass m_H. The Higgs particle is composite on the cut-off scale but pointlike in the critical region. In the limit of infinite Higgs correlation length ($\xi = \Lambda/m_H \rightarrow \infty$) the famous mass relation $m_H = 2m$ exists for large N.

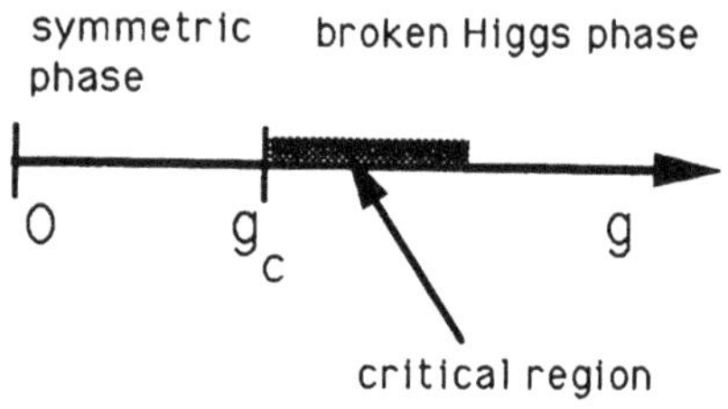

Fig. 1. The phase diagram of the NJL model is separated by the critical value g_c of the dimesionless coupling constant. The shaded area in the broken phase corresponds to continuum physics with a large Λ/m_H ratio.

Historically the phase transition and the mass spectrum were calculated in Hartree-Fock approximation [5] which becomes exact in the large-N limit. The Hartree-Fock approximation is expected to capture the relevant physical properties of the model at finite N and controlled quantitative information can be obtained from large scale computer simulations [6].

3. Equivalent Higgs-Yukawa Model

A simple universality argument suggests that the NJL model defined by Eq.(1) is equivalent to the Higgs-Yukawa sector of a simplified SM with $U(1)$ chiral symmetry:

$$S^{\Lambda}_{SM} = \int d^4x\{\frac{1}{2}\partial_{\mu}\varphi_i\partial_{\mu}\varphi_i + \frac{1}{2}\mu_0^2\varphi_i\varphi_i + \frac{\lambda_0}{4!}(\varphi_i\varphi_i)^2 + \overline{\psi}\gamma^{\mu}\partial_{\mu}\psi + y_0(\varphi_1\overline{\psi}\psi + i\varphi_2\overline{\psi}\gamma_5\psi)\},\ i = 1, 2\ . \quad (2)$$

The model in Eq.(2) and other related Higgs-Yukawa models have been the subject of detailed analytic and numerical investigations over the last few years [9]. The lattice version of Eq.(2) reads

$$S_{SM}^{latt} = -\kappa \sum_{x,\mu} \phi_i(x)(\phi_i(x+\mu) + \phi_i(x-\mu))$$
$$+ \sum_x \phi_i(x)\phi_i(x) + \frac{\lambda_0^{latt}}{4!} \sum_x [\phi_i(x)\phi_i(x) - 1]^2$$
$$+ \frac{1}{2} \sum_{x,\mu} \overline{\psi}(x)\gamma^\mu(\psi(x+\mu) - \psi(x-\mu)) \qquad (3)$$
$$+ y_0^{latt} \sum_x [\phi_1(x)\overline{\psi}(x)\psi(x) + i\phi_2(x)\overline{\psi}(x)\gamma_5\psi(x)],$$

where, on the classical level, the lattice parameters and fields are related trivially to those of Eq.(2) [#1].

Fig.2 shows the sketch of the phase diagram for the lattice Higgs-Yukawa model of Eq.(3) at $N = 20$ with the particular choice $\lambda_0^{latt} = 0$ (the scaled Yukawa coupling of Fig.2 is defined by $\tilde{y} = \sqrt{N} \cdot y_0^{latt}$). The second order phase transition line relevant for us is the one separating the symmetric (SYM) and spontaneously broken ferromagnetic (FM) phases originating from the point ($y_0^{latt} = 0, \kappa = \frac{1}{8}$) (this part of the phase diagram remains qualitatively the same at $\lambda_0^{latt} \neq 0$). Along the SYM-FM second order phase transition line there is no sign of a non-trivial fixed point. The points of the critical line between the SYM and FM phases belong to the domain of attraction of the Gaussian fixed point ($\kappa = \frac{1}{8}, y_0^{latt} = 0, \lambda_0^{latt} = 0$). Approaching this critical line from the broken phase one obtains an (almost) cut-off independent theory with small low energy renormalized couplings. The scalar field develops a non-zero vacuum expectation value, there are massive fermions and

[#1] In the continuum limit Eq.(3) describes $N' = 16N$ fermions (doubling) which is, however, not an issue in the following considerations; the lattice spacing a is set to 1 throughout the paper.

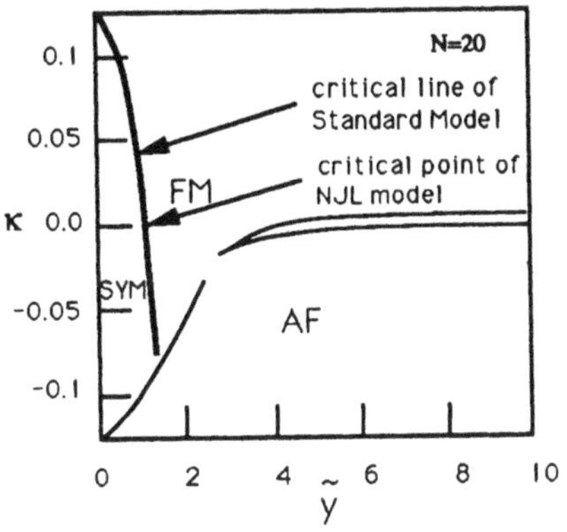

Fig. 2. The phase diagram of the lattice Higgs-Yukawa model. The antiferromagnetic phase (AF) and a special symmetric phase in the narrow strip between the horizontal lines are not relevant for the continuum limit.

a massive scalar (Higgs boson) together with a massless Goldstone boson. Every point close to the second order line in the FM-phase corresponds to the Higgs-Yukawa sector of the simplified Standard Model.

The renormalization properties of the Higgs-Yukawa theory close to the SYM-FM critical line follow the triviality scenario. For any given set of small low energy couplings y_R, λ_R one can tune the bare parameters in such a way that over a certain range of the cut-off the physical predictions become cut-off independent (up to small inverse power cut-off corrections). The cut-off can not be completely removed, however. Or, saying differently, for any value of the bare couplings y_0 and λ_0 (≥ 0), as μ_0^2 is tuned towards the critical surface (Λ is increased), the corresponding low energy couplings y_R^2 and λ_R go to zero. Asymptotically they behave as $\sim (ln\Lambda)^{-1}$ and the prefactors are determined by the 1-loop β-function.

If we approach the critical surface along the $\kappa = 0$ line in Fig.2 for $\lambda_0^{latt} = 0$, the action in

Eq.(3) is quadratic in the scalar fields which can be integrated out to obtain

$$S^{latt}_{NJL} = \frac{1}{2}\sum_{x,\mu}\overline{\psi}(x)\gamma_\mu(\psi(x+\mu)-\psi(x-\mu))$$
$$-\frac{1}{2}G^{latt}\sum_x[(\overline{\psi}(x)\psi(x))^2-(\overline{\psi}(x)\gamma_5\psi(x))^2], (4)$$

where $G^{latt}=1/2(y_0^{latt})^2$ in cut-off units. Eq.(4) is the lattice regularized version of the NJL model in Eq.(1). It follows then that the critical properties of the NJL model are the same as the Higgs-Yukawa sector of the simplified SM at a special point of the critical line. The equivalence with a generalized NJL model is anticipated to hold everywhere along the critical line in the absence of some unexpected singularity.

4. Equivalence Proof in the 1/N Expansion

The equivalence of the Higgs-Yukawa sector of the simplified SM and the generalized NJL model with identical renormalized Green's functions can be shown analytically in the 1/N expansion [6]. Let us consider the following chiral symmetric generalization of the NJL-model in Eq.(1)

$$\begin{aligned}S^\Lambda_{GNJL} = &\int d^4x\,\{\overline{\psi}\gamma^\mu\partial_\mu h(\partial^2)\psi\\ &-\frac{1}{2}\frac{g}{\Lambda^2}\Big[(\overline{\psi}\psi+\frac{d}{\Lambda^2}\partial_\mu\overline{\psi}\partial_\mu\psi)f(\partial^2)\\ &\times(\overline{\psi}\psi+\frac{d}{\Lambda^2}\partial_\nu\overline{\psi}\partial_\nu\psi)\\ &-(\overline{\psi}\gamma^5\psi+\frac{d}{\Lambda^2}\partial_\mu\overline{\psi}\gamma^5\partial_\mu\psi)f(\partial^2)\\ &\times(\overline{\psi}\gamma^5\psi+\frac{d}{\Lambda^2}\partial_\nu\overline{\psi}\gamma^5\partial_\nu\psi)\Big]\Big\},\end{aligned} \qquad (5)$$

where $f(\partial^2)=1+(f_1/\Lambda^2)\partial^2+...$ and $h(\partial^2)=1+(h_1/\Lambda^2)\partial^2+...$ are *finite* polynomials (therefore the model is local), while $d, f_1, ..., g, h_1, ...$ are dimensionless couplings. We work in euclidean space and Λ is a spherical momentum space cut-off. The action in Eq.(5) contains a large number of parameters. This is a redundant set and the physical predictions depend on the same number of parameters as those of the Higgs-Yukawa sector of the simplified SM in Eq.(2).

Assuming that the couplings $d, f_1, ..., h_1, ...$ are independent of N, while $g=O(1/N)$, the generalized NJL-model can be solved exactly in the large-N limit. The physical predictions depend on $d, f_1, ..., h_1, ...$ through two independent functions $\alpha=\alpha(d, f_1, ..., h_1, ...)$ and $\beta=\beta(d, f_1, ..., h_1, ...)$ which are the relevant additional parameters of the generalized NJL-model. It was shown [6] that the renormalized Green's functions in the two models are *identical* for all physical momenta if the following identification is made

$$y_0^2N_c=\frac{1}{\beta},\qquad \lambda_0N_c=3\frac{\alpha}{\beta^2}\ . \qquad (6)$$

The same equivalence is true for the exact effective potential of these two theories. We can add to these results the general expectation (based on renormalization group arguments) that the low energy behaviour of any model with the given particle content and symmetries can be described by Eq.(2). The conclusion is that up to a trivial redefinition of the parameters the generalized NJL-model *is* the simplified Standard Model. The only missing point is to demonstrate that as $d, f_1, ..., h_1, ...$ run over their allowed range, the couplings α and β (or y_0 and λ_0) cover their full allowed range. For that it is enough to consider a simple example:

$$d\neq 0, f_1\neq 0, f_2=...=h_1=...=0\ . \qquad (7)$$

We obtain for this case the result

$$\alpha=\frac{1}{2\pi^2}(4d+3d^2+\frac{4}{3}d^3+\frac{1}{4}d^4)\,,$$

$$\beta = \frac{1}{8\pi^2}\left[2f_1(1+d+\frac{1}{3}d^2)+3d+\frac{3}{4}d^2\right] \quad . \quad (8)$$

The functions α and β have to be positive, which determines the allowed range of the parameters d and f_1. It is clear that y_0 and λ_0 run over the full parameter range as d and f_1 run over their allowed range.

5. Hill's Toy Model and the NJL Condensate

Hill introduced a toy gauge model [8] as a minimal extension of the Standard Model to illustrate the role of the NJL condensate in the top quark mode. The model Lagrangean has the gauge symmetry $U(1)\otimes SU(2)_L\otimes SU(3)_1\otimes SU(3)_2$ before symmetry breaking is introduced at some very high energy scale $\Lambda \sim 10^{15}$ GeV. The assignments of quarks and leptons under the gauge transformations $SU(2)_L$, $SU(3)_1$, $SU(3)_2$ are as follows:

$$\begin{aligned}
&(u,d)_L;\ (c,s)_L && \rightarrow (2,3,1)\\
&(\nu_e,e);\ (\nu_\mu,\mu);\ (\nu_\tau,\tau) && \rightarrow (2,1,1)\\
&u_R, d_R, c_R, s_R, b_R && \rightarrow (1,3,1)\\
&e_R, \mu_R, \tau_R, (\nu_{iR}) && \rightarrow (1,1,1).
\end{aligned}$$

The top quark is assigned

$$(t,b)_L \rightarrow (2,1,3); \quad t_R \rightarrow (1,1,3).$$

The assignment is not anomaly free, but an additional electroweak singlet quark which otherwise will play no role in the analysis provides the necessary anomaly cancellations.

The gauge symmetry $SU(3)_1\otimes SU(3)_2$ is broken at some scale Λ by a heavy scalar (Higgs) field ϕ^a_b which transforms as $(1,3\bar{3})$. The VEV which has the form $\langle\phi\rangle = \Lambda\cdot\delta^a_b$ breaks $SU(3)_1\otimes SU(3)_2$ to a massless gauge group $SU(3)_c$ with gluons A_μ and a residual global $SU(3)'$ with degenerate, massive gauge bosons ("colorons") B_μ. The original gauge fields $A_{1\mu}$ and $A_{2\mu}$ of $SU(3)_1\otimes SU(3)_2$ are related to A_μ and B_μ by

$$\begin{aligned}
A_{1\mu} &= \cos\theta\cdot A_\mu - \sin\theta\cdot B_\mu\ ,\\
A_{2\mu} &= \sin\theta\cdot A_\mu + \cos\theta\cdot B_\mu\ ,
\end{aligned}$$

where $\tan\theta = h_1/h_2$; h_1, h_2 designate the gauge coupling constants of $SU(3)_1$ and $SU(3)_2$, respectively. The QCD coupling constant at scale Λ is given by $g_3 = h_1\cos\theta$ and the mass of the degenerate octet of heavy colorons is $M_B = \sqrt{h_1^2+h_2^2}\,\Lambda$.

The $SU(3)'$ current has the form

$$h^A_\mu = g_3\cot\theta\left(\bar{t}\gamma_\mu\frac{\lambda^A}{2}t + \bar{b}_L\gamma_\mu\frac{\lambda^A}{2}b_L\right) + \cdots$$

where $h_2 \gg h_1$ is assumed and terms $\propto g_3\tan\theta$ are only indicated by ellipsis. The contribution of the electroweak singlet quark is not shown.

The exchange of a heavy coloron between fermions after Fierz rearrangement leads to the effective top quark Lagrangean

$$\frac{g^2}{\Lambda}\left(\bar{\psi}^a_L t_{Ra}\right)_i\left(\bar{t}_{Rb}\psi^b\right)^i$$

which is of the NJL type with large-N indices (a,b) and an $SU(2)_L$ index i; the effective NJL coupling is given by

$$\frac{g^2}{\Lambda^2} = \frac{g_3^2\cot^2\theta}{\Lambda^2}\ .$$

Radiative corrections coming from higher order diagrams, like the exchange of two heavy colorons generate new higher dimensional terms in the effective Lagrangean. Hill estimates the leading correction from the box diagram to be

$$L = L_{NJL} + \chi\frac{g}{M_B^2}\left(D^\mu\bar{\psi}^a_L D_\mu t_{Ra}\right)^i\left(\bar{t}_{Rb}\psi^b\right)^i \quad (9)$$

which has the form we discussed earlier in the generalized NJL framework. The estimate on the size of the correction term is $\chi \approx -\frac{5}{24}$ which is argued to have negligible effect on the low energy predictions of BHL for m_t (or m_H), shifting the mass values by less than one percent [11].

6. Conclusions

The attractive idea to replace the Higgs sector with an effective four-fermion interaction and its top quark condensate can be implemented in the 1/N expansion in a well-defined and systematic way. We have shown that the number of independent parameters in the general NJL framework is identical to that of the minimal SM. The two representations can be made completely equivalent. Specifically, for a given cut-off Λ, any m_H/m_t ratio which can be obtained in the Standard Model will be reproduced by the generalized NJL-model. The "compositeness" of the Higgs particle does not lead to a predictable mass relation and the value of the top quark mass is independent of the scale of the top quark condensate. Our conclusions are not completely unexpected. In a recent independent paper Suzuki considered a specific modification of the NJL-model and also found that the physical predictions are sensitive to this change [10].

Based on the top color gauge model Hill argues that "generic" values for the added terms generated by any sensible theory which drives the top quark condensate will not change the original predictions. In addition to the difficulty of deriving the effective NJL condensate from the original gauge model, the 1-loop Renormalization Group had to be used at strong coupling to obtain estimates.

Speculation about the effects of the various terms in the generalized NJL Lagrangean when used as an effective theory at and below some high energy scale Λ will continue. Before the true meaning of the NJL effective approximation is further clarified in a quantitative non-perturbative fashion, the predictive power of the top quark condensate idea should be taken with some reservations.

Acknowledgements

The work presented here has been done in collaboration with Anna Hasenfratz, Peter Hasenfratz, Karl Jansen, and Yue Shen. I am greatful to them and the organizers for giving me the opportunity to present our results at this conference. I am indebted to Bill Bardeen, Howard Georgi and Christopher Hill for very interesting discussions. I would also like to thank the organizers for the stimulating atmosphere at the conference. This work was supported by DOE grant #DE-FG-03-90ER40546.

References

[1] Y. Nambu, Enrico Fermi Institute preprints, 88-39(1988), 88-62(1988) and 89-08(1989).

[2] W. A. Bardeen, C. T. Hill and M. Lindner, Phys. Rev. D41 (1990) 1647; C. T. Hill, FNAL preprint, FERMI-CONF-90/170-T(1990).

[3] V. A. Miransky, M. Tanabashi and K. Yamawaki, Mod. Phys. Lett. A4 (1089) 1043; Phys. Lett. 221B (1989) 177.

[4] W. J. Marciano, Phys. Rev. Lett. 62 (1989) 2793.

[5] Y. Nambu and G. Jona-Lasinio, Phys. Rev. 122 (1961) 345.

[6] A. Hasenfratz, P. Hasenfratz, K. Jansen, J. Kuti, and Y. Shen, Nucl. Phys. B365 (1991) 79.

[7] J. Zinn-Justin, Nucl. Phys. B367 (1991) 105.

[8] C. T. Hill, Phys. Lett. B266 (1991) 419.

[9] A.M. Thornton, Phys. Lett. 214B (1988) 577; A. Hasenfratz, W. Liu and T. Neuhaus, Phys. Lett. 236B (1990) 339. For a recent review see M.F.L. Golterman, Nucl. Phys. (Proc. Suppl.) B20 (1991) 528.

[10] M. Suzuki, Mod. Phys. Lett. A5 (1990).

[11] W. A. Bardeen, FNAL preprint 90/269/T (1990).

VARIOUS TOPICS

THE CONFINEMENT AND LOCALIZATION OF QUARKS*

Janos Polonyi †

Center for Theoretical Physics
Laboratory for Nuclear Science
and Department of Physics
Massachusetts Institute of Technology
Cambridge, Massachusetts 02139 U.S.A.

ABSTRACT

It is shown that the symmetry with respect to the center of the global gauge group requires non-polynomial terms in the renormalized QCD lagrangian. This new interaction generates linear potential between static quarks and breaks the chiral symmetry of the vacuum independently of the renormalized three and four gluon vertices. This confinement mechanism is similar to the localization of electrons observed in solid state physics.

I. INTRODUCTION

Effective theories are used when the dynamics of the fundamental degrees of freedom is either (i) unknown or (ii) too complicated to treat in its complexity. The well known example for (ii) is QED or QCD where the interactions for the short distance vacuum fluctuations or the long range modes, respectively are strong. Thus certain quasiparticles (bound states of virtual e^+e^- pairs for QED and hadrons in QCD) are introduced with some effective interactions. It is believed that by starting with the bare Yang-Mills lagrangian

$$L_b = \frac{1}{2g_b^2} \operatorname{tr} F_{\mu\nu}^2, \tag{1.1}$$

* This work is supported in part by funds provided by the U. S. Department of Energy (D.O.E.) under contract #DE-AC02-76ER03069.

† Supported in part by the Alfred P. Sloan Foundation and the NSF-PYI grant PHY-8958079. On leave from CRIP, Budapest, Hungary

with

$$F_{\mu\nu} = \partial_\mu A_\nu - \partial_\nu A_\mu + [A_\mu, A_\nu] \tag{1.2}$$

and

$$A_\mu = A^a_\mu \frac{\lambda^a}{2i} \tag{1.3}$$

at the ultraviolet cutoff we shall ultimately recover the confining long distance behavior after the resummation of the perturbation expansion. Since this program goes well beyond our present analytical capabilities we have to be content with some kind of effective theory for the long distance phenomena which displays the desired features in a more accessible manner.

One expects nothing surprising in the case (ii). In fact, QED or QCD can be considered to be correct in the appropriate range of length scales set by some underlying grand unified scheme. The detailed manner the grand unified model cuts off these models in the ultraviolet is actually irrelevant if the "sub"-theories are renormalizable. The requirement of renormalizability combined with other general considerations based on symmetry properties and locality restricts the class of allowed interactions in a drastic manner. The unmanageable richness of the infrared physics is constrained by considering only such interactions which have controllable behavior at short distances. The consistency of this scheme is that the renormalizable interactions are just the relevant ones in the infrared. One can verify that the effects of the non-renormalizable vertex is $O((\frac{p}{\Lambda})^{\omega-4})$ in the infrared, where p and Λ are the infrared momentum scale and the ultraviolet cutoff, respectively and ω is the sum of the canonical and the anomalous dimensions.

Our goal is to demonstrate that this scheme gives a richer Yang-Mills theory then originally thought. It is well known that the asymptotically free perturbation expansion gives (1.1) as the only relevant operator. But a symmetry argument presented below suggests that the action of the confining Yang-Mills theory involves a non-polynomial term L_{np},

$$L_b = \frac{1}{2g_b^2} \operatorname{tr} F_{\mu\nu}^2 + L_{\text{np}}, \tag{1.4}$$

in addition to the conventional quartic part. The theory should remain renormalizable in the presence of this interaction. The renormalized part of L_{np} alone generates the expected long range structure of the vacuum without the three and four gluon vertices [1]and it appears as another relevant operator of the Yang-Mills system. We shall derive a low energy effective theory for (1.4). If the non-perturbative renormalization of the complete QCD lagrangian were understood then the quantities which characterize the long distance features of the vacuum, such as the string tension and the chiral condensate could be obtained analytically by the help of this effective theory. Though we are not yet at that stage we believe that the virtue of this approach is that it shifts the difficulties of the analytic solution of QCD from the infrared to the ultraviolet domain where we have more chances to succeed ultimately.

The quarks scatter off certain localized potentials, called haarons when they propagate through the vacuum build by L_{np}. The quark mass acquires an imaginary part in this process and its asymptotic states are removed from the theory. This mechanism is similar to the localization in solid state physics, except that it takes place in the space-time rather than the space. This difference arises because the haaron potential is localized in the time as well.

The organization of the paper is the following. In Section II. we show that the invariance under the center of the global gauge group requires a non-polynomial piece in the Yang-Mills action. Section III. approaches this problem from a different point of view, by arguing that the gauge field is not smooth even in the renormalized theory. The renormalization properties and the low energy effective theory is discussed in Section IV. We mention briefly the computation of the string tension and the derivation of a simpler finite effective theory in Section V. Section VI. contains the description of the haaron gas expansion for the quark Green function. Finally, Section VII. is for the summary.

II. SYMMETRIES OF THE YANG–MILLS SYSTEM

We point out first the presence of a discrete symmetry for the Yang-Mills system which plays a key role in the confinement mechanism [2]. It has been isolated in lattice gauge gauge theory and should contain the right physics unless quarks play fundamental role in the formation of the confining forces. The gauge invariance of the vacuum should be taken seriously when the origin of the confining forces are sought. In fact, the gauge-dependent component of the vacuum describes background charges. These may break the electric flux tube between test charges and shield the confining potential when Gauss' law is not exactly implemented.

The starting point is the hamiltonian formalism in temporal gauge. The canonical variables are $\mathbf{A}^a(\mathbf{x})$ and $\mathbf{E}^a(\mathbf{x}) = \delta/i\delta\mathbf{A}^a(\mathbf{x})$. They are collected into anti-Hermitean matrices,

$$\mathbf{A} = g\mathbf{A}^a\frac{\lambda^a}{2i}, \tag{2.1}$$

and

$$\mathbf{E} = \mathbf{E}^a \frac{\lambda^a}{2ig}. \tag{2.2}$$

The hamiltonian is

$$H = -\mathrm{tr} \int d^3x \{g^2 \mathbf{E}^2(\mathbf{x}) + \frac{1}{g^2} \mathbf{B}^2(\mathbf{x})\}, \tag{2.3}$$

where

$$B_i = \epsilon_{ijk}(\partial_j A^k + A^j A^k) \; . \tag{2.4}$$

The temporal gauge condition $A_0 = 0$ allows time independent gauge transformations $\mathcal{G}$ which act on the coordinates as

$$\mathbf{A} \rightarrow \mathbf{A}^h = h(\partial + \mathbf{A})h^\dagger, \tag{2.5}$$

and are generated by the operators

$$G(\mathbf{x}) = G^a(\mathbf{x})\frac{\lambda^a}{2i} = \mathbf{D} \cdot \mathbf{E}^a(\mathbf{x})\frac{\lambda^a}{2i} = \boldsymbol{\partial} \cdot \mathbf{E}(\mathbf{x}) + [\mathbf{A}(\mathbf{x}), \mathbf{E}(\mathbf{x})] \; . \tag{2.6}$$

in the Hilbert space.

The basis vectors of the Hilbert space corresponding to (2.3) can be classified by constructing the irreducible representations of $\mathcal{G}$. These states are labeled by the eigenvalues of the commuting combinations of the generators. This set of eigenvalues specify the static local background charge and states sharing the same charge density form a charge superselection class. We shall work in the vacuum sector which corresponds to the singlet representation of $\mathcal{G}$.

The basic building block for the quantum theory is the matrix element of the time evolution operator between coordinate eigenstates,

$$< \mathbf{A}'|e^{-itH}|\mathbf{A} >_{\mathbf{phys}}, \tag{2.7}$$

where the subscript is the remainder that the gauge singlet component is kept only. (2.7) has manifest gauge symmetry,

$$< \mathbf{A}'^h|e^{-itH}|\mathbf{A} >_{\mathbf{phys}} = < \mathbf{A}'|e^{-itH}|\mathbf{A} >_{\mathbf{phys}}, \tag{2.8}$$

which can be used to generate the Ward identities by using

$$\frac{\delta}{\delta h} < \mathbf{A}'^h|e^{-itH}|\mathbf{A} >_{\mathbf{phys}} = 0. \tag{2.9}$$

Though the symmetry leads directly to the Ward identities, the opposite is not true. It is well known that the the spontaneous breakdown of a global symmetry makes the observables non-symmetrical but preserves the Ward identities. We shall see below that the eventual breakdown of the global part of $\mathcal{G}$ destroys the confinement mechanism. Thus extra care is needed to ensure that the approximations we ultimately have to make do not themselves break $\mathcal{G}$.

Can the global component of $\mathcal{G}$ be broken "spontaneously" in (2.7) ? It is sufficient to look into the center of the global gauge transformations. The center of a group consists of elements which commute with the whole group. The global center gauge transformations for gauge group SU(N) is Z_N which is formed by the matrices

$$(z_k)_{ij} = \delta_{ij} e^{i\frac{2\pi k}{N}}, \tag{2.10}$$

with $k = 0, \cdots, N-1$. It is clear that the breakdown of the center symmetry destroys the symmetry with respect to the continuous part of the global subgroup as well.

The center symmetry is broken dynamically for the short time matrix elements, $t \sim 0$. In order to understand the way this happens we digress into the details of the construction of the gauge singlet (2.7). The singlet component is selected by averaging over the gauge transformations

$$< \mathbf{A}'|e^{-itH}|\mathbf{A} >_{\mathbf{phys}} = \int \mathcal{D}[g] < \mathbf{A}'^{g}|e^{-itH}|\mathbf{A} > . \tag{2.11}$$

This procedure can formally be written as

$$< \mathbf{A}'|e^{-itH}|\mathbf{A} >_{\mathbf{phys}} = < \mathbf{A}'|\mathcal{P}e^{-itH}|\mathbf{A} >, \tag{2.12}$$

where $\mathcal{P}$ is a projection operator

$$\mathcal{P} = \int \mathcal{D}[g]\hat{g} = \int \mathcal{D}[\alpha(\mathbf{x})] e^{i\int d^3y \alpha^a(\mathbf{y}) G^a(\mathbf{y})}. \tag{2.13}$$

Consider now the functional integral

$$< \mathbf{A}'|e^{-itH}|\mathbf{A} >_{\mathbf{phys}} = \int \mathcal{D}[g] e^{-\mathcal{H}[g;\mathbf{A}',\mathbf{A},t]}, \tag{2.14}$$

with

$$\mathcal{H}[g; \mathbf{A}', \mathbf{A}, t] = -\log < \mathbf{A}'^{g}|e^{-itH}|\mathbf{A} > . \tag{2.15}$$

It is instructive to follow $\mathcal{H}$ along the trajectory of the global gauge transformations, $g_0(x) = g$. $\mathcal{H}(g_0)$ is a periodic function since

$$\mathcal{H}[gz_k] = \mathcal{H}[g]. \tag{2.16}$$

This relation holds because the global gauge transformations from the center leave the gauge field unchanged and the matrix element in (2.15) is invariant in a trivial manner. Consider the case $\mathbf{A}' = \mathbf{A}$ for simplicity. Then

$$\mathrm{Re}\Big(\mathcal{H}(g_0) - \mathcal{H}(z_k)\Big) = O(\frac{\mathbf{A}^2 V}{t}) \tag{2.17}$$

for $t \sim 0$ and $g_0 \neq z_k$ where V is the three volume of the system. This can be seen by approximating (2.15) with the contribution of the straight trajectory connecting the initial and the final coordinates and recalling that only states with non-localized field contribute in the large volume limit due to their entropy. The "hamiltonian" $\mathcal{H}$ develops deep minima and the "classical statistical mechanical system" (2.14) is in the center-symmetry broken phase for the short time processes.

Such a symmetry breaking is supported by numerical results in lattice gauge theory [3] where the theory is considered at finite temperature. The partition function is

$$Z = \int \mathcal{D}[\mathbf{A}] < \mathbf{A}|e^{-\beta H}|\mathbf{A} >_{\mathrm{phys}}, \tag{2.18}$$

with $\beta = 1/T$ and the center symmetry is found to be broken at high temperature, short Euclidean time. The color component of the Euclidean heavy quark propagator $G(\mathbf{x}, \beta, \mathbf{x}, 0)$ is just the gauge transformation $g(\mathbf{x})$. In fact, this is the gauge transformation in (2.14) which is performed between the initial and the final time. This propagator is vanishing in the center symmetrical phase since

$$\sum_k z_k = 0. \tag{2.19}$$

Thus the center symmetry is equivalent with the absence of the quarks.

The lesson of the deconfined phase matrix elements at short time is that the Ward identities are not sufficient to recover the confining long distance behavior in general. We have to make sure in addition that the center symmetry is manifest in the vacuum. This in turn involves large gauge fields and is a non-perturbative issue.

In order to make contact with the path integral formalism we note that the matrix element

$$< \mathbf{A}'^g|e^{-itH}|\mathbf{A} >, \tag{2.20}$$

with $g(\mathbf{x}) = e^{i\alpha^a(\mathbf{x})\lambda^a}$ can be written as

$$\int \mathcal{D}[\mathbf{A}](\mathbf{x},t)e^{iS[\mathbf{A},\frac{1}{t}\alpha(\mathbf{x})]}, \tag{2.21}$$

where

$$S[\mathbf{A},A_0] = \frac{1}{2g^2}\int d^4x \, \mathrm{tr}\, F^2_{\mu\nu}. \tag{2.22}$$

The boundary conditions on the vector field in the path integration are the usual ones,

$$\mathbf{A}(\mathbf{x},0) = \mathbf{A}(\mathbf{x}), \tag{2.23}$$

and

$$\mathbf{A}(\mathbf{x},t) = \mathbf{A}'(\mathbf{x}). \tag{2.24}$$

Thus the integral variable of the projection operator acts like a static A_0 field. The complete matrix element (2.7) can now be written as

$$< \mathbf{A}'|e^{-itH}|\mathbf{A} >_{\mathrm{phys}} = \int \mathcal{D}[tA_0]\mathcal{D}[\mathbf{A}]e^{iS[A_\mu]}. \tag{2.25}$$

The important point is that the integration measure $\mathcal{D}[tA_0]$ is the invariant Haar measure for the gauge group. In fact, the gauge invariance (2.8) is exactly the condition

$$\int dg f(g) = \int dg f(gh) \tag{2.26}$$

for the integration measure over the gauge group. The invariant measure for SU(N) is

$$d_{\mathrm{inv}}\alpha = dv \prod_i d\lambda_i \sum_n \delta(\sum_i \lambda_i - 2\pi n) \prod_{i<j} \sin^2(\frac{\lambda_i - \lambda_j}{2}), \tag{2.27}$$

where the group element g is written as $g = vav^\dagger$ with a, a diagonal matrix, $v \in SU(N)$ and the eigenvalues of g are $e^{i\lambda_j}, j = 1, \cdots, N$. It is instructive to see the simplest case, $N = 2$. Using the parameterization

$$g = e^{i\boldsymbol{\alpha}\cdot\frac{\boldsymbol{\sigma}}{2}}, \tag{2.28}$$

the invariant measure is

$$d\phi d(\cos\theta)d\alpha \sin^2\alpha = d^3\alpha(\frac{\sin\alpha}{\alpha})^2, \tag{2.29}$$

with

$$\boldsymbol{\alpha} = \alpha(\sin\theta\cos\phi, \sin\theta\sin\phi, \cos\theta). \tag{2.30}$$

In the conventional path integral formalism one uses the flat integration measure for the gauge fields. Since the invariant Haar measure is flat in the vicinity of the identity the difference between the usual path integral formulae and (2.25) becomes important in the large, non-perturbative field domain. In fact, the center symmetry is the invariance of (2.25) under the change $A_0 \to A_0^z$ where

$$e^{tA_0^z} = e^{i\frac{2\pi}{N}} e^{tA_0}. \tag{2.31}$$

Note that the flat measure misses the periodic structure and brakes the center symmetry explicitly. It leads to the "deconfined phase" realization of the matrix elements and is the source of an additional infrared problem in conventional perturbative QCD.

III. SINGULAR TRAJECTORIES IN THE PATH INTEGRAL

The necessity of the invariant measure for tA_0 is obvious in the static temporal gauge

$$\partial_0 A_0 = 0, \tag{3.1}$$

according to (2.8) and (2.26). But one wonders whether by relaxing the gauge condition (3.1) and allowing the time dependence for A_0 we could avoid this complication. The standard argument to support the simple flat integration measure in the path integral is that the transformation

$$A_0 \to g(\partial_0 + A_0)g^\dagger \tag{3.2}$$

actually leaves the flat integration measure and the quartic action (2.22), (1.2) invariant.

The need of the invariant measure for non-static A_0 arises by a subtle mechanism which is characteristic of the quantum systems. The point is that there is no reason to expect that only smooth configurations enter in the path integral. In fact, we shall see that the center symmetry leads to Dirac delta singularities on the gauge field. The conventional formalism ceases to be gauge invariant for such non-smooth configurations.

It is well known that the path integral for the free one dimensional non-relativistic particle is dominated by nowhere differentiable trajectories. The hand-waving argument is that the typical trajectories of the integral

$$\prod_i \int dx_i e^{-\frac{1}{2}\Delta t \sum_i \left(\frac{x_{i+1}-x_i}{\Delta t}\right)^2} \tag{3.3}$$

have an O(1) contribution in the exponent for each i. This gives

$$\frac{dx}{dt} = O(\frac{1}{\sqrt{\Delta t}}). \tag{3.4}$$

This singular structure is essential to reproduce the canonical commutation relations. The ultraviolet singularities become worse in higher dimensions and the configurations contributing to the path integral formulae of a field theory are nowhere smooth [4]. The non-smooth fluctuations around a smooth background field are negligible only in the weak coupling limit [5]. One would naively think that the singularities which generate an ultraviolet divergence in the classical action must drop out from the path integration during the renormalization process. But this is not necessarily so for sufficiently strong coupling when the entropy factor may make such configurations relevant in the infrared [6], [7].

In order to see the general features of this phenomenon we shall briefly mention some aspects of the compactified sine-Gordon model in two dimensions [8]. This is a simple model with compact variables, the feature which lends additional importance to the singular configurations. The ordinary sine-Gordon model is described by the lagrangian

$$L = \frac{1}{2}(\partial_\nu \phi)^2 + \frac{\mu}{\lambda^2}\cos(\lambda\phi). \tag{3.5}$$

There are kinks in the model whose stability is due to the conservation of the topological current

$$J_\mu = \frac{1}{2\pi}\epsilon_{\mu\nu}\partial_\nu\phi. \tag{3.6}$$

Note that $\partial_\mu J_\mu$ may be non-vanishing at point singularities. In order to allow such configurations in the path integral we compactify the model by introducing

$$z(x) = e^{i\lambda\phi(x)} \tag{3.7}$$

as the fundamental variable subject to the constraint $|z|^2 = 1$ and write the lagrangian as

$$L_c = \frac{1}{2\lambda}[\partial_\nu z^\dagger \partial_\nu z + \mu(z + z^\dagger)]. \tag{3.8}$$

The lagrangians (3.5) and (3.8) obviously generate the same perturbation expansion but they differ in the non-perturbative domain. The difference is partially due to the existence of point singularities, called vortices in the compact version. A configuration contains a vortex at the space-time location y^μ if it can be transformed into the form

$$z(x) = \frac{(x-y)^1 + i(x-y)^2}{\sqrt{(x-y)^2}} \tag{3.9}$$

for $x \sim y$ by the application of smooth deformations. It is easy to show that the kink number jumps by one at the vortices,

$$\partial_\nu J_\nu(x) = m(x). \tag{3.10}$$

The right hand side is the vortex charge density, whose volume integral gives the total vortex number. Thus the point singularities make the kinks unstable and generate anomalous contributions to the divergence of the topological current.

The important but difficult question is whether the vortex configurations survive the renormalization procedure. The lattice regularization of (3.8) gives the two dimensional X-Y model with an external field. Fortunately the renormalization group trajectories are known for this system in the dilute vortex gas approximation [6]. It turns out that this model has two phases. The weak coupling phase is identical to the original model (3.5) where the kinks are stable and the semiclassical expansion works. The renormalized model in the strong coupling phase has non-vanishing vortex density in the vacuum, anomalous topological current, develops kink-condensate and its spectrum is fundamentally different from the semiclassical solution. In addition, a new relevant coupling constant, the fugacity for vortices appears which controls the density of the singularities. The renormalizable model

$$L_c = \frac{1}{2\lambda}[\partial_\nu z^\dagger \partial_\nu z + \mu(z + z^\dagger)] + \frac{\log\xi}{2\pi\lambda^2}(\epsilon_{\mu\nu} z^\dagger \partial_\mu \partial_\nu z)^2 \tag{3.11}$$

has three relevant coupling constants in this phase. Note that our present analytical methods are not sufficient to deal with the bare field theory in the continuum.

The non-linear O(3) sigma model in three dimensions and the four dimensional compact U(1) gauge theory display similar structure. In the latter case the disordered phase is populated by Dirac long fluctuating strings which generate the confining forces. Unfortunately the renormalization properties of these models are far from being trivial due to the possible appearance of new relevant coupling constants.

We now return to Yang-Mills theories and quote the path integral expression for (2.7) without the gauge fixing (3.1). The actual derivation goes by inserting the projection operator (2.13) at each time slice in the derivation of the path integral expressions. The field A_0 becomes time dependent and the final result is

$$< \mathbf{A}'|e^{-itH}|\mathbf{A} >= \int \mathcal{D}[\Delta t A_0]\mathcal{D}[\mathbf{A}]e^{iS_c[A_\mu]}, \tag{3.12}$$

where the invariant Haar measure is used for the gauge transformations

$$\omega(x) = e^{\Delta t A_0(x)}. \tag{3.13}$$

This is the gauge rotation performed between the consecutive time slices and $\Delta t \to 0$. The center symmetry requires that configurations where (3.13) is close to the global center $\omega(x) \sim z_k$ remain present in the renormalized theory. This in turn requires the finiteness of the integral

$$\int_t^{t+\Delta t} dt' A_0(\mathbf{x}, t') \tag{3.14}$$

when $\Delta t \to 0$. Thus the gauge field have Dirac delta singularities in the renormalized confining gauge theory.

The kinetic energy for the gauge field is proportional to the square of

$$\frac{1}{\Delta t}\Big(\mathbf{A}(\mathbf{x}, t+\Delta t) - \mathbf{A}(\mathbf{x}, t) - \Delta t \mathbf{D} A_0(\mathbf{x}, t)\Big). \tag{3.15}$$

The covariant gradient results from the parallel transport of the coordinate $\mathbf{A}$ between the time slices in computing the velocity. When finite gauge transformation may occur during the time Δt then the covariant gradient term is inappropriate because it is valid for infinitesimal gauge transformations only. The center symmetry can be preserved by keeping the effect of the parallel transport exactly,

$$\Delta t \mathbf{E}(\mathbf{x}, t) = \mathbf{A}^{\omega(\mathbf{x},t)}(\mathbf{x}, t+\Delta t) - \mathbf{A}(\mathbf{x}, t). \tag{3.16}$$

The action S_c in (3.12) is constructed with (2.3) and (3.16).

The explicit Lorentz or O(4) symmetry is lost in the temporal gauge but the gauge invariant observables remain covariant. When only gauge invariant and Lorentz or O(4) covariant approximations are made then the path integrals (2.25) and (3.12) yield covariant gauge invariant observables. But it might be useful to have a more covariant formalism in order to have more freedom in choosing the approximation schemes later. Thus one may introduce similar compactifications in the space directions, too. This requires local ultraviolet cutoff in space-time and results a bare theory which is a close variant of Wilson's lattice gauge model [9]. We shall use the simpler version (3.12) in what follows and the issue of the covariance of our result will be addressed briefly later.

Note that the Dirac delta singularities of the Yang-Mills theory have even more reason to be present in the renormalized theory than the vortices in the sine-Gordon model since they have no suppression factor from the action. In order to make the configurations smoother one can impose a gauge fixing condition involving the derivatives of the gauge field.

IV. RENORMALIZATION AND THE EFFECTIVE THEORY

The path integral (3.12) yields an apparently non-renormalizable bare lagrangian. The integration measure for A_0 is

$$\begin{aligned} d_{\rm inv} A_0(\mathbf{x},t) &= d\Omega(\mathbf{x},t) du(\mathbf{x},t) \sin^2(u(\mathbf{x},t)) \\ &= d\Omega(\mathbf{x},t) du(\mathbf{x},t) e^{\ln \sin^2(\Delta t u(\mathbf{x},t))} \end{aligned} \tag{4.1}$$

in SU(2) gauge theory where $d\Omega$ stands for the integration over the direction of A_0^a in the color space and $u = \sqrt{A_0^a A_0^a}$. The last multiplicative factor in (4.1) can be added to the action,

$$S \to S - \frac{1}{a^4} \int d^4x \ln \sin^2 \sqrt{-2\mathrm{tr}[aA_0(\mathbf{x})]^2}. \tag{4.2}$$

The parameter a is the ultraviolet cutoff in a local regularization procedure. We can take it as the lattice spacing, though the actual form of the regularization is not used. This factor is to make the additional term in the action dimensionless.

In order to assess the role of the new piece in the action we compare the following bare theories:

(i) The conventional Yang-Mills theory with the action (2.22), (1.2) and flat integration measure in dimensional regularization.

(ii) The same action and integral measure with momentum space cutoff.

(iii) Wilson's lattice gauge theory [9].

(iv) The theory (2.25) or its manifestly covariant version in "any" local regularization scheme.

The renormalizability of the theory (i) can be proven by induction in the order of the loop expansion. The inductive step proceeds by the combination of the power counting and the BRST invariance and defines consistent short distances properties. The long range properties and the very existence of the theory is left unclear in this perturbative approach.

The cutoff theory (ii) is perturbatively non-renormalizable due to the lack of gauge invariance. At each order of the loop expansion one has to introduce new counterterms to ensure the finiteness. These counterterms are taken into account automatically in (i).

(iii) has exact gauge invariance and infinitely many vertices. The power counting argument [10] and the BRST symmetry can be generalized for this regularization.

By the help of these ingredients the renormalizability can be proven for any finite order of the loop expansion in an analogous manner to (i) [11]. The irrelevant lattice vertices lead no new overall divergences and the overlapping divergences can be shown to be absent by using the BRST symmetry. The invariant measure and the compact functions of the Wilson action generate just that infinite tower of irrelevant vertices which are needed to restore the perturbative gauge invariance at short distances in this regularization. The renormalized action turns out to be (2.22) with (1.2) and the model is asymptotically free.

Finally, the case (iv) can be formulated in "any" local, regularization. We know no other local regularization than the lattice, thus this case is actually equivalent to Wilson's theory. When the discreteness of the three-space is kept explicitly in (2.25) then the Ward identities are violated and the systematic treatment of this problem would reintroduce the link variables in spatial directions. In the case of the time dependent A_0 we end up with compact variables in each directions.

Thus the role of the irrelevant vertices in perturbation expansion is to maintain the Ward identities for all order of the loop expansion. One may conclude that (iv) would reduce to (i) if the regulator had as "nice" counterterm structure as in the dimensional regularization. But the discussion of the center symmetry above makes clear that the perturbative Ward identities are not enough for confinement. The role of the irrelevant terms of the invariant measure is not only to correct the gauge dependence of the regulator but the whole periodic structure is required for confinement. The confining Yang-Mills theory what we find in the Monte Carlo computations needs this periodicity in the large field, non-perturbative domain.

This suggests the following decoupling of the relevant and the irrelevant operators. The relevant, renormalizable part of the action is responsible mainly for the asymptotically free ultraviolet behavior and the irrelevant pieces fix up the symmetry properties. They may correct the perturbative ultraviolet behavior if the regulator is not manifestly gauge invariant. The most important role of gauge invariance in the infrared is to preserve the center symmetry. This is achieved by the irrelevant terms, too. Our effective theory is based on an attempt to split the perturbative and the non-perturbative regions. On the one hand, we assume that the model (iv) is renormalizable, asymptotically free and one can use simple continuum regulator in the perturbation expansion. On the other hand, since the renormalized theory displays unbroken center symmetry the renormalized action must have some left over piece from the invariant Haar measure. This term is completely missed in the perturbative scheme. Since the bare measure term is local and periodic the renormalized lagrangian should contain a finite periodic local potential for the gauge field. The detailed shape of this potential should be such that its vacuum wave functional is invariant under the shift of the gauge field by the period length of the potential.

Rather than deriving the renormalized lagrangian we only outline the way it could be computed. This simplified argument is still useful because the the leading infrared properties of the renormalized theory can be studied without the detailed knowledge of the periodic potential. The first complication arises at the gauge fixing. The center symmetry $A_0 \rightarrow A_0^z$ in (2.31) transforms different Gribov copies of the path integral (2.25) into each other. This can be seen easily by recalling that the gauge fixing conditions for (2.25) are (2.23), (2.24) and (3.1). The center symmetry requires the summation over these copies. One should avoid the usual relativistic gauge because this summation is too complicated there. Furthermore, that gauge breaks the center symmetry explicitly. Instead, we choose the gauge where A_0 is diagonal. The remaining Abelian gauge invariance is removed by adding the term

$$S_{\rm gf} = -\frac{1}{\xi}\int d^4x \,{\rm tr}(\partial_\mu A_0)^2 \tag{4.3}$$

to the lagrangian. The center symmetry is respected in this gauge and the summation over the corresponding copies is made automatically. The diagonality condition for A_0 removes the integration over the direction in (2.27) and (4.1) and we are left with a one component field

$$A_0^a(x) = \delta^{a,3}u(x) \tag{4.4}$$

in the SU(2) case.

We can now state our main assumption which is the claim that the bare lagrangian

$$L_B = -\frac{1}{2g_B^2}\,{\rm tr}\,F_{\mu\nu}^2 - \frac{1}{a^4}\ln\sin^2 A_0 \tag{4.5}$$

gives rise to the renormalized lagrangian

$$L_R = -\frac{1}{2g^2}\,{\rm tr}\,F_{\mu\nu}^2 + V_R(A_0) \tag{4.6}$$

in our gauge. The gauge fixing terms are suppressed in (4.5) and (4.6). The local potential is periodic function

$$V_R(u) = \lambda \sum_{\sigma=-\infty}^{\infty} v_\sigma e^{i\sigma\ell u} \tag{4.7}$$

where $v_{-n} = v_n$ and $v_0 = 0$. The value of the parameter ξ in (4.3) is chosen in such a manner that the kinetic term for A_0 is O(4) invariant in the renormalized action.

In order to demonstrate the relevance of the periodic potential we now set the vector gluon field zero, $\mathbf{A} = 0$ in the renormalized theory. The remaining part of the lagrangian is

$$L_{\text{eff}} = \frac{1}{g^2}\text{tr}[\partial_\mu u(x)\partial_\mu u(x)] + V_R(u(x)) \tag{4.8}$$

which describes a four dimensional Euclidean sine-Gordon theory. It is easy to see that the remnant of the center symmetry

$$u(x) \to u(x) + \frac{2\pi}{g\ell} \tag{4.9}$$

prevents the mass generation and requires the presence of several terms in the Fourier expansion (4.7) with different sign. We shall investigate this model by the help of a partial resummation of the perturbation expansion in λ. The asymptotically free gluon exchange contributions can be taken into account order by order at some later stage.

V. THE EFFECTIVE QUARK DYNAMICS

One can compute the Euclidean generator functional

$$< e^{ig\int d^4x\rho(x)u(x)} >= \int \mathcal{D}[u]e^{-\int d^4x L_{\text{eff}}[u]}e^{ig\int d^4x\rho(x)u(x)} \tag{5.1}$$

by expanding in λ. We distinguish three classes of the connected graphs in this expansion. (i) The propagator

$$G(x) = \int \frac{d^4p}{(2\pi)^4}\frac{1}{p^2}e^{-ipx} = \frac{1}{4\pi^2x^2}. \tag{5.2}$$

connects source locations in the space-time. No other loops are present. (ii) One leg of each propagator is at the source and the other is connected to a sine-Gordon vertex, arising from the expansion of the exponential function in (4.7). (iii) There are loops made of the sine-Gordon vertices and the source is connected to the loops by the propagator. The graphs from (iii) give the radiative corrections to the the class (i) and (ii). The contributions of these two classes can be resummed. The result is [1]

$$< e^{ig\int d^4x\rho(x)u(x)} >= I_{(i)}I_{(ii)}, \tag{5.3}$$

where

$$I_{(i)} = e^{-\frac{1}{2}g^2\int d^4xd^4y\rho(x)G(x-y)\rho(y)} \tag{5.4}$$

is the sum of the diagrams of the class (i) and

$$I_{(ii)} = e^{-\int d^4x V_R(ig\int d^4y\rho(y)G(x-y))} \tag{5.5}$$

comes from the resummation of the class (ii). We shall replace $I_{(i)}$ by a constant because it plays no role in the leading infrared contributions. The perturbative gluon exchange which can be included later gives similar result, too.

We mention two applications of the generator functional. First we choose

$$\rho(\mathbf{x},t) = \delta(\mathbf{x}) - \delta(\mathbf{x}+\mathbf{L}) \tag{5.6}$$

for $0 < t < T$ and zero otherwise. This source corresponds to a heavy quark-anti quark pair with separation L. (5.3) can be written as

$$-\int d^4y V_R(\frac{i\tilde{g}\ell}{|\mathbf{y}|} - \frac{i\tilde{g}\ell}{|\mathbf{y}-\mathbf{L}|}) \tag{5.7}$$

with $\tilde{g} = \frac{g}{4\pi}$ which gives for the exponent

$$4\pi T\lambda \sum_\sigma \left(\sigma^2 v_\sigma \tilde{g}^2\ell^2 L + \sum_{n=0}^{\infty} c_{\sigma,n}\frac{1}{L^n}\right) \tag{5.8}$$

where the coefficients $c_{\sigma,n}$ are ultraviolet divergent. The first term, the leading infrared contribution is finite and gives the string tension

$$\sigma = 2\pi\lambda \sum_{n=-\infty}^{\infty} (n\tilde{g}\ell)^2 v_n = -\frac{g^2}{8\pi} V_R''(0) \; . \tag{5.9}$$

This can be non-vanishing in spite of the massless behavior of the field $u(x)$ due to the symmetry (4.9). The potential V_R has maximum at $u = 0$ and we expand around a tachion. But it is just the center symmetry, (4.9) which stabilizes the theory.

Another application of (5.3) is when the current $\rho(x)$ is a dynamical variable. We now start with the action

$$S = \int d^4x \{\frac{1}{2}(\partial_\mu u(x))^2 + V_R(u) + \bar{\psi}(x)[i\partial\!\!\!/ - g\gamma_0\frac{\sigma^z}{2}u(x)]\psi(x)\} \tag{5.10}$$

for the dynamical quark field ψ and u. The elimination of the field $u(x)$ in the same manner as before gives the effective theory

$$S_{\text{eff}} = \int d^4x \{\bar{\psi}(x)i\partial\!\!\!/\psi(x) + V(g\ell\int d^4y\bar{\psi}(y)\gamma^0\frac{\sigma^z}{2}\psi(y)G(y-x))\} \tag{5.11}$$

for the quarks. The leading infrared contribution can be obtained by keeping the renormalizable part of the potential

$$V_R(u) \sim V_0 + \frac{1}{2}u^2 V_2 \tag{5.12}$$

This reduces (5.11) to a non-local Nambu-Jona Lasinio type model,

$$\begin{aligned} S_{\text{eff}} = & \int d^4x \bar{\psi}(x) i \partial\!\!\!/ \psi(x) \\ & - 4\pi\sigma \int d^4x d^4y \bar{\psi}(x)\gamma^0 \frac{\sigma^z}{2}\psi(x) G_2(x-y)\bar{\psi}(y)\gamma^0\frac{\sigma^z}{2}\psi(y), \end{aligned} \tag{5.13}$$

where σ is the string tension (5.9) and

$$G_2(x-y) = \int d^4z G(x-z)G(z-y) = \int \frac{d^4p}{(2\pi)^4}\frac{1}{p^4}e^{-ip(x-y)} \ . \tag{5.14}$$

A simpler form of this effective theory can be obtained by introducing an auxiliary field $\sigma(x)$ representing a collective mode of the original A_0 field:

$$\begin{aligned} S_{\text{eff}} = & \int d^4x \bar{\psi}(x)[i\partial\!\!\!/ - 8\pi\sigma\gamma^0\frac{\sigma^z}{2}\sigma(x)]\psi(x) \\ & - \frac{1}{2}\int d^4x d^4y \sigma(x)(\partial_\mu\partial_\mu)^2\sigma(y) \ . \end{aligned} \tag{5.15}$$

The model (5.15) is clearly not covariant. The covariance has been lost during the approximations in the renormalization procedure and the subsequent resummation. But it is easy to find a simple covariant extension of (5.15). It involves a vector field $u_\mu(x)$ and the lagrangian is

$$L = -\frac{1}{4}F_{\mu\nu}\partial^2 F_{\mu\nu} + \frac{1}{2}(\partial_\mu\partial_\nu u_\nu)^2 + \bar{\psi}i\gamma_\mu[\partial_\mu + i8\pi\frac{\sigma^z}{2}u_\mu]\psi. \tag{5.16}$$

The similarity of this lagrangian with the dual superconductor model [12] is noteworthy. Another interesting feature is that this is a finite model in the zero charge sector.

VI. LOCALIZATION BY THE HAARON GAS

Another application of the resummation employed in the derivation of the effective potential is the direct computation of an arbitrary quark Green function. This computation provides an important insight into the mechanism which confines charges in the effective model. A rather lengthy but straightforward manipulation results

$$< \psi(x_1)\psi(x_2)...\bar{\psi}(y_1)\bar{\psi}(y_2)... >= \sum_{m=0}^{\infty} \frac{\lambda^m}{m!} \prod_{j=1}^{m} \Big(\int d^4x_j \sum_{\sigma(j)} v_{\sigma(j)} \Big) \\ < \psi(x_1)\psi(x_2)...\bar{\psi}(y_1)\bar{\psi}(y_2)... >_{iU} \ . \tag{6.1}$$

The Green function in the right hand side corresponds to free fermions propagating on the pure imaginary background field

$$U(x) = \ell \sum_{j=1}^{m} \sigma(j) \frac{1}{4\pi^2 (x - x_j)^2}. \tag{6.2}$$

The expression (6.1) is surprisingly similar to the dilute instanton gas approximation except that it is not a saddle point expansion and is consistent in the infrared domain. We called this method the haaron-gas representation because the scattering off the singular peaks generates the periodic effects of the Haar measure. We can recover the same static potential in the haaron gas picture by the help of the heavy quark propagators as in the computation of the string tension above.

The result (6.1) shows interesting similarities with the localization phenomenon observed in solid state physics [13]. To make this point clearer we simplify (6.1) by replacing the $\frac{1}{x^2}$ singularities by a Dirac delta and consider a non-relativistic charged particle described by the hamiltonian

$$H = \frac{\mathbf{p}^2}{2m} + \lambda \sum_{j=1}^{N} \delta(\mathbf{x} - \mathbf{x}_j)\delta(t - t_j). \tag{6.3}$$

The additional rule is that the potential stands for "quenched impurities". In other words, the location of the Dirac delta peaks in the potential is distributed uniformly throughout space-time and the physical observables are averages of the expectation values

$$< O >_{\text{phys}} = \prod_{j=1}^{N} \int d^3x_j dt_j < O >, \tag{6.4}$$

where the density of the scattering centers in space-time, $\frac{N}{VT}$, is finite. The haaron appears as a quenched potential in the quenched approximation, i.e. when the fermion loop effects are neglected. This is believed to be a reasonable approximation for the long range phenomena in the QCD vacuum. The average potential due to the haarons is $O(\lambda)$. The partial re-summation of the perturbation expansion in λ generates poles at imaginary energies for the resolvent

$$< \frac{1}{E-H} >_{\mathrm{phys}} . \tag{6.5}$$

This is just the model for Anderson's localization when the potential in (6.3) is time independent. The extended states are removed from the system by the destructive interference due to the random phase shifts suffered during the collisions off the impurity potential. The time-dependent potential of (6.3) eliminates the localized zero momentum states as well. Confinement appears as a localization phenomenon in space-time.

It is instructive to consider the propagator for two particles which are coupled to the potential with the opposite strength, λ and $-\lambda$. Due to the finite extent of the haarons the locations of the Dirac delta are correlated. This correlation can be taken into account in a rudimentary manner by replacing the Dirac delta by a step function which is constant in a region of size L. The two-particle propagator is now written as the sum over the particle trajectories. This sum is dominated by the trajectories where the two particles traverse the same step function domains. In fact, whenever the particles propagate through different potential domains, the cancellation due to the stochastically independent potentials suppresses the contribution. Thus the particles do not separate more than L during the propagation, which indicates that the confinement radius is the correlation length of the potential.

Note that the important ingredients of this model are the large fluctuations and the finiteness of the correlation length of the quenched potential. These two properties are contradicting in a usual renormalizable field theoretical model for the potential. In fact, the finite correlation length should lead to finite mass, which in turn suppresses the large values of the field. It is the periodic Haar measure which makes these ingredients compatible by maintaining the center symmetry in a renormalizable manner.

Finally, few words about the chiral symmetry breaking. The confining potential is chiral invariant which indicates that the chiral symmetry should be broken dynamically in the vacuum [14]. The breakdown of the chiral invariance can be derived either by the Bethe-Salpeter type resummation in the Jona-Lasinio model [15] or directly from the haaron-gas representation. This latter argument proceeds similarly to the case of dilute instanton gas : one can show that there are localized zero modes for the Dirac operator on a haaron background. These zero modes describe quarks bound by the instantaneous haaron potential and the average haaron density is $O(<\bar{\psi}\psi>)$. The average distance of the haarons gives the distance scale which separates the physics of our effective theory from perturbative QCD. It is interesting to speculate whether the resonance effects of the bound states created by the haarons could be observed after the averaging procedure for the haaron potential. This might lead to some non-trivial scattering phenomenon on the vacuum, similarly to the Kondo-effect [16].

VII. SUMMARY

We presented arguments that the center invariance requires periodic terms in the QCD lagrangian. These new interactions generate linear static quark potential and break the chiral symmetry dynamically without the conventional three- or four-gluon vertices. The conventional gluon vertices are supposed to play role in the dimensional transmutation which generates the finite scale parameters of the periodic renormalized potential in the lagrangian.

We believe that it will be possible to compute the quark Green functions analytically on a haaron-background. This will open the way for an analytical hadron model: The haaron-gas representation generates the appropriate long distance structure and the inclusion of the perturbative gluon vertices provides the asymptotically free short distance behavior. Another way to address the long and short distance features in a coherent manner is to combine the leading effective theory (5.16) with perturbative QCD. The gluon propagator is the sum of the long range, $\frac{1}{p^4}$ type and the perturbative one. We plan to explore these possibilities in the near future.

ACKNOWLEDGEMENT

I thank Kenneth Johnson and Laurent Lellouch for stimulating discussions.

REFERENCES

[1] K. Johnson, L. Lellouch and J. Polonyi, "An Effective Theory for the QCD Vacuum", to appear in *Nucl. Phys.*

[2] J. Polonyi, *Phys. Lett.* **213B** (1988) 340; in the *Quark-Gluon Plasma*, ed.. R. Hwa (World Scientific, 1990).

[3] L. McLerran and B. Svetitsky, *Phys. Lett.* **98B** (1981) 195; J. Kuti, J. Polonyi and K. Szlachanyi, *Phys. Lett.* **98B** (1981) 199.

[4] L. S. Schulman, *Techniques and Applications of Path Integration*, (John Wiley, 1981).

[5] J. Polonyi, in the Proceedings of *LATTICE'90* (Tallahassee, Florida, 1990).

[6] J. M. Kosterlitz and D. J. Thouless, *J. Phys.* **C6** (1973) 118; J. M. Kosterlitz, *J. Phys.* **C7** (1974) 1046; J. V. Jose, L. P. Kadanoff, S. Kirkpatrik and D. R. Nelson, *Phys. Rev.* **B16** (1977) 1217.

[7] C. Callen, R. Dashen and D. Gross, *Phys. Rev.* **D17** (1978) 2717.

[8] K. Huang and J. Polonyi, *Int. J. of Mod. Phys.* **A6** (1991), 409.

[9] K. Wilson, *Phys. Rev.* **D14** (1974) 2445.

[10] T. Reisz, *Comm. of Math. Phys.* **116** (1988) 81.

[11] T. Reisz, *Nucl. Phys.* **B318** (1989) 417.

[12] V. P. Nair and C. Rosenzweig, *Phys. Rev.* **D31** (1985) 401; C. Rosenzweig, *Phys. Rev.* **D38** (1988) 1934;

[13] P. W. Anderson, *Phys. Rev.* **109** (1958) 1492.

[14] A. Casher, *Phys. Lett.* **83B** (1979) 395.

[15] R. Delburgo and M. D. Scadron, *J. Phys.* **G5** (1979) 1621; J. Finger and J. Mandula, *Nucl. Phys.* **B199** (1982) 168; A. Amer, A. Le Yaouanc, L. Oliver, O. Péne and J-C. Raynal, *Phys. Rev.* **D28** (1983) 1530; A. Le Yaouanc, L. Oliver, O. Péne and J-C. Raynal, *Phys. Rev.* **D29** (1984) 1233; A. Le Yaouanc, L. Oliver, S. Ono, O. Péne and J-C. Raynal, *Phys. Rev.* **D31** (1985) 137; S. Adler and C. Davis, *Nucl. Phys.* **B244** (1984) 469; S. Adler, *Progr. Theor. Phys. Suppl.* **86** (1986) 12; M. Hirata, *Progr. Theor. Phys.* **77** (1987) 939; R. Alkofer and P. A. Amundsen, *Nucl. Phys.* **B306** (1988) 305.

[16] J. Kondo, *Progr. Theor. Phys.* **32** (1964) 37.

CLASSICAL STRUCTURES IN THE ELECTROWEAK THEORY

JÜRGEN BAACKE
Fachbereich Physik, Universität Dortmund, Postfach 500500
D-4600 Dortmund 50 , Germany

ABSTRACT

We consider the existence and stability of classical solutions which are supposed to play a rôle in the electroweak theory: the sphaleron, nonabelian monopoles and electroweak skyrmions.

1 Introduction

Among the nonperturbative effects that may play a rôle in the electroweak theory are the classical solutions obtained in the limit of vanishing weak mixing angle Θ_W: the sphaleron [1,2] which can catalyse baryon number violating processes at high temperatures [3] and in the scattering of W bosons [2,4], and the electroweak skyrmeon [5-7] which could possibly exist due to an effective "low" energy chiral action a la Skyrme [8] in analogy to the proton in the low energy approximation to QCD. Both objects have masses in the TeV region and could be accessible in forthcoming experiments.

The sphaleron solution represents an unstable state that describes the point in function space that has the highest energy on a minimal energy path between two ground states with different winding number, i. e. it is a saddle point with (at least) one unstable direction. It has recently been found [9,10] that for a ratio ξ between Higgs and W mass larger than 12 a second unstable mode develops, and more such modes appear for even higher values of ξ. The implications for possible new minimal energy paths are still under investigation [11]. If we consider the stability equation as a Schrödinger equation these instabilities correspond to bound states with negative energy. All of those states have been found in the $K = 0$ partial wave.Here K denotes the sum of spin and isospin which is a good quantum number in the sphaleron background which is K-spin symmetric. If the "potential" of this Schrödinger equation is so strong that several negative energy S - wave bound states appear one might expect that also higher partial waves contain such states. This possibility has been analyzed recently [13] using the Jacobi criterion, however no instabilities besides the known $K = 0$ ones have been found. The general aspects of such a stability analysis are presented in section 2. The specific application to the sphaleron is described in section 3.

The same technique had been used previously [12] also for an analysis of stability of the 't Hooft Polyakov monopole [14,15]. The result that the monopole ist stable with respect to spherical and nonspherical fluctuations was expected but had indeed never established. We report on this analysis in section 4.

For the electroweak skyrmeon a similar stability analysis was planned but the classical solution for finite Higgs mass has meanwhile met a different fate which makes such an analysis superfluous. The classical object was established [5-7] in the limit of infinite Higgs mass for the electroweak theory modified by adding an effective Skyrme [8] term . Subsequently the theory with finite Higgs mass was investigated [16] and the object continued to exist at even smaller masses (few TeV). The form of the Skyrme term used there was criticized recently by Carson [17] and his critique seems justified. No classical solutions were found anymore by Carson. We have repeated his analysis and come (unfortunately) to the same conclusion. This matter is discussed in section 5. We end with sum concluding remarks in section 6.

2 Analysis of Stability

If one has found a classical solution of the equations of motion of some field theory it remains to be investigated whether this solution is stable. Usually the classical Ansatz has a (generalized) spherical symmetry . The usual numerical methods will end up with converging to a local minimum, but this will be a local minimum only within the chosen Ansatz. A stability analysis starts with expanding the field into fluctuations around the classical solution and determining the corrections to the Lagrangian up to second order in the field fluctuations. The linear term in this expansion vanishes because we are in a local minimum, the quadratic one leads to a linear set of equations of motion, the stability equations. These can be written in the general form

$$H_{ij}\phi_j^\alpha = \omega_\alpha^2 \phi_i^\alpha = E_\alpha \phi_i^\alpha \tag{1}$$

Here ϕ_i are the fluctuation fields in a general notation, i labels the components, α the solutions, ω is the frequency of the fluctuation and the notation H_{ij} and E_α stresses the similarity to a Schrödinger equation. H is an operator of second and lower order in the spatial derivatives and couples the different fluctuation fields. Typically it is of the form

$$H_{ij} = \delta_{ij}\left(-\frac{1}{r^2}\frac{d^2}{dr^2} - \frac{2}{r}\frac{d}{dr} + \frac{l_i(l_i+1)}{r^2}\right) + V_{ij}(r) \tag{2}$$

An instability with respect to small fluctuations is signalized by the occurence of imaginary frequencies or , in the Schrödinger equation language, by a negative eigenvalue E_α of H. One way of analysing stability is to determine the "bound state spectrum" of H (which has a continuum if ω^2 is bigger than the minimal $mass^2$ of the theory) and scanning it for a negative eigenvalue. This method was followed in [9,10]. Another possibility is to apply the Jacobi criterion (see e.g. [18]). This criterion can be rephrased in the quantum mechanical language as a consequence of the node rule (which is not a "rule of thumb" but can be proven riguorously [19]). Applied to the present case it says that the number of zeros of the wave function regular at $r = 0$, or , for a system of n coupled wave functions , of the determinand of a set of n independent solutions regular at $r = 0$, increases with energy. Now the lowest eigenstate of H has one such zero at $r \to \infty$. Therefore an instability is signalized by the occurence of (at least one) node in the solution of Eq.1 for $\omega = 0$ at some finite r, which is precisely the Jacobi criterion.

This simplifies the investigation considerably. The main labour consists in finding the operator H resp. the Lagrangian of second order in the fluctuation fields. The subsequent analysis of Eq.1 can only be numerical since in most of the relevant cases already the classical solutions are known only numerically. The fluctuations can be labelled by some angular momentum type quantum number ("K- Spin") and of course only the lowest partial waves can be investigated. However, in the quantum mechanical language, the increasing "centrifugal barrier" prevents the occurence of deeply bound states for higher partial waves and so the restriction to low ones is not a serious limitation of the method. Such an investigation can anyway not be considered as a mathematical proof though it gives certainly strong evidence for or against stability.

A special situation occurs in the generalized " p waves " which contain rotational and translational zero modes. In this case one has to extend the analysis fourth order in the fluctuations. We will not come back on this point below, we just state that this has been investigated [12,13] but has not lead to the discovery of instabilities.

We would like to add some remarks about the numerics here. The test of stability requires the classsical solution to be known with high precision and with a large (200-400) number of mesh points. This requirement is stringent especially in those channels where one has zero modes. They have to vanish as $r \to \infty$ but this node can easily appear at finite r due to a slight inaccuracy of the classical profiles. If one does not like to have recourse to some software package a method designed by Bais and Primack [20] for determining the 't Hooft Polyakov monopole profiles proves to be superior to shooting or relaxation methods. Let the equations of motion be written in the form

$$D^0_\alpha \Phi_\alpha = F_\alpha(\Phi) \tag{3}$$

Here Φ_α are the classical fields are D^0_α are the appropriate free Klein Gordon operators (e.g. $r^2d^2/dr^2 + 2rd/dr - m^2r^2$). The terms on the right hand side are the remaining - in general nonlinear - terms occuring in the equation of motion. Let the Green function G^0_α of D^0_α be defined by

$$D^0_\alpha G^0_\alpha(r, r') = \delta(r - r'). \tag{4}$$

Then Eq. 3 can be rewritten in the form of an integral equation

$$\Phi_\alpha = A_\alpha(r) + \int_0^\infty dr' G^0_\alpha(r,r')(F_\alpha(\Phi)(r') - D^{0'}_\alpha A_\alpha(r')) \tag{5}$$

Here the function $A_\alpha(r)$ has to to satisfy the - in general nontrivial - boundary conditions at $r = 0$ and for $r \to \infty$ and is needed in order to impose these boundary conditions on $\Phi_\alpha(r)$. The iteration of this equation converges typically within 20 iterations to an accuracy of the energy of 10^{-8}. It can even be used on a PC.

3 The Sphaleron

The model considered here is the lagrangian of the standard electroweak theory in the limit where $sin^2\Theta_W = 0$. All fermions are decoupled and the $U(1)$ sector of the theory is ignored. One is then left with the Lagrangean for a triplet of vector (gauge) bosons and a complex Higgs doublet:

$$\mathcal{L} = -\frac{1}{4}W^a_{\mu\nu}W^{\mu\nu a} + (\mathcal{D}_\mu\Phi)^\dagger(\mathcal{D}^\mu\Phi) + \lambda(\Phi^\dagger\Phi - \frac{1}{2}v^2)^2 \tag{6}$$

where, as usual,

$$W^a_{\mu\nu} = \partial_\mu W^a_\nu - \partial_\nu W^a_\mu + g\epsilon^{abc}W^b_\mu W^c_\nu \tag{7}$$

and

$$\mathcal{D}_\mu\Phi = \partial_\mu\Phi - \frac{i}{2}g\tau^a W^a_\mu\Phi. \tag{8}$$

The W-boson and the Higgs mass are given by $M_W = gv/2$ and $M_H = \sqrt{2\lambda}v$ respectively.

The sphaleron Ansatz [2] is then

$$\begin{aligned} W^{cl}_{0a} &= 0 \\ W^{cl}_{ja} &= \frac{f_A - 1}{gr}\epsilon_{jam}\hat{x}_m \\ \Phi^{cl} &= vH(r)\begin{pmatrix} 0 \\ 1 \end{pmatrix}. \end{aligned} \tag{9}$$

The resulting classical equations of motion for $f_A(r)$ and $H(r)$

$$f_A'' = m_W^2 H^2(f_A - 1) + \frac{f_A(f_A^2 - 1)}{r^2} \tag{10}$$

$$H'' = \frac{1}{2}m_H^2 H(H^2 - 1) - \frac{2H'}{r} + \frac{(f_A - 1)^2}{2r^2}H \tag{11}$$

are to be solved with the boundary conditions

$$f_A(0) = -1 \quad f_A(\infty) = 1$$
$$H(0) = 0 \quad H(\infty) = 1$$

For its solution the method of Bais and Primack sketched in the previous section has been used.

The stability analysis requires now the Lagrangian to be expanded up to second order in the fluctuations of the fields around the classical solution. This requires first a choice for the parametrisation of the field fluctuations. The background field is symmetric under the K-spin, the sum of angular momentum and isospin. The expansion will therefore carried out in a basis of K spin tensor spherical harmonics. We have used a cartesian tensor basis used previously for the analysis of pion nucleon scattering in Skyrme type models [21]

Explicitly this reads

$$\begin{aligned} W_{0a} &= W_{0a}^{(1)} \\ W_{ja} &= W_{ja}^{cl} + W_{ja}^{(1)} \\ \Phi &= \Phi^{cl} + \Phi^{(1)} \end{aligned} \tag{12}$$

where the fluctuation fields $W_{0a}^{(1)}, W_{ja}^{(1)}$ and $\Phi^{(1)}$ are expanded as

$$\begin{aligned} W_{0a}^{(1)} &= \frac{1}{g} \sum_{K,M} \sum_{\alpha=1}^{3} s_\alpha^K(r) \mathbf{J}_a^\alpha Y_{KM} \\ W_{ja}^{(1)} &= \frac{1}{g} \sum_{K,M} \sum_{\alpha=1}^{9} t_\alpha^K(r) \mathbf{I}_{ja}^\alpha Y_{KM} \\ \Phi^{(1)} &= v \sum_{K,M} \{ h_1^K(r) + i\tau^a \sum_{\alpha=1}^{3} p_\alpha^K(r) \mathbf{J}_a^\alpha \} Y_{KM} \begin{pmatrix} 0 \\ 1 \end{pmatrix} \end{aligned} \tag{13}$$

A complete operator basis is given by

$$\begin{aligned} \mathbf{J}_a^1 &= \hat{x}_a \\ \mathbf{J}_a^2 &= r\nabla_a \\ \mathbf{J}_a^3 &= \epsilon_{abc} x_b \nabla_c = \Lambda_a \end{aligned} \tag{14}$$

and

$$\begin{aligned} \mathbf{I}_{ja}^1 &= \hat{x}_j \hat{x}_a & \mathbf{I}_{ja}^2 &= \hat{x}_j r\nabla_a & \mathbf{I}_{ja}^3 &= r\nabla_j \hat{x}_a \\ \mathbf{I}_{ja}^4 &= \hat{x}_j \Lambda_a & \mathbf{I}_{ja}^5 &= \Lambda_j \hat{x}_a & \mathbf{I}_{ja}^6 &= r\nabla_j \nabla_a \\ \mathbf{I}_{ja}^7 &= r\nabla_j \Lambda_a & \mathbf{I}_{ja}^8 &= \Lambda_j r\nabla_a & \mathbf{I}_{ja}^9 &= \Lambda_j \Lambda_a \end{aligned} \tag{15}$$

The resulting equations of motion for the fluctuations are too lengthy to be reproduced here and we refer the reader to [13]. They are of the general form of Eqs.

1 and 2 except for those of t_1^K, t_2^K and t_4^K in which no derivatives with respect to r appear. These amplitudes can therefore be expressed by the remaining ones. Up to now the gauge has not yet been fixed. In order to avoid the associated zero modes we have used the temporal gauge. One is then left with the spatial amplitudes t_α^K for the W field and the amplitudes p_α^K for the Higgs field. For the analysis of stability we have to solve these equations for frequency $\omega = 0$.

There are, for each K-spin, two sets of coupled amplitudes corresponding do a definite parity,

i) the magnetic modes MK containing the amplitudes $t_1^K, t_2^K, t_3^K, t_6^K, t_9^K, p_1^K$ and p_2^K with a regular threshold behaviour r^K or r^{K-2}. In view of the constraint for t_1^K and t_2^K there are five independent amplitudes.

ii) the electric modes EK containing the amplitudes $t_4^K, t_5^K, t_7^K, t_8^K, p_8^K$ and h_1^K with a regular threshold behaviour $r^{K\pm1}$. Here the amplitude t_4^K is constrained so that there are again five independent amplitudes.

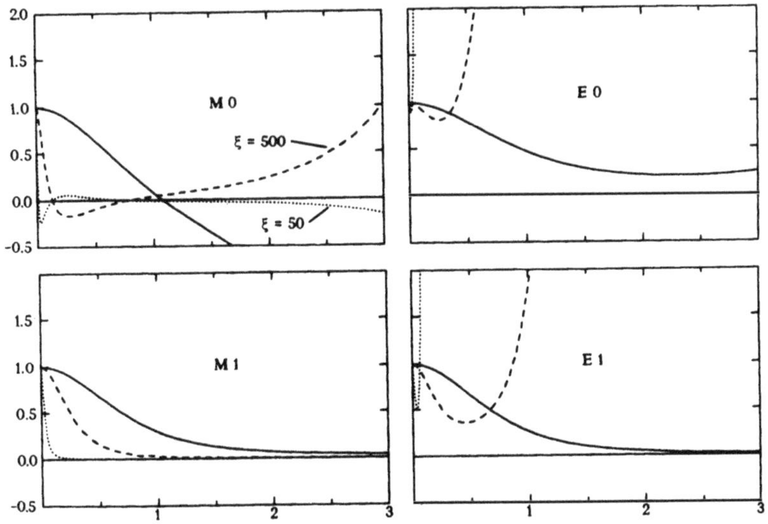

Figure 1 : The Jacobi criterion applied to the lowest fluctuation modes of the sphaleron. The determinand of independent solutions is plotted versus $x = M_W r$ for $\xi = M_H/M_W = 1$ (solid line), 10 (dashed line) and 100 (dotted line) unless indicated otherwise.

For each of these sets there is a fundamental system of five independent solutions. If the determinands of these solutions have zeros as functions of r then according to the Jacobi criterion the classical solution is unstable with respect to fluctuations in the corresponding channel. We have taken out all threshold factors so that the determinands start at $r = 0$ as constants, normalized to 1. The graphs of the determinands are displayed in Figure 1 for the lowest modes and for different ratios of $\xi = M_H/M_W$.

The magnetic $K = 0$ mode $M0$ is the only one for which the determinands have zeros. There is for all values of ξ at least one zero which corresponds to the genuine sphaleron instability. Above $\xi = 12.038$ a second one develops (we display the graph for $\xi = 50$) and a third one appears at $\xi = 138.32$. These instabilities are the ones discovered previously [9,10]. Our analysis shows that the $M0$ mode is the only one to display such instabilities.

4 The 't Hooft-Polyakov Monopole

The model considered by 't Hooft and Polyakov [14,15] is an $SO(3)$ gauge theory with spontaneous symmetry breaking associated with a Higgs field in the triplet representation. It may exist as a classical object in Grand Unified Theories embedding the electroweak theory [22] . The Lagrangean is given by

$$\mathcal{L} = -\frac{1}{4} W^a_{\mu\nu} W^{\mu\nu}_a + \frac{1}{2} D_\mu \Phi_a D^\mu \Phi^a - \mathcal{V}(\vec{\Phi}) \tag{16}$$

The gauge fields $W^{\mu\nu}_a$ are defined as previously in Eq. 7. The covariant derivative of the Higgs field is given by

$$D^\mu \Phi_a = \partial^\mu \Phi_a - g\epsilon_{abc} W^\mu_b \Phi_c \tag{17}$$

The Higgs self coupling is described by the potential $\mathcal{V}$ as

$$\mathcal{V}(\vec{\Phi}) = \frac{\lambda}{4}(\vec{\Phi}^2 - v^2)^2 \tag{18}$$

The masses of the massive vector bosons, M_W, and of the Higgs particle , M_H are

$$\begin{aligned} M_W &= gv \\ M_H &= \sqrt{\beta} M_W \end{aligned} \tag{19}$$

with $\beta = 2\lambda/g^2$. The Ansatz for the classical solution is

$$\begin{aligned} W^{cl}_{ai} &= \frac{M_W}{g} W(x) \epsilon_{aib} \hat{x}_b \\ \Phi^{cl}_a &= \frac{M_W}{g} V(x) \hat{x}_a \end{aligned} \tag{20}$$

Inserting this Ansatz into the Euler-Lagrange equations yields the following equations of motion for the radial profiles $V(x)$ and $W(x)$

$$x^2 W'' + 2xW' = (xV^2 + W(2 + xW))(1 + xW) \tag{21}$$

and

$$x^2V'' + 2xV' = 2V(1 + xW)^2 + \frac{1}{2}\beta x^2 V(V^2 - 1) \tag{22}$$

The basic monopole solution is determined by these equations and the following boundary conditions :

$$W(x) \propto x \quad V(x) \propto x \tag{23}$$

as $x \to 0$ and

$$xW(x) \to -1 \quad V(x) \to 1 \tag{24}$$

as $x \to \infty$.

Again we have to expand around this classical solution. The Ansatz for the fluctuations of the gauge fields is again given by Eq. 13 while the Higgs field is expanded as

$$\Phi_a = \Phi_a^{cl} + M_W \sum h_\alpha^K \mathbf{J}_a^\alpha Y_{KM} \tag{25}$$

Previously [12] we used the radial gauge, here we present the results of an analoguous calculation in the temporal gauge. Then the amplitudes s_α^K vanish. One obtains for each value of the K spin two systems of coupled amplitudes:

i) the magnetic modes MK with $t_1^K, t_2^K, t_3^K, t_6^K, t_9^K$ and h_3^K; again the amplitudes t_1^K and t_2^K can be expressed by the other ones;

ii) the electric modes EK with $t_4^K, t_5^K, t_7^K, t_8^K, h_1^K$ and h_2^K, where t_4^K can be expressed by the other ones.

We have again determined the r dependence of the determinands of independent solution for both systems. They are displayed in Fig. 4 for the lowest modes. The actual calculation was extended up to $K = 5$ and Higgs masses of $M_H = 20 M_W$. No instability was found.

5 The Electroweak Skyrmeon

The Higgs Lagrangean of the standard model can be rewritten in the form

$$\mathcal{L}_{Higgs} = \frac{h^2}{4} Tr(\mathcal{D}_\mu U)^\dagger(\mathcal{D}^\mu U) + \frac{1}{2}\partial_\mu h \partial^\mu h - \frac{\lambda}{4}(h^2 - v^2)^2 \tag{26}$$

where the Higgs dublet Φ has been reparametrized by introducing the unitary matrix $U(x)$ and the modulus $h(x)$ of the Higgs field via

$$M(x) = \begin{pmatrix} \Phi_2^* & \Phi_1 \\ -\Phi_1^* & \Phi_2 \end{pmatrix} = \frac{h(x)}{\sqrt{2}} U(x) \tag{27}$$

In this form of the Lagrangean resembles to the low energy pion Lagrangean with its chiral $SU(2)_L \times SU(2)_R$ symmetry. As in this case it may be expected [23] that

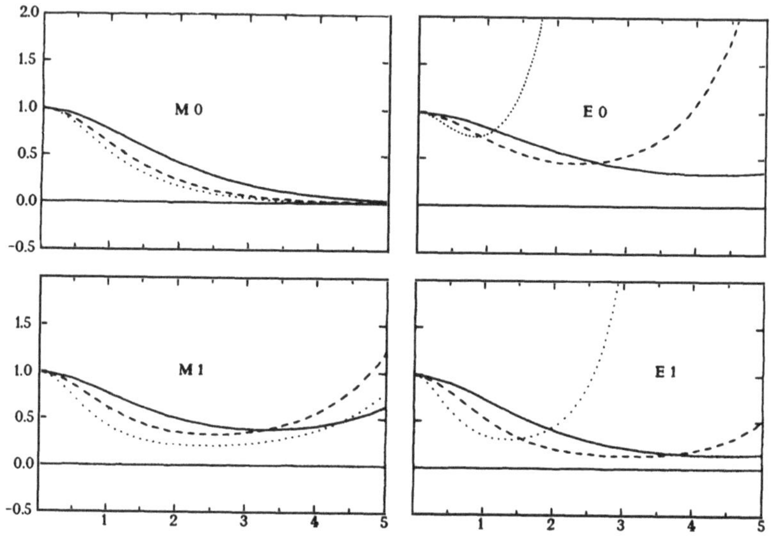

Figure 2 : The Jacobi criterion applied to the stability of the 't Hooft - Polyakov monopole. The determinand of independent solutions is plotted versus $x = M_W r$ for the lowest modes and for $M_H/M_W = 0$ (solid line), 1 (dashed line) and 10 (dotted line).

one loop quantum corrections introduce an effective Skyrme term and that solitons may exist in analogy to the Skyrme nucleon.

This possibility was investigated first [7] for infinite Higgs mass. The Skyrme term was added in the form

$$\mathcal{L}_{Skyrme} = \frac{1}{32e^2} Tr[R_\mu, R_\nu]^2 \tag{28}$$

where

$$R_\mu = \mathcal{D}_\mu U U^\dagger \tag{29}$$

The Ansatz for the classical solution was chosen, after using the gauge degrees of freedom, as

$$\begin{aligned} U &= \cos\Theta(r) + i\tau\hat{\mathbf{x}}\sin\Theta(r) \\ W_{0a} &= 0 \\ \frac{g}{2}W_{ia} &= \frac{\alpha(r) - \frac{1}{2}}{r}\epsilon_{ila}\hat{x}_l - \frac{\delta(r)}{r}\hat{x}_i\hat{x}_a \end{aligned} \tag{30}$$

and the modulus of the Higgs field $h(r)$ was frozen to the vacuum expectation value v. Stable electroweak skyrmeons were found with masses of a few TeV. The system was then investigated also for finite Higgs mass [16] where one has now $h(r)$ as an additional degree of freedom. Again classical solutions were found with energies somewhat lower than those obtained at infinite Higgs mass.

The form of the Skyrme term, Eq. 28, had been taken over unchanged. This latter point has been criticized recently by Carson [17]. Indeed if the Skyrme term is induced as an effective one loop potential one expects it to be regular in the Higgs field. The matrix U is however obtained from the matrix M by dividing by the modulus. A regular Skyrme term would take the form [17]

$$\mathcal{L}'_{Skyrme} = \frac{1}{8e^2v^4}Tr((\mathcal{D}_\mu M)^\dagger(\mathcal{D}_\nu M) - (\mu \leftrightarrow \nu))^2 \tag{31}$$

The modification with w.r.t. the previous definition can be made more explicit by rewriting $\mathcal{L}'$ as

$$\mathcal{L}'_{Skyrme} = (h/v)^4\mathcal{L}_{Skyrme} + \frac{\partial_\mu h^2\partial_\nu h^2}{16e^2v^4}Tr((g^{\mu\nu}R^\lambda R_\lambda - R^\mu R^\nu) \tag{32}$$

The $M_H = \infty$ Skyrme term Eq.28 appears here multiplied with a factor $(h(r)/v)^4$. $h(r)$ tends to a finite value as $r \to 0$. For the electroweak skyrmions of Baacke et. al. [16] based on Eq. 28 this value was in the range $0 < h(0) < v$. $h(r)$ reaches the asymptotic value v exponentially within an interval of order $r = O(M_H^{-1})$. Therefore the additional factor $(h(r)/v)^4$ appearing in Eq. 32 suppresses the first term of the new Skyrme Lagrangean very effectively within this range of r; this term is however essential for the stability of skyrmeons, it provides a potential that prevents the classical solution from collapsing to zero size. If - in an iterative procedure solving the classical equations of motion - the skyrmeon has shrunk to a size smaller than M_H^{-1} this term becomes now ineffective and the solution shrinks further to vanishing size and energy. This is precisely what is observed when following e.g. the Bais-Primack iteration [24].

So besides the topological argument given by Carson [17] and the calculations mentioned in his work [25] we have and independent evidence for the nonexistence of electroweak skyrmeons at finite Higgs mass.

6 Conclusions

We have discussed the existence and stability of classical objects of the electroweak theory and its extensions.

The conclusions are:

i) There is no instability of the sphaleron in higher partial waves, the ones in the $K = 0$ magnetic mode, that led to the concept of deformed sphalerons [9], are the only ones. This investigation was performed for $M_H/M_W \leq 500$ and for $K \leq 5$

ii) There is no instability in any partial wave for the 't Hooft Polyakov monopole for $M_H/M_W \leq 20$ and $K \leq 5$. This result found first in the radial gauge [12] is confirmed here using the temporal gauge.

iii) There is no evidence for the existence of an electroweak skyrmeon at finite Higgs mass. Own numerical calculations [24] confirm the general argument given by Carson [17] and numerical calculations reported by him [25].

The limitations we mentioned in the first two of these statements are not supposed to be serious ones. Centrifugal barriers in higher K modes tend prevent instabilities more and more. For both objects there is also no visible tendency to become more unstable at values of M_H/M_W higher than the ones we have investigated.

7 References

1. R. Dashen, B. Hasslacher and A. Neveu, *Phys.Rev.* **D10** (1974) 4138
2. F. Klinkhamer and N. Manton, *Phys. Rev.* **D30** (1984) 2212
3. V. Kuzmin, V. Rubakov and M. Shaposnikov, *Phys. Lett.* **B155** (1985) 369
4. H. Aoyama and H. Goldberg, *Phys. Lett.***B188** (1987) 506
5. J. Ambjørn and V.A. Rubakov, *Nucl. Phys.***B256** (1985) 434
6. G.Eilam, D.Klabucar and A.Stern, *Phys. Rev. Lett.* **56** (1985) 133
7. G. Eilam and A. Stern, *Nucl. Phys.* **B294**(1987) 775
8. T.H.R. Skyrme, *Proc. Roy. Soc.* **A260** (1961)
9. J. Kunz and Y. Brihaye, *Phys. Lett.*
10. L. Yaffe, *Phys. Rev.***D40**(1989) 3463 **216B** (1989) 353
11. Y. Brihaye, S. Giler, P. Kosinski and J. Kunz, *Phys. Rev.***D42** (1990) 2846
12. J. Baacke, *Fluctuations and Stability of the 't Hooft-Polyakov Monopole* , University of Dortmund preprint DO-TH 90/17, *Z. Phys.* to appear
13. J. Baacke and H. Lange, University of Dortmund preprint DO-TH 91/17 (Nov. 1991), to be published in *Mod. Phys. Lett*
14. G. 't Hooft, *Nucl. Phys.* **B79** (1974) 276, ibid. **B105** (1976) 1156

15. A. M. Polyakov, *JETP Letters* **20** (1974) 194

16. J. Baacke, G. Eilam and H. Lange, *Phys. Lett.*(1988) 127

17. L. Carson, *Are there electroweak skyrmions ?*, University of Alabama Preprint UAHEP9012, December 1990

18. I.M. Gel'fand and S.V. Fomin, *Calculus of Variations*, (Prentice-Hall Inc., Englewood Cliffs, N. J. , 1963)

19. see e.g. A. Messiah, *Quantenmechanik* (de Gruyter, Berlin 1976)Vol.1, section 3.2.5

20. F. A. Bais and J. R. Primack, *Phys. Rev.* **D13** (1975) 819

21. B. Schwesinger, H. Weigel, G. Holzwarth and A. Hayashi, *Phys. Rep.* **173** (1989) 173

22. see e.g. J. Ellis in *Magnetic Monopoles* eds. R.A. Carrigan,Jr. and W.P. Trower (Plenum Press, N.Y., 1983) p. 17

23. T. Appelquist and C. Bernard, *Phys. Rev.* **D22** (1980) 200

24. J. Baacke and H. Lange, unpublished

25. L. Carson and A. Stern, unpuplished

Status of Four-fermion Interactions in Low-energy QCD

A.A.Andrianov and V.A.Andrianov

Department of Theoretical Physics, University of Sankt-Peterburg, 1,ul. Ulyanovskaya,198904 Sankt-Peterburg, Russia

Abstract

We examine conditions for quark models with four-quark interaction to be a reasonable approximation to the QCD at low energies. In accordance with effective-field theory approach the independence of effective potential on the cutoff is exploited in the quark sector to establish the universality law for effective coupling constants. On the contrary the insensitivity of effective potential in respect to "intrinsic"field dilatations leads to a selection rule for parameters of quark models to make a truncation of QCD.

1 Introduction

The field-theory approach to description of particle interactions relies substantially on the concept of effective field theory that includes a certain set of physical states and is applicable reliably in a limited energy region. The influence of other degrees of freedom, for instance, of particles with large enough masses and energies is then taken into account by means of effective-interaction vertices. Thereby in any of physical models the presence of a scale Λ is implied that allows to divide all degrees of freedom into essential or dynamical ones and inessential or virtual ones. That scale Λ roughly marks the range of applicability of an effective field theory.

From the viewpoint of a fundamental theory which presumably includes the entire spectrum of physical states the physical observables have not to depend on a choice of above scale. In the case that a fundamental theory is unknown but still implied to exist the property of universality or indifference in a scale may be implemented with a choice of dependence of effective coupling constants on an energy-spectrum cutoff Λ. We stress that the universality condition does not inevitably entail the renormalizability of a theory in the common sense, i.e. with a finite number of coupling constants.

Typically the effective field theory incorporates the infinite set of coupling constants of a different importance.Yet in practice, for a selected energy region we retain the finite subset of vertices to achieve a reasonable precision. In other words we engage a different field theory to approach to the original one. However the truncated theory itself might be no more equivalent to the initial fundamental theory in the entire range of energies though being a satisfactory approximation for selected energies.

To our mind the QCD at low energies ($< \Lambda \sim 1$GeV) delivers a relevant example of that when being adapted for by a quark model with four-fermion interaction (the Nambu-Jona-Lasinio model [1] and its generalizations[2,3]) in the dynamical symmetry-breaking regime(DSB). In applications to the low-energy hadron physics the above quark models deal with fields of a bounded energy spectrum with momenta $p < \Lambda \sim 1$GeV. Henceforth the problem arises to control the correlation of both theories by adjusting their main scales.

In our paper we propose the selection rule which promotes to find a particular 4-fermion model related to QCD in a better way. This rule we establish in examination of a set of 4-fermion models with different scales Λ though with fixed, Λ-independent coupling constants. Due to universality it is equivalent to consider the models with a given Λ and (specifically) variated coupling constants.

The variation of the cutoff $\Lambda \to \Lambda e^{-\sigma}$ can be induced by a dilatation of fermion fields (due to their "radial" fluctuations) $\psi \to \psi e^{\sigma/2}$. Such a dilatation of low-energy fields only does not affect coupling constants generated presumably by high-energy fields. As a selection rule we draw the variational principle in dilatational mode, i.e. the intrinsic stability principle. Namely we demand that in the DSB regime the (intrinsic) vacuum energy be weakly dependent on a choice of cutoff Λ.

With the universality principle together that intrinsic stability criterion allows to determine an optimal value for Λ and moreover to fix unambiguously coupling constants as functions of Λ and of order parameters (condensates).

In Sect.2 we remind the basic properties of quark models with 4-fermion interaction. Their effective potential is obtained in the mean-field (or large-N_c) approach and applied to describe the DSB regime when the non-vanishing average value of scalar mean-field yields a dynamical mass for quarks.

In Sect.3 the principles of universality and stability are formulated and their role in modelling the low-energy QCD is illuminated. The pertinent anomalous scale equation for the trace of quark energy-momentum tensor happens to be essential for QCD-inspired models.

In Sect.4 the universality criterion is implemented for the 4-fermion model without gluons. It is shown how the intrinsic stability principle (in a weak form) supplemented by scale equations of QCD leads to the optimal value of a so-called chiral-symmetry breaking scale Λ_{CSB}.

Possible extensions of our approach are shortly discussed in Sect.5.

2 Models with four-quark interaction and their effective potential

Let us adopt two schemes of derivation of effective quark lagrangian suitable to treat low-energy phenomena of QCD. The common but formal approach is to absorb (integrate out) all gluon interaction into effective many-fermion vertices[3] and then to make a truncation of that infinite set of vertices retaining the lowest-order ones, i.e. the 4-fermion coupling. Furthermore when selecting the soft-momentum region for quark fields one may expect to have a local limit for the 4-fermion vertex function and eventually to come to the NJL-type model without explicit gluon fields. On the other hand, one could apply the effective action method and take into account the high-energy gluon interaction only by means of perturbation theory. Then the quark effective action to lowest non-trivial order will incorporate both the four-fermion interaction and the coupling to low-energy ("non-perturbative")gluons[4,5]. We postpone the comparative derivation of two approximations from QCD. Rather let us treat them as different theories which might be adjusted to reproduce main features of QCD in a selected energy region.

In this report we consider quark models with local 4-fermion interaction and without explicit gluon coupling. When their coupling is large enough they reveal the dynamical breaking of chiral symmetry. In order to examine this DSB regime let us make a simplification and retain the scalar vertex only which is responsible for DSB. Then the basic lagrangian density can be presented in two forms (the euclidean-space formulation is taken here),

$$\mathcal{L} = z(\Lambda)\,\bar{\psi}\hat{D}\psi + \frac{g^2 z^2}{4N_c}(\bar{\psi}\psi)^2 = z(\Lambda)\,\bar{\psi}(\hat{D} + im(x))\psi + \frac{N_c}{g^2}m^2(x), \qquad (1)$$

where $\hat{D} = i\gamma_\mu\partial_\mu$. We simplify our analysis further on and set up the number of flavours $N_F = 1$ and the current quark mass $m_q = 0$. In Eq.1 the scalar mean-field $m(x)$ is introduced in order to make the DSB phenomenon more transparent (for large N_c).

Let us implement the (gauge invariant) cutoff Λ for quark spectrum in accordance with the QCD effective-action concept. Namely we suppose the cutoff is defined before auxilliary mean-fields are inserted in Eq.1. Hence one should start from the basic quark spectrum, $\hat{D}\psi_\lambda = \lambda\psi_\lambda$, $|\lambda| < \Lambda$. Thus low-energy quark fields are $\psi_l = \Theta(\Lambda^2 - \hat{D}^2)\psi$ where $\Theta(x)$ is a step-function. In that way the projected mean field $\hat{m} = \Theta m \Theta$ is non-local.

The regularized fermion vacuum functional Z_F^Λ is now described by the following relations,

$$Z_F^\Lambda(m) = \left\langle \exp\left(-\int d^4x\, \mathcal{L}_F(m(x))\right)\right\rangle_{\bar{\psi}\psi} \qquad (2)$$

or[6]

$$\ln Z_F^\Lambda(m) = \frac{N_c}{2}Tr\left\{\ln\frac{\hat{D}^2 + i[\hat{D},\hat{m}] + \hat{m}^2}{\mu^2}\Theta(\Lambda^2 - \hat{D}^2)\right\} \qquad (3)$$

where the γ_5-symmetry of Θ is exploited, $[\Theta, \gamma_5] = 0$.

The constant $\mu = \mu_0 z^{-1/2} = \mu(\Lambda)$, $\mu_0 = const \cdot \Lambda$ is a normalization scale for fermion fields which includes the renormalization scale of QCD $\mu_0 \neq \Lambda$ and the effective constant $z(\Lambda)$ characteristic to a preparation procedure of low-energy effective action. From physical viewpoint μ may be thought of as a typical scale at which the experiments involving quark fields are supposed to make.

The total effective action S^{Λ}_{eff} for $m(x)$ fields reads

$$S^{\Lambda}_{eff} = -\ln Z^{\Lambda}_{F} + \frac{N_c}{g^2}\int d^4x\, m^2(x) \simeq \int d^4x\, V_{eff}. \tag{4}$$

In the mean-field approach (at large N_c) there is a saddle point $m(x) = const$ for S_{eff}. In the adiabatic approximation the effective potential can be easily obtained,

$$\begin{aligned} V_{eff}(\Lambda, \mu, m, g^2) \quad = \quad & \frac{N_c}{8\pi^2}\Big\{\frac{\Lambda^4}{2}\big(\frac{1}{2} - \ln\frac{\Lambda^2+m^2}{\mu^2}\big) \\ & -\frac{m^2\Lambda^2}{2} + \frac{m^4}{2}\ln\frac{\Lambda^2+m^2}{m^2} + \frac{8\pi^2 m^2}{g^2}\Big\} \end{aligned} \tag{5}$$

Herein V_{eff} is derived by calculating the quark determinant.

The main feature of above models consists in the existence of nontrivial solution of the mass-gap equation,

$$R(m) \equiv \frac{4\pi^2}{N_c}\cdot\frac{\partial V_{eff}}{\partial m} = 0 \tag{6}$$

which yields the dynamical mass for quarks $m = m_0$ when the 4-fermion coupling constant is strong enough.

In the large N_c approach the mass-gap equation can be expressed as follows

$$R(m) \simeq m\Big(\frac{8\pi^2}{g^2} + m^2\ln\frac{\Lambda^2+m^2}{m^2} - \Lambda^2\Big) = 0 \tag{7}$$

The nontrivial solution of Eq.7 exists in the case that $g^2 > g^2_{crit} = 8\pi^2/\Lambda^2$. It means that the nonlinear 4-fermion interaction is attractive $g^2 > 0$ and strong enough in order to generate the mass for composite scalar meson. We remark that there is no dependence in Eq.7 on a normalization scale of fermion fields μ. The minimum of effective potential V_{eff} is provided when

$$\frac{2N_c}{g^2_{eff}} \equiv \frac{\partial^2 V_{eff}}{(\partial m)^2} = \frac{N_c}{2\pi^2}m_0^2\Big(\ln\frac{\Lambda^2+m_0^2}{m_0^2} - \frac{\Lambda^2}{\Lambda^2+m_0^2}\Big) > 0. \tag{8}$$

The corresponding coefficient in m-field lagrangian is related to the mass of scalar meson.

3 Universality and stability principles for effective action

Let us analyze the four-fermion models from two standpoints. At first one can treat them as a certain approximation to the QCD-effective action. Then the constants of a model should depend on a scale Λ as well as on a normalization point $\mu = \mu_0$ so that the insensitivity of physical observables to a choice of Λ and μ be assured. The latter property we label as the universality. In that case the functional dependence of coupling constants on Λ and μ appears due to high-energy fluctuations of quarks and gluons and it is adjusted to compensate the dependence on the scale which bounds the low-energy quark spectrum.

Thus the effective potential for quarks V_{eff} as being derived from QCD is a function of $g(\Lambda,\mu)$, $z(\Lambda,\mu)$ and also depends on Λ,μ explicitly. In order to obtain the consequences of renorminvariance in Λ and μ for the quark sector we have then to separate the flavour part of QCD vacuum energy and compare it with effective potential of 4-fermion model.

The second standpoint is to pay attention to the intrinsic dynamics of 4-fermion models assuming the coupling constants to be fixed ($\partial_\Lambda g_i = \partial_\Lambda z = 0$). Then we examine the low-energy dependence on Λ in the DSB regime. In combination with the universality it allows to make a selection between 4-fermion models with different cutoffs or, equivalently, between those models with a given cutoff but sliding coupling constants. The main goal of this trial is to find a model which reproduces the low-energy QCD in a better way.

In order to relate the vacuum energy of quarks $\simeq V_{eff}(m = m_0)$ to an appropriate part of QCD-vacuum energy one has to separate a quark contribution to the latter one. In the QCD we expect that:

1. the total vacuum energy is determined by the trace of energy-momentum tensor in a renorminvariant way and has a canonical dimension 4,

$$\langle T^{\nu,tot}_{\nu}\rangle = -4\,E^{tot} = d_\tau E^{tot} \tag{9}$$

   where d_τ denotes an infinitesimal dilatation of all scales[7], $(\Lambda_{QCD},\mu) \to e^{-\tau}(\Lambda_{QCD},\mu)$;

2. due to the scale anomaly[8] Eq.9 specifies the value of (non-perturbative) vacuum energy,

$$E^{tot} = -\frac{\beta(g_c^2)}{4g_c^2}\langle Tr(G^2_{\rho\sigma})\rangle \simeq \frac{\beta}{8}\langle (G^a_{\rho\sigma})^2\rangle \tag{10}$$

   where the total β-function is

$$\beta(g_c^2) \equiv \frac{d\ln g_c^2}{d\ln\mu^2} \simeq -\frac{11N_c - 2N_f}{48\pi^2}g_c^2 \tag{11}$$

   in the one-loop approach.

3. Eqs.9 - 11 result in zero-momentum sum rules (one-loop approach),

$$\langle tr(G^2_{\rho\sigma}(x))\rangle \simeq \left\langle T\left\{tr(G^2_{\rho\sigma}(x))\cdot\int d^4y\,\frac{\beta(g)}{4g^2}\,tr(G^2_{\rho\sigma}(y))\right\}\right\rangle \tag{12}$$

Let's expand E^{tot} in powers of N_F (i.e. the number of flavours), $E^{tot}\simeq E_0+N_fE_1$. Then one can obtain from the basic definition of QCD that

$$\begin{aligned} E_1 &= \partial_{N_f}E^{tot} = \langle\partial_{N_f}(\mathcal{L}_{QCD})\rangle \\ &= -\left\langle\frac{1}{2}\partial_{N_F}\left(\frac{1}{g_c^2}\right)tr(G^2_{\rho\sigma})+\frac{1}{2}Tr\langle x|\,ln\,\frac{\hat{D}^2}{\mu_F^2}\,|x\rangle_{reg}\right\rangle_G \end{aligned} \tag{13}$$

where a gauge-invariant regularization is applied to make sense for fermion determinant (for instance, finite-mode regularization[6]). The dilatation of Eq.13 leads to sum rules for the flavour part of vacuum energy ($\sim N_F$) when Eq.12 is taken into account,

$$\begin{aligned} \left\langle tr\left[\langle x|\ln\frac{\hat{D}^2}{\mu^2}|x\rangle\right]_{reg}\right\rangle &\simeq \frac{1}{48\pi^2}\left\langle tr(G^2_{\rho\sigma}(x))\right\rangle + \\ &+\left\langle T\left\{tr\left[\langle x|\ln\frac{\hat{D}^2}{\mu^2}|x\rangle\right]_{reg}\cdot\int d^4y\,\frac{\beta(g)}{4g^2}tr(G^2_{\rho\sigma}(y))\right\}\right\rangle \end{aligned} \tag{14}$$

We compare now these relations with Eqs.1-5 which provide V_{eff} in the mean-field approach and assign that the vacuum energy of 4-fermion model $V_{eff}(m_0)$ should be referred to the QCD-fermion energy in Eq.13.

In order to make their fitting more transparent let's accept that:

1. the QCD coupling constant is normalized at different scales μ_B for gluons and μ_F for quarks and the main QCD-scale is a function of N_f,

$$\Lambda_{QCD}(N_f)\simeq\Lambda_B\,\xi^{-b},\qquad b=\frac{2N_f}{11N_c-2N_f}$$

2. respectively a fundamental fermion scale is attributed $\Lambda_F\equiv\xi\Lambda_B,\quad \xi\geq 1$; then the one-loop QCD coupling constant reads

$$\frac{1}{g_c^2}\simeq\frac{1}{48\pi^2}\left(11N_c\ln\frac{\mu_B^2}{\Lambda^2_{B,QCD}}-2N_f\ln\frac{\mu_F^2}{\Lambda^2_{F,QCD}}\right) \tag{15}$$

3. there exists a normalization point $\mu_F\equiv\exp(-\eta)\Lambda_F$ such that the correlator between one-fermion loop operator and gluon density in Eq.14 be suppressed.

Evidently the second QCD-scale can be implanted by a choice of relevant normalization point for quark momenta,

$$\partial_{N_f}\left(\frac{1}{g_c^2}\right)_{\mu_F=\Lambda_{F,QCD}} = 0 \tag{16}$$

The introducing of the second, fermion QCD-scale Λ_F helps to arrange dilatations of fermion scales in QCD and thereby to establish the correspondence between quark vacuum energies in QCD and in 4-fermion models.

One can see from Eqs.13, 15 that the variation of fermion scales in the QCD concerns Λ_F and μ_F in both terms of Eq.13 or equivalently μ_F only in the second one. This variation leads to the anomaly equation for fermion energy-momentum tensor,

$$\langle T_\mu^{\mu,F}\rangle = d_\sigma E^F = -\frac{\beta^F}{2}\langle (G^a_{\rho\sigma})^2\rangle \simeq -\frac{N_f}{12}C_g \tag{17}$$

where $\beta^F \simeq 2N_f g_c^2/48\pi^2$ represents the quark part of β-function in the one-loop approximation and the conventional gluon condensate[8], $C_g \equiv g^2/4\pi^2\langle (G^a_{\rho\sigma})^2\rangle$ is inserted for vacuum average. Eq.17 is approximately renorminvariant under variation of fermion scales (up to the next order in $1/N_c$), $d_\sigma^2 E^F \simeq 0$.

From the assumption 3 one can fix the value of fermion part of vacuum energy,

$$E^F = \frac{N_f}{48}C_g(1+2\eta), \quad d_\sigma E^F \simeq -\frac{4}{1+2\eta}E^F \tag{18}$$

To some extent the latter relation may be thought of as a definition of anomalous dimension in respect to fermion-scale dilatations. One can exploit the negative value of total vacuum energy, $E^{tot} \simeq E^{gl} + E^F < 0$ in order to derive the upper bound for $1 + 2\eta < 11N_c/2N_f$ or $\eta < 9/4$ for $N_f = 3$.

Now let us specify the scale dependence of V_{eff} in the quark model and consider the intrinsic scale fluctuations of low-energy quark fields $\psi_l \to \psi_l e^{\sigma/2}$. These "radial" fluctuations induce the dilatation of the effective Dirac operator $\hat{D} + im \to e^{\sigma/2}(\hat{D} + im)e^{\sigma/2} \equiv \hat{D}^\sigma$ (conformal Weyl transformations) which in turn leads to rescaling the spectral parameters $\Lambda \to \Lambda e^{-\sigma}$, $\mu \to \mu e^{-\sigma}$ in Eq.5. It follows from our definition of the lagrangian and low-energy quark fields (Eqs.1,3). When the mean-field is introduced we imply its participation in above dilatations due to rescaling of quark fields.

In the quark model (Sect.2) employed as a truncation of QCD-effective action the rescaling involves both intrinsic dilatations of Λ, μ and induced variations of effective constants g and z, $d_\sigma = \partial_\sigma + g'_\sigma \partial_g + z'_\sigma \partial_z$. In order to approach the QCD by means of a quark model we suppose that the flavour part of QCD-vacuum energy is related to $V^0_{eff} \equiv V_{eff}(m = m_0)$ and thereby from Eqs.17, 18 one has

$$d_\sigma V^0_{eff} \simeq d_\sigma E^F = -\frac{N_f}{12}C_g \tag{19}$$

Respectively the approximate scale invariance should hold for higher-order dilatations if $V^0_{eff} \simeq E^F$,

$$d^n_\sigma V^0_{eff} \simeq 0, \quad n > 1 \tag{20}$$

The next equation we need for V^0_{eff} is achieved from β-function in Eq.10 for $\mu^F = \exp(-\eta)\Lambda^F$ and displayed in Eq.18, i.e.

$$V_{eff} \simeq \frac{N_f}{48} C_g(1 + 2\eta) \tag{21}$$

This relation cannot be valid for constant η and arbitrary choice of normalization point that is evident from its compatibility with Eq.20. However we assume the existence of an appropriate normalization scale which supports Eq.21. The latter equation we shall use for the normalization of fermion scales in a quark model.

Thus we expect that Eqs.19 - 21 are to determine correct scaling properties of a quark-model truncation. As a direct consequence one obtains the independence of dynamical mass m_0 on fermion scales. Indeed, the effective potential in the vicinity of $m \sim m_0$ has the following expansion,

$$V_{eff}(m) = V_{eff}(m_0) + \frac{1}{2} V''_{eff}(m_0)(m - m_0)^2 + \cdots \tag{22}$$

due to the mass-gap Eq.7. Respectively its scale invariance,

$$d^2_\sigma V_{eff}|_{m=m_0} = d^2_\sigma V^0_{eff} + V''_{eff}(m_0)(d_\sigma m_0)^2 = 0 \tag{23}$$

can be provided iff $d_\sigma m_0 = 0$ since in the DSB regime (Eq.9) $V''_{eff}(m_0) > 0$. Therefore the scaling behavior of 4-fermion coupling constant g is uniquely determined by the mass-gap Eq.8 for $m = m_0$, $d_\sigma m_0 = 0$.

Let's analyze now the intrinsic scaling properties of quark models with low-energy cutoff. The dilatational potential V^{dil}_{eff} for radial fluctuations of quark fields, $\psi \to \psi e^{\sigma/2}$ can be derived from conformally non-invariant part of quark vacuum functional Z^Λ_F by means of a group-averaging ansatz[9]

$$Z^\Lambda_F = \int d\sigma \frac{Z_F(\hat{D} + im)}{Z_F(\hat{D}^\sigma)} \cdot Z_{inv} \equiv Z_{inv} \cdot \int d\sigma \exp\Big[-S_{dil}(\sigma, \Lambda, \mu, m)\Big]. \tag{24}$$

The effective action S_{dil} is generated by a conformal anomaly and can be calculated along the scheme of conformal bosonization[9]. For $\sigma \simeq const$ (the adiabatic approach) one has

$$V^{dil}_{eff}(\sigma) \equiv \frac{S_{dil}(\sigma)}{(vol)} = V_{eff}(\sigma = 0) - V_{eff}(m, g, z, \Lambda e^{-\sigma}, \mu^{-\sigma}). \tag{25}$$

This is our definition of the effective potential in the dilatational mode. In Eq.24 the constants g, z are fixed ($d_\sigma g, d_\sigma z = 0$) since the intrinsic dependence on Λ as a low-energy cutoff is under examination.

We select the appropriate quark model (satisfying Eqs.19, 20, 23) in searching for its intrinsic stability against field dilatations. Namely we demand that the free vacuum energy be weakly dependent om a choice of Λ in the DSB regime for $\Lambda \simeq \Lambda_{CSB}$, i.e.

$$\partial_\sigma V^{dil}_{eff}(m = m_0) = 0 \tag{26}$$

The strong form of stability principle might be the requirement to reach a minimum for V^{dil}_{eff} for a selected scale Λ_{CSB}, i.e. $\partial^2_\sigma V^{dil}_{eff} > 0$. However it is not realized in simple quark models without explicit gluon coupling (see in[5] a counterexample for the gauged quark model).

4 Approaching QCD with four-fermion interaction without gluons

In the quark model with 4-fermion interaction we have three basic parameters (Eqs.1, 3): $\Lambda, \mu(\Lambda)$ and $g(\Lambda)$. The coupling constant,

$$\frac{8\pi^2}{g^2} = \Lambda^2 - m_0^2 \ln \frac{\Lambda^2 + m_0^2}{m_0}$$

where m_0 is conceivably controlled by the gluonic scale Λ_B. The universality condition inspired by QCD (Eqs.19, 20) implies the indifference to a choice of scale Λ. It allows to obtain the function $\mu^2 \equiv \Lambda^2 \exp f(\Lambda)$. The differential equation for $f(\Lambda)$ takes the following form,

$$\begin{aligned} \tfrac{8\pi^2}{N_c} d_\sigma V_{eff}(m_0) &= 2\Lambda^4 \Big[\ln(1 + \frac{m_0^2}{\Lambda^2}) - f(\Lambda) - \frac{1}{2}\Big] + m_0^2 d_\sigma \Big(\frac{8\pi^2}{g^2}\Big) + \frac{\Lambda^4}{2} d_\sigma f \\ &= -\frac{2\pi^2}{3N_c} C_g \equiv -\gamma m_0^4 \end{aligned} \tag{27}$$

The explicit solution can be easily found,

$$f = -\frac{1}{2} + \ln(1 + \frac{m_0^2}{\Lambda^2}) - \frac{m_0^2}{\Lambda^2} + \frac{m_0^4}{\Lambda^4}\Big\{\ln \frac{\Lambda^2 + m_0^2}{m_0^2} + \gamma\Big(\ln \frac{\Lambda^2}{\rho m_0^2} + \frac{1}{2} + \eta\Big)\Big\} \tag{28}$$

It contains an arbitrary constant ρ which determines the value of "physical" cutoff $\Lambda = \Lambda_{CSB}$ if Eq.21 for V_{eff} is fulfilled. When substituting Eqs.5, 27, 28 into Eq.21 one obtains the simple relation

$$\Lambda^2_{CSB} = \rho m_0^2 \tag{29}$$

In order to fix the constant ρ let's exploit the stability condition, Eq.26. It can be realized in a weak form,

$$\frac{8\pi^2}{N_c} \partial_\sigma V^{dil}_{eff}|_{\sigma=0} = 0 = 2\Lambda^4 \Big(\frac{1}{2} - ln\Big(1 + \frac{m^2}{\Lambda^2}\Big) + f(\Lambda)\Big)_{m=m_0} \tag{30}$$

which selects out the value of ρ, and furthermore Λ in accordance to Eq.29,

$$\rho - \ln(\rho + 1) = \gamma(\frac{1}{2} + \eta) \tag{31}$$

For the expecting value of dynamical mass, $m_0 \simeq 300$ MeV and the conventional value of $C_g \simeq (360MeV)^4$ one has $\gamma \simeq 4.6$. Then for $\eta = 0$ (canonical dimension for V_{eff}) and $\eta = 1$ we find from Eq.31 $\rho \simeq 4$ and $\simeq 9$ respectively. It gives $\Lambda \simeq 600$MeV or $\simeq 900$MeV.

The strong form of stability condition (the minimum of V_{eff}^{dil})is not available since,

$$\partial_\sigma^2 V_{eff}^{dil}\Big|_{\sigma=0} = -\frac{N_c \Lambda^4 m^2}{2\pi^2(\Lambda^2 + m^2)} < 0 \tag{32}$$

It may be shown [5] that the supplementing of above quark model with explicit gauge coupling to low-energy gluons makes it possible to realize the intrinsic stability in the strong form.

5 Discussion

In our report we have outlined the scheme of approaching the QCD with a different quark model which is based on the concept of effective field theory[10] and related scaling laws of QCD. Our aim has been to decsribe principal features of approximate low-energy dynamics and to explain the correspondence between main scales of QCD and its truncation by a quark model.

Insofar as a second scale for quarks has been introduced one may inquire on its necessity in our construction. In fact, its role is convenient in the selection of fermion part of QCD vacuum energy though main scaling relations for E^F, i.e. Eqs.13, 14, 17, are insensitive to a choice of fermion scale Λ_F.

We have shown that the intrinsic stability of truncating model allows to determine the fitting cutoff $\Lambda \simeq \Lambda_{CSB}$ uniquely. Still we are aware of all simplifications that have been done and guess that a more sophisticated model with more derivatives[11](and more fermion legs) in vertices might serve better for a reasonable approximation of QCD. In particular, the intrinsic dilatational mode might be implemented as a true collective variable for QCD effective action.All these interesting options will be explored elsewhere.

Certainly the realistic model might happen to be different. However we are sure that the common treatment and applications of QCD-inspired quark models[2,3] when using the physical cutoff have to be improved with taking into account true scaling laws of QCD. Just the way of how to make it we have presented.

6 Acknowledgements

We are very much indepted to the Organizing Committee of the Workshop on Effective Field Theories (August 1991, Hungary) for a support. It is our pleasure to thank the participants of the Workshop for fruitful discussions and especially Prof. A.Patkos for informal encouragement.

7 References

1. Y. Nambu and G. Jona-Lasinio, *Phys. Rev.* **122** (1961) 345.

2. M. K. Volkov, *Ann. Phys. (N.Y.)* **157** (1984) 282;
T. Hatsuda and T. Kunihiro, *Phys. Lett.* **B145** (1984) 7;
A. Dhar, R. Shankar and S. R. Wadia, *Phys. Rev.* **D31** (1985) 3256;
D. Ebert and H. Reinhard, *Nucl. Phys.* **B271** (1986) 188;
V. Bernard, R. L. Jaffe and U.-G. Meissner, *Nucl. Phys.* B308 (1988) 753;
M. Wakamatsu and W. Weise, *Z. Phys.* **A331** (1988) 173.

3. T. Eguchi, *Phys. Rev.* **D14** (1976) 2755;
N. Karchev and A. A. Slavnov, *Sov. J. Theor. Math. Phys.* **65** (1985) 192;
D. McKay and H. Munczek, *Phys. Rev.* **D32** (1988) 266;
T. R. Cahill, J. Praschifka and C. D. Roberts, *Int. J. Mod. Phys.* **A4** (1989) 1681; 4929;
R. D. Ball, *Int. J. Mod. Phys.* **A5** (1990) 4391.

4. W. A. Bardeen, C. N. Leung and S. T. Love, *Nucl. Phys.* **B273** (1986) 649;
K.-I. Kondo, S. Shuto and K. Yamawaki, *Nagoya preprint* **DPNU-91-14/ CHIBA-EP-46** (1991) 10p.

5. A. A. Andrianov and V. A. Andrianov, *Amsterdam preprint* **ITFA-90-30** (1990) 8p.;
in *Proceeding of Steklov Math.Inst., St.Peterburg branch*, ed. P. P. Kulish and V. N. Popov (Nauka, St.Peterburg, 1992), to be published.

6. A. A. Andrianov and L. Bonora, *Nucl. Phys.* **B233** (1984) 232.

7. C. G. Callan, S. Coleman and R. Jackiw, *Ann. Phys. (N.Y.)* **59** (1970) 42.

8. J. C. Collins, A. Duncan and S. D. Joglekar, *Phys. Rev.* **D16** (1977) 438;
V. A. Novikov, M. A. Shifman, A. I. Vainstein and V. I. Zakharov, *Nucl. Phys.* **B191** (1981) 301;
J. M. Cornwall and A. Soni, *Phys. Rev.* **D29** (1984) 1424.

9. A. A. Andrianov, V. A. Andrianov, V. Yu. Novozhilov and Yu. V. Novozhilov, *Phys. Lett.* **186B** (1987) 401.

10. H. Georgi, *Harvard preprint* **HUTP-90/A077** (1991) 18p.

11. A. Hasenfratz, P. Hasenfratz, K. Jansen, J. Kuti and Y. Shen, *San-Diego preprint* **UCSD/PTH 91-06** (1991) 18p.

How strong can the Yukawa coupling be?

Christoph Frick
HLRZ at KFA Jülich, P.O.Box 1913, D-5170 Jülich, Germany

Abstract

The renormalized Yukawa coupling is investigated in the weak coupling region of the broken phase of a lattice regularized $SU(2)_L \otimes SU(2)_R$ fermion-scalar model by nonperturbative numerical Monte Carlo simulation. We did not find any indication for the existence of a nontrivial fixed point. The renormalized Yukawa coupling appears to be bounded from above. The maximal possible values for a fermionic correlation length above 3 seems not to exceed the perturbative unitarity bound.

Introduction. The scalar field selfcoupling and the Yukawa couplings in the electroweak theory are believed to vanish in the limit of infinite cut-off, as suggested by the signs of the perturbative β-functions. But the confirmation of this 'triviality' of these couplings and a reliable determination of its consequences require nonperturbative methods, because one has to control the dynamically most difficult regime – when the values of the renormalized couplings have maximal possible values which could be, in principle, large. The lattice regularization allows to use numerical Monte Carlo simulation techniques which proved to be appropriate in such situations.

Though the accumulating phenomenological evidence that the top quark mass does not exceed 200 GeV suggests that the Yukawa coupling of the top quark is weak enough to be treated reliably by perturbative methods, for example a fourth family with fermions much heavier than the top quark might exist. Thus it is of interest to investigate the limits of the Standard Model in the regime of maximal renormalized quartic and Yukawa couplings.

In spite of the fact that presently we are only able to study rather simplified versions of the Standard Model – the problems of regularizing a chiral theory and having the correct fermion content of the spectrum beeing still unsolved [1, 2] – we hope that it is possible to address some of the general quantum field theoretical aspects of a strong Yukawa coupling: Are the Yukawa theories in 4 dimensions really trivial or do some nontrivial fixed points exist? If they are trivial, how far and in what form Yukawa theories can be used as effective quantum field theories? What are their relations to the purely fermionic theories as for example the ones with four-fermion coupling (see in particular the contribution of J. Kuti in these proceedings)? Among the most important phenomenological questions (e.g. what would be the effect of a strong Yukawa coupling on the triviality bound for the Higgs boson mass [3]? – and what is the lower bound on the Higgs boson mass which follows in the case of a strong Yukawa coupling from the vacuum stability requirement [3, 4]?) we want to concentrate in this short contribution on the investigation of how strong the Yukawa coupling can actually be, which means to ask how large the fermion mass can be when generated by a Yukawa interaction [5].

The SU(2)$_L$⊗SU(2)$_R$ Yukawa model with naive fermions. The naive Yukawa model on the euclidean lattice is defined by the action

$$S = -\kappa \sum_{x\mu} \tfrac{1}{2}\mathrm{Tr}\left\{\Phi_x^\dagger \Phi_{x+\hat\mu} + \Phi_{x+\hat\mu}^\dagger \Phi_x\right\} + \sum_x \tfrac{1}{2}\mathrm{Tr}\left\{\Phi_x^\dagger\Phi_x + \lambda\left(\Phi_x^\dagger\Phi_x - 1\!\!1\right)^2\right\}$$
$$+ \sum_{x\mu} \frac{1}{2}\left(\overline{\Psi}_x\gamma_\mu\Psi_{x+\hat\mu} - \overline{\Psi}_{x+\hat\mu}\gamma_\mu\Psi_x\right) + y\sum_x \overline{\Psi}_x(\Phi_x P_R + \Phi_x^\dagger P_L)\Psi_x \,. \qquad (1)$$

Here Φ is a 4-component scalar field with quartic selfcoupling and Ψ is a fermionic SU(2) doublet field coupled to the scalar field by the Yukawa interaction with the bare Yukawa coupling y. P_R and P_L are the right and left handed chiral projectors. Because both fermions in the doublet couple with the same strength y to the scalar field, the action has a global SU(2)$_R$ flavour symmetry. Together with the global SU(2)$_L$ symmetry, which would turn into a local symmetry when gauge fields are included, the action is invariant under the global chiral SU(2)$_L$⊗SU(2)$_R$ transformations

$$\Psi_0 \to (\Omega_L P_L + \Omega_R P_R)\Psi_0, \quad \overline{\Psi}_0 \to \overline{\Psi}_0(\Omega_L^\dagger P_R + \Omega_R^\dagger P_L), \quad \Phi_0 \to \Omega_L \Phi_0 \Omega_R^\dagger, \qquad (2)$$

where $\Omega_{L,R} \in \mathrm{SU}(2)_{L,R}$. In the following we will set the bare quartic coupling λ of the scalar field to infinity, which freezes the radial mode of the scalar field, $\frac{1}{2}\mathrm{Tr}\,\Phi_x^\dagger\Phi_x = 1$, so that Φ_x can be represented by an SU(2) matrix. The hopping parameter of the scalar field κ and the bare Yukawa coupling y remain as free parameters.

This model using 'naive' lattice fermions represents a simplification of the Standard Model with respect to the fermion spectrum: due to the problem of fermion doubling on the lattice [6] it describes 16 degenerate fermions and thus does not reproduce the chiral character of the Standard Model. As none of the approaches to achieve a more realistic fermion spectrum has been successful yet [1, 2] we are restricted to this approximation. For the technical reason to be able to use the Hybrid Monte Carlo algorithm [7] in the numerical simulation we have to further double the number of fermions by squaring the fermion matrix. Thus we are actually simulating a model with 32 fermion doublets in the hope that qualitative aspects of Yukawa models are not changed by a large fermion number.

Search for the strongest renormalized Yukawa coupling. First important insights in the general properties of the Yukawa interaction can be obtained by looking to the phase diagrams of Yukawa models. In particular, the fact that various fermion-scalar models, e.g. models with different symmetry groups and/or alternative formulations of fermions on the lattice (naive, Wilson, mirror, staggered fermions; see the review [1] and the references therein), have the most important features of the phase structure in common, can be seen as an argument in favour of our hope to get results which are also relevant for the Yukawa sector in the Standard Model.

The phase diagram of our SU(2)$_L$⊗SU(2)$_R$ Yukawa model (1) in the space of the bare parameters κ and y is shown in Fig. 1 [8, 9]. It consists of several phases and phase regions with ferromagnetic (FM), paramagetic (PM), antiferromagnetic (AM) and ferrimagnetic (FI) ordering of the scalar field. A comprehensive discussion is given in Ref. [8]; in the present context let us just emphasize the most important general features of the phase structure which are shared by various other fermion-scalar models:
(i) in addition to the weak coupling phases or phase regions FM(W), PMW and AM(W) where the fermion mass obeys the perturbative relation $am_F \simeq y\,\langle\Phi\rangle$ there are at large values of y the strong coupling phases or phase regions FM(S), PMS and AM(S) with non-perturbative behaviour of several fermionic observables;

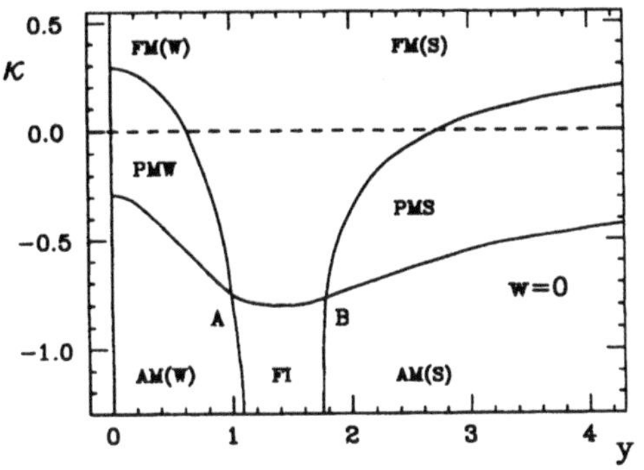

Figure 1: *Schematic phase diagram of the Yukawa model with naive fermions (1).*

(ii) at intermediate values of y the ferromagnetically ordered FM phase continues into the negative κ region, which means that there the spontaneous symmetry breaking is entirely due to the Yukawa interaction;
(iii) there exist points where several phase transition lines meet, here the points labeled by A and B in Fig. 1.

These points have been speculatively considered as candidates for nontrivial fixed points. Therefore we have performed a detailed investigation of the scaling laws for the fermion mass am_F and the scalar field expectation value $\langle\Phi\rangle$ in the whole scaling region near the FM(W)-PMW phase transition line including the near vicinity of our point A. In Fig. 2 on the left we display the numerical results for the three κ values which are closest to the point A. For direct comparison the results for an exemplary positive κ are also included. Everywhere the scaling behaviour is found to be consistent with the scaling laws of the trivial Gaussian fixed point situated at $\lambda = y = 0$ and $\kappa = 1/8$. Furthermore we have translated all our results on the fermion mass into a plot of the lines of constant $(am_F)^2$ which is shown in the right plot of Fig. 2. It can be clearly concluded that the lines of constant $(am_F)^2$ do not flow into the point A, thus diminishing the probability for point A to be a nontrivial fixed point. In all, none of our numerical results can be interpreted as an indication for the existence of a nontrivial fixed point in our Yukawa model. This gives some nonperturbative support to the perturbative scenario of the triviality of the Yukawa coupling.

The assumtion that Yukawa theories are trivial requires the interpretation of Yukawa models as effective theories for a finite cut-off. This framework implies the existence of an upper bound for the fermion mass generated by Yukawa interaction in analogy to the upper bound for the Higgs boson mass. This upper bound has to be determined in the limit of strongest renormalized quartic and Yukawa couplings. As in the purely scalar sector, the maximal renormalized quartic couplings are expected in the limit of infinite bare quartic coupling λ and therefore we restrict ourselves to $\lambda = \infty$. For the Yukawa coupling the situation is different. The nonperturbative fermion mass behaviour found in the strong coupling regions for $y \gtrsim 1.4$ yields a complete decoupling of the fermions from the spectrum when ap-

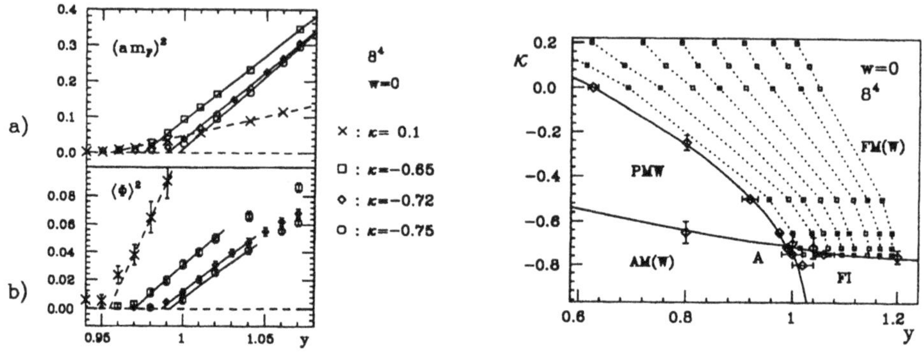

Figure 2: *Evidence for the absence of a nontrivial fixed point. Left: Scaling of* $(am_F)^2$ *and* $\langle\Phi\rangle^2$ *as a function of* y *for the three values of* κ *which are closest to the point A and for one exemplary positive* κ*, where the latter data are shifted by an amount of* $\Delta y = 0.44$ *in the positive* y *direction. Right: The lines of constant* $(am_F)^2$ *in the FM(W) phase region. From left to right* $(am_F)^2$ *changes from 0.1 to 0.8 in steps of 0.1.*

proaching the FM(S)-PMS phase transition line leaving the pure scalar sector behind. Thus the region of strongest renormalized Yukawa couplings with scaling fermions is expected for the maximal bare y which are still in the weak coupling FM(W) region. One naturally ends up in the region with negative κ at intermediate y. This conclusion is the motivation to concentrate in the following on the vicinity of the FM(W)-PMW phase transition line at negative κ.

The ratio $(am_F)/\langle\Phi\rangle$ and the wavefunction renormalization constant Z_π. Here we use the tree level definition of the renormalized Yukawa coupling,

$$y_R = \frac{(am_F)}{v_R} = \frac{am_F}{\langle\Phi\rangle}\sqrt{Z_\pi}\,, \tag{3}$$

where Z_π is the Goldstone wavefunction renormalization constant. Recently we could confirm these results by an alternative definition related to the 3-point-function [10].

It turns out to be interesting to look first separately to the ratio $(am_F)/\langle\Phi\rangle$ and to the wavefunction renormalization constant Z_π before combining these in the full definition (3). In Fig. 3 in the left diagram the full squares represent the results for the ratio $(am_F)/\langle\Phi\rangle$ as a function of κ for the fixed fermionic correlation length $\xi_F := \Lambda/m_F = 1/(am_F) = 3.2$. This value of ξ_F corresponds to the line with $(am_F)^2 = 0.1$ in Fig. 2. The ratio $(am_F)/\langle\Phi\rangle$ increases approximately linearly as the hopping parameter κ is decreased approaching the vicinity of point A along a line of constant cut-off to fermion mass ratio.

But this ratio has to be renormalized by the Goldstone wavefunction renormalization constant Z_π which is to be extracted from the propagator of the Goldstone boson. A detailed discussion of the scalar propagators in the presence of fermions, in particular in the weak coupling region at negative κ of the Yukawa model, can be found elsewhere [10, 11]; here let us only mention that their analysis is made quite difficult by massive lattice artefacts causing a very complex structure of the scalar propagators which make them different from a simple

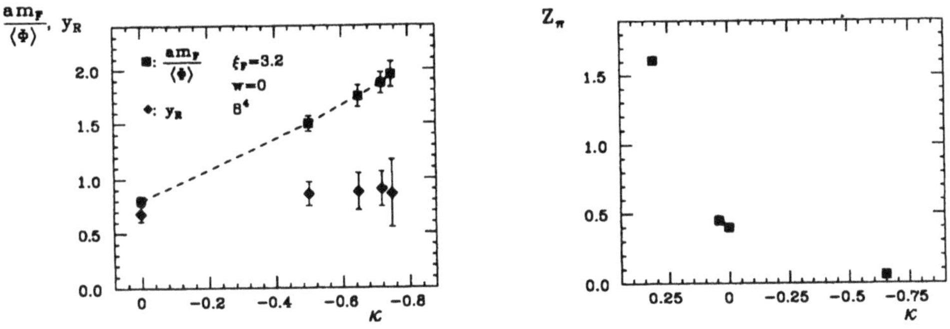

Figure 3: *Left: The ratio $am_F/\langle\Phi\rangle$ and the renormalized Yukawa coupling y_R as a function of κ for the fixed fermionic correlation length $\xi_F = 3.2$. Right: The Goldstone wavefunction renormalization constant Z_π as a function of κ close to the FM(W)-PMW phase transition.*

free particle propagator. However, these phenomena can be quantitatively understood and are thus completely under control. Some of the resulting values for Z_π at typical parameter points are shown in the right plot in Fig. 3. For decreasing κ along a line of constant fermionic correlation length we observe a rapid monotonic decrease of Z_π which just compensates the increase of the ratio $(am_F)/\langle\Phi\rangle$. This is demonstrated in Fig. 3 where the resulting values for the renormalized Yukawa coupling y_R from definition (3) represented by the full diamonds are compared to the unrenormalized ratios given by the full squares. The obtained values of y_R are within the error bars practically independent of κ for a fixed fermionic correlation length.

The renormalized Yukawa coupling. First let us mention that the numerical data for y_R at positive κ are in good agreement with 1-loop perturbation theory. All our results in the negative κ region are collected in Fig. 4 where y_R is plotted as a function of the fermionic correlation length $\xi_F = \Lambda/m_F = 1/am_F$. Fig. 4 demonstrates that for negative κ the renormalized Yukawa coupling y_R at fixed ξ_F is almost independent of κ. As the fermionic correlation length grows within the interval $0 < \xi_F \lesssim 3$ the renormalized Yukawa coupling decreases rapidly. As ξ_F is increased above ~ 3 the renormalized Yukawa coupling y_R flattens out and becomes within the error bars approximately a constant as a function of ξ_F. Of course, the expected logarithmic decrease of y_R cannot be seen in our results because of the large error bars and the smallness of the available interval for the fermionic correlation length.

It can be seen clearly from our data that in the FM(W) phase the renormalized Yukawa coupling is bounded from above. All our results of y_R for different values of κ and y form roughly a universal curve as a function of ξ_F. Interestingly, the dotted curve $y_R = \frac{1}{\sqrt{2\beta_0|\ln(am_F)|}}$ which is obtained from the perturbative 1–loop formula by performing formally the limit $y_0 \to \infty$ matches this universal behaviour of y_R quite well.

The values for y_R for $\xi_F > 3$ are well compatible with the perturbative tree level unitarity bound given in this model by $y_R \lesssim 0.63$ [9] as shown by the dashed line in Fig. 4. In

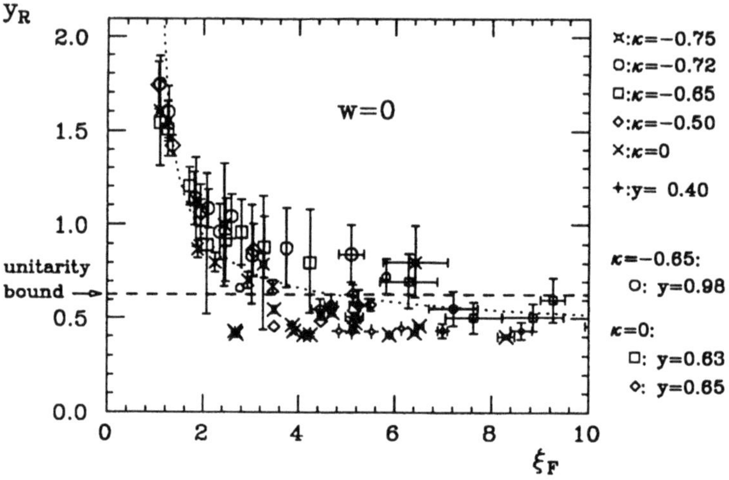

Figure 4: *The renormalized Yukawa coupling y_R as a function of the fermionic correlation length $\xi_F = 1/(am_F)$ for scans in the negative κ region. The dotted curve represents 1-loop perturbation theory at $\kappa = 0$.*

particular the data on larger lattice volumes lie mostly below this bound. Note that in Φ^4–theory the upper bound for the renormalized quartic coupling determined by means of analytical and numerical calculations is significantly smaller than the perturbative unitarity bound [12, 13, 14].

We conclude that the scaling behaviour obtained for the vacuum expectation value of the scalar field and for the fermion mass and the results for the renormalized Yukawa coupling are completely consistent with the triviality of the Yukawa theory for the whole FM(W) scaling region near the phase transition to the PMW phase. The likelihood of point A beeing a nontrivial fixed point is therefore very low. Even more, the renormalized Yukawa coupling from this nonperturbative investigation agrees nicely with 1-loop perturbation theory, the values beeing compatible with the unitarity bound.

Acknowledgements. This work was done in collaboration with Wolfgang Bock, Asit K. De, Karl Jansen, Jiří Jersák and Thomas Trappenberg. The numerical simulations were performed on the Cray Y-MP/832 at HLRZ Jülich and on the VP200-EX at RWTH Aachen. I would like to thank the organizers of the workshop for the possibility to present these results and for the very nice atmosphere at Dobogókö.

References

[1] A.K. De and J. Jersák, HLRZ-91-83, to be published in *Heavy Flavours*, eds. A. Buras and M. Lindner (World Scientific, Singapore).

[2] J. Smit, Nucl. Phys. B (Proc. Suppl.) 17 (1990) 3;
J. Shigemitsu, Nucl. Phys. B (Proc. Suppl.) 20 (1991) 515;
M.F.L. Golterman, Nucl. Phys. B (Proc. Suppl.) 20 (1991) 528;
W. Bock, A.K. De and J. Smit, HLRZ-91-81;
W. Bock, Conference Lattice '91, Tsukuba, Japan, to be published in the proceedings;
M.F.L. Golterman, Conference Lattice '91, Tsukuba, Japan, to be published in the proceedings.

[3] M. Lindner, Z. Phys. C31 (1986) 295;
M. Lindner, M. Sher and H.W. Zaglauer, Phys. Lett. 228B (1989) 139.

[4] M.J. Duncan, R. Philippe and M. Sher, Phys. Lett. B153 (1985) 165;
M. Sher and H.W. Zaglauer, Phys. Lett. B206 (1988) 537;
M. Sher, Phys. Rep. 179 (1989) 273.

[5] M.B. Einhorn and G.J. Goldberg, Phys. Rev. Lett. 57 (1986) 2115.

[6] L.H. Karsten and J. Smit, Nucl. Phys. B183 (1981) 103;
H. Nielsen and M. Ninomiya, Nucl. Phys. B185 (1981) 20.

[7] S. Gottlieb, W. Liu, R.L. Renken, R.L. Sugar and D. Toussaint, Phys. Rev. D35 (1987) 2531.

[8] W. Bock, A.K. De, K. Jansen, J. Jersák, T. Neuhaus and J. Smit, Nucl. Phys. B344 (1990) 207.

[9] W. Bock, A.K. De, C. Frick, K. Jansen und T. Trappenberg, HLRZ 91-21, to be published in Nucl. Phys. B.

[10] W. Bock, A.K. De, C. Frick, J. Jersák and T. Trappenberg, HLRZ-91-82, Amsterdam ITFA 91-33.

[11] C. Frick, Conference Lattice '91, Tsukuba, Japan, to be published in the proceedings.

[12] A. Hasenfratz, K. Jansen, C.B. Lang, T. Neuhaus and H. Yoneyama, Phys. Lett. B199 (1987) 531;
A. Hasenfratz, K. Jansen, J. Jersák, C.B. Lang, T. Neuhaus and H. Yoneyama, Nucl. Phys. B317 (1989) 81;
T. Neuhaus, Nucl. Phys. B (Proc. Suppl.) 9 (1989) 21.

[13] J. Kuti, L. Lin and Y. Shen, Phys. Rev. Lett. 61 (1988) 678.

[14] M. Lüscher and P. Weisz, Phys. Lett. B212 (1988) 472; Nucl. Phys. B318 (1989) 705.

Critical Exponents for Wetting of Hot Gluons

Thomas Trappenberg

Institute of Theoretical Physics E, RWTH Aachen, D-5100 Aachen, Germany

and

HLRZ at KFA Jülich, D-5170 Jülich, Germany

Abstract

When a hot deconfined quark-gluon plasma is cooled it undergoes a phase transition to the confined hadron-glueball phase. The features of this transition have crucial cosmological consequences and are important in the context of heavy ion collisions. Until now detailed studies have been performed mainly in the quenched approximation [1, 2], i.e. the quarks have been neglected and only a pure SU(3) gauge theory has been considered. But also in this idealized situation there are still open questions. We sketch different scenarios of the SU(3) finite temperature phase transition and calculate the critical exponents in the case of the wetting scenario.

Bubble nucleation and the transition in the presence of domain walls. The phase transition of the SU(3) gauge theory at finite temperature is related to the spontaneous breakdown of the Z(3) center symmetry. The order parameter is the Polyakov loop, which is a complex scalar field in three dimensions,

$$\Phi(\vec{x}) = Tr(\mathcal{P} \exp \int_0^\beta dt A_4(\vec{x}, t)) . \qquad (1)$$

Here $A_4(\vec{x}, t)$ is the SU(3) gauge field in euclidean time direction and $\mathcal{P}$ denotes path ordering. At high temperatures Φ has a nonvanishing vacuum expectation value, whereas at low temperatures $< \Phi >= 0$. Color is *deconfined* in the ordered high temperature phase and the disordered low temperature phase is the *confined phase.*

The phase transition is known to be of first order and therefore the different phases coexist at the transition temperature T_c. When the temperature falls below T_c some domains of the confined phase form. The bubbles of the confined phase have a surface, which costs extra free energy. Therefore supercooling is necessary to stabilize the bubbles. It is clear that the reduced interface tension, which is the free energy F of the surface per area A normalized to the temperature T

$$\sigma_{cd} = \frac{F}{AT} \qquad (2)$$

is an important thermodynamic quantity. The interface tension is labeled with the index '*cd*' in order to denote that it corresponds to the interfaces between the confined and a deconfined phase. The interface tension σ_{cd} also sets the scale for the average distance between the confined bubbles generated at T_c. If the average distance of bubbles is small in comparison to their average size they can easily fuse and increase their size, forming

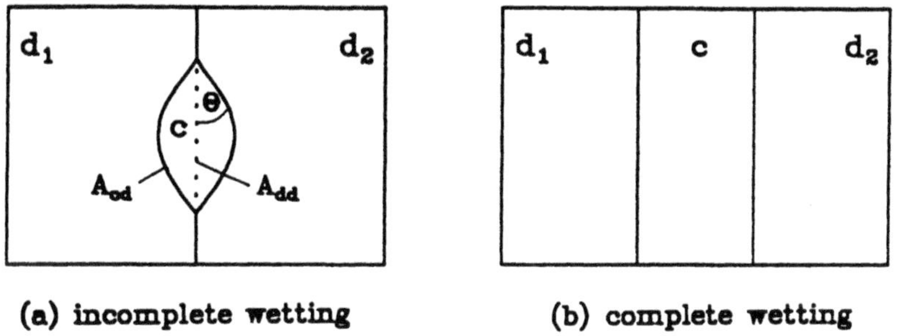

Figure 1: Scenarios of the phase transition in the presence of a domain walls

large domains of the confined phase. If, on the other hand, the average distance is large in comparison to their size, the bubbles shrink to zero because of the cost of the extra surface free energy. Then strong supercooling is necessary to stabilize the confined bubbles. After forming large domains at $T < T_c$ the confined bubbles quickly increase their size building up the pure confined phase. This is the bubble nucleation scenario and its consequences are discussed in the literature in great detail (see for example [3, 4]).

The situation is rather different when domain walls between different deconfined phases are present above T_c. In principle there are three different deconfined phases which correspond to the three different vacuum expectation values of the Polyakov loop when the Z(3) center symmetry is spontaneously broken. If domain walls are generated at $T > T_c$ there are two possible scenarios for the phase transition which are illustrated in fig.1.

Let us simplify the discussion by considering a single domain wall between two different deconfined phases d_1 and d_2 with a reduced interface tension σ_{dd}. Then it is energetically favourable to generate confined bubbles at this interface, because it costs only the free energy $(2\sigma_{cd}A_{cd} - \sigma_{dd}A_{dd})T$ instead of $2\sigma_{cd}A_{cd}T$ for a confined bubble surrounded by one deconfined phase with the same surface area. In the situation illustrated in fig.1a the contact angle Θ is given by

$$\sigma_{dd} = 2\sigma_{cd} \cos \Theta \ . \tag{3}$$

Therefore if $2\sigma_{cd} > \sigma_{dd}$ the confined phase forms a lens at the deconfined-deconfined domain wall.

In thermal equilibrium $2\sigma_{cd}$ cannot be smaller than σ_{dd}. However a special situation occurs if $2\sigma_{cd} = \sigma_{dd}$. In this case illustrated in fig.1b the contact angle Θ is zero and the confined phase forms a film separating the two deconfined phases. The confined phase completely wets the deconfined phase instead of the incomplete wetting of the previous scenario. In the complete wetting scenario the creation of the confined film does not cost extra surface free energy and therefore a confined layer can be created already at $T > T_c$. This is qualitatively different from the bubble creation scenario. When the transition temperature is reached the width of the confined film increases and actually diverges at T_c. The divergence of the width is characterized by a critical exponent ψ, which is known in condensed matter physics. It should be mentioned that in the wetting scenario no new scale is generated at the transition temperature. The relevant scale for the creation of the confined phase is now provided by the structure of the deconfined-deconfined domain wall network.

It was suggested by Frei and Patkós [5] that complete wetting might occur at the SU(3) finite temperature transition and recent simulations of quenched QCD support this expectation [1, 2]. Together with U.-J. Wiese we have calculated the critical exponents for the wetting scenario which is described in the next section.

The instanton solution of the Z(3) symmetric Φ^4 theory. Of course, at the transition temperature QCD is not tractable by analytic methods. Because the phase transition is first order there are no universal bulk quantities allowing a connection to simpler models. However, in a wetting situation there exist universal wetting properties (see [6] for a detailed collection of wetting phenomena). Now I sketch our calculculation of the critical exponents of complete wetting described in detail in ref. [7].

Formally integrating out all degrees of freedom except the Polyakov loops leads to an effective action for a 3-d theory of Polyakov loops

$$\exp(-S_{eff}[\Phi]) = \int \mathcal{D}A\, \delta[\Phi(\vec{x}) - Tr(\mathcal{P}\exp\int_0^\beta dtA_4(\vec{x},t))]\exp(-S[A])\,, \tag{4}$$

where $S[A]$ is the action of the SU(3) gauge theory. Of course this cannot be done in practice. However to calculate critical exponents it is possible to study a model in the same universality class, for example a local action of the form

$$S[\Phi] = \int d^3x[\frac{1}{2}\partial_i\Phi^*\partial_i\Phi + V(\Phi)]\,. \tag{5}$$

The requirements on the potential $V(\Phi)$ are that $S[\Phi]$ should have the same symmetries as $S_{eff}[\Phi]$ which are the

$$\text{charge conjugation symmetry:} \quad V(\Phi^*) = V(\Phi) \tag{6}$$
$$\text{and the Z(3) symmetry:} \quad V(\Phi z) = V(\Phi) \tag{7}$$

with $z \in$ Z(3). In addition we demand that the corresponding quantum theory should be perturbatively renormalizable. The last requirement restricts the potential to terms up to sixth order in Φ. Decomposing Φ into $\Phi = \Phi_1 + i\Phi_2$ the simplest model is given by (5) with the potential

$$V(\Phi) = a|\Phi|^2 + b\Phi_1(\Phi_1^2 - 3\Phi_2^2) + c|\Phi|^4\,. \tag{8}$$

The interfaces between the different vacua are then described to lowest order of a quasiclassical expansion by the solution of the equations of motion [8]. The equations of motion are given by

$$\begin{aligned}\Phi_1'' &= \frac{\partial V}{\partial \Phi_1} = \Phi_1(2a + 4c|\Phi|^2) + 3b(\Phi_1^2 - \Phi_2^2)\,,\\ \Phi_2'' &= \frac{\partial V}{\partial \Phi_2} = \Phi_2(2a + 4c|\Phi|^2) - 6b\Phi_1\Phi_2\,.\end{aligned}$$

Close to the critical temperature, where the 4 minima of the potential (8) are nearly degenerate, the solution connecting the two minima with nonvanishing imaginary parts is given by [7]

$$\Phi_1(z) = -\frac{1}{4}\Phi_0[2 + \tanh\frac{\mu}{2}(z - z_0) - \tanh\frac{\mu}{2}(z + z_0)] ,$$
$$\Phi_2(z) = \frac{\sqrt{3}}{4}\Phi_0[\tanh\frac{\mu}{2}(z - z_0) + \tanh\frac{\mu}{2}(z + z_0)] ,$$

with $\mu = -\frac{b}{\sqrt{2c}}$ and $z_0 = -\frac{1}{2\mu}\ln(\frac{1}{2} - \frac{2ac}{b^2})$. The critical temperature corresponds to $b^2 = 4ac$. Hence, the thickness z_0 of the confined layer diverges and the interface magnetization $\Phi_1(0)$ vanishes at T_c. This is the complete wetting. A path of the average Polyakov loop connecting different deconfined phases always runs through the origin in the complex Φ plane, which corresponds to the confined phase. The scaling laws and hence the critical exponents can be read off from the instanton solution and are given by

$$z_0 \propto (T - T_c)^{-\psi} \quad \text{with} \quad \psi = 0 \tag{9}$$
$$\Phi_1(0) \propto (T - T_c)^{\beta} \quad \text{with} \quad \beta = \tfrac{1}{2} . \tag{10}$$

If the complete wetting occurs at the SU(3) finite temperature phase transition and if the model (5) with the potential (8) is in the same universality class for critical interface effects, then these are also the scaling laws in the SU(3) gauge theory.

Wetting in other models. In the following I discuss wetting behaviour of other perturbatively renormalizable models. The most general potential leading to a renormalizable theory in three dimensions with the symmetries (6) and (7) is a sixth-order polynomial in Φ_1 and Φ_2 given by

$$\begin{aligned} V(\Phi) &= a|\Phi|^2 + b\Phi_1(\Phi_1^2 - 3\Phi_2^2) + c|\Phi|^4 + d|\Phi|^2\Phi_1(\Phi_1^2 - 3\Phi_2^2) \\ &+ e|\Phi|^6 + f(\Phi_1^6 - 15\Phi_1^4\Phi_2^2 + 15\Phi_1^2\Phi_2^4 - \Phi_2^6) . \end{aligned} \tag{11}$$

To examine if the higher order terms destroy the complete wetting, a special case is discussed first. If we consider $b = d = f = 0$ the potential is U(1) symmetric. In this case there is no potential barrier isolating the different Z(3) symmetric minima of the broken phase and hence there is no wetting. However is was tested [7], that complete wetting is still present in other regions of the phase transition sheet in the 6-dimensional phase diagram. To give some illustrative examples the equipotential lines of the potential of eq.(11) are depicted in three different situations in fig.2. The first example (fig.2a) is close to the U(1) symmetric case and no wetting occurs. The last example (fig.2c) is in a region, where complete wetting is present. The other example (fig.2b) is in the region where the transition between complete and incomplete wetting occurs.

The discussion in this section shows, that not all renormalizable Z(3) symmetric Φ^6 models have the feature that complete wetting occurs. The calculation of the critical exponents for wetting in the SU(3) gauge theory has to be done in a model with the same wetting behaviour. If complete wetting occurs in the SU(3) gauge theory, the exponents calculated in the Φ^4 model in the last section are the relevant ones. Whether complete wetting occurs in the SU(3) gauge theory, however, can be decided only in the SU(3) gauge theory itself.

Recent results. There is a numerical evidence for complete wetting from exploring the average path connecting different deconfined phases as well as from a determination of the reduced interface tensions σ_{cd} and σ_{dd}. However the interface tension is difficult to obtain in Monte Carlo simulations.

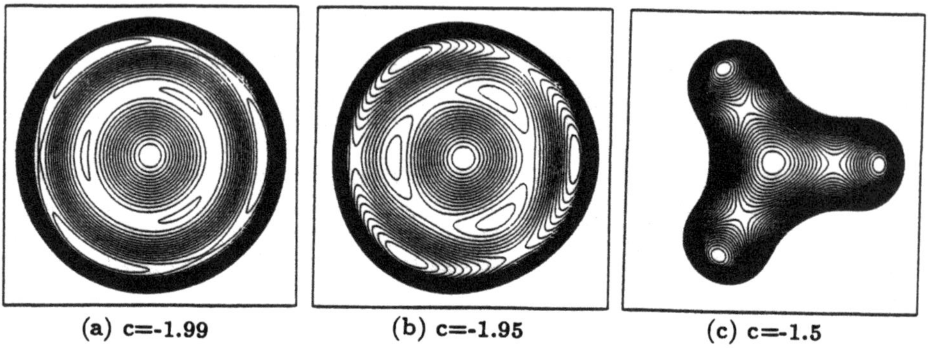

Figure 2: The equipotential lines of $V(\Phi)$ in the complex Φ plane at the critical temperature. The parameters are $a = 1$, $b = b_c$, $d = 0$, $e = 1$, $f = 0$ and three different values of c.

On the other hand interface dominated field configurations influence the spectrum of the transfer matrix in a finite volume. It has been found [7] that complete wetting leads to a different energy splitting compared to incomplete wetting. A recent investigation of the spectrum on a finite lattice [9] is consistent with complete wetting. This supports the expectation, that in the presence of domain walls the dynamics of the phase transition is qualitative different from the bubble nucleation scenario.

Acknowledgement. This work has been done in collaboration with U.-J. Wiese. I like to thank A. Patkós for his inspiring comments in Dobogókö.

References

[1] K. Kajantie, L. Kärkkäinen and K. Rummukainen, Nucl. Phys. B357 (1991) 693 ; L. Kärkkäinen and K. Rummukainen, Nucl. Phys. B(Proc. Suppl.)20 (1991) 306.

[2] J. Potvin and C. Rebbi, Nucl. Phys. B(Proc. Suppl.)20 (1991) 317.

[3] C. Hogan, Phys. Lett. 133B (1983) 172.

[4] K. Kajantie and H. Kurki-Suonio, Phys. Rev. 34 (1986) 1719.

[5] Z. Frei and A. Patkós, Phys. Lett. B229 (1989) 102.

[6] R. Lipowsky, Critical behavior of interfaces: wetting, surface melting and related phenomena, Jül-Spez-438 (1988).

[7] T. Trappenberg and U.-J. Wiese, HLRZ-91-52, to be published in Nucl. Phys. B[FS].

[8] S. Coleman, Phys. Rev. D (1977) 2929.

[9] B. Großmann, M.L. Laursen, T. Trappenberg and U.-J. Wiese, in preparation.

SUMMARY

WORKSHOP ON EFFECTIVE FIELD THEORIES SUMMARY TALK *†

HOWARD GEORGI
Lyman Laboratory of Physics, Harvard University
Cambridge, MA 02138

I am not sure how to go about summarizing this workshop. I will begin by dividing the talks into three sets:

- talks that I could clearly identify as being about effective field theory;
- talks that I could clearly identify as being about ineffective field theory;
- talks that I could not clearly identify.

Ineffective Field Theory

The ineffective field theory talks included discussions of Skyrmions, the spin content of the proton and the anomaly at finite temperature. I won't talk much about the ineffective field theory talks, but will just hit one highlight. I very much liked the talk by Dr. Smilga about the anomaly. This was a very ambitious (though not entirely successful) attempt to provide a physical picture of baryon number violation due to the anomaly at high temperature. This is an important problem, in my opinion. Apart from the problems that Smilga outlined, it seems to me that there was another questionable step in his analysis of the four dimensional case. I think that it is really not obvious that reducing the four dimensional case to a two dimensional one by turning on a uniform magnetic field gets the four dimensional physics right. This needs a separate proof in my opinion.

Among the talks that I could not clearly identify were Trappenberg's and Polonyi's. Trappenberg's was my first introduction to the idea of complete wetting, in which at the confining phase transition, layers of confined phase form between the different regions of deconfined phase. This is a pretty idea, although it was not clear to me what this had to do with effective field theory. In any event, Trappenberg got the prize for the prettiest transparencies at the workshop with his multicolor renderings of the effective potential.

I found Polonyi's talk a little frustrating. I had the feeling that it should have been possible to formulate the problem he addressed as an effective field theory problem,

---

*Research supported in part by the National Science Foundation under Grant #PHY-8714654. Harvard preprint #HUTP-91/A052

†Research supported in part by the Texas National Research Laboratory Commission, under Grant #RGFY9106.

to justify his title. However, I do not believe that he actually did it. Somehow, the point is that in QCD with no quarks, the response of the theory to color excitations in the limit of infinitely long wavelength is insensitive to all the details of the theory except for the confining structure of the QCD vacuum. Unfortunately, I could not find a realization of this idea in what Polonyi actually said.

We now come to primary subject of the workshop. I will further subdivide the effective field theory talks into talks about what one might call "conventional" chiral perturbation, and everything else.

Unconventional Effective Field Theory

One of the talks outside conventional chiral perturbation theory was Espriu's talk on electroweak interactions. [1, 2, 3, 4] One reason to use the effective field theory formalism in this context is to organize the possible effects of physics at high energies in the most economical way and to properly include the constraints of $SU(2) \times U(1)$ gauge symmetry. I am sure that you are all as tired as I am of papers on "phenomenology" that focus on one piece of physics at high energies without paying any attention to how it fits together with the tremendous body of information we have about the electroweak interaction. A typical crazy strategy is to take some piece of a nonrenormalizable interaction, and consider it in isolation. What is worse, the term is always allowed to have a very large coefficient (so that the effects are measurable) that would produce enormous renormalizations of many other terms that are not considered. The formalism discussed by Espriu avoids these pitfalls.

I want to point out that there is another reason to use the effective field theory formalism.[5] It actually contains information about the physics that goes beyond conventional perturbation theory. A commonplace example of this is the renormalization group itself, where we all know that if we renormalize the theory at the wrong scale, we can mess up the convergence of perturbation theory. There is another nice example of the principle in the electroweak interactions if $m_t \gg M_W$. The t couplings to the Goldstone bosons break the custodial $SU(2)$ symmetry. In this instance, when you integrate out the t, as you all know, you get a large contribution to the mass splitting between the W and Z, that translates into a contribution to the ρ parameter that grows as m_t^2. In an effective field theory language, this contribution is incorporated into the parameters of the effective Lagrangian at $\mu = m_t$. When this effective Lagrangian is run down to the W and used to calculate the electroweak parameters, the large contribution from the t loop is incorporated **nonperturbatively** into the calculation through the dependence of the $W - Z$ mass splitting. This was certainly understood many years ago by Cohen, Grinstein and me. [6] It was probably understood by Veltman in the last century. However, it has recently been rediscovered in a very cumbersome, ineffective language by the precise radiative correction community, where it goes by the name of "resummation" of m_t dependence. [7] As you see, it is a trivial consequence of effective field theory.

Manohar and Jenkins gave a pair of talks about a slightly unconventional chiral perturbation theory description of baryons.[8] This work is unconventional in two ways. It is based on the heavy fermion effective field theory. [9] And, in Jenkins's talk, it includes the spin 3/2 decuplet as well as the spin 1/2 octet. I found these fascinating because I am working on the subject myself with my student Chris Carone. Particularly impressive was Liz Jenkins's talk, summarizing an interesting calculation of chiral logs. An interesting question arose in connection with this talk. After the talk, Ulf Meissner asked about chiral logs in the baryon masses of the form

$$\Delta m\, m_K^2 \ln \frac{m_K}{\mu} \tag{1}$$

where Δm is the splitting between the decuplet and the octet in the chiral limit. The trouble with (1) is that the derivative with respect to m_K^2 diverges as $m_K^2 \to 0$ for fixed Δm. If this were the whole story, then the expectation value of $\bar{s}s$ would not be finite in the baryon state. In fact, as various of us realized (and Jenkins had said specifically in her talk), this is an interchange of limits problem. (1) is valid only for $\Delta m \ll m_K$. Jenkins argued in her talk that this may be a reasonable approximation in the real world. Thus the "chiral limit" in this case requires $\Delta m \to 0$ as well as $m_K \to 0$.

In fact, this is the point. Chiral logs are infrared effects. Transitions between different $SU(3)$ multiplets can only produce chiral in the limit in which the multiplets are degenerate, so that zero momentum Goldstone bosons can be exchanged in the chiral limit.

While Jenkins did not find a reasonable fit to the p-wave nonleptonic hyperon decays, she did give a reasonable explanation for the failure. The leading contributions to the p-wave decays are accidentally (?) small because of cancellations between different terms. Thus higher order contributions are unusually important. An interesting question is whether a full second order calculation is useful. In the general case, it seems that there are so many additional terms that one will be able to fit anything.

Finally, in the unconventional category, there were the talks by Kuti and Yamawaki on the "top-mode standard model." [11] There is an English phrase that describes the top-mode standard model rather well. It is a "tempest in a teapot." In fact, as Kuti showed in the limit of zero gauge coupling (about which I will say more later), the top-mode standard model is nothing but the standard model with a relatively heavy Higgs boson and t quark.

My own way of saying what is going on in the top-mode standard model is the following. Simply adopt as your definition of what it means for the Higgs to be a bound state of $\bar{t}t$ the condition that there is some scale Λ at which the Higgs self-coupling and its coupling to $\bar{t}t$ are both large. That is all there is too it! From there on, the renormalization group automatically gives you all the reliable results of Bardeen, Hill and Lindner (BHL). [12]

The top-mode people have actually done slightly LESS than this. The only thing they have added is an unreliable analysis (the ladder approximation, valid only in large N_c) of a nonsensical model (the NJL model). This is clearly a negative contribution to our understanding of what is going on.

When I say that the NJL model is nonsensical, I mean not just that, as Kuti explained, it only makes sense as a cut-off theory. Worse, the crucial dynamics that binds the spinless state in this model is happening at a scale of the order of the cut-off. A sensible theory may give rise to the NJL model as an approximate theory far below the cut-off. But at the cut-off, there is no reason to ignore the infinite set of higher dimension terms that appear as well.

Yamawaki's talk was a good example of what not to do. He presented his analysis of the NJL model plus color gauge interaction. Kuti, in his talk, distinguished between the NJL model with no color, which he showed was equivalent to the standard model, and the NJL model with color, which is more complicated. Unfortunately, for Yamawaki, the complications associated with a color gauge interaction occur at strong coupling. In the physically interesting case of color $SU(3)$, the color interaction is not strong enough to do anything very interesting. Thus Yamawaki model makes no more sense than the simple NJL model.

At scales far below the cut-off, Λ, Yamawaki's analysis is an exercise in self-torture. He is working far harder than he should, and all he is doing (if he does it correctly), is reproducing the renormalization group analysis in the standard model. This is the world's most complicated way of doing the Veltman analysis of the ρ parameter.

At scales of the order of the cut-off, Yamawaki's analysis is an exercise in self-delusion. Here the theory just doesn't make sense.

I should say that one of these objections does not apply to attempts such as the recent model of Chris Hill to build the NJL four-fermion interactions out of spontaneously broken gauge boson exchange. [13] These models are really much more consistent theories. However, even here, the question remains whether the approximation schemes used to analyze the very strong interactions in the model have any validity at all. And of course, none of these models solves or even addresses the real problem of the standard model with a fundamental Higgs boson, which is why is there such a large gap between the $SU(2) \times U(1)$ breaking scales and the much higher scales that seem to exist in the world. There is just as much fine tuning as in the conventional standard model.

Conventional Effective Field Theory

Now we are getting to the central topic of the workshop. Most of the talks were in this category. I will have to further subdivide them in three parts, tests, applications and attempts to calculate parameters.

Tests

The tests were of two kinds, systematic and tricky. The systematic tests were discussed by Kaisser and Kambor, for meson-meson scattering and meson weak decays respectively. These talks represent the continuation of the program of Gasser and Leutwyler. They are the reason that this workshop exists, at least in this place and with this set of people. To summarize such a large body of hard, careful work is presumptuous. But I would say that the main moral is that chiral perturbation theory works very well up to energies at which there are large contributions from vector meson exchange. Beyond that point, the vector mesons must be put into the theory explicitly. This is still more of an art than a science. I will return to this below.

By tricky tests, I mean examples in which loop effects in chiral perturbation theory are less sensitive than naively expected to the higher order terms in the chiral Lagrangian, so that reliable results can be obtained in leading nontrivial order. A beautiful example of this was discussed by Donoghue in his introduction to the workshop, $\gamma\gamma \to \pi^0\pi^0$. This process does not occur in tree level in the lowest order chiral Lagrangian. The one-loop Feynman graphs that contribute are finite and order p^2 (in the chiral limit). What is happening is an infrared effect. There is no order p^2 term in the Lagrangian that can contribute, just because of the symmetries. But the graph is order p^2 by power counting. Effectively, there is a $1/p^2$ infrared behavior.

This kind of thing happens frequently in the interactions of γs and π^0s. Two different examples of similar effects were discussed by Meissner, $\gamma p \to \pi^0 p$ and $\gamma p \to \gamma p$.

Applications

A number of talks at the workshop used chiral perturbation theory to address other issues in physics. The two talks that we have just heard were in this category. Saino, gave us an update on the σ term in $\pi - N$ scattering. The long standing puzzle of the anomalously large contribution of the s quark mass to the mass of the proton may be going away. It would be very interesting to compare Saino's results to the value of the σ term that comes from Jenkins's analysis. I hope that she will have the result soon.

Heiri Leutwyler assured me that his talk about finite size effects would be self-summarizing. I will assume that it was a sufficient summary of Eletzky's talk as well.

The other application we heard about was to the subject of quark-gluon plasma. We had two talk about this, from Gerber and from Goity on the subject of hot pions and other hadrons. Goity explained very honestly some of the difficulties with this subject. His conclusion was that while chiral perturbation theory is consistent with the observations of the results of nucleus-nucleus collisions at high energies, these observations cannot really be called a test of chiral perturbation theory. I believe that this is unlikely to change, ever. There is just too much going on. We will never be able to have enough theoretical control over all the parts of this process to be able to make reliable predictions. I suspect that this field will always remain an application rather than a test of chiral Lagrangian techniques.

Attempts to Calculate Parameters

Finally we come to the things that Donoghue called models in his introductory talk. These are attempts to calculate the coefficients in the chiral Lagrangian. This is a tough job.

Let me first discuss Dr. Andrianov's talk. I have a theory about this one. You see (goes my theory) at the Leningrad University in Saint Petersburg, seminars seldom last less than three hours. So when Andrianov ran out of time after 45 minutes, we had only heard 15% of his talk, making it hard to understand what he was trying to do. I asked Dr. Nagy to inquire further about this, and got some more information. The talk was about the NJL model as a low energy representation of QCD. The basic idea of Andrianov's talk was that this by itself doesn't work. If you try to represent low energy QCD with only quarks and four quark interactions, the theory doesn't tell you at what scale to match onto the chiral Lagrangian and the matching cannot reproduce the physics of the underlying QCD theory. But when gluons are also included in the NJL model, then it works.

I don't understand the details of this, but it seems sensible to me, because it fits into the general picture of matching that I discussed in my own talk. The high energy theory must tell you where to match onto the low energy theory.

Several talks discussed the Espriu, DeRafael, Taron [16] (EDT) quark loop "calculations" of coefficients in the chiral Lagrangian. I am thinking particularly about the talks by Pich on the $\Delta I = 1/2$ rule, and Bijnens in the anomaly in next to leading order. Of course, I am being somewhat unfair to these speakers, each of whom had an interesting topic. But in both cases, what I found most interesting about the talks were the discussions of the EDT scheme. I confess that I do not understand why this scheme should work at all. It seems to me to be an expansion in the QCD coupling constant in a region in which it is large (since it is causing the chiral symmetry breakdown). I would like to encourage people that work on these calculations to spend less time on the mathematic and more time on the physics. In particular, to explain why the manipulations make any sense at all.

One justification that we heard a few time is "the chiral quark model." [18] As one of the originators of this idea, let me say that I am not convinced that it hangs together very well. The idea is that at the chiral symmetry breaking scale, QCD is really still rather weak. If this were really correct, then EDT would probably work. What I have never been able to quite is get straight is how to deal consistently with the gluons in this picture. More discussion is needed. The chiral quark model by itself is not an explanation.

The funny thing is that EDT seems to work. It is possible, of course, that the success of EDT is accidental, and that as we probe more terms in the chiral Lagrangian, the apparent agreement will look less impressive. We shall see.

The other model that was extensively discussed was what might be called the vector meson dominance model, in which the higher derivative terms in the chiral Lagrangian are assumed to arise primarily from the exchange of vector mesons. A number of comments are worth making here.

My first comment is that attempts at unitarization alone are misguided. If the effective field theory is working, then final state interaction should be under control. You can calculate them. When they stop working, you need to add something else. In this case, the ρ and the over vector mesons. The process of putting in the vector mesons consistently, that is building an effective theory including the vector mesons has been begun, as we heard from Kaisser, but not finished. It has not yet been put on the same systematic footing as the chiral Lagrangian describing just the Goldstone bosons. Including the vector mesons is more complicated for a number of reasons. For one thing, the chiral symmetry does not constrain the transformation properties of the matter fields so much as it does those of the Goldstone bosons. There are many different representations of the vector meson fields with the same transformation properties under Gell-Mann's $SU(3)$. If we could write down the full effective field theory, we could use any of these representations and go back and forth from one to another by making nonlinear field redefinitions. However, in practice, some representations are better than others, because each representation corresponds to a different set of higher order terms in the momentum expansion. We should choose the representation for the vector mesons that makes the momentum expansion as good as possible. I do not know how to do this for sure, except in one limit. If the vector mesons were really much lighter than everything else except the Goldstone bosons, then it would be important to incorporate them using the formalism of hidden local symmetry, pioneered by Dr. Yamawaki and collaborators. [19] The reason is that this approach avoids any hidden factors of $1/m_\rho$ in the couplings of the vector mesons to nonconserved currents.

The second comment is that just because you incorporate the ρ doesn't mean that there are no remaining higher order terms in the chiral Lagrangian. In fact, I am somewhat surprised that explicit calculations of the higher order term work as well as they do assuming that the vector mesons are the whole story. I don't see why this should work. This is a mystery like EDT.

The third comment is that even if you included the vector mesons in the optimum possible way, and even if you allow for reasonable higher order terms in the momentum expansion, it doesn't have to work. You know that at some momentum, the effective field theory picture of the QCD interactions in terms of the hadron states is going to break down. At sufficiently high energies, the right description is quarks and leptons, not hadrons. The nice paper by Bob Jaffe[20] (whose title, "Ineffective Field Theory," we have taken in vain at various times during the workshop) shows in detail how this can happen in the example of heavy quark bound states. The only question is whether this happens at 2 GeV, or 1 GeV, or 500 MeV. I hope that it is not until some rather high energy like 2 GeV where we will be forced to abandon the effective field theory of hadrons once and for all. But if it is 500 MeV instead, then we have to pack it in, and be content with refining the analysis of really low energy phenomena.

Final Remarks

Finally, let me thank the physicists from Hungary and Switzerland for organizing this workshop. I would particularly like to thank Andreas Patkos and his students for all their help. It has been a terrific experience.

References

[1] E. Jenkins and A. Manohar, Phys. Lett. **237B** (1989) 259.

[2] M. Golden and L. Randall, "Radiative Corrections to Electroweak Parameters in Technicolor Theories".

[3] B. Holdom and J. Terning, Phys. Lett. **247B** (1990) 88.

[4] G. Altarelli,

[5] H. Georgi, Nucl. Phys. **B** (1991) .

[6] A. Cohen, H. Georgi and B. Grinstein, Nucl. Phys. **B232** (1984) 61.

[7] M. Consoli, *et al.*, in Phys. Lett. **227B** (1989) 167.

[8] E. Jenkins and A. Manohar, Phys. Lett. **255B** (1990) 558; Phys. Lett. **259B** (1990) 353.

[9] H. Georgi, Phys. Lett. **240B** (1990) 447.

[10] H. Georgi and I. MacArthur, unpublished; D. Kaplan and A. Manohar.

[11] V.A. Miransky, *et al.*, Phys. Lett. **221B** (1989) 177, and Mod. Phys. Lett. **A4** (1989) 1043; Y. Nambu, Phys. Rev. **D42** (1990) 3918.

[12] W. Bardeen, C. Hill and M. Lindner, Phys. Rev. **D41** (1990) 1647.

[13] C. Hill, Phys. Lett. **266B** (1991) 419.

[14] J. Gasser and H. Leutwyler, Ann. Phys. **158** (1984) 142.

[15] J. Gasser and H. Leutwyler, Nucl. Phys. **B250** (1985) 465.

[16] D. Espriu, E. De Rafael and J. Taron, Nucl. Phys. **B345** (1990) 22.

[17] M. Chanowitz, M. Golden and H. Georgi, Phys. Rev. **D36** (1987) 1490.

[18] A. Manohar and H. Georgi, Nucl. Phys. **B234** (1984) 189.

[19] See M. Bando, T. Kugo and K. Yamawaki, Physics Reports **164** (1988) 217-314.

[20] R. L. Jaffe, Phys. Lett. **245B** (1990) 221.

LIST OF PARTICIPANTS

V. Andrianov (St. Petersburg)
J. Baacke (Dortmund)
J. Balog (Munich–Budapest)
J. Beringer (Bern)
V. Bernard (CRN Strasbourg)
J. Bijnens (CERN)
H. Bock (Aachen)
P. Büttiker (Bern)
J.F. Donoghue (Amherst)
G. Ecker (Vienna)
V.L. Eletsky (ITEP)
D. Espriu (CERN)
P. Frick (Jülich)
Z. Fodor (Budapest)
P. Forgács (Munich–Budapest)
H. Georgi (Harvard)
P. Gerber (Bern)
J. Goity (PSI)
P. Hasenfratz (Bern)
Z. Horváth (Budapest)
E. Jenkins (San Diego)
J. Kambor (Munich)
N. Kaiser (CRN Strasbourg)
J. Kuti (San Diego)
H. Leutwyler (Bern)
A. Manohar (San Diego)
U.-G. Meißner (Bern)
H. Meyer-Ortmanns (Heidelberg)
P. Minkowski (Bern)
J. Missimer (PSI)
M. Nagy (Bratislava)
T. Nagy (Budapest)
H. Neufeld (Vienna)
F. Niedermayer (Munich–Budapest)
A. Patkós (Budapest)
T.N. Pham (Palaiseau)
A. Pich (CERN)
G. Pócsik (Budapest)
J. Polónyi (MIT)
M. Sainio (Helsinki)
J. Schechter (Syracuse)
A. Smilga (Bern)
T. Trappenberg (Aachen)
T.N. Truong (Palaiseau)
Z. Trócsányi (Debrecen)
G. Valencia (CERN)
A. Wirzba (NORDITA)
D. Wyler (Zürich)
K. Yamawaki (Nagoya)

www.ingramcontent.com/pod-product-compliance
Lightning Source LLC
LaVergne TN
LVHW082254241026
850861LV00008B/45
9789810210014